Mechanistic Bioinorganic Chemistry

ADVANCES IN CHEMISTRY SERIES 246

Mechanistic Bioinorganic Chemistry

H. Holden Thorp, EDITOR
University of North Carolina at Chapel Hill

Vincent L. Pecoraro, EDITOR
University of Michigan

Developed from a symposium sponsored
by the Division of Inorganic Chemistry, Inc.,
at the 205th National Meeting
of the American Chemical Society,
Denver, Colorado,
March 28–April 2, 1993

American Chemical Society, Washington, DC 1995

Library of Congress Cataloging-in-Publication Data

American Chemical Society. Meeting. (205th: 1993: Denver, Colo.)

Mechanistic bioinorganic chemistry / [editors] H. Holden Thorp, Vincent L. Pecoraro; developed from a symposium sponsored by the Division of Inorganic Chemistry, Inc., at the 205th National Meeting of the American Chemical Society, Denver, Colorado, March 28–April 2, 1993.

p. cm.—(Advances in chemistry series, ISSN 0065–2393; 246)

Includes bibliographical references and indexes.

ISBN 0–8412–3062–5 (clothbound)

1. Bioinorganic chemistry—Congresses.

I. Thorp, H. Holden, 1964– . II. Pecoraro, Vincent L. III. American Chemical Society. Division of Inorganic Chemistry. IV. Title. V. Series.

QP531.A395 1993
574.19′214—dc20 95–15128
CIP

This book is printed on acid-free paper. ∞

PRINTED IN THE UNITED STATES OF AMERICA

1995 Advisory Board

Advances in Chemistry Series

FOREWORD

The ADVANCES IN CHEMISTRY SERIES was founded in 1949 by the American Chemical Society as an outlet for symposia and collections of data in special areas of topical interest that could not be accommodated in the Society's journals. It provides a medium for symposia that would otherwise be fragmented because their papers would be distributed among several journals or not published at all.

Papers are reviewed critically according to ACS editorial standards and receive the careful attention and processing characteristic of ACS publications. Volumes in the ADVANCES IN CHEMISTRY SERIES maintain the integrity of the symposia on which they are based; however, verbatim reproductions of previously published papers are not accepted. Papers may include reports of research as well as reviews, because symposia may embrace both types of presentation.

ABOUT THE EDITORS

H. HOLDEN THORP started his career as an undergraduate with Thomas J. Meyer at the University of North Carolina at Chapel Hill, where he studied the electrocatalytic reduction of carbon dioxide. He completed his Ph.D. on the photochemistry of dioxorhenium(V) with Harry B. Gray at the California Institute of Technology. His postdoctoral work at Yale University with Gary W. Brudvig centered on the redox chemistry of oxomanganese clusters in aqueous solution. He is currently assistant professor of chemistry at the University of North Carolina at Chapel Hill. His research is directed toward understanding the fundamental redox pathways in DNA oxidation, developing electrochemical sensors for DNA hybridization, elucidating the mechanisms of excited-state proton transfer to multiply bonded ligands, and activating small molecules of biological significance. He has received the Camille Dreyfus Teacher–Scholar Award, a Presidential Young Investigator Award from the National Science Foundation, and a fellowship in science and engineering from the David and Lucile Packard Foundation. Thorp has appeared throughout the Southeast as the bassist for Nick Demos and the Greek Islanders, and he and his wife Patti enjoy watching college basketball and their new son, John.

VINCENT L. PECORARO received his B.S. degree in biochemistry from the University of California—Los Angeles and was granted the Ph.D. in chemistry, working on microbial and mammalian iron metabolism under the mentorship of Kenneth N. Raymond at the University of California—Berkeley. He reverted to biochemistry again when he joined the laboratory of W. W. Cleland in the biochemistry department at the University of Wisconsin—Madison. In Madison, he was an NIH postdoctoral fellow, using chiral metallonucleotides to probe the mechanism of phosphoryl group transfer in kinases and

ATP synthases. He joined the faculty in the chemistry department at the University of Michigan—Ann Arbor as an assistant professor and was promoted to professor. He spent a term at the University of North Carolina at Chapel Hill as visiting associate professor of chemistry. He is a member of the metallobiochemistry study section for the National Institutes of Health and was appointed an associate editor of *Inorganic Chemistry*. His awards include the G. D. Searle Biomedical Research Scholarship and an Alfred P. Sloan Fellowship. His research interests span a wide range of disciplines that include the biological chemistry of manganese and vanadium, the development and exploitation of metallacrowns for new materials, and the de novo synthesis of metallopeptides. He has been involved in undergraduate curricular reform and was a recipient of the University of Michigan Literature, Sciences, and the Arts Award for Excellence in Undergraduate Instruction and the University of Michigan Faculty Recognition Award for his contributions in science, education, and professional service. Vince and his wife Peggy Carver enjoy traveling, cooking, and especially eating.

CONTENTS

INDEXES

PREFACE

THE FIELD OF BIOINORGANIC CHEMISTRY provides a common ground for practitioners of macromolecular biochemistry, inorganic model chemistry, and a host of spectroscopic techniques ranging from magnetic resonance to time-resolved optical methods. The maturation of these areas has placed the field in a position to move beyond asking questions solely about structure and electronic properties and into asking detailed questions about the mechanisms of metal-catalyzed reactions and how structure and electronic properties influence these pathways. Probably the most important factor causing the increased popularity of a mechanistic emphasis is the rapid recent progress in large-molecule crystallography that has allowed for a detailed static picture of the active-site structure of more and more proteins. With better defined structures as a framework, the challenge for bioinorganic chemists shifts from answering the question, "What is the structure of the active center?" to answering the question, "How does such a center facilitate catalysis?"

In this volume, we attempt to bring together topics that span the breadth of metal-catalyzed reactions in biological systems and have accordingly selected chapters not only on catalytic reaction of metalloproteins, but also on other reactions of metal ions in biological systems, such as electron transfer and nucleic acid scission. Of course, an emphasis on mechanisms does not mean that structure and electronic properties can be ignored. In fact, a mechanistic emphasis requires a detailed knowledge of how structure and electronic properties influence reactivity, and these subjects are also explored here.

In surveying the roles that metal ions can play in biological systems, a number of functions are apparent. Many metal ion requirements are purely structural, as illustrated by Ca^{2+} mediation of signal transduction with calmodulin or the recent discovery of Zn^{2+} gene regulation by zinc finger proteins. Metal ions can also be present to carry electrons. In this capacity, the metal ions need to exist in two adjacent, stable redox forms, as seen in cytochromes and iron–sulfur proteins. Hydrolytic reactions can be catalyzed by metal ions that can coordinate water and hydroxide ligands that participate in the hydrolysis. This catalysis is observed in reactions that range from nucleic acid hydrolysis by ribozymes to small-molecule hydrolysis by carbonic anhydrase. Finally, metal ions can catalyze demanding redox reactions, such as the oxidation of water or methane, by activating the oxidant or the substrate by both redox chemistry and coordination. All these roles are discussed in this volume, which we hope will serve as a valuable resource for practitioners in the field and as a tantalizing introduction for those contemplating questions that fall under the rubric that is bioinorganic chemistry.

Acknowledgments

We thank Lederle Laboratories; Mallinckrodt Medical, Inc.; the Burroughs-Wellcome Company; the ACS Division of Inorganic Chemistry, Inc.; and Mitch's Tavern for financial support.

H. HOLDEN THORP
Department of Chemistry
Venable and Kenan Laboratories, 045A CB 3290
University of North Carolina at Chapel Hill
Chapel Hill, NC 27599–3290

VINCENT L. PECORARO
Department of Chemistry
University of Michigan
Ann Arbor, MI 48109

Understanding the Mechanisms in Bioinorganic Chemistry

H. Holden Thorp

Department of Chemistry, University of North Carolina, Chapel Hill, NC 27599-3290

The field of bioinorganic chemistry has experienced tremendous growth in scope and level of understanding. Advances in physical methods and the constant discovery of new biological systems that rely on metal ions have created a field that is evolving to be both broader and more detailed. As a result, bioinorganic chemists are increasingly concerned with a broad range of different types of mechanisms. In this review, the broad nature of the subject of bioinorganic chemistry is illustrated by discussing examples of important mechanisms in which metal ions mediate electron transfer, hydrolytic catalysis, redox catalysis, and gene expression.

THE FIELD OF BIOINORGANIC CHEMISTRY has grown dramatically over the past 20 years (*1*). From beginnings in studies of heme proteins (*2*) and platinum antitumor compounds (*3*) to recent advances in such new areas as Zn finger proteins (*4*), metal-responsive gene expression (*5*), RNA catalysis (*6*), and metalloprotein electron transfer (*7*), the field has become a haven for practitioners of such diverse areas as enzymology, coordination chemistry, spectroscopy, and molecular biology. What attracts these scientists from such seemingly diverse areas to bioinorganic chemistry? The answer can lie only in the unique abilities of transition metals to complement and mediate biological processes and the chemical transformations of biological molecules.

Transition metals can exceed the constraints of organic functionalities and assume coordination numbers greater than four in a wide array of structural geometries. For this reason, many transition metals are present in biological systems to serve (at least partly) a structural role, allowing a macromolecule to assume a particular conformation that would be impossible without a transition-metal structural element. Another useful

0065-2393/95/0246-0001$08.00/0

feature of transition metals in biological systems is that these elements may exist in multiple oxidation states and often are found as ions or inorganic clusters that facilitate multielectron oxidations or reductions of substrates. Because the number of suitable donors to metals is quite large, the chemical potential of a system can be fine tuned over an impressive range. Biological systems fully exploit transition-element chemistry by coupling this tunable (in terms of *both* stoichiometry and potential) redox activity with the capacity to coordinate nucleophiles. In this way a remarkable range of chemical transformations, from nitrogen fixation to water oxidation, are undertaken in a seemingly effortless process.

Another useful feature of transition metals in biological systems is the unique battery of physical techniques that can be used to study their structural and electronic properties (*8*). Many of these techniques are dependent on the unusual geometric and electronic properties of transition-metal centers and allow for study of the metal ion without interference from other material in the system.

In assuming functional roles, transition metals use one or both of the abilities of existing in multiple oxidation states and coordinating nucleophiles. In electron-transfer enzymes, the ability of transition metals to exist in multiple oxidation states allows for the storage of an electron at a localized site in the enzyme, creating the need for transfer of electrons across relatively large distances (*9*). These electron-transfer reactions are responsible for the transduction of biochemical energy, initiation of photosynthesis and other catalytic processes, and maintenance of proton gradients (*10, 11*). In hydrolytic enzymes and ribozymes, the ability of transition metals to coordinate water and hydrolyzable substrates is responsible for catalytic activity (*6, 12*). These two unique features are combined in redox catalysis enzymes, because in this process metals bind substrates of interest and change the oxidation state of the metal center to effect a net activation of the bound small molecule, which can then be released in an altered form.

Bioinorganic chemistry has been called a "maturing frontier" (*1*). With this maturing, the level of understanding of the structures and reactivity of metal ions in biological systems has been raised considerably. The availability of X-ray absorption spectroscopy and other spectroscopic techniques and the increasing database of results from macromolecular X-ray crystallography have provided an increasingly detailed picture of the structural properties of metal ions in biological systems. Whereas many issues of structure remain unresolved, key structural features of many metalloenzyme systems are now known. As a result, bioinorganic chemists find themselves contemplating mechanistic issues more carefully. The mechanistic emphasis in physical studies of metalloenzymes is increasing (*8*); the development of synthetic model complexes is progressing from initial goals of mimicking spectroscopic or structural features to more sophisticated goals of functional modeling (*13*); the chemistry of DNA cleavage has progressed toward achieving an understanding of the precise molecular mechanism of cleavage reactions on both the metal and DNA sides of the reaction, including the precise sites of DNA reactivity (*14–17*). As a result of this new interplay, the lines between structural and mechanistic studies are becoming blurred.

In this chapter, the unique features of transition metals in biological systems are discussed from the point of view of structural roles, spectroscopic properties, electron transfer, hydrolytic and redox catalysis, and metal-responsive gene expression. The following chapters provide more detail on these subjects. Several important examples not discussed elsewhere in this volume will be presented. The goal of this chapter (and this volume) is to acquaint the reader with the wide range of roles played by metal ions in biological systems and thereby to demonstrate why metals are such useful cofactors and why scientists from such broad disciplines are drawn to study their properties.

Structural Roles for Transition Metals

The ability of transition metals to assume a variety of coordination geometries is largely responsible for their unique biological function. Most metal centers in biology are either four-, five-, or six-coordinate; however, the recent discovery of unusual three-coordinate iron centers in nitrogenase (*18*) illustrates the apparently limitless diversity of coordination environments for metal centers in biological systems. In the four-coordinate environments, tetrahedral and square-planar geometries are most common. Examples of the tetrahedral geometry include iron centers in iron–sulfur clusters and many copper centers, all of which are discussed in this volume.

A large number of zinc-binding enzymes called Zn-finger proteins have been discovered (*19*). In these enzymes, the Zn^{2+} center is a structural element that organizes two histidine and two cysteinate residues on either end of a peptide strand to form a cylindrical structure that has been termed a zinc finger (*20, 21*). These proteins typically contain multiple fingers that bind to DNA cooperatively, as shown in Figure 1. In the absence of Zn^{2+}, these peptides do not bind to DNA; however, once the finger structure is imposed by coordination to the metal center, the residues in the finger section are positioned properly for DNA recognition. Other zinc–binding domains, including ones involving dinuclear zinc complexes (*4*), have been identified in related transcriptional factors (*22*). The discovery of the zinc-binding transcription factors shows that nature has taken advantage of the ability of metal ions to organize relatively small peptides into highly specific structures.

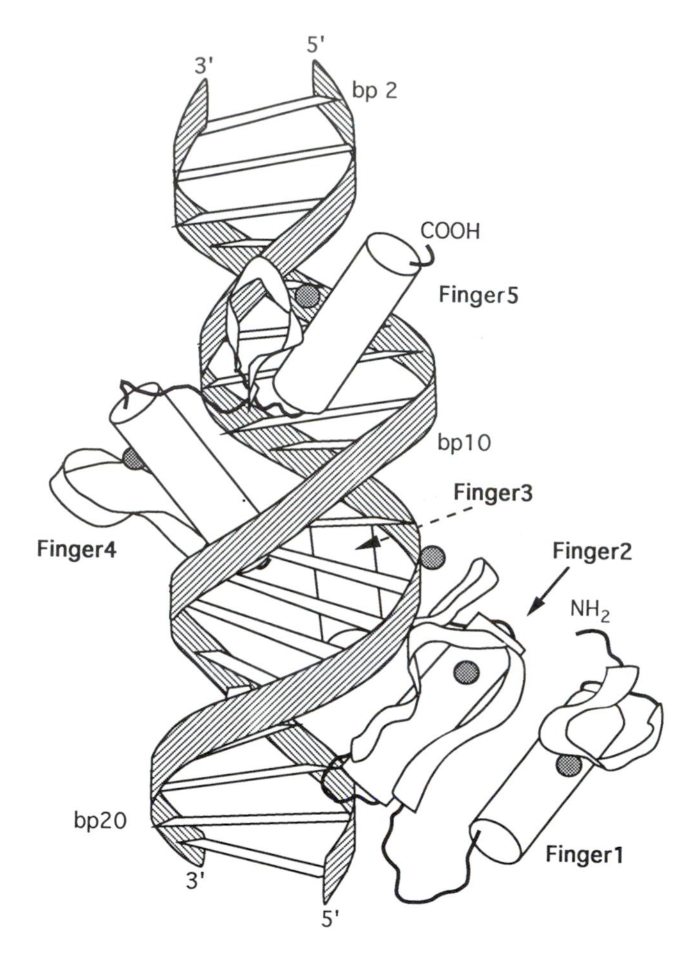

Figure 1. Sketch of the complex of the five-finger human GLI protein with a 21-base pair DNA fragment. (Reproduced with permission from reference 20. Copyright 1993 American Association for the Advancement of Science.)

The square-planar geometry is less common than tetrahedral geometry among four-coordinate metal ions in biological systems. The anticancer drug cisplatin, *cis*-$Pt(NH_3)_2Cl_2$, is a square-planar complex and is an effective agent against testicular and ovarian cancers (*3*). The complex binds primarily to two adjacent guanine bases on a single DNA strand of a double helix. Because of the square-planar geometry of the platinum center, the guanines are forced out of the usual stacked, parallel arrangement to one where they are disposed at right angles (Figure 2) (*23*). The coordination structure of this intrastrand adduct induces a kink in the DNA that is recognized by structure-specific recognition proteins (SSRPs), suggesting that these proteins play a role in cisplatin

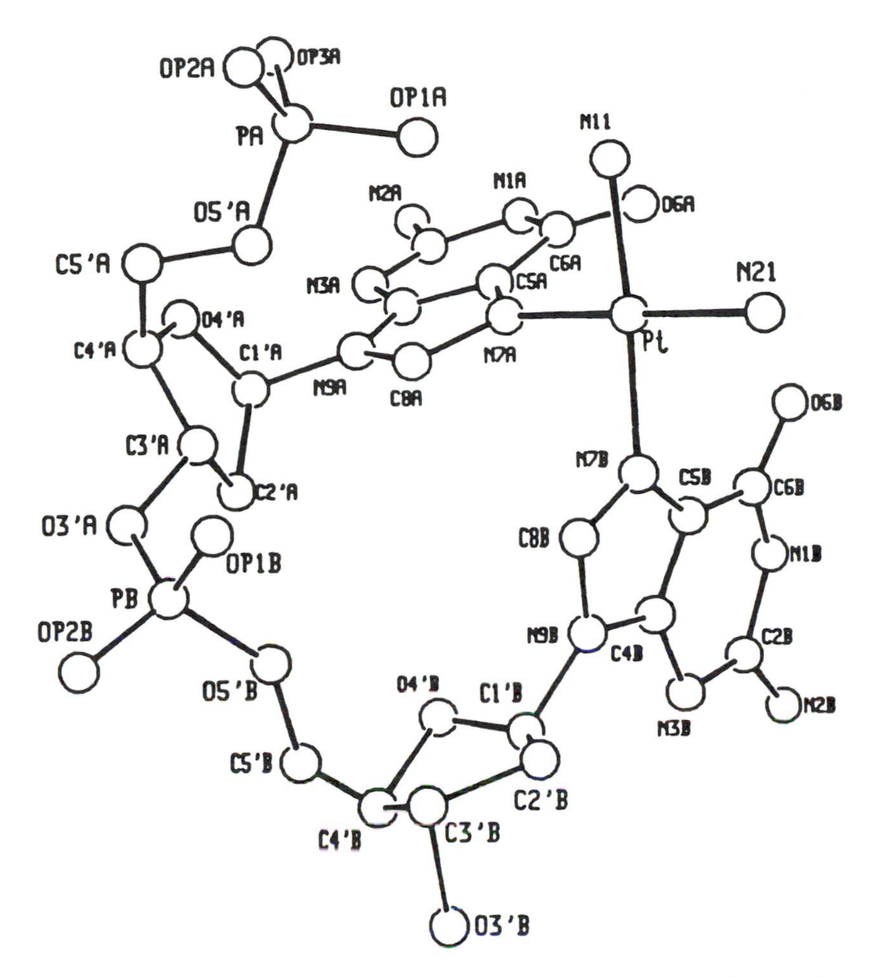

Figure 2. X-ray crystal structure of the cis-$\{Pt(NH_3)_2[d(pGpG)]\}$ *complex showing the displacement of the adjacent guanines from the usual stacked arrangement. (Reproduced with permission from reference 23. Copyright 1985 American Association for the Advancement of Science.)*

cytotoxicity (*24, 25*). These proteins do not recognize adducts of inactive platinum complexes, such as *trans*-$Pt(NH_3)_2Cl_2$; thus, the anticancer properties of cisplatin are a direct result of its stereochemistry and square-planar geometry, which is characteristic of a number of transition metals in the appropriate oxidation state. The recent findings on the role of SSRPs in cisplatin cytotoxicity demonstrate the breadth of challenges in bioinorganic chemistry: Cisplatin started out as an exciting problem in coordination chemistry and is now an equally exciting problem in molecular biology.

Spectroscopic Signatures of Transition Metals

An alluring feature of transition metals in biological systems is the wide array of physical techniques that are unique to the characterization of metal centers. In particular, a number of methods allow the metal center to be uniquely examined in the presence of other optically transparent or diamagnetic material in the system (*8*). Perhaps the most obviously unique feature of systems that contain transition metals is that many of them are strongly colored. This trait arises because of the unique electronic structures of transition-metal chromophores. For example, the characteristic optical properties of blue copper proteins, heme cofactors, and iron–sulfur clusters are all discussed in this volume. Detailed study of these optical properties has been particularly revealing with regard to understanding the geometric and electronic structures of transition metals in biological systems.

Continuous-wave and pulsed electron paramagnetic resonance (EPR) methods permit the characterization of only paramagnetic metal centers without complications arising from other diamagnetic centers elsewhere in the system (*8*). When the metal center or centers can be cycled through different oxidation states, signals can be turned off and on in a controlled manner, permitting detailed characterization of individual signals and their associated paramagnetic centers (*26*). The use of pulsed EPR methods permits in many cases the determination of the number and nature of ligands bound to the metal center that gives a particular signal (*27*). The use of EPR methods to characterize metalloenzymes is described in many chapters in this volume, and the combination of information from EPR with that from optical spectroscopy is often particularly powerful.

X-ray absorption spectroscopy (XAS) has also had a profound impact on bioinorganic chemistry (*28*). The advantage of XAS is that it is element-specific because using modern synchrotron radiation, the XAS spectrum of any element of interest can be obtained without interference from other types of elements in the system (*29*). This result means that even diamagnetic, optically inactive metal centers can be characterized independently for each transition element in a given system. Thus, the properties of one element can be studied independently in the presence of another element that may have optical or magnetic properties. Analysis of the so-called absorption edge is revealing with regard to the electronic structure of the metal center, which can permit many conclusions regarding the oxidation state and coordination environment to be drawn. Even more powerful is the analysis of extended X-ray absorption fine structure (EXAFS). This analysis permits determination of bond lengths to particular donor atoms and estimation of coordination numbers for each of these donors. The special advantage of EXAFS is that the bond lengths determined are often of quite high precision. Calculations involving the bond valence sum method show that EXAFS bond lengths can be used to estimate either coordination numbers or oxidation states if one of the two is known (*30, 31*).

All of the special techniques used for metalloenzymes are most useful when analogous small molecules have been prepared by inorganic synthesis and characterized by X-ray crystallography. The preparation of these model complexes allows a given physical property to be associated with a particular coordination structure or oxidation state. Synthesis of model complexes with structural, functional, or spectroscopic properties relevant to a particular metalloenzyme is a vital aspect of bioinorganic chemistry (*13*) and is described in numerous chapters in this volume.

Hydrolytic Catalysis

Metal ions are vital to the function of many enzymes that catalyze hydrolytic reactions. Coordination of a water molecule to a metal ion alters its acid–base properties, usually making it easier to deprotonate, which can offer a ready means for catalyzing a hydrolytic reaction. Also, the placement of a metal center in the active site of a hydrolytic enzyme could permit efficient delivery of a catalytic water molecule to the hydrolyzable substrate. In fact, the first enzyme discovered, carbonic anhydrase, is a metalloenzyme that requires a Zn^{2+} center for its catalytic activity (*32*). The function of carbonic anhydrase is to catalyze the hydrolysis of carbon dioxide to bicarbonate:

$$CO_2 + H_2O \rightleftharpoons HCO_3^- + H^+ \quad (1)$$

which plays an important role in regulating respiratory processes in animals, plants, and bacteria (*12*). The X-ray crystal structure shows that the zinc center is coordinated to three histidine residues and a water molecule (*33*), and the catalytic mechanism has been proposed to involve formation of a ZnOH complex and insertion of CO_2 into the Zn–O bond:

$$(His)_3Zn\text{–}OH_2 \rightarrow (His)_3Zn\text{–}OH + H^+ \quad (2)$$

$$(His)_3Zn\text{–}OH + CO_2 \rightarrow (His)_3Zn\text{–}OCO_2H \quad (3)$$

$$(His)_3Zn\text{–}OCO_2H + H_2O \rightarrow (His)_3Zn\text{–}OH_2 + HCO_3^- \quad (4)$$

Looney and co-workers prepared a hydroxo complex of zinc that reversibly binds CO_2 (*34, 35*):

$$[HB(3\text{-}t\text{-Bu-5-Mepz})_3]ZnOH + CO_2 \rightleftharpoons [HB(3\text{-}t\text{-Bu-5-Mepz})_3]ZnOCO_2H \quad (5)$$

where pz is pyrazolyl.

Hydrolytic catalysis by metal ions is also important in the hydrolysis of nucleic acids, especially RNA (*36*). Molecules of RNA that catalyze hydrolytic reactions, termed ribozymes, require divalent metal ions to effect hydrolysis efficiently. Thus, all ribozymes are metalloenzymes (*6*). There is speculation that ribozymes may have been the first enzymes to evolve (*37*), so the very first enzymes may have been metalloenzymes! Recently, substitution of sulfur for the 3′-oxygen atom in a substrate of the tetrahymena ribozyme has been shown to give a 1000-fold reduction in rate of hydrolysis with Mg^{2+} but no attenuation of the hydrolysis rate with Mn^{2+} and Zn^{2+} (*38*). Because Mn^{2+} and Zn^{2+} have stronger affinities for sulfur than Mg^{2+} has, this feature provides strong evidence for a true catalytic role of the divalent cation in the hydrolytic mechanism, involving coordination of the metal to the 3′-oxygen atom. Other examples of metal-ion catalyzed hydrolysis of RNA involve lanthanide complexes, which are discussed in this volume.

Electron Transfer

The ability to exist in more than one oxidation state allows transition-metal complexes to serve as the active site of enzymes whose function is to transfer electrons (*39*). A great deal of effort has been directed at understanding the mechanisms of electron transfer in metalloproteins, such as cytochromes and blue copper proteins (*40*). Of particular interest is the mechanism by which an electron can tunnel from a metal center that is imbedded in a protein matrix to a site on the outer surface of the protein (*7*). A discussion of current theories is given in this volume.

In natural systems, redox proteins such as cytochrome c (cyt c) function not only to transfer electrons, but to transfer electrons *specifically* to a particular redox partner, usually another macromolecule. Transfer of electrons between subunits of modified hemoglobins and within complexes of cyt c with cyt b_5 and cyt c with cyt c peroxidase (Ccp) have therefore been studied extensively (*41, 42*). These studies have revealed the fundamental requirements for the recognition process leading to the formation of the protein–protein complex as well as the thermodynamic features of the electron-transfer reaction itself. This reaction, outlined in equation 6, consists of three fundamental processes: recognition to form a complex (K_1), electron transfer within the complex, and dissociation of the redox-altered complex (K_2). For the cyt c–Ccp complex, Fe(II) cyt c corresponds to ${P_2}^{red}$ and oxidized Ccp corresponds to ${P_1}^{ox}$.

$$ {P_1}^{ox} + {P_2}^{red} \overset{K_1}{\rightleftharpoons} {P_1}^{ox} \cdot {P_2}^{red} \rightarrow {P_1}^{red} \cdot {P_2}^{ox} \overset{K_2}{\rightleftharpoons} {P_1}^{red} + {P_2}^{ox} \qquad (6) $$

The rate of electron transfer within the complex is a function of the thermodynamic driving force of the reaction, the conformation of the protein–protein complex, and the intervening medium (*41*, *42*). Originally, an invariant phenylalanine residue in cyt c was thought to mediate its electron-transfer reactions, but (in one of the early applications of the technique) site-directed mutagenesis of the Phe residue to Ser, Gly, and Tyr showed that the Phe was not essential for function (*43*). The recent crystal structure of the cyt c–Ccp complex shows that electron-transfer pathways do indeed exist that do not involve the Phe residue (Figure 3) (*7*, *44*).

Individual residues from each protein that are vital for complex formation have also been identified through site-directed mutagenesis (9). Thus, formation of the protein–protein complex occurs through a series of site-specific interactions of amino acids on the surface of one protein with particular amino acids on the surface of the other protein. The crystal structures of the complexes of Ccp with cyt c from yeast and horse confirm the site-specific interactions (*44*). Hake et al. (*45*) prepared mutants of Ccp in which key aspartate residues required for complex

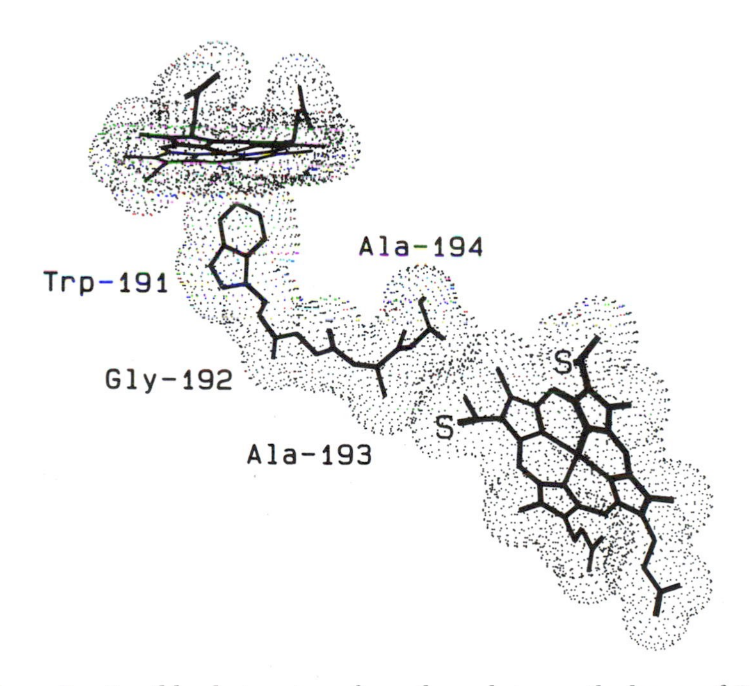

Figure 3. Possible electron-transfer pathway between the hemes of CCp and yeast cyt c determined from the X-ray crystal structure of the protein–protein complex. (Reproduced with permission from reference 44. Copyright 1992 American Association for the Advancement of Science.)

formation were changed to lysine, disrupting any hydrogen bonds formed by the aspartate in the complex. Affinity chromatography experiments confirm the important role played by these residues and also show that the reactant in equation 6, Fe(II) cyt c, binds more strongly to Ccp than the product, Fe(III) cyt c, at physiological ionic strength. Of course, this situation is desired for efficient catalytic electron transfer in the natural system because strong binding of the Fe(III) cyt c product to Ccp would inhibit further reaction. Surprisingly, the order of binding affinities for the various mutants was different for Fe(III) cyt c and Fe(II) cyt c. Thus, the binding probably occurs at different sites on the protein surfaces for the two oxidation states. These experiments show that the simple presence or absence of an electron on a single metal center buried in a large protein can dramatically affect the interactions of the entire macromolecule with its protein partner. This result underlines the profound impact that fundamental inorganic chemistry, such as the oxidation state of a single metal ion, can have on the behavior of complex biological systems.

Redox Catalysis

Enzymes that catalyze the redox reactions of small molecules, such as the reduction of nitrogen to ammonia, the oxidation of water to oxygen, or the oxidation of methane to methanol, have long captured the fascination of inorganic chemists and biochemists, and these and other related processes are discussed elsewhere in this volume. Perhaps these systems are so attractive because the reactions catalyzed are difficult to achieve using either small inorganic catalysts or a protein without a metal center. Therefore, only through the marriage of protein and transition-metal chemistry can these complex transformations occur efficiently with high catalytic turnover. These systems thus present the tantalizing possibility that a detailed understanding of structure and mechanism will lead to logical means of synthesizing small molecules capable of mediating the same transformations. As mentioned above, this interplay of enzyme chemistry and inorganic synthesis has led to several new systems capable of acting as functional models for a number of enzyme reactions (*13*), many of which are discussed here.

In addition to catalyzing the oxidative or reductive transformation of small molecules, redox metalloenzymes can also effect the translocation of protons against a chemical gradient across a membrane. Conceptually, this movement can be understood from the point of view of the fundamental coordination chemistry of metal–oxo complexes (*46*, *47*). Reversible reduction of metal–oxo complexes often occurs concomitantly with protonation, as shown in equations 7 and 8.

$$[M(n + 2)O]^{n+} + H^+ + e^- \rightarrow [M(n + 1)OH]^{(n-1)+} \quad (7)$$

$$[M(n + 1)OH]^{(n-1)+} + H^+ + e^- \rightarrow [M(n)OH_2]^{(n-2)+} \quad (8)$$

These reactions can occur with terminal oxo complexes or bridging oxo complexes (*48*), as well as with complexes of other ligands that can accept protons, such as thiolates or alkoxides. This process is driven by the coupling of the reduction of the metal center to an increase in the basicity of the oxo ligand. From this elementary model, a simple proton pump can be devised if protons are taken up from one side of the membrane upon metal reduction and released to the other side of the membrane upon reoxidation. To achieve this requirement, the pump must exist in two different states, each of which is in protonic contact with a different side of the membrane. In this manner, cycling of the redox state of the metal center leads to proton translocation, and electron transfer can be used to drive the pumping of protons against an electrochemical gradient. This process has been discussed in terms of an eight-state cubic model, in which the eight states correspond to the four forms (two redox and two protonic) of both states of accessibility to each side (*49*).

The respiratory enzyme complex cytochrome c oxidase (Cco) catalyzes the oxidation of cyt c by dioxygen (*10, 11, 50*):

$$4 \text{ cyt } c^{2+} + O_2 + 4H^+ \rightarrow 4 \text{ cyt } c^{3+} + 2H_2O \quad (9)$$

The protein resides in the inner mitochondrial membrane and receives electrons from the cytosolic side while consuming protons from the matrix side. In this way, a proton electrochemical gradient is generated. In addition, the enzyme pumps as many as four additional protons for each dioxygen molecule reduced. This latter mode of proton pumping, in which proton translocation is not coupled to substrate reduction, has been termed "vectorial" in contrast to the "scalar" proton pumping resulting from dioxygen reduction (equation 9).

Four redox-active metal centers are present in Cco: two copper centers, Cu_A and Cu_B, and two hemes from cytochromes a and a_3 (*10, 11, 50*). The heme from cyt a_3 and the Cu_B center are strongly ($-J \geq 200$ cm^{-1}) antiferromagnetically coupled and exist as an $S = 2$ binuclear complex (*51*), which is the probable site of dioxygen reduction. To achieve the strong antiferromagnetic coupling, the Cu_B and heme a_3 centers probably interact via some type of bridging ligand, possibly a μ-oxo ligand. Two laboratories have reported the syntheses of μ-oxo copper–heme complexes (*52, 53*), which are shown in Figure 4. In these complexes, the iron and copper centers are strongly antiferromagnetically coupled, as in Cco. Models for how this site might reduce dioxygen have been proposed (Figure 5) (*10*), and the next step for the modeling

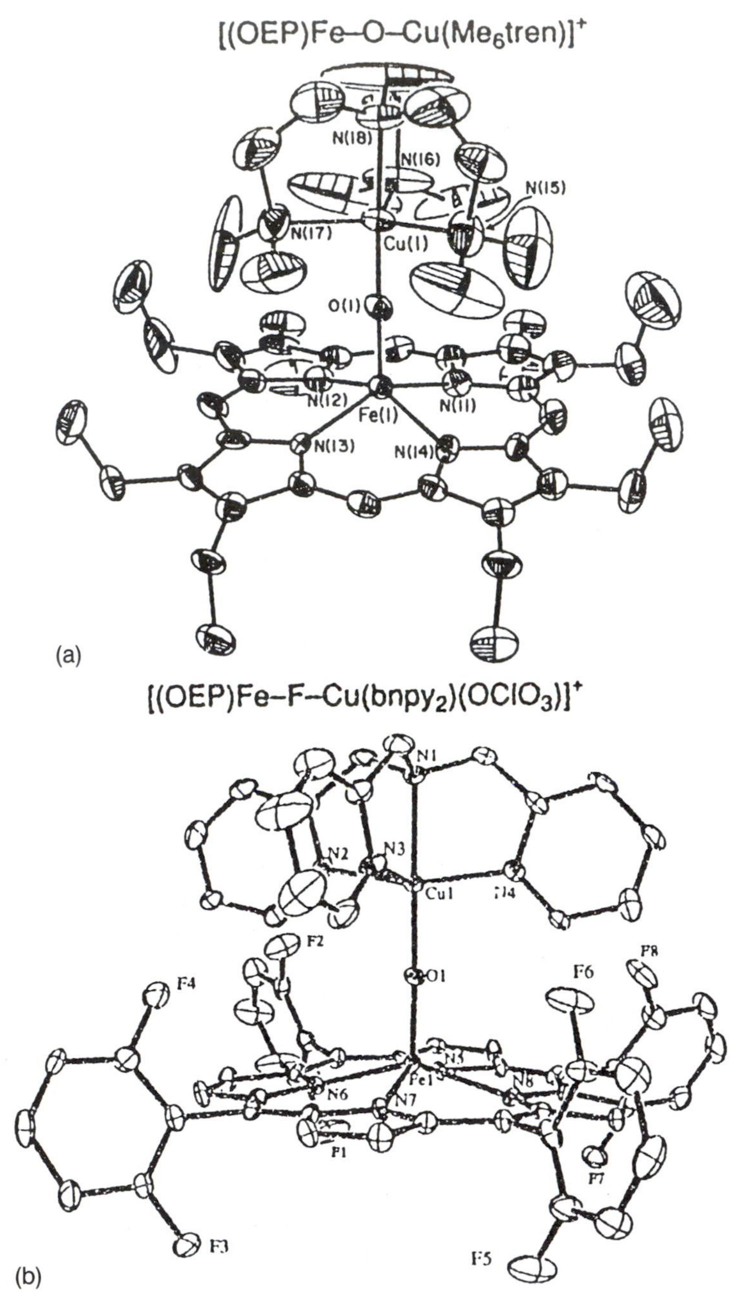

Figure 4. a, X-ray crystal structure of an oxo-bridged heme–copper complex synthesized by Lee and Holm. (Reproduced from reference 52. Copyright 1993 American Chemical Society.) b, X-ray crystal structure of an oxo-bridged heme–copper complex synthesized by Nanthakumar and co-workers. (Reproduced from reference 53. Copyright 1993 American Chemical Society.)

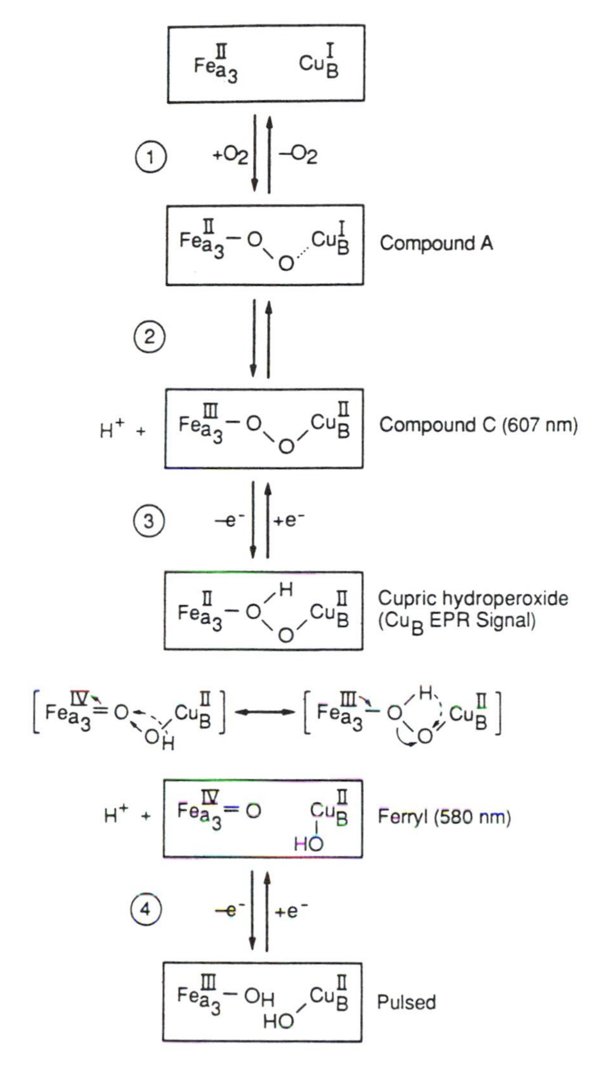

Figure 5. Possible mechanism for reduction of O_2 in cytochrome oxidase involving the binuclear heme–copper site. (Reproduced from reference 10. Copyright 1990 American Chemical Society.)

chemistry will be to achieve dioxygen reduction using the coupled copper–heme models.

In addition to the mechanism of dioxygen reduction, an understanding of how Cco pumps protons is also desirable. Models have been proposed that allow for linkage of the proton pumping to the dioxygen reduction reaction (*50*). One attractive model involves the Cu_A site and is shown in Figure 6 (*10*). In this mechanism, the Cu_A center is ligated to two histidines and two thiolates and receives the initial electron from

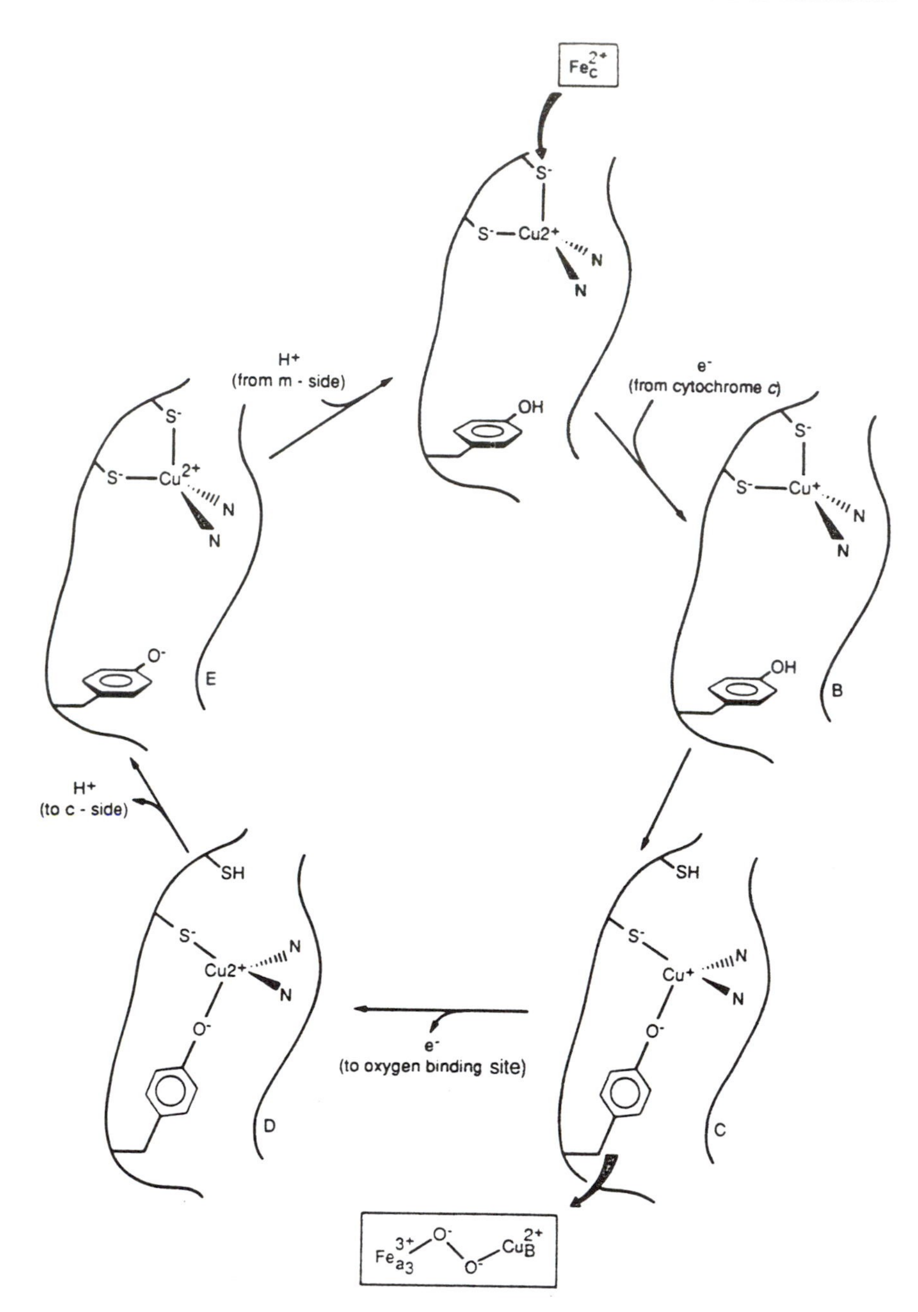

Figure 6. Model for redox-linked proton pumping in cytochrome oxidase involving the Cu_A site. (Reproduced from reference 10. Copyright 1990 American Chemical Society.)

cyt c. The reduced complex releases one thiolate and coordinates a nearby tyrosine ligand, while the proton from the tyrosine is concurrently transferred to the released thiolate. The reduced phenoxide complex then transfers an electron to the oxygen-bound Cu_B–heme a_3 site, and a proton is lost from the sulfhydryl to the cytosol side of the membrane. The resulting free thiolate then displaces the alkoxide ligand on the oxidized copper center, and the free phenoxide is then protonated from the matrix side. This model is attractive because it uses simple coordination chemistry, such as that seen in metal–oxo and related complexes with acid–base active ligands, to effect proton translocation.

Metal-Responsive Gene Expression

As discussed here, biological systems rely on metals for numerous catalytic and structural functions. Living systems therefore have a tremendous need for a regulatory system that will ensure that a particular metal ion is available when it is needed to fulfill one of these catalytic or structural roles. On the other hand, high concentrations of metal ions can be toxic, and the regulatory system must also maintain the concentrations of free metal ions below toxic levels. These regulatory systems must therefore be very finely tuned, and most systems rely on proteins, termed metalloregulatory proteins, that specifically bind the metal ion of interest (*5*). The metal-bound metalloregulatory protein then binds to DNA and induces transcription of messenger RNA (mRNA) that codes for a protein that can catalyze the removal or storage of the excess metal ion. For example, when Hg^{2+} concentrations reach toxic levels, the metalloregulatory protein merR binds a single mercuric ion (*54, 55*). The Hg^{2+}–merR protein distorts the DNA in the promoter region and induces transcription of the gene for mercuric ion reductase (Figure 7), which then renders Hg^{2+} harmless by catalyzing its reduction to the volatile Hg^0 form. A related system regulates copper concentrations by activating transcription of copper metallothionein, a copper storage protein (*56, 57*).

As with copper and mercury ions, concentrations of iron must also be carefully regulated (*58, 59*). The chief iron storage protein is ferritin, which is capable of storing up to 4500 iron atoms in an iron–oxo mineral lattice (*60*). When intracellular iron concentrations are high, synthesis of ferritin is required, as seen with copper metallothionein (*56, 57*). However, the mechanism by which ferritin synthesis is regulated is unique in metalloregulation. Rather than by activation of transcription, in which high metal ion concentrations induce activity of RNA polymerase and production of mRNA, ferritin synthesis proceeds via regulation at the *translational* level, that is, synthesis of protein from mRNA is regulated by iron concentration. The mRNA for ferritin contains a

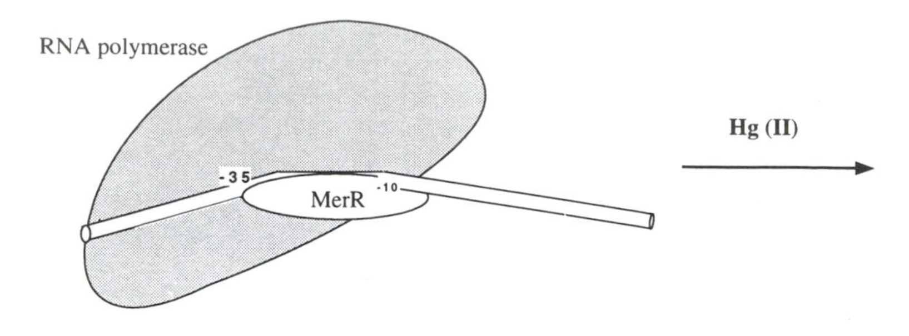

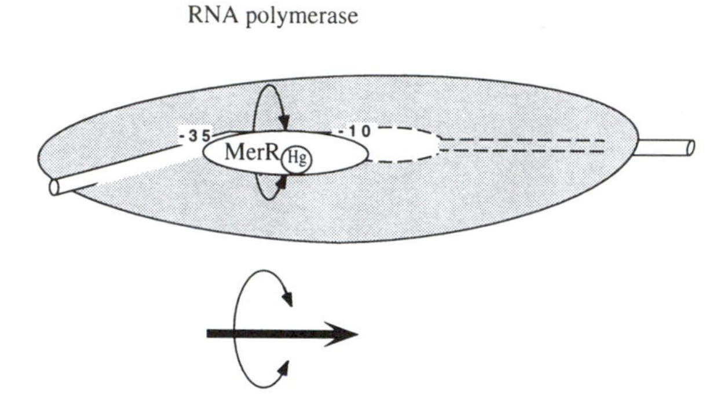

Figure 7. Potential mechanism for Hg^{2+}-responsive gene expression. Binding of Hg^{2+} to merR causes local underwinding of the DNA that optimizes contact of the DNA with RNA polymerase. (Reproduced with permission from reference 5. Copyright 1993 American Association for the Advancement of Science.)

30-nucleotide stem-loop structure, termed the iron recognition element (IRE) (*61*). IREs are present in mRNAs for other iron-storage and -transport proteins, such as transferrin and transferrin receptor, and also play a role in regulating synthesis of these enzymes (*58*). Regulation of ferritin is achieved via interaction of the IRE with the IRE-binding protein (IREBP), which inhibits translation by binding to the IRE. The IREBP binds to the IRE at low iron concentrations, but when iron levels are increased, an iron–sulfur cluster is assembled in the IREBP, inducing a structural change that prohibits RNA binding and permits ferritin synthesis (*62*). Thus, ferritin is synthesized as desired when iron concentrations are high.

The biochemistry of the IRE regulation system for ferritin has a number of unique features. As already stated, the IRE system is unusual in operating at the translational level instead of the transcriptional level. In addition, regulation of ferritin synthesis involves *negative* regulation because binding of the regulatory protein in the absence of the metal ion stimulates gene expression. The other metalloregulatory systems discussed here operate via positive regulation, because binding of the metal ion to the metalloregulatory protein stimulates gene expression. Because the system operates at the translational level, it relies on the structure of an RNA segment, which is much more heterogeneous than the DNA structures that mediate transcriptional activation. Recent studies show that the structure of the IRE is strongly correlated with the presence of highly conserved base pairs in so-called flanking regions on either side of the actual IRE (*63*). An understanding of the complex structure of the IRE is an important goal in unraveling the mechanisms of iron homeostasis, and structural studies using transition-metal cleavage agents appear particularly fruitful (*64*).

In addition to a unique biochemistry, the inorganic chemistry of ferritin regulation is also unusual. The coordination chemistries of the metalloregulatory protein and the regulated protein are quite different (*5*). Assembly of the ferritin core involves synthesis of a metal–oxo core, whereas the active site of the IREBP is an iron–sulfur cluster, which involves a significantly different type of coordination chemistry. Related metalloregulatory systems exhibit similar chemistry for both the regulated protein and the metalloregulatory protein. Also, when the iron–sulfur cluster is assembled in the IREBP, the protein can function as an aconitase enzyme that catalyzes the conversion of citrate to isocitrate (*62*). This example is the only one known of a separate enzymatic activity for a metalloregulatory protein. The role (if there is one) of this enzymatic activity in the iron regulation process is unclear; however, the coupling of metalloregulation to metalloenzymatic catalysis by a single enzyme is an exciting possibility.

Conclusions

The future of bioinorganic chemistry is filled with promise and excitement. As more sophisticated physical methods are developed for the study of biological systems, even higher levels of structural characterization will become available and more detailed mechanistic studies will be possible. Greater understanding of enzyme mechanisms will be accompanied by the development of synthetic model complexes that not only mimic spectroscopic and structural features but also act as functional models. Significant progress in this direction is already being made (*13*). This research will provide an increased understanding of enzyme mechanisms and a host of synthetic catalysts with new reactivities and efficiencies. The new understanding of the mechanisms of hydrolytic and oxidative cleavage of nucleic acids will provide new techniques for understanding the structures of complex nucleic acids and clues for manipulating DNA and RNA in living systems (*6, 65–67*). As the fundamental properties of metal ions in biological systems become apparent, the ability of living systems to regulate metal ion concentrations becomes an important issue (*5*). An understanding of these regulatory mechanisms represents an important new frontier that may ultimately allow scientists to manipulate the concentrations of metal ions in vivo and to prepare pharmaceutical products with desirable properties (*68*).

Acknowledgments

I thank the National Science Foundation for a Presidential Young Investigator award and the David and Lucile Packard Foundation for a Fellowship in Science and Engineering.

References

1. Lippard, S. J. *Science (Washington, D.C.)* **1993,** *261*, 699–700.
2. Ibers, J. A.; Holm, R. H. *Science (Washington, D.C.)* **1980,** *209*, 223–235.
3. Lippard, S. J. *Acc. Chem. Res.* **1978,** *11*, 211–217.
4. Coleman, J. E. *Annu. Rev. Biochem.* **1992,** *61*, 897–946.
5. O'Halloran, T. V. *Science (Washington, D.C.)* **1993,** *261*, 715–725.
6. Pyle, A. M. *Science (Washington, D.C.)* **1993,** *261*, 709–714.
7. Beratan, D. N.; Onuchic, J. N.; Winkler, J. R.; Gray, H. B. *Science (Washington, D.C.)* **1992,** *258*, 1740–1741.
8. Solomon, E. I.; Lowery, M. D. *Science (Washington, D.C.)* **1993,** *259*, 1575–1581.
9. McLendon, G.; Hake, R. *Chem. Rev.* **1992,** *92*, 481–490.
10. Chan, S. I.; Li, P. M. *Biochemistry* **1990,** *29*, 1–12.
11. Malmström, B. G. *Chem. Rev.* **1990,** *90*, 1247–1260.
12. Vallee, B. L. In *Zinc Enzymes;* Spiro, T. G., Ed.; Wiley: New York, 1983; pp 1–24.
13. Karlin, K. D. *Science (Washington, D.C.)* **1993,** *261*, 701–708.
14. Kozarich, J. W.; Worth, L., Jr.; Frank, B. L.; Christner, D. F.; Vanderwall, D. E.; Stubbe, J. *Science (Washington, D.C.)* **1989,** *245*, 1396–1399.
15. Sitlani, A.; Long, E. C.; Pyle, A. M.; Barton, J. K. *J. Am. Chem. Soc.* **1992,** *114*, 2303–2312.
16. Hecht, S. M. *Acc. Chem. Res.* **1986,** *19*, 83–89.
17. Guajardo, R. J.; Hudson, S. E.; Brown, S. J.; Mascharak, P. K. *J. Am. Chem. Soc.* **1993,** *115*, 7971–7977.
18. Chan, M. K.; Kim, J.; Rees, D. C. *Science (Washington, D.C.)* **1993,** *260*, 792–794.
19. Berg, J. M. *Prog. Inorg. Chem.* **1989,** *37*, 143–185.
20. Pavletich, N. P.; Pabo, C. O. *Science (Washington, D.C.)* **1993,** *261*, 1701–1707.
21. Pavletich, N. P.; Pabo, C. O. *Science (Washington, D.C.)* **1991,** *252*, 809–817.

22. Harrison, S. C. *Nature (London)* **1991,** *353,* 715–719.
23. Sherman, S.; Gibson, D.; Wang, A. H. J.; Lippard, S. J. *Science (Washington, D.C.)* **1985,** *230,* 412–415.
24. Brown, S. J.; Kellett, P. J.; Lippard, S. J. *Science (Washington, D.C.)* **1993,** *261,* 603–605.
25. Pil, P. M.; Lippard, S. J. *Science (Washington, D.C.)* **1992,** *256,* 234–237.
26. Brudvig, G. W. In *Metal Clusters in Proteins;* Que, L., Jr., Ed.; ACS Symposium Series 372; American Chemical Society: Washington, D.C., 1988; pp. 231–237.
27. Abragam, A.; Bleaney, B. *Electron Paramagnetic Resonance of Transition Ions;* Clarendon: Oxford, 1970.
28. Cramer, S. P.; Hodgson, K. O. *Prog. Inorg. Chem.* **1979,** *25,* 1–40.
29. Teo, B. K. *EXAFS: Basic Principles and Data Analysis*; Springer-Verlag: New York, 1986.
30. Liu, W.; Thorp, H. H. *Inorg. Chem.* **1993,** *32,* 4102–4105.
31. Thorp, H. H. *Inorg. Chem.* **1992,** *31,* 1585–1588.
32. Banci, L.; Bertini, I.; Luchinat, C.; Donaire, A.; Martinez, M.-J. Maratal Mascarelli, J. M. *Comments Inorg. Chem.* **1990,** *9,* 245–262.
33. Erikkson, A. E.; Jones, T. A.; Liljas, A. *Proteins: Struct. Funct. Genet.* **1988,** *4,* 274–282.
34. Looney, A.; Parkin, G.; Alsfasser, R.; Ruf, M.; Vahrenkamp, H. *Angew. Chem. Int. Ed. Engl.* **1992,** *31,* 92–93.
35. Looney, A.; Han, R.; McNeill, K.; Parkin, G. *J. Am. Chem. Soc.* **1993,** *115,* 4690–4697.
36. Cech, T. R. *Science (Washington, D.C.)* **1987,** *236,* 1532–1539.
37. Piccirilli, J. A.; McConnell, T. S.; Zaug, A. J.; Noller, H. F.; Cech, T. R. *Science (Washington, D.C.)* **1992,** *256,* 1420–1424.
38. Piccirilli, J. A.; Vyle, J. S.; Caruthers, M. H.; Cech, T. R. *Nature (London)* **1993,** *361,* 85–88.
39. Marcus, R. A.; Sutin, N. *Biochim. Biophys. Acta* **1985,** *811,* 265–316.
40. Wuttke, D. S.; Gray, H. B. *Curr. Opin. Struct. Biol.* **1993,** *3,* 555–563.
41. Hoffman, B. M.; Natan, M. J.; Nocek, J. M.; Walin, S. A. *Struct. Bonding* **1991,** *75,* 85–108.
42. McLendon, G. *Struct. Bonding (Berlin)* **1991,** *75,* 159–174.
43. Pielak, G. J.; Mauk, A. G.; Smith, M. *Nature (London)* **1985,** *313,* 152–154.
44. Pelletier, H.; Kraut, J. *Science (Washington, D.C.)* **1992,** *258,* 1748–1755.
45. Hake, R.; McLendon, G.; Corin, A.; Holzchu, D. *J. Am. Chem. Soc.* **1992,** *114,* 5442–5443.
46. Thorp, H. H. *Chemtracts: Inorg. Chem.* **1991,** *3,* 171.
47. Meyer, T. J. *J. Electrochem. Soc.* **1984,** *131,* 221C–231C.
48. Thorp, H. H.; Sarneski, J. E.; Brudvig, G. W.; Crabtree, R. H. *J. Am. Chem. Soc.* **1989,** *111,* 9249–9250.
49. Wikström, M. *Biochem. Biophys. Acta* **1979,** *549,* 177–222.
50. Babcock, G. T.; Wikström, M. *Nature (London)* **1992,** *356,* 301–309.
51. Brudvig, G. W.; Morse, R.; Chan, S. I. *J. Magn. Reson.* **1986,** *67,* 189–201.
52. Lee, S. C.; Holm, R. H. *J. Am. Chem. Soc.* **1993,** *115,* 5833–5834.
53. Nanthakumar, A. S. F.; Murthy, N. N.; Karlin, K. D.; Ravi, N.; Huynh, B. H.; Orosz, R. D.; Day, E. P.; Hagen, K. S.; Blackburn, N. J. *J. Am. Chem. Soc.* **1993,** *115,* 8513–8514.
54. Wright, J. G.; Natan, M. J.; MacDonnell, F. M.; Ralston, D. M.; O'Halloran, T. V. *Prog. Inorg. Chem.* **1990,** *38,* 323–412.

55. Schiering, N.; Kabsch, W.; Moore, M. J.; Distefano, M. D.; Walsh, C. T.; Pai, E. F. *Nature (London)* **1991,** *352,* 168–172.
56. Fürst, P.; Hamer, D. *Proc. Natl. Acad. Sci. U.S.A.* **1991,** *86,* 5267–5271.
57. Thiele, D. J. *Nucleic Acids Res.* **1992,** *20,* 1183.
58. Theil, E. C. *Biofactors* **1993,** *4,* 87–93.
59. Klausner, R. D.; Rouault, T. A.; Harford, J. B. *Cell* **1993,** *72,* 19–28.
60. Taft, K. L.; Papaefthymiou, G. C.; Lippard, S. J. *Science (Washington, D.C.)* **1993,** *259,* 1302–1305.
61. Harrell, C. M.; McKenzie, A. R.; Patino, M. M.; Walden, W. E.; Theil, E. C. *Proc. Natl. Acad. Sci. U.S.A.* **1991,** *88,* 4166–4170.
62. Haile, D. J.; Rouault, T. A.; Harford, J. B.; Kennedy, M. C.; Blondin, G. A.; Beinert, H.; Klausner, R. D. *Proc. Natl. Acad. Sci. U.S.A.* **1992,** *89,* 11735–11739.
63. Dix, D. J.; Lin, P.-N.; McKenzie, A. R.; Walden, W. E.; Theil, E. C. *J. Mol. Biol.* **1993,** *231,* 230–240.
64. Theil, E. C. *New J. Chem.* **1994,** *18,* 435–441.
65. Pyle, A. M.; Barton, J. K. *Prog. Inorg. Chem.* **1990,** *38,* 413–476.
66. Neyhart, G. A.; Grover, N.; Smith, S. R.; Kalsbeck, W. A.; Fairley, T. A.; Cory, M.; Thorp, H. H. *J. Am. Chem. Soc.* **1993,** *115,* 4423–4428.
67. Stubbe, J.; Kozarich, J. W. *Chem. Rev.* **1987,** *87,* 1107–1143.
68. Abrams, M. J.; Murrer, B. A. *Science (Washington, D.C.)* **1993,** *261,* 725–730.

RECEIVED for review November 8, 1993. ACCEPTED revised manuscript March 15, 1994.

2

Insights into the Role of Nickel in Hydrogenase

Michael J. Maroney,[1] Michelle A. Pressler,[1] Shaukat A. Mirza,[1] Joyce P. Whitehead,[1] Ryzard J. Gurbiel,[2] and Brian M. Hoffman[2]

[1]Department of Chemistry, University of Massachusetts, Amherst, MA 01003
[2]Department of Chemistry, Northwestern University, Evanston, IL 60280

The nickel site in Thiocapsa roseopersicina *hydrogenase was examined in the five redox forms defined by electron paramagnetic resonance spectroscopy at 77 K by use of X-ray absorption spectroscopy. These studies show that the nickel site is remarkably insensitive to changes in the redox state of the enzyme, a result that is inconsistent with nickel-centered redox chemistry. Model studies of a series of nickel complexes with ligands of the type* $RN(CH_2CH_2S)_2$ *show that the products of one-, two-, and four-electron oxidations all reflect sulfur-centered chemistry. The role of the nickel site in binding hydrogen was explored by using a combination of electron nuclear double resonance and X-ray absorption spectroscopic techniques. These studies do not completely rule out a role for the nickel site, but they point to the possibility that nickel is not the hydrogen-binding site. These results are discussed within the context of biological and inorganic chemical literature pertaining to nickel thiolate complexes.*

HYDROGENASES (H_2ases) are a widely distributed class of enzymes found in both prokaryotes and eukaryotes that catalyze the reversible two-electron oxidation of molecular hydrogen (eq 1) (*1–3*). Thus, H_2ases may function to provide reducing equivalents for energy production via the uptake and oxidation of H_2 or may reduce H^+ in the production of H_2. Hydrogen oxidation (uptake) is generally coupled with phosphorylation and ultimately with the reduction of inorganic substrates such as SO_4^{2-} (*Desulfovibrio* species), CO_2 (*Methanobacterium* species), NO_3^- (e.g., *Paracoccus denitrificans*), or O_2 (*Alcaligenes* and *Nocardia*). H_2ases

0065–2393/95/0246–0021/$11.60/0

may also play a role in cycling hydrogen produced in other systems (e.g., nitrogenase) or in generating a proton gradient (*1–4*).

$$H_2 \rightleftharpoons 2H^+ + 2e^- \quad (1)$$

Hydrogenases have been grouped into three classes (*1*) that are based on the inorganic content of the enzymes and are immunologically and biochemically distinct (*5*). With the possible exception of a recently purified enzyme, N^5,N^{10}-methylenetetrahydromethanopterin dehydrogenase from *Methanobacterium thermoautotrophicum*, that possesses H_2ase activity but does not contain Fe (*6, 7*), all H_2ases contain Fe,S clusters. Four enzymes have been rigorously shown to contain only Fe and S^{2-} (*1*). With the exception of a H_2ase from *D. vulgaris*, these Fe-only enzymes are monomeric proteins of 60-kDa molecular weight that are extremely O_2 sensitive ($t_{1/2} \approx$ a few minutes in air) and irreversibly deactivated. Because they catalyze both H_2 oxidation and H_2 production at high rates in vitro (V_m (H_2 evolution) $\approx$ 6000 μmol min^{-1} mg^{-1}); V_m (H_2 oxidation) $\approx$ 20,000 μmol min^{-1} mg^{-1}), they are frequently referred to as bidirectional H_2ases.

Since the discovery of Ni as a biological component of *Methanobacterium bryantii* in 1980 (*8*) and the subsequent identification of the Ni-containing component as a H_2ase (9), dozens of examples of H_2ases that require a single Ni atom as well as Fe,S clusters have been characterized. H_2ases belonging to the Ni,Fe class are generally associated with hydrogen oxidation in vivo and are widely distributed; examples are known from fermentative (*10*), SO_4-reducing (*5, 11*), methanogenic (*12*), photosynthetic (*4, 13*), facultative (*14*), and aerobic bacteria (*3, 15*). The Ni,Fe H_2ases are often $\alpha\beta$ dimers with subunit molecular weights of approximately 60 and 30 kDa. In enzymes containing more than two subunits (e.g., in *Alcaligenes eutrophus*, an $\alpha\beta\gamma\delta$ tetramer (*16*)), two subunits with the characteristic molecular weights are usually associated with the H_2ase activity. The Ni,Fe H_2ases have activities that are only 1–10% as large as those typical of the Fe-only enzymes (V_m (H_2 evolution) $\approx$ 450 μmol min^{-1} mg^{-1}; V_m (H_2 oxidation) $\approx$ 1500 μmol min^{-1} mg^{-1}) (*5*) and are more oxygen tolerant ($t_{1/2}$ in air varies from several hours to several weeks (*3*)). Furthermore, the deactivated enzymes may be reductively reactivated.

Among the H_2ases containing Ni are a class of enzymes that also contain Se: the Ni,Fe,Se H_2ases (*5*). The Se is generally present as a single selenocysteine residue, most notably in *D. baculatus*. The selenocysteine residue has been shown to be encoded by an internal TGA codon (*17*), to constitute a conservative replacement for a cysteine residue in enzymes lacking selenocysteine (*17*), and to be one of the Ni ligands (*18, 19*). However, examples of enzymes that contain labile Se are also known in which the H_2ase gene does not contain the TGA codon

(*20*). Although selenocysteine-containing enzymes are clearly related to the Ni,Fe enzymes (*21*), they are even more O_2 tolerant than are the Ni,Fe enzymes and are typically isolated in air in a form that does not require reductive activation (*5*). The catalytic activities of the Ni,Fe,Se enzymes are even lower than those found in the Ni,Fe enzymes (V_m (H_2 evolution) $\approx$ 450 μmol min^{-1} mg^{-1}; V_m (H_2 oxidation) $\approx$ 100 μmol min^{-1} mg^{-1}) (*5*).

In general, the trends in decreasing H_2ase uptake and production activity are paralleled by increasing affinities for H_2 and higher H_2/HD ratios in proton–deuterium exchange assays (*5*). Clearly, the presence of Ni and the addition of selenocysteine have an effect on the catalytic activities and oxygen sensitivities of the H_2ases, although the physiological significance of these differences is not known. It appears that the only feature characteristic of all H_2ases may be metal–sulfur bonds. One view of the role played by Ni is that nature varies the metal composition of the active site in the enzyme in order to influence the reactivity of the metals toward H_2 or O_2. Another view is that the role of Ni is to modify the active site, which might involve chemistry that occurs at Fe or at metal ligands. We used spectroscopic techniques in combination with a synthetic model approach to investigate the role of the Ni center in hydrogenase. These studies are summarized here within the context of the biological and inorganic literature pertaining to Ni thiolate complexes.

One of the most interesting aspects of the H_2ase Ni site is its association with redox chemistry involving unusual formal oxidation states of Ni. In contrast with Ni(II), which has an even number of electrons, the presence of Ni in H_2ases is often detected by the appearance of characteristic rhombic electron paramagnetic resonance (EPR) signals (g = 2.4–2.0) in oxidized and reduced samples of the enzyme (*2, 3*). These signals have been associated with an S = ½ Ni species, from the observation of hyperfine splitting arising from ^{61}Ni-labeled samples (*22*) and have provided the principal biophysical probe of the Ni site. The "Ni EPR signals" in the enzyme can be distinguished from those arising from Fe,S clusters because they can be observed at 77 K, whereas those from the Fe,S clusters require temperatures below 30 K to become observable. The EPR signals observed at 77 K have been used to monitor the redox state of the enzyme (*2, 3, 23–25*) and to infer an interaction of the Ni site with inhibitors (e.g., CO) (*26*), H_2 (*27*), and another paramagnet (e.g., an Fe,S cluster) (*3*).

A model for the redox chemistry of the Ni site based on redox titrations of H_2ase is summarized in Figure 1. Ni,Fe H_2ases are generally isolated in air as a combination of two fully oxidized and inactive forms that can be distinguished by their Ni EPR spectra and their kinetics of activation. Form A (g = 2.31, 2.23, and 2.02) requires extensive incu-

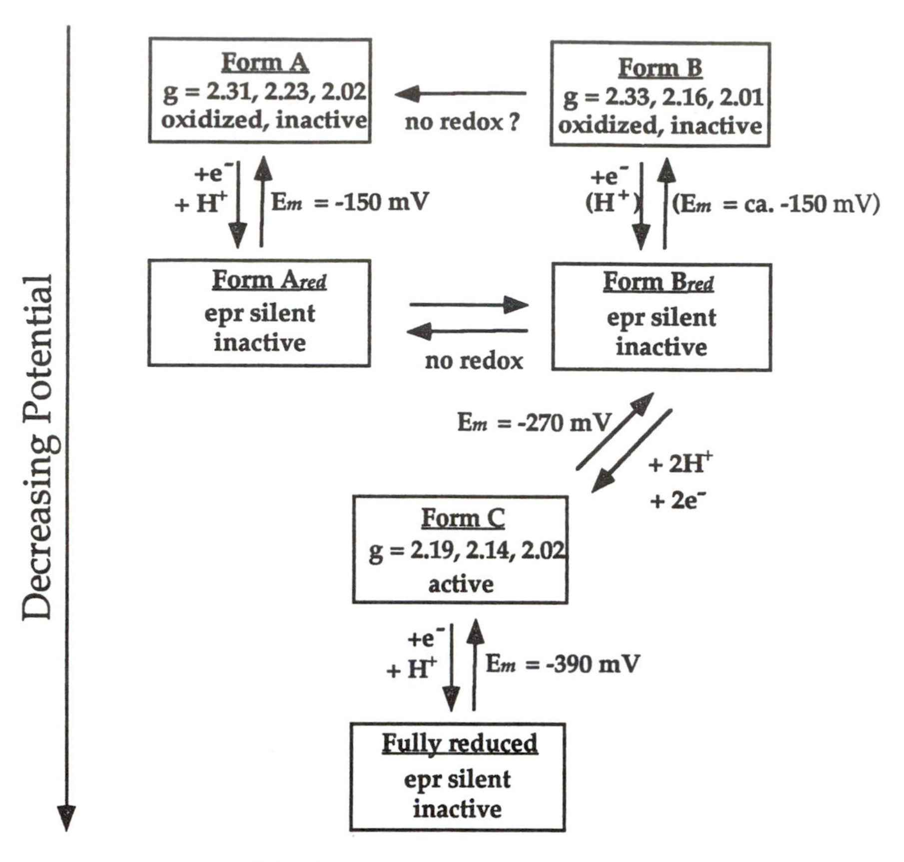

Figure 1. A model for the interconversion of Ni,Fe hydrogenase, based on the work of Cammack et al. (3) and Albracht et al. (28). The example potentials and pH dependencies are those determined for D. gigas *hydrogenase at pH 7 vs. NHE (3).*

bation under H_2 or treatment with strong reducing agents to be activated, whereas form B (g = 2.33, 2.16, and 2.01) is instantaneously activated by exposure to H_2. Both forms are initially reduced by H_2 to form an EPR silent state of the Ni center. Recently obtained evidence indicates that form B may be converted to form A, but form A is not converted to form B prior to reduction. This data implies that the EPR silent state may be an equilibrium mixture of two forms, one that is oxidized to form A, the other to form B (28). In any event, studies of the magnetic properties of the Ni center in the EPR silent intermediate form of H_2ase from *D. baculatus* (an enzyme that contains a selenocysteine Ni ligand) indicate that the Ni is diamagnetic and must therefore be Ni(II) (29).

Further exposure to H_2 results in the formation of a third EPR active form, form C (g = 2.19, 2.14, and 2.02). Redox titrations show that the intensity of this signal reaches a maximum at the same potentials associated with catalytic activity; thus, form C has been attributed to a catalytically active form of the enzyme (*3*). Form C may be further reduced to a fully reduced form that is EPR silent at 77 K. Redox titrations of several H_2ases have been performed to determine the potentials associated with the transformations of the Ni EPR spectra (*2, 3*). The potentials associated with the redox processes of Ni in H_2ase generally lie between ~0 and –414 mV (vs. NHE), the latter being the potential of the H^+/H_2 couple at pH 7. The potentials determined vary somewhat with the source of the enzyme used; those determined for *D. gigas* are shown in Figure 1.

The nature of the ligand environment of the Ni center has been addressed by use of a combination of spectroscopic techniques. X-ray absorption spectral data are now available for Ni,Fe H_2ases from *D. gigas* (*30–32*), *M. thermoautotrophicum* (*33*), and *Thiocapsa roseopersicina* (*34, 35*) and for the Ni,Fe,Se enzyme from *D. baculatus* (*19*). In general, these studies indicate that the Ni site is 5–6 coordinate and contains at least two S-donor ligands at a distance of 2.2 Å. Because of the lack of visible hyperfine interactions in the enzyme spectra, EPR has not been of much value in probing the ligand environment of the Ni. However, the use of isotopes with magnetic nuclei has proved useful. Studies using bacteria grown on ^{33}S-enriched media reveal that the Ni signal interacts with 1–2 sulfur atoms (*36*). EPR has also shown that ^{13}CO interacts strongly with the unpaired spin, consistent with binding of this inhibitor to the Ni center (*26*). On the other hand, studies using $^{17}O_2$ in the deactivation of the enzyme (formation of forms A and B) revealed only weak hyperfine interactions, leading to the conclusion that O_2 does not interact directly with Ni but is bound in the vicinity of the Ni in both oxidized forms (*26*). Electron spin echo envelope modulation (ESEEM) studies reveal an interaction with a N atom in many (*37–39*), but not all (*39*) cases; it is not clear, however, whether this interaction represents ligation by a N-donor ligand.

More information regarding possible Ni-binding ligands has been obtained from an examination of homologies in 17 amino acid sequences, coupled with site-directed mutagenesis on *Escherichia coli* H_2ase-1 and studies of the Ni binding capacity and catalytic activity of the mutants (*21*). These studies indicated that the large subunit of Ni,Fe H_2ases contains the Ni-binding site and is highly conserved. Within the amino acid sequence of the large subunits are two fully conserved sequences: R-X-C-X-G-C near the amino terminus and D-P-C-X-X-C near the carboxyl terminus. The only exceptions are the case of two Ni,Fe,Se enzymes, in which the first cysteine residue in the carboxyl terminal region is substituted by the known Ni ligand, selenocysteine, and in the case of *M. thermoautotrophicum* F_{420}-reducing H_2ase, in which the conserved glycine is substituted by cysteine. The conserved sequences contain six possible Ni ligands: four cysteines, an aspartate, and an arginine. The mutagenesis and biochemical studies of the mutants indicate that these six amino acids are potential Ni ligands. Histidines flanking the conserved regions were less well conserved, and mutants lacking these residues were active H_2ases.

The structure that emerges from the physical and biochemical studies is one that contains a 5–6 coordinate Ni atom in a mixed ligand environment, featuring at least two cysteinate ligands and one site that is available for binding exogenous ligands.

Ni Redox Chemistry

Various schemes using formal Ni oxidation states IV–0 have been used to account for the appearance and disappearance of EPR signals associated with the Ni site (*3, 24, 40*). Two of these schemes are summarized in Table I. Proposal A simply assigns oxidation states of III–0, with odd oxidation states corresponding to EPR active species. Such redox chemistry within a 400-mV potential range is unprecedented in Ni chemistry. Alternatively, proposal B uses only one-electron redox chemistry for Ni but suggests that somehow the Ni site becomes reoxidized at a *lower* potential, implying that the structure and the protonation state of the Ni site have changed.

In the absence of data regarding the redox role of the Ni ligands, the Fe,S clusters, or other groups nearby, or data that have a direct bearing on changes in the electron density and charge density of the Ni, schemes assigning formal oxidation states to the Ni atom offer no chemical or mechanistic insight. To assess the role of Ni-centered redox chemistry involving widely differing oxidation states for Ni, we have examined the Ni K-edge X-ray absorption spectrum of samples of *T. roseopersicina* H_2ase poised in each of the five states defined by EPR spectra observed at 77 K (*35*). The EPR spectra obtained from samples

Table I. Redox Schemes

Observed Phenomenon	E_m *(mV vs. NHE at pH 7.0)*	*Proposal A*	*Proposal B*
EPR of form A	−150	Ni(III)	Ni(III)
EPR of form B		Ni(III)	Ni(III)
EPR silent intermediate		Ni(II)	Ni(II)
Appearance of form C signal	−270	Ni(I)	Ni(III)
Disappearance of form C signal	−390	Ni(0)	Ni(II)
Reductive activation	−310		
Oxidative deactivation	−133		

frozen in sample holders for the X-ray absorption spectroscopy (XAS) experiment are shown in Figure 2. These spectra demonstrate that ≥80% of the Ni present in the enzyme is poised in the desired form. They also illustrate the sequential change in the EPR spectrum that is observed as the enzyme is reduced.

The Ni K-edge X-ray absorption spectra corresponding to the EPR spectra are shown in Figure 3 and summarized in Table II. The most striking feature of the H_2ase Ni K-edge XAS spectra is the lack of sensitivity to the oxidation state of the enzyme as determined by the EPR spectrum of the Ni center. The edges do not exhibit a significant shift to lower energy upon reduction of the enzyme. Values for the edge energy are within 0.2 eV of each other with the exception of Form A (the same formal oxidation state as for Form B), which has an edge energy that is ≤1 eV higher in energy.

The X-ray absorption edge energy is a sensitive measure of the charge residing on the metal center and is therefore a good probe of metal-centered redox processes. K-edge energy shifts have been widely used to monitor redox chemistry in metalloenzymes. The results from several metalloprotein studies are shown in Table III. In general, shifts of <1 eV indicate that the redox chemistry involved is not localized on the metal center. This is clearly the case for the Ni site in *T. roseopersicina* H_2ase.

This approach was used to examine the redox chemistry of the Cu site in galactose oxidase (*41*), which had been proposed to contain an unusual Cu(III) center (*52*). The lack of a significant Cu K-edge energy shift between the oxidized and reduced forms of the protein demonstrated that the redox chemistry was not metal-centered and implicated another redox active site. The crystal structure of the protein subsequently revealed a novel thioether composed of a cysteine and a tyrosinate ligand of the Cu site that is likely to be involved in the redox process (*53*).

The pre-edge region of the spectra obtained for redox-poised samples of *T. roseopersicina* show no evidence of a peak or shoulder near 8338 eV that has been assigned to a $1s \rightarrow 4p_z$ transition (with shakedown contributions) (*54–57*) and is observed only in the XAS spectra of planar four-coordinate complexes and pyramidal five-coordinate complexes (*32, 58*). In several cases, it is possible to resolve a weak peak near 8332 eV that has been assigned to a $1s \rightarrow 3d$ transition (Figure 3) (*32, 59, 60*). The $1s \rightarrow 3d$ transition is symmetry-forbidden in centrosymmetric point groups, but is expected to gain intensity in geometries that allow p–d mixing to occur (*58, 61*). In general, the $1s \rightarrow 3d$ peaks range from 0 to 0.015(5) eV in the enzyme, in ratios of 0–0.13 relative to the area of the $(Et_4N)_2[NiCl_4]$ pre-edge peak (*58*). The low intensity of this feature in the H_2ase spectra is consistent with either a planar four-coordinate

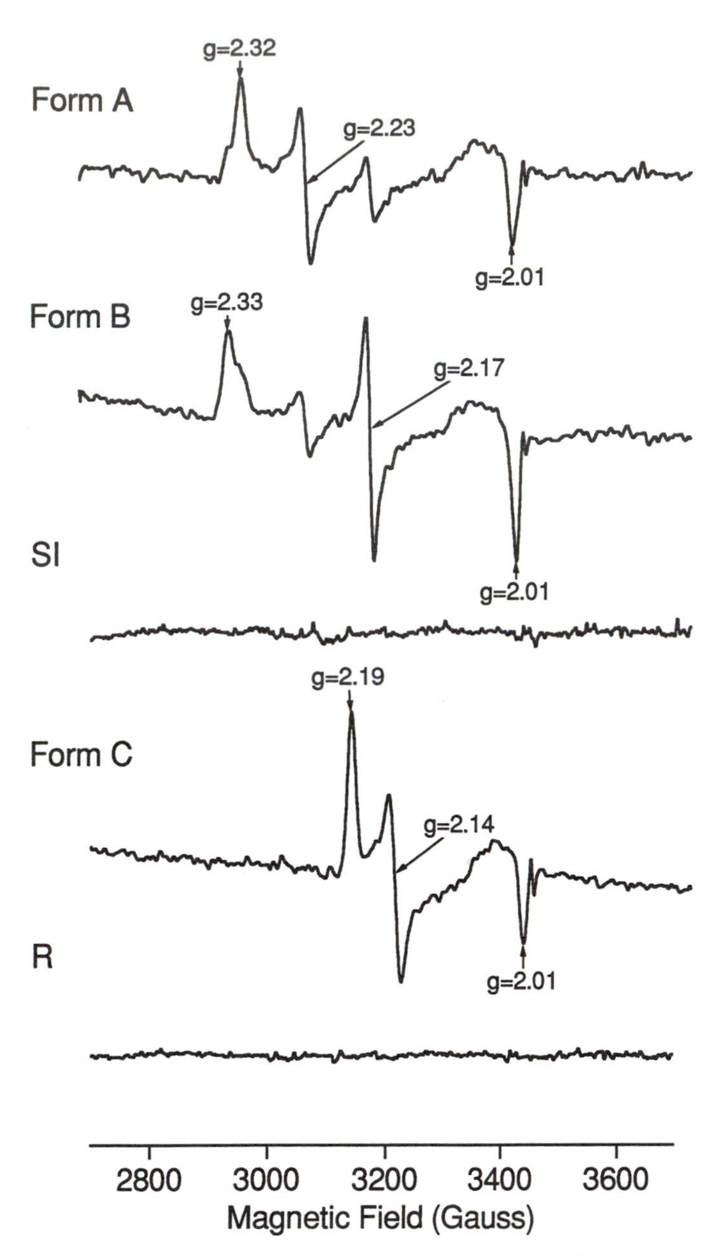

Figure 2. EPR spectra from Thiocapsa roseopersicina *hydrogenase obtained on samples used in XAS experiments at 77 K. The spectra are arranged in order of decreasing redox potential top to bottom. Forms A and B correspond to oxidized enzyme, SI is an EPR silent intermediate, form C is an active form of the enzyme that is also EPR-active, and R is the fully reduced enzyme. (Reproduced from reference 35. Copyright 1993 American Chemical Society.)*

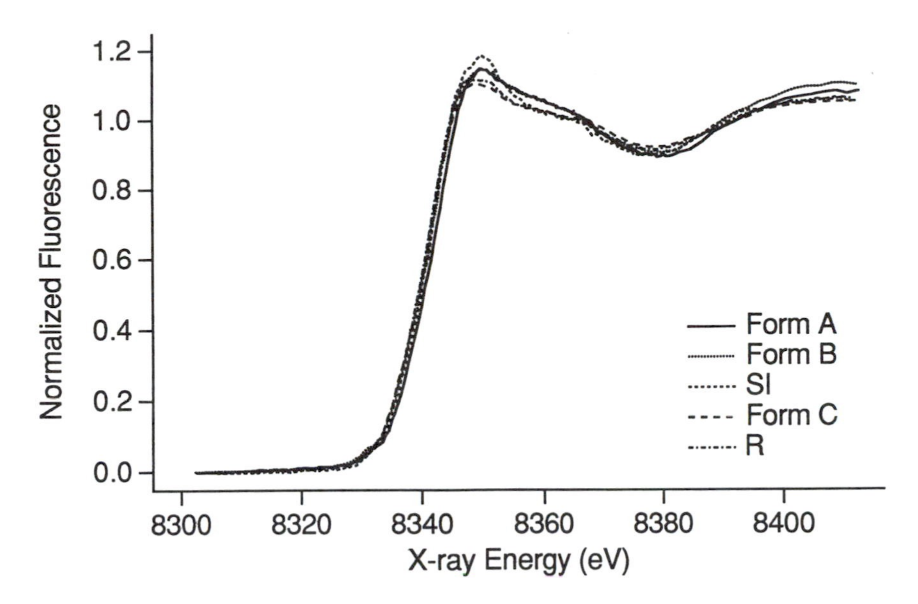

Figure 3. Nickel K-edge X-ray absorption spectra obtained on samples of Thiocapsa roseopersicina *hydrogenase poised in the five forms defined by the EPR spectra at 77 K. (Reproduced from reference 35. Copyright 1993 American Chemical Society.)*

geometry [peak areas of 0–0.029(5) eV (58)] or a six-coordinate geometry [peak areas of 0.006(5)–0.040(5) eV (58)]. Given the absence of a $1s \rightarrow 4p_z$ transition that is expected for the planar and pyramidal geometries (shoulder), the spectra obtained from all of the H_2ase samples are most consistent with a six-coordinate or a five-coordinate trigonal-bipyramidal Ni site.

The post-edge X-ray absorption near-edge structure (XANES) observed for the five redox states remains nearly constant, also suggesting that the structure of the Ni site does not change during reduction of the enzyme. The lack of a significant change in the structure of the Ni site is also seen from an analysis of the first coordination sphere extended X-ray absorption fine structure (EXAFS) data for the Ni site (Figure 4), which can be fit with 2–3 S,Cl donors at 2.23 ± 0.03 Å and 3 ± 1 N,O-donors at 2.00 ± 0.06 Å regardless of oxidation state. This result can be contrasted with expectations for metal-centered redox chemistry based on the structures of $[Ni^{III}(pdtc)_2]^-$ and $[Ni^{II}(pdtc)_2]^{2-}$, which show a decrease of 0.14 Å in the average Ni–S bond length upon oxidation of Ni(II) to Ni(III) (62). A logical conclusion is that the redox chemistry associated with the processes that give rise to the EPR spectra characteristic of H_2ase are largely not centered in Ni orbitals.

Table II. EPR and Ni K-Edge Data

Enzyme Redox State	*Ni K-Edge Energy ± 0.2 (eV)*	*1s→3d Peak Area ± 0.005 eV (relative to $NiCl_4^{2-}$)*[a]	*% Ni EPR Detectable*[b]	*% of EPR Active Ni Poised*[c]
Form A	8340.4	0.015 (0.132)	91	78
Form B	8339.4	0.040 (0.351)	90	85
SI	8339.8	0.014 (0.123)	0	(100)
Form C	8339.6	<0.001 (0)	80	100
R	8339.5	0.007 (0.061)	0	(100)
Form C + light	8339.4	0.012 (0.140)	65	100

[a] Data are from ref. 38.

[b] Based on protein concentration determination and 1.0 Ni/protein (35).

[c] Percentage of EPR active Ni poised in the desired form. A value of 100% indicates that a single EPR active species is present.

Table III. XAS Edge Analysis in Metalloproteins

Protein	*Formal Oxidation State Change*	*Edge Shift (eV)*	*Ligand vs. Metal Redox*	*References*
Galactose oxidase	Cu(III)→Cu(I)	~0	ligand	41
CO dehydrogenase	Ni(III)→Ni(II)	~0	Fe,S cluster	42
Sulfite oxidase	Mo(VI)→Mo(V)	0.5	?	43
	Mo(V)→Mo(IV)	0.5	?	
Mn OEC				
S_1→S_2		0.8	metal	44, 45
S_2→S_3		~0	ligand	
Cobalamin	Co(II)→Co(I)	1.0	metal	46
	Co(III)→Co(II)	1.5		
Cytochrome oxidase	Cu_{a3}(II)→Cu_{a3}(I)	1.3	metal	47
CO dehydrogenase	Ni(III)→Ni(II)	1.42	metal	48
Cytochrome oxidase	Cu(II)→Cu(I)	1.5	?	49
Cytochrome oxidase	Fe(III)→Fe(II)	2.2	metal	49
Stellacyanin	Cu(II)→Cu(I)	2.0	metal	50
Plastocyanin	Cu(II)→Cu(I)	2.2	metal	49
Xanthine oxidase	Mo(VI)→Mo(IV)	3.1	metal	51

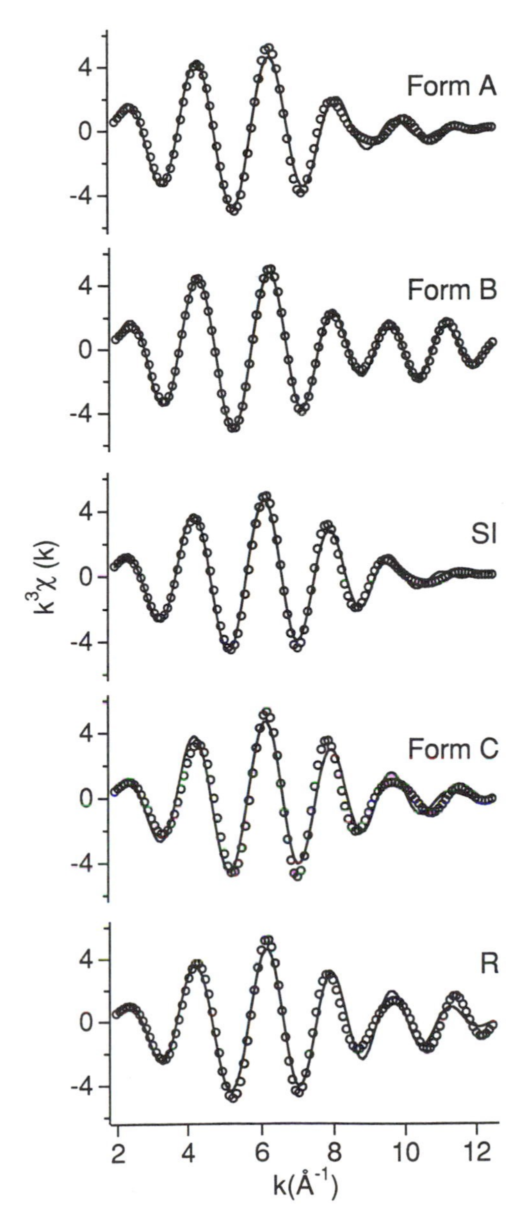

Figure 4. First coordination sphere (backtransform window = 1.1–2.7 Å) Fourier-filtered Ni K-edge EXAFS spectra from redox-poised Thiocapsa roseopersicina *hydrogenase samples (○) and fits (——): form A, (3 ± 1)N,O at 1.97(2) Å + (2 ± 1)S at 2.23(2) Å; form B, (3 ± 1)N,O at 1.95(2) Å + (2 ± 1)S at 2.24(2) Å, SI, (2 ± 1)N,O at 2.03(2) Å + (3 ± 1)S at 2.24(2) Å; form C, (2 ± 1)N,O at 2.01(2) Å + (2 ± 1)S at 2.25(2) Å; R, (2 ± 1)N,O at 2.03(2) Å + (2 ± 1)S at 2.24(2) Å. (Reproduced from reference 35. Copyright 1993 American Chemical Society.)*

The most appropriate description of the Ni oxidation state in all forms of the enzyme appears to be Ni(II), based on a valence bond sum analysis of the best EXAFS model (Thorp, H. H., University of North Carolina, unpublished data). This model is in agreement with studies of the magnetic properties of the EPR silent intermediate of the H_2ase from *D. baculatus* that demonstrated that the Ni site was diamagnetic, albeit in an enzyme that has a selenocysteinate Ni ligand (*29*).

Although Ni-centered redox chemistry does not appear to be involved, redox chemistry clearly occurs in the enzyme and cannot be attributed solely to either the Fe,S clusters present or the Ni center. Detailed studies of the redox chemistry associated with the Fe,S clusters have been performed on the Ni,Fe enzyme from *D. gigas* (*63*, *64*) and on the Fe,Ni,Se enzyme from *D. baculatus* (*40*). These studies used redox titrations and EPR spectroscopy to vary and monitor the redox state of the enzymes and magnetic field-dependent Mössbauer experiments to examine the cluster types, spin states, and redox states of the Fe,S clusters present in each sample. In the oxidized *D. gigas* enzyme, these studies identified three clusters: an oxidized, paramagnetic ($S = \frac{1}{2}$, $g = 2.02$) $[3Fe,4S]^+$ cluster and two diamagnetic $[4Fe,4S]^{2+}$ clusters. The $[3Fe,4S]^+$ cluster is reduced at −70 mV to an integer spin cluster ($S = 2$, $g = 12$) that is not reduced further, but its EPR spectrum is sensitive to the redox states of the other clusters. The two $[4Fe,4S]^{2+}$ clusters are reduced at different potentials (−290 and −340 mV) and display distinct and unusual spectral properties in their reduced forms. Only broad, ill-defined EPR signals from one $[4Fe,4S]^+$ cluster are observed, suggestive of spin–spin interactions with other paramagnets. In contrast, the *D. baculatus* enzyme was shown not to contain a [3Fe,4S] cluster in either an oxidized or reduced state. This enzyme has two [4Fe,4S] clusters that exhibit Mössbauer properties similar to those characterized for the *D. gigas* enzyme and have similar midpoint redox potentials of −315 mV. However, these clusters exhibit EPR signals ($g = 1.94$) that are typical of reduced [4Fe,4S] clusters, indicating that interaction with the [3Fe,4S] cluster in *D. gigas* H_2ase is responsible for the unusual EPR properties of the [4Fe,4S] clusters in that enzyme. The potentials determined for the Fe,S clusters in these two enzymes indicate that the oxidized forms of H_2ase (forms A and B) will contain an oxidized [3Fe,4S] cluster (if present) and that it will be reduced in the EPR silent forms. The potentials determined for [4Fe,4S] clusters indicate that the active form (form C) has at least one reduced [4Fe,4S] cluster and that the fully reduced state contains only reduced clusters.

The potentials determined for reduction of the Fe,S clusters indicate that they are not directly responsible for all of the changes in the EPR spectrum that have been associated with Ni. Furthermore, because three electrons appear to be involved in the redox chemistry associated with

Ni, many enzymes (e.g., *D. baculatus* and *T. roseopersicina*) do not contain the number of clusters that would be required to account for three one-electron redox processes. Clearly, other redox centers must be present. Frequently, high formal oxidation state intermediates in biology involve oxidation of some other group rather than the metal center (e.g., galactose oxidase (*41*), Mn O_2-evolving complex (OEC) (*65*), horseradish peroxidase compound I (*66, 67*)). In this regard, it is worth noting that the only Ni-containing functional model reported that features a mixed-donor ligand environment composed of N-, O-, and S-donors uses only the Ni(II/I) couple to catalyze the H–D exchange reaction (*68*). It is possible that protein chemistry, such as that observed for galactose oxidase, might explain some of the redox chemistry that occurs in H_2ase. Another possibility is that the cysteinates bound to the Ni site might account for some of the redox chemistry of H_2ase, and if the changes were localized to a large extent on the S-donor ligands, they might not affect the Ni site. We have explored this possibility using Ni complexes of alkyl thiolate ligands.

Redox Chemistry of Nickel Thiolates

The similarity of the oxidized enzyme EPR spectra (forms A and B) to those obtained from Ni(III) coordination complexes (*69–73*) led to the assignment of the EPR signals to low-spin tetragonal Ni(III) centers with the unpaired spin in the d_{z^2} orbital ($g_z < g_{x,y}$). This assignment, coupled with the identification of S-donor ligands in the coordination sphere of Ni and the low oxidation potential of the biological Ni site, has motivated a number of studies aimed at producing Ni(III) complexes with thiolate ligands, and several examples are now known (*74*). Alternatively, the $S = 1/2$ EPR signals could be assigned to Ni(I) species (particularly appropriate for the reduced enzyme), and Ni(I) model compounds with thiolate ligation and rhombic EPR spectra have been reported (*75, 76*). Many of these models were developed with the goal of understanding the factors that lead to stabilization of Ni(III) or Ni(I), rather than to explore the redox chemistry of Ni thiolates like those that might exist in H_2ase.

Nickel(II) coordination compounds with simple N- or O-donor ligands generally lead to redox potentials of $\geq +1$ V for the oxidation of Ni(II) to Ni(III), and of ≤ -1 V for the reduction of Ni(II) to Ni(I), potentials that are well beyond the relevant range (*69, 70, 73*). However, ligands featuring deprotonated amides, thiocarboxylates, and oxime ligands, which are known to stabilize higher oxidation states of metals but have little biological relevance, have been shown to lower the oxidation potential of Ni(II) (*74*). The oxidation of one alkyl thiolate complex points to the stabilization of formally Ni(III) centers by thiolate ligands, although it is not yet clear which centers are primarily involved in the oxidation. The sterically encumbered complex $Ni(nbdt)^{2-}$ undergoes a reversible one-electron oxidation to a formally Ni(III) species that exhibits an axial EPR spectrum ($g_{\parallel}$ = 2.14, $g_{\perp}$ = 2.05) at potentials as negative as −0.76 V (*77*).

Reductive chemistry is not readily accessible in $Ni(nbdt)^{2-}$, presumably because of the negative charge on the complex. Only one system has been reported that produces stable Ni(III) and Ni(I) complexes (*76*), and no system with redox potentials for the Ni(III/II) and Ni(II/I) couples differing by 400 mV has been characterized. $Ni(terpy)(SAr)_2$ can be reduced by $Na_2S_2O_4$ to give a Ni(I) complex with an axial EPR spectrum ($g_{\parallel}$ = 2.25, $g_{\perp}$ = 2.13) that is capable of reversibly binding CO and H^-. The CO adduct and the H^- adduct both give rhombic EPR spectra (CO adduct: g_1 = 2.24, g_2 = 2.14, g_3 = 2.05; H^- adduct: g_1 = 2.24, g_2 = 2.19, g_3 = 2.05). Chemical oxidation of these complexes is difficult and leads to decomposition of the complex (*76*). However, if the terpy ligand is changed to 2,6-bis[(1-phenylimino)ethyl]pyridine (DAPA), both oxidized and reduced complexes may be prepared (*76*). This change is attributed to the presence of a less extensive π-system in the DAPA ligand. Reduction of the Ni(II) complex $Ni(DAPA)(SPh)_2$ by $Na_2S_2O_4$ leads to chemistry similar to that observed for the terpy complexes. The Ni(I) complex that forms has a rhombic EPR spectrum (g_1 = 2.26, g_2 = 2.14, g_3 = 2.09) and reacts with CO to form a adduct that also has a rhombic EPR spectrum (g_1 = 2.20, g_2 = 2.15, g_3 = 2.02). Oxidation of $Ni(DAPA)(SPh)_2$ with $Fe(CN_6)^{3-}$ leads to the formation of a formally Ni(III) species with an axial EPR spectrum ($g_{\parallel}$ = 2.03; $g_{\perp}$ = 2.21). The oxidation product does not bind CO, but does bind CN^- to give a complex with a rhombic EPR spectrum (g_1 = 2.26, g_2 = 2.21, g_3 = 2.04). The reduced complex also binds CN^-, but CO is capable of displacing the CN^- ligand. Because CO binds to active H_2ase (form C) to form a Ni–CO adduct (*26*), this model chemistry points to a more reduced Ni center in the active protein.

We have focused our modeling efforts on a system that contains alkyl thiolate ligands, in analogy with cysteinate coordination in the enzyme. Reaction of $Ni(OAc)_2$ with a series of tridentate ligands $[RN(CH_2CH_2SH)_2]$ leads to the formation of a series of dimeric complexes, $\{Ni[RN(CH_2CH_2S)_2]\}_2$ [R = $CH_2CH_2SCH_3$ (**1**), $CH_2CH_2SCH_2Ph$ (**2**), CH_3 (**3**), CH_2Ph (**4**), $CH_2CH_2CH(Ph)_2$ (**5**)], containing distorted planar Ni centers coordinated by three thiolate donors and a tertiary amine (*78*, *79*) (Pressler, M. A. and Maroney, M. J., University of Massachusetts, unpublished results). The structure of these dimers is represented by **1** in Figure 5. Two of the thiolate ligands bridge between the Ni centers giving a butterfly-shaped cluster, with the two planar Ni complexes joined along an edge with an angle of 105° between the $Ni(S_{br})_2$ planes in the example shown. This type of structure is charac-

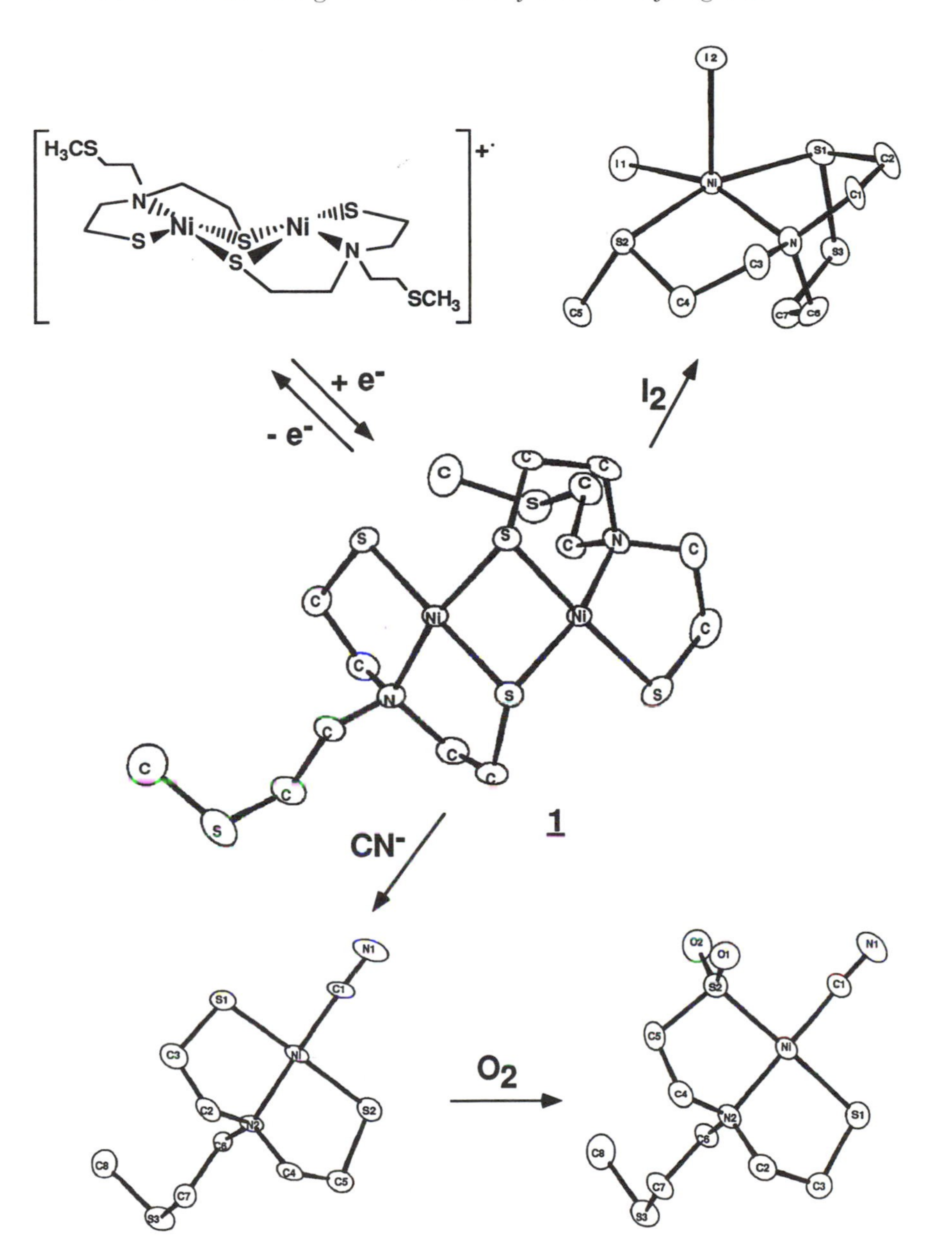

Figure 5. The oxidative chemistry of a series of Ni thiolate complexes. Crystallographically characterized structures are shown as ORTEP diagrams.

teristic of this class of ligands (*80*) and has been observed for the *formally* Ni(III,II) mixed-valent one-electron oxidation product of the Ni(II) complex of a tertiary phosphine ligand with three pendant arylthiolates ($P(o\text{-}C_6H_4S)_3$) (*81*).

The oxidative chemistry that is characteristic of the dimers is summarized in Figure 5. As discussed below, the most striking feature is that the products of oxidation reveal S-centered chemistry in every case. Despite the obvious shortcomings of such complexes as models for a mononuclear five- or six-coordinate biological Ni site, the redox chemistry of this series of compounds provides insight into the possible roles of the Ni thiolate ligands in the redox chemistry exhibited by H_2ase.

One of these complexes (**1**) provided the first example of a Ni(II) thiolate complex that undergoes a reversible one-electron oxidation to a formally Ni(III)-containing complex (*78*). Data from cyclic voltammetric studies of the redox chemistry of **1** are shown in Figure 6. Scans that proceed first in the cathodic direction (a) show no reductions of the Ni(II) dimer within the limits provided by the solvent or electrolyte used (0.1 M $n\text{-}Bu_4N(ClO_4)/CH_2Cl_2$). Scanning to increasing anodic potentials reveals two distinct oxidations, the first of which is quasireversible (Figure 6b) and has an $E_{1/2} \sim 0$ V versus NHE (Table IV). The fact that the process is reversible suggests that the product is still dimeric, an expectation that is given credence by the isolation and structural characterization of a mixed-valent Ni(II/III) dimer with a core structure like that observed in **1** (*81*). Coulometric measurements show that the electrochemical process is a one-electron oxidation, and the oxidation product yields a rhombic EPR spectrum ($g_1 = 2.17$, $g_2 = 2.11$, $g_3 = 2.08$) appropriate for an $S = 1/2$ species (Figure 7). This spectrum is reminiscent of those obtained from H_2ases in that it is a rhombic spectrum with *g*-values between $g = 2.3$ and 2.0 and displays no obvious ligand hyperfine splittings despite the presence of an N-donor ligand. The first oxidation has an Ip_a/Ip_c ratio near 1 only at high scan rates. The fact that this ratio becomes >1 at slower scan rates suggests that the oxidation product is unstable, an observation that is confirmed by EPR spectroscopy.

If the voltammetric scan is not reversed after the first oxidation, a second oxidation (Ep_a = 0.2–0.3 V) is observed (Figure 6c). This oxidation renders the lower potential oxidation irreversible (Ep_a = 0–0.1 V) and also leads to the observation of an irreversible reduction at negative potentials (Ep_c = −0.8 to −0.9 V). It has not been possible to measure accurately the number of electrons involved in either the second oxidation or in the coupled reduction process by coulometry due to the formation of films on the electrode surfaces. However, in comparison with that observed for the first oxidation, the peak currents are consistent with one- and two-electron processes, respectively. In spite of the irreversible nature of the cyclic voltammogram, continuous cycling be-

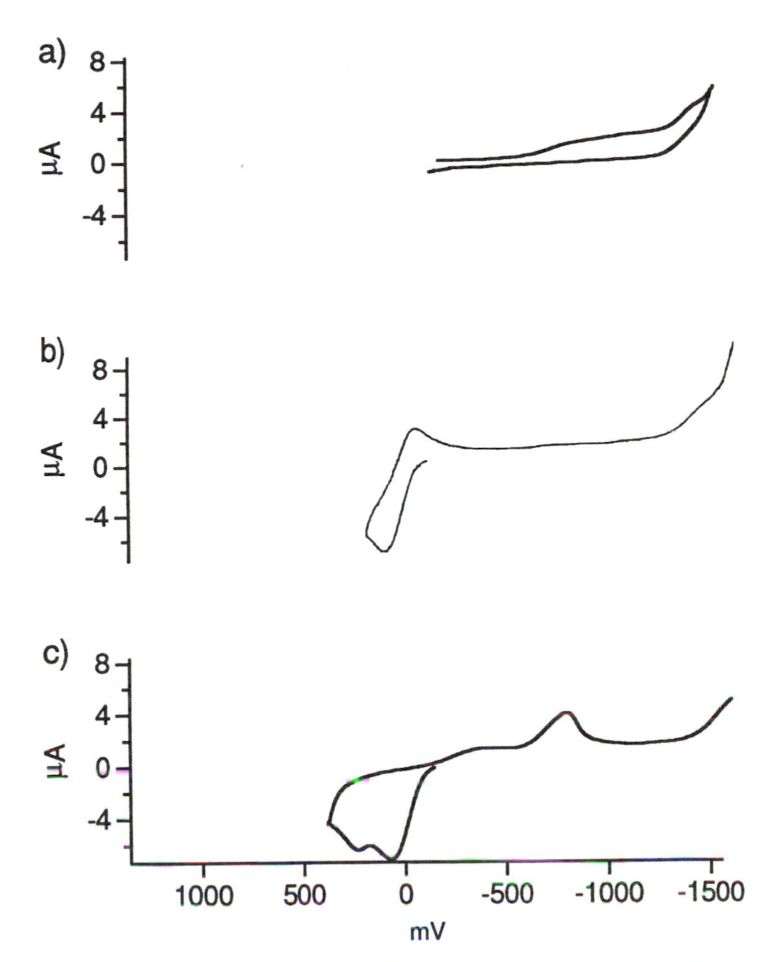

Figure 6. Voltammetric studies of $\{Ni[RN(CH_2CH_2S)_2]\}_2$, R = CH_2CH_2SMe. *A, initial scan in the cathodic direction. B, initial scan in the anodic direction, reversing the direction of the scan following the first oxidation. C, initial scan in the anodic direction, reversing the direction of the scan following the second oxidation.*

tween +1.5 and −1.5 V does not lead to any changes (i.e., no new products are formed, and the dimer is not consumed on the cyclic voltammetric time scale). Cyclic voltammograms obtained on the one-electron oxidation product were identical to those obtained from the starting material.

Ligands lacking a pendant thioether donor show very similar oxidative chemistry (Figure 8). The largest difference is in the first oxidation, which is irreversible and occurs at a more positive potential (Ep_a = 0.4–0.5 V). The oxidation remains a one-electron process that leads

Table IV. Electrochemical and EPR Data for the Oxidation of Complexes 1–5

Complex	Ep_a *(mV vs. NHE)*	*Characteristics*	*EPR g-values*[a]	Ep_a *(mV vs. NHE)*	Ep_c *(mV vs. NHE)*
1	100	quasireversible $E_{1/2} = +30$ $n = 1.0$	2.17, 2.11, 2.08	260	−755
2	10	quasireversible $E_{1/2} = +30$ $n = 1.1$	2.16, 2.11, 2.07	220	−900
3	400	irreversible $n = 0.8$	2.21, 2.14, 2.03	880	−960
4	470	irreversible $n = 0.9$	2.20, 2.14, 2.02	700	−950
5	430	irreversible $n = 0.8$	2.21, 2.14, 2.02	800	−1200

NOTE: Potentials are taken from 250 mV/s cyclic voltammetric scans at a Pt button working electrode on 0.5 mM solutions of the complexes in 0.1 M n-$Bu_4N(ClO_4)$ at −30 °C, and converted to NHE using the Fc/Fc+ couple (= +440 mV vs. NHE); the value of n was determined by controlled potential electrolysis at a Pt gauze electrode under the same conditions.

[a] Samples were generated by electrolysis and measured as frozen solutions at 77 K.

to the formation of an EPR-active $S = ½$ species (Figure 7). Although the g-values characteristic of ligands with N-alkyl substituents ($g_1 = 2.20$, $g_2 = 2.14$, $g_3 = 2.02$) are distinct from those that feature pendant thioethers (Table IV), they are virtually identical to those observed for the active form (form C) of H_2ase ($g_1 = 2.19$, $g_2 = 2.14$, $g_3 = 2.02$). One possibility that could account for the differences between these two classes of dimers is that the pendant thioether donor becomes coordinated in solution, a possibility that is suggested by the solid-state structures of a Ni(II) complex with a similar ligand (*82*) and of the mixed-valence Ni(III/II) dimer (*81*).

As is the case for the complexes with pendant thioether donors, the N-alkyl complexes show a second oxidation (Ep_a = 0.7–0.9 V) that is coupled to the formation of a new species with an irreversible reduction (Figure 8c); the potential can be continuously cycled between +1.5 and −1.5 V without changing the voltammogram.

The redox chemistry observed for these compounds by using electrochemical methods is similar to that observed for thiolates in the absence of metal ions (*83, 84*) and strongly suggests the formation of disulfides. This behavior has also been noted in the electrochemical study of the oxidation of $[Ni(pdmt)SPh]^-$ (*85*). The largest difference observed

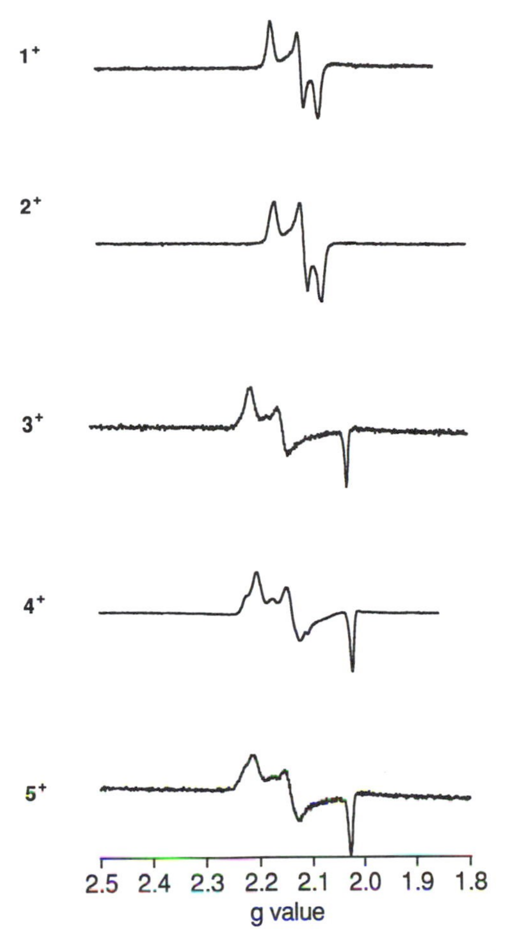

Figure 7. X-band EPR spectra of the one-electron oxidation products of complexes **1–5**.

for the dimers would appear to be that the oxidation of the two thiolates involved in the formation of the disulfide occur at different potentials.

Although there is no proof that the two-electron oxidation process leads to a product that is still dimeric, the fact that the cyclic voltammogram does not reveal the formation of new species even during repeated cycling is consistent with either a dimeric product or mononuclear complexes that rapidly re form the starting dimer (*85*).

Disulfide formation in the two-electron oxidation products is also supported by chemical oxidations of **1**. The two-electron oxidation of each Ni center in the dimer by I_2 leads to the formation of a five-coordinate mononuclear Ni complex with a disulfide ligand (*78*). The struc-

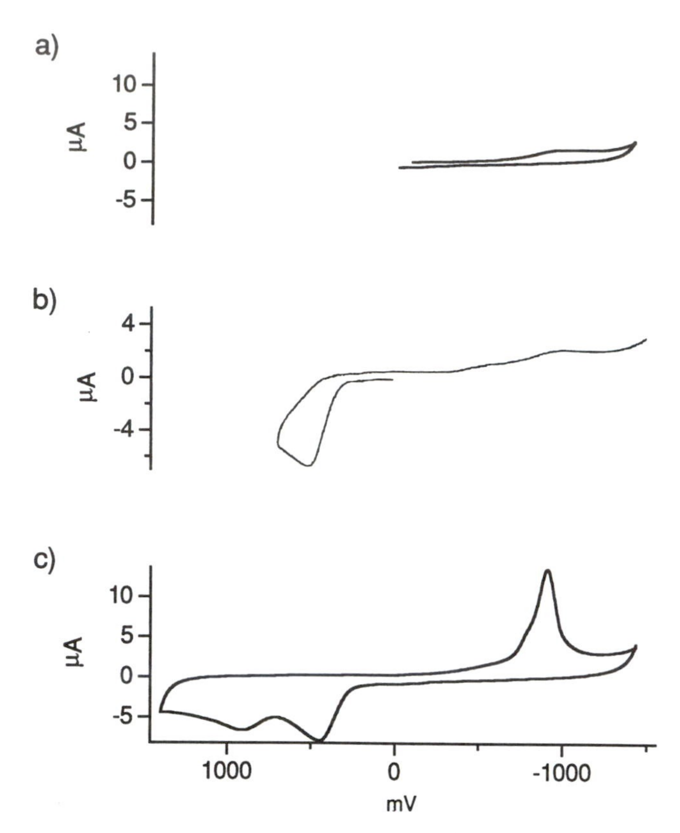

Figure 8. Voltammetric studies of [Ni(RN(CH₂CH₂S)₂)], R = CH₃. A, initial scan in the cathodic direction. B, initial scan in the anodic direction, reversing the direction of the scan following the first oxidation. C, initial scan in the anodic direction, reversing the direction of the scan following the second oxidation.

ture of this complex, shown in Figure 5, features a high-spin ($S = 1$) five-coordinate Ni(II) center in a distorted pyramidal environment composed of a thioether ligand, a tertiary amine donor, two iodide ligands, and one sulfur of a disulfide. Thus, the two-electron oxidation of each Ni(II) center in the dimer results in the one-electron oxidation of all of the thiolates.

The frozen solutions containing the one-electron oxidation products reported in Figure 7 may be thawed, whereupon they give isotropic spectra with $g_{iso} = g_{ave}$. At room temperature, the EPR signals are rapidly lost due to the formation of an EPR silent product. This chemistry, which occurs at a slower rate following the one-electron oxidation, likely accounts for the increasing value of Ip_a/Ip_c with decreasing scan rate. The rate of the loss of the EPR signal can be followed by incubating the samples at subambient temperatures and measuring the EPR spectra as a function of time. The loss of the EPR signal is first-order in [dimer] (Figure 9), and the slower rate of signal loss exhibited by the derivative containing a pendant thioether is consistent with the stabilization of the oxidation product via coordination of this ligand. Two mechanisms for the decomposition of *formally* Ni(III) thiolates to disulfides have been suggested (*85*). The first mechanism involves dimerization of the oxidized complexes; the second involves dissociation of thiyl radicals and the subsequent formation of a disulfide. The first-order decay of the EPR signal suggests that coupling of two complexes does not occur in the rate-determining step. The data are consistent with a rate-limiting dissociation of a thiyl radical, followed by the rapid formation of a disulfide. Hypothetical mechanisms for the chemistry associated with the products of one- and two-electron electrochemical oxidations are summarized in Figure 10.

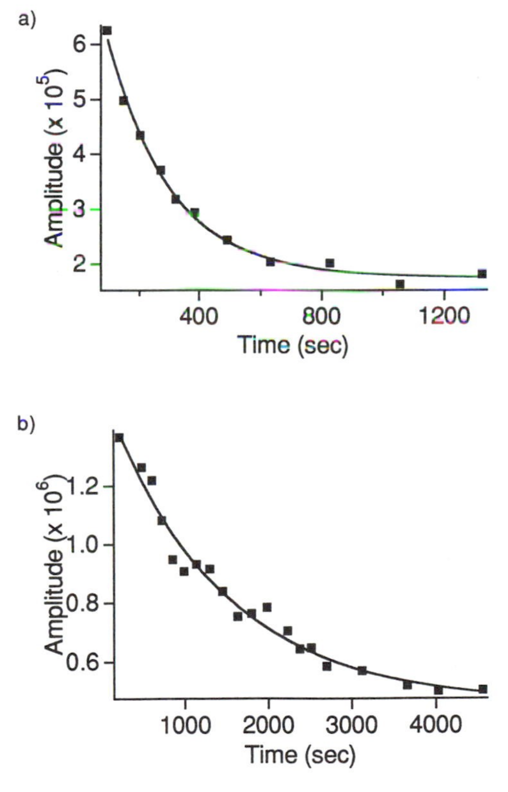

Figure 9. First-order kinetic plots of the decay of the EPR signals of $\mathbf{1}^{+\bullet}$ *at 25° (a) and* $\mathbf{4}^{+\bullet}$ *at −25° (b). Solid lines represent fits obtained for* $k = 4.7 \times 10^{-3}\ s^{-1}$ *and* $k = 7.2 \times 10^{-4}\ s^{-1}$*, respectively.*

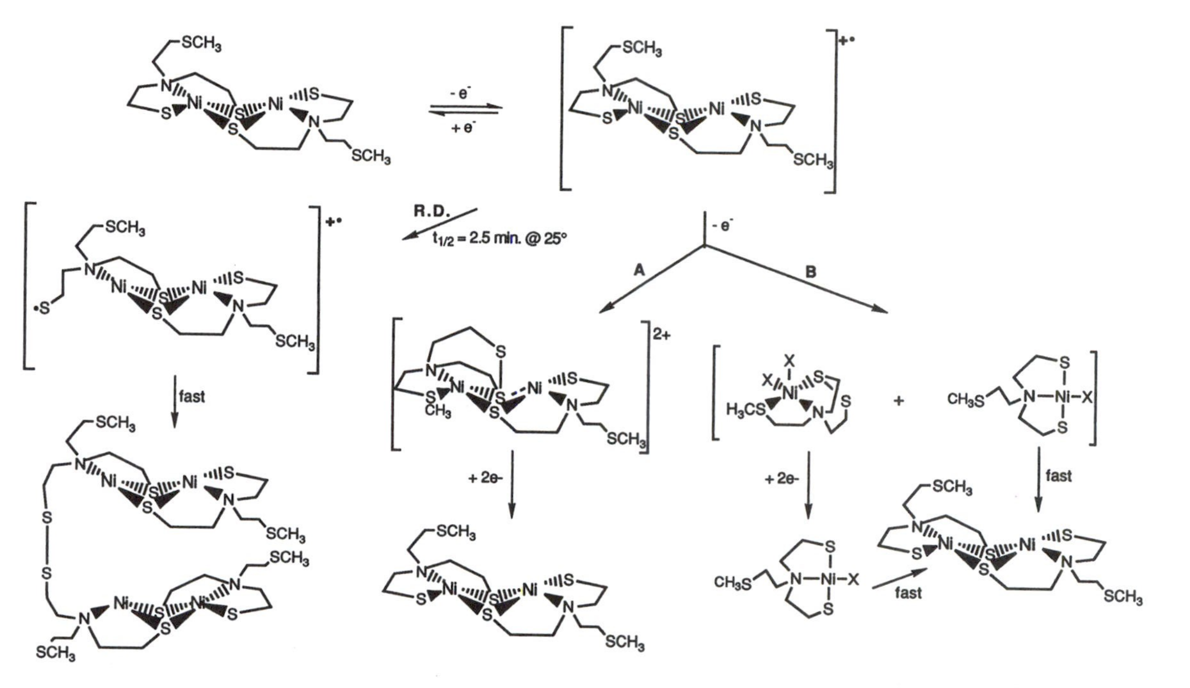

*Figure 10. Hypothetical mechanisms for the oxidative chemistry of **1** and the reaction of the one-electron oxidation product, $\mathbf{1}^{+\bullet}$. Left from $\mathbf{1}^{+\bullet}$: The process corresponding to the slow loss of the EPR signal and two paths for the formation of disulfides upon a second one-electron oxidation. Path A: Retaining a dimeric structure, the two-electron reduction of the disulfide regenerates **1**. Path B: The second one-electron oxidation cleaves the dimer leading to the formation of a mononuclear disulfide complex, in analogy with that shown in Figure 5 and a mononuclear Ni(II) complex that rapidly dimerizes to form **1**, in analogy with chemistry known for similar complexes (85). Reduction of the disulfide leads to production of the same mononuclear Ni(II) complex.*

The electronic structure of the one-electron oxidation products has also been explored. At first glance, the observation of g-values > 2.0 suggests that a Ni(III) center has been formed and that spin–orbit coupling from metal-centered orbitals is responsible for the observed g-values. However, electrons localized on S also experience spin-orbit coupling of sufficient magnitude to give rise to the observed g-values. EPR spectra of thiyl radicals (e.g., cysteinyl radical) taken in H-bond donor solvents (e.g., MeOH) reveal axial spectra with $g_{||} = 2.3$ and $g_{\perp} = 2.0$ ($g_{ave} = 2.1$) (*86*). The spectra reported for the dimers (and many other *formally* Ni(III) thiolates) have $g_{ave} = g_{iso} = 2.1$. Presumably, the spectra of thiyl radicals are typically axial due to the degeneracy of the p-orbitals that are not involved in the S–C s-bond. This degeneracy could be lifted in a metal complex, giving rise to rhombic spectra with $g_{ave} \sim 2.1$. Thus, analysis of g-values does not provide unequivocal information regarding the nature of the molecular orbital that contains the unpaired spin, particularly because the orbital likely contains contributions from both the sulfur and the metal atom.

Additional information regarding the metal character of the orbital containing the unpaired spin can be obtained from the observation of metal hyperfine interactions. When ^{61}Ni ($I = 3/2$) is used to prepare the dimer, line broadening with no resolved hyperfine coupling is observed. The fact that only line broadening is observed is consistent with only a small amount of spin density interacting with the Ni centers. The effects of ^{61}Ni substitution are illustrated in Figure 11 and are similar in dimers with and without pendant thioethers. Although the problem is not a simple one, attempts were made to simulate the EPR spectra and estimate the maximum value of hyperfine coupling constants. The simulations shown in Figure 11 were obtained for two Ni centers without constraining the magnitude of the hyperfine interaction to be the same for both centers. In fact, the values obtained were essentially the same and lead to average values for the two Ni centers of $|A_1|$, $|A_2|$, and $|A_3| = 2.1$, 10.1, and 10.0 G, and 10.5, 0.0, and 4.1 G for compounds **1** and **4**, respectively. The fact that equal participation of two Ni centers is required to simulate the EPR spectra suggests that the radical cation dimers are examples of delocalized mixed-valence complexes. This notion is consistent with the structure obtained from a similar Ni(II/III) complex, in which the Ni centers are structurally indistinguishable (*81*).

When the magnitudes of the dipolar hyperfine interactions in the dimers are compared with a theoretical value of 67.5 G for an unpaired spin localized in a Ni 3d orbital (*87*), it can be seen that the molecular orbital containing the unpaired spin has a relatively small contribution from Ni (≤30%). A similar situation is observed in H_2ases. When bacteria are raised on a source of ^{61}Ni, hyperfine is observed in the EPR signal originating from H_2ase (*22*). (In fact, observation of this hyperfine was

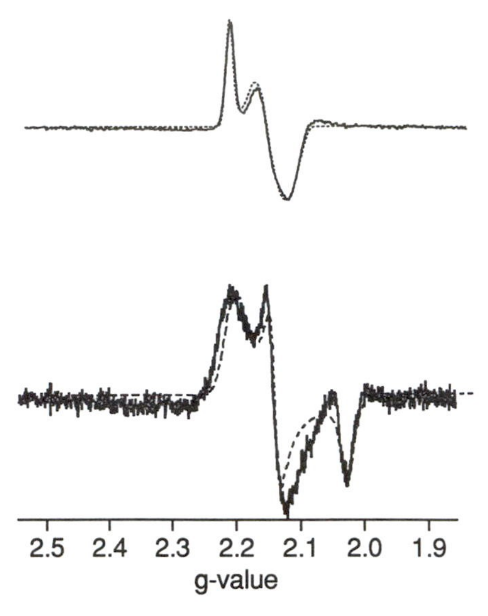

Figure 11. X-band EPR spectra of ^{61}Ni derivatives of $\mathbf{1}^{+\bullet}$ *(top) and* $\mathbf{4}^{+\bullet}$ *(bottom) at 77 K. Simulated spectra assuming Lorentzian line shapes are shown as dashed lines. Simulations were obtained by using the program NEWSIM. Initial g-values and linewidths for the simulations were obtained from the complexes with natural abundance Ni and were used as a starting point for the refinement of the hyperfine components. The final refinement allowed the g-values, linewidths, and hyperfine values to vary, although this did not result in significant changes in the g-values or linewidths. The best fits shown above result from using two independently refined sets of hyperfine tensor elements for each of the Ni centers in the dimer, although the hyperfine parameters obtained for each Ni center are nearly the same and were subsequently averaged.* $\mathbf{1}^{+\bullet}$: *(g_1 = 2.168, g_2 = 2.111, g_3 = 2.077; G_1 = 8.83 G, G_2 = 8.95 G, G_3 = 9.14 G; $|A_1|$ = 2.1 G, $|A_2|$ = 10.0 G, $|A_3|$ = 9.7 G).* $\mathbf{4}^{+\bullet}$: *(g_1 = 2.209, g_2 = 2.141, g_3 = 2.028; G_1 = 6.21 G, G_2 = 14.29 G, G_3 = 12.06 G; $|A_1|$ = 10.8 G, $|A_2|$ = 0.0 G, $|A_3|$ = 4.5 G).*

the basis for assigning the EPR signals to Ni-containing species (*8*).) In the case of active H_2ase (form C), the hyperfine is highly anisotropic, and resolved hyperfine coupling of ~20 G is observed only on the highest field feature; the hyperfine observed for the other two g-values is very small. If the value observed at g_3 is used to estimate the spin density on Ni, a value of ~30% is obtained. This value is very close to that observed for the dimers and suggests that the orbital containing the unpaired spin is not localized on Ni centers in either case.

Evidence of spin density on the S-donor atoms has been obtained from ^{1}H-ENDOR spectroscopy (Gurbiel, R. and Hoffman, B. M., Northwestern University, unpublished results). In the case of radical cation of model compound **4**, a resonance with a coupling constant of 12–14

MHz is observed (Figure 12A). Given that the only protons that would be expected to give rise to significant coupling with the spin are the methylene protons on the carbon atom a to the S-donors, this resonance must arise from these protons (or a subset of these protons), analogues of cysteine β-CH_2 protons. In fact, nearly identical ^{1}H-ENDOR spectra are observed for *T. roseopersicina* H_2ase (form C) (Figure 12B) (*88*). This result suggests that the covalent interaction between the cysteine donors and Ni in the enzyme is similar to that existing between the Ni and the thiolate ligands in **4**.

Another oxidation reaction of Ni thiolate complexes is of potential relevance to the deactivation of H_2ases upon exposure to O_2. Reaction of the Ni(II) dimeric complexes with two equivalents of CN^- leads to the formation of planar, *trans*-dithiolato complexes. These mononuclear complexes react with O_2 or with air under ambient conditions (*79, 89*). Manometric measurements show that the reaction stoichiometry is 1 Ni:1 O_2. The reaction products are planar, diamagnetic Ni(II) complexes featuring one thiolato and one sulfinato ligand. This reaction corresponds to a formal four-electron oxidation of a Ni(II) complex to give a product reflecting oxidation of a thiolate ligand. Structurally characterized examples of the mononuclear *trans*-dithiolate obtained from **1** and its oxidation product are shown in Figure 5 (*79, 89*).

The oxidation can be followed spectrophotometrically (Figure 13). The spectra of the *trans*-dithiolato complexes are characterized by ab-

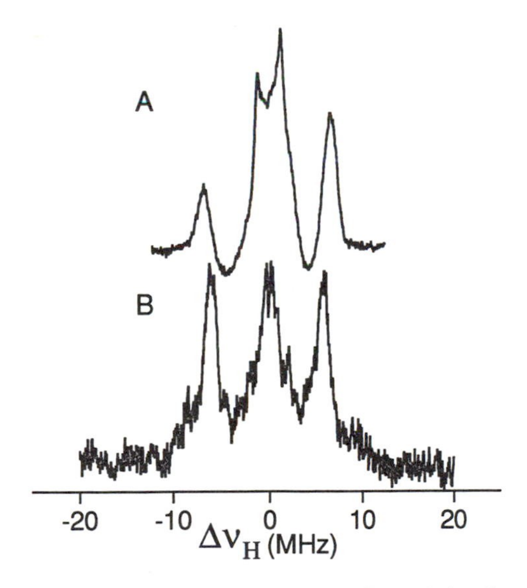

Figure 12. ^{1}H-ENDOR spectra (35 GHz) of $4^{+\bullet}$ (A) taken at g = 2.208 and of the nonexchangeable protons in Thiocapsa roseopersicina *hydrogenase-form C (B) taken at g = 2.19.*

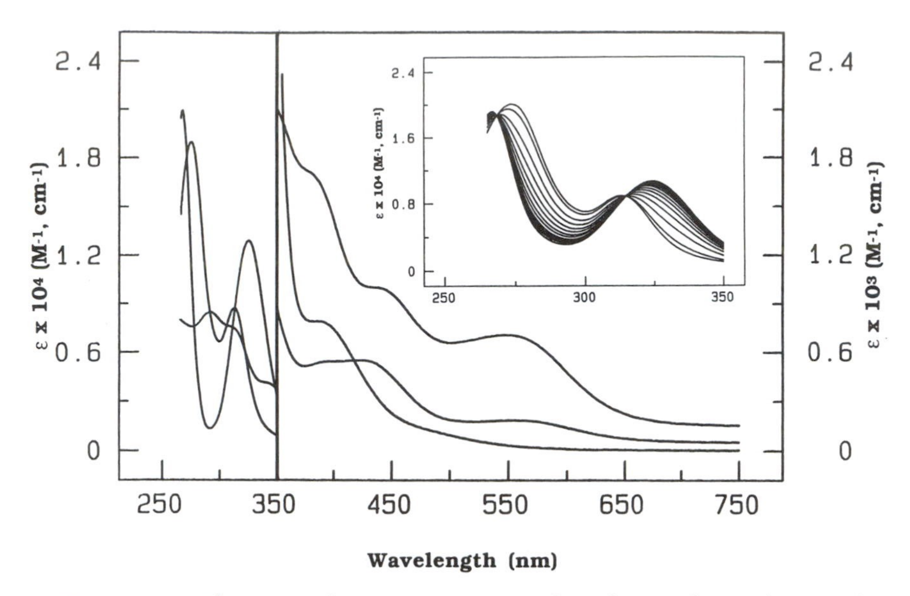

*Figure 13. Electronic absorption spectra taken during the oxidation of $Ni(MeSCH_2CH_2N(CH_2CH_2S)_2CN)^-$ by O_2 in DMF. Spectrum 1 is 0.30 mM **1**. Spectrum 2 is following the addition of 2 equiv of $Et_4(CN)$ under N_2. Spectrum 3 is following the oxidation of the CN^- complex by O_2. The inset describes the reaction with O_2 at 30 °C at times t = 7, 30, 90, 150, 210, 270, 330, 390, 450, 510, 570, 630, 720, and 750 min. Extinction coefficients are for a constant [Ni] = 0.60 mM. (Reproduced from reference 89. Copyright 1989 American Chemical Society.)*

sorption maxima near 290 and 310 nm in their electronic absorption spectra. Upon exposure to O_2, new bands associated with the formation of the monosulfinato complex appear near 265 and 325 nm. The spectral changes proceed with the formation of isosbestic points, indicating that no stable intermediates are formed in the oxidation process. Because the final spectra obtained are identical to isolated samples of the monosulfinates, there is no evidence that a bis-sulfinate, a disulfide, or any other oxidation product forms.

The spectral changes observed during oxidation provide a means for monitoring the kinetics of the oxidation process. The reactions follow a rate law that is first order in [Ni] and first order in $[O_2]$: Rate $= k[Ni(L)CN]^-[O_2]$ (79). The reaction is relatively slow and not very sensitive to the nature of the N-substituent ($k = 1.4\text{–}3.1 \times 10^{-2}\ M^{-1}\ s^{-1}$ in dimethylformamide (DMF) at 30 °C). The reaction rates are independent of the presence of a singlet oxygen scavenger or radical traps. The activation parameters $\Delta H^{\ddagger}$ and $\Delta S^{\ddagger}$ were measured by using the temperature dependence of the second-order rate constants in DMF and have values of 13.1–14.7 kcal/mol and −24.2 to −18.7 eu, respectively.

Reactions that use isotopically labeled O_2 and mass spectrometry reveal that the sole source of the O atoms in the sulfinate product is O_2 (*89*) and that both atoms of a single molecule of O_2 are incorporated into ~80% of the product (*79*).

In light of these results, hypothetical mechanisms for the oxidation of Ni thiolate complexes by O_2 may be discussed. The only well-characterized mechanism for the oxidation of thiolates to sulfinates in transition metal complexes involves O_2^{2-} as an oxidant and proceeds via the stepwise formation of sulfenates (Scheme 1) (*90, 91*).

The reaction mechanism has been interpreted as involving nucleophilic attack of the coordinated thiolate S atom on the peroxide (H_2O_2, $H_3O_2^+$) and follows the rate law: Rate = k_2[Nuc][H_2O_2]. Although the rate law and the kinetic barriers determined for this mechanism in a number of complexes are similar to those observed for the oxidation of $[Ni(L)CN]^-$ (e.g., for $[(en)_2Cr(SCH_2CH_2NH_2)]$: $\Delta H^{\ddagger}$ = 9.7(2) kcal/mol, $\Delta S^{\ddagger}$ = −26 eu), the stoichiometry of the reactions studied here (1Ni: $1O_2$) rules out a stepwise mechanism for the oxidation of $[Ni(L)CN]^-$ complexes. Such a mechanism would also be expected to give rise to complete scrambling of labeled oxygen in reactions involving $^{16}O_2$ and $^{18}O_2$, in contrast to the observation that the oxidation of $[Ni(L)CN]^-$ proceeds mostly with the incorporation of both atoms of a single O_2 molecule.

Despite the differences in mechanism, the expectation that thiolate ligands will act as nucleophiles appears to be a feature of the oxidation of $[Ni(L)CN]^-$ by O_2. The strong tendency of Ni thiolates to form dimers and higher polymers (*80, 92–94*) and the fact that the presence of a tightly bound anionic ligand appears to be required to cleave the dinuclear complexes (e.g., CN^- or thiolate (*85*)) is evidence of the nucleophilicity of terminal thiolate ligands in planar Ni(II) complexes.

Because reaction of thiyl radicals with O_2 is known to form sulfinyl radicals (*95*), mechanisms involving radicals were also considered. Although such a mechanism would have the correct stoichiometry and account for the lack of scrambling in the studies involving labeled O_2, reactions involving the formation of free radicals were ruled out because

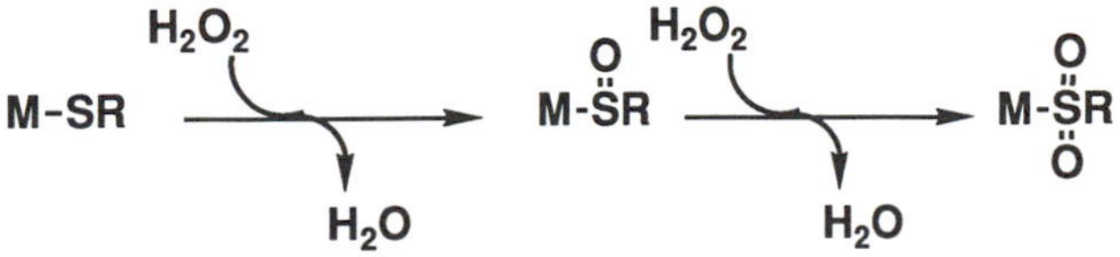

$$\text{M-SR} \xrightarrow[\ -H_2O\ ]{H_2O_2} \text{M-S(=O)R} \xrightarrow[\ -H_2O\ ]{H_2O_2} \text{M-S(=O)_2R}$$

Scheme 1

no EPR active species were observed and the rates of reaction are not affected by the presence of radical scavengers.

The ability of transition metals to form complexes with O_2 is well known. Ni(II) complexes are known to reversibly bind O_2 (*96, 97*). The Ni-O_2 complexes are powerful oxidizing agents and have been described as involving formal Ni(III) and O_2^- centers. The formation of a Ni–O_2 complex as a precursor to the ligand oxidation of $[Ni(L)CN]^-$ is unlikely because no intermediates are detected in the oxidation process and the spectrum of $[Ni(L)CN]^-$ taken under 1 atm of O_2 at 15 °C, a temperature that is below that required to halt the ligand oxidation, is identical to those of anaerobic samples of $[Ni(L)CN]^-$ and does not change in 6 h. Mechanisms involving the production of free O_2^- via oxidation of the metal center, followed by oxidation of thiolate ligands by the O_2^- (or O_2^{2-} from reaction with a second Ni center) are inconsistent with the lack of EPR signals from Ni(III) and the absence of an effect on the rates of reaction in the presence of radical traps. Furthermore, these reactions would be expected to proceed stepwise, as in Scheme 1, leading to the formation of intermediates and to complete scrambling of labeled O_2.

A mechanism that is consistent with the data gathered in this study for the oxidation of $[Ni(L)CN]^-$ by O_2 is shown in Scheme 2. This mechanism accounts for the stoichiometry of the reaction, has a major pathway involving incorporation of both atoms of a single O_2 molecule, and features a rate-determining step that involves one Ni complex and one O_2 molecule.

The proposed mechanism involves the formation of persulfoxide or thiadioxirane intermediates. These intermediates have been suggested to be involved in the oxidation of thioethers by $^1O_2$ (*98–100*) and have been postulated for the oxidation of Ni dithiolenes to bis-sulfinates (*101*). The rate-determining step in the proposed mechanism is the cleavage of the O–O bond in the thiadioxirane intermediate. This conclusion is based on the similarity between the enthalpy and entropy of activation for a number of oxidations in which O–O bond cleavage is rate determining (e.g., Scheme 1).

The mechanism of thioether oxidation is described as involving electrophilic attack of $^1O_2$ on the sulfur atom (or conversely nucleophilic attack of thioether S on $^1O_2$) leading to the formation of sulfoxides and

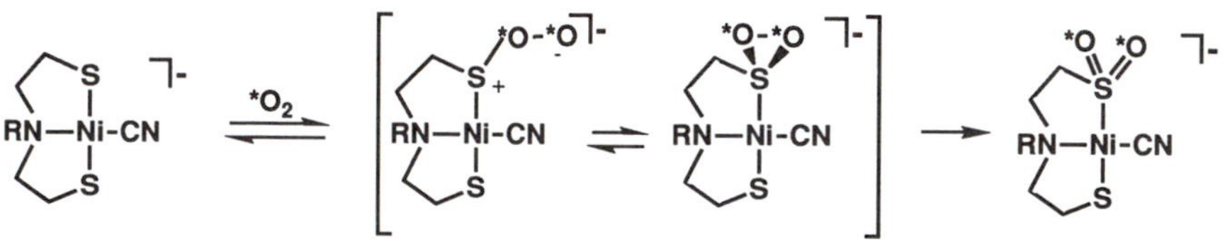

Scheme 2

sulfones. A recent study using $^{18}O_2$ revealed that the product sulfone contains two atoms of oxygen derived from the same O_2 molecule (*98*). Because addition of the $^1O_2$ scavenger 1,4-diazabicyclo[2.2.2]octane has no effect on the rate of reaction, singlet oxygen is not directly involved. However, nucleophilic attack of the coordinated thiolates on $^3O_2$ does account for the stoichiometry of the reaction, the derivation of the major product from a single O_2 molecule, and the absence of radical intermediates, and does not involve Ni-centered redox chemistry. It is possible that the increased nucleophilicity of thiolate sulfur versus thioether sulfur is enough to cause the reaction to occur at a slow rate with the weaker electrophile.

Nickel-containing H_2ases are deactivated upon exposure to air in a process that is dependent on O_2 and results in a weak interaction of the $S = 1/2$ center in the enzyme with $^{17}O_2$. This deactivation suggests that O_2 binds near the Ni site (*26*). In addition to differing in terms of their EPR spectrum, the reductive activation kinetics under H_2 also differ for forms A and B. Exposure to H_2 results in the rapid reduction of form B, whereas form A exhibits a lag that has been associated with the removal of O_2 from the sample (*3*). Several observations point to a possible role for S-oxidation in the deactivation process. First, in addition to the compounds discussed here, the oxidation of Ni thiolate ligands to sulfinate ligands has been observed in other compounds (*102*). This oxidation of Ni thiolate ligands suggests that the oxidation is typical of nickel thiolates and there is no apparent reason why it would not occur in an enzyme site that is accessible to O_2. Second, in many cases form B has been shown to react in air to produce form A. Furthermore, the reaction is slow, with $t_{1/2}$ on the order of several hours to days depending on the enzyme involved and the temperature (*3*). This is a time scale similar to what is observed for the O_2 oxidation in Ni thiolate complexes. Last, the substitution of Se for the S center that reacts with O_2 might be expected to lead to an enzyme that is more oxygen tolerant, because oxides of selenium are more difficult to form than sulfur oxides and are more readily reduced (*103*). This appears to be the case for H_2ases that contain a selenocysteine ligated to the Ni center. These enzymes can be isolated in air in an EPR silent state without oxidation to form A or B (*5*).

Is Ni the H_2-Binding Site?

Another potential role for the Ni center in H_2ases is as a site involved in H_2 activation. A catalytically viable form of the enzyme (form C) is characterized by a rhombic EPR signal, Ni–C ($g_{1,2,3}$ = 2.19, 2.15, 2.02), that has been attributed to a Ni(III) or Ni(I) complex of H^- or H_2, based on EPR and ENDOR spectroscopic studies (*24, 26, 27*). In *D. gigas* H_2ase, three sets of ^{1}H ENDOR signals have been observed. A set of nonexchangeable protons with a coupling constant of 12 MHz was observed and attributed to cysteine β-CH_2 protons. The observation of protons with the same coupling constant in both oxidized (forms A and B) and reduced enzyme (form C) is direct evidence of delocalization of spin density onto Ni cysteine ligands and suggests that the electronic structure of the EPR active species is similar in both the oxidized and reduced forms of the enzyme (i.e., that Ni does not change oxidation state). In addition to the nonexchangeable protons, two sets of solvent exchangeable protons were observed. The first is weakly coupled (4 MHz) and is assigned to a protonated Ni ligand (e.g., H_2O). The second set of protons, more strongly coupled (17 MHz), is due to a proton that is derived from H_2. Although the coupling constant for this proton is relatively large, it does not resemble the couplings typical of known paramagnetic Ni-H hyperfine interactions, which are in the 300–500-MHz range and are visible in EPR spectra (*104, 105*). The nature of the proton giving rise to the 17-MHz coupling in H_2ase is not clear, although a hydride bound to Ni using an orbital containing the unpaired electron is ruled out by the EPR and ENDOR spectra. It was suggested that this proton could be a hydride ligand bound to a Ni(III) center through a d orbital in the *x,y* plane that does not contain spin density (*27*). This possibility is bolstered by EPR spectra obtained on $Ni^{III}(CN)_4(H_2O)_2^-$, in which no hyperfine splitting due to equatorial $^{13}CN^-$ is observed (*106*). The possible involvement of a Ni dihydrogen ligand (*107*) remains an open question.

Form C is light-sensitive and is known to be converted to an EPR active photoproduct by exposure to visible light at low temperature (*3, 24, 108*). The photoproduct exhibits a unique EPR signal, Ni–L ($g_{1,2,3}$ = 2.29, 2.13, 2.05) (Figure 14). The photochemistry is reversible, and annealing the sample at a higher temperature regenerates the Ni–C EPR spectrum. In the case of *T. roseopersicina* H_2ase, the transition from Ni–L to Ni–C is very sharp and occurs at 194 K (Figure 14) (*88*). It has been suggested that form C corresponds to a hydride (or dihydrogen) Ni adduct (*26, 27*) and that the photochemistry involves dissociation of the hydride from the Ni center (*24*). We examined this proposal using a combination of ENDOR and XAS measurements and found that although the photochemistry does indeed correspond to the loss of a bound proton(s), there is no conclusive evidence to support the assignment of Ni as the binding site.

The ^{1}H-ENDOR spectra obtained on *T. roseopersicina* H_2ase reveal resonances attributable to two of the three types of protons observed for the *D. gigas* enzyme, the nonexchangeable cysteine β-CH_2 protons (A = 12 MHz) and the more strongly coupled exchangeable proton (A = 21.5 MHz). The weakly coupled exchangeable proton (4 MHz) is not

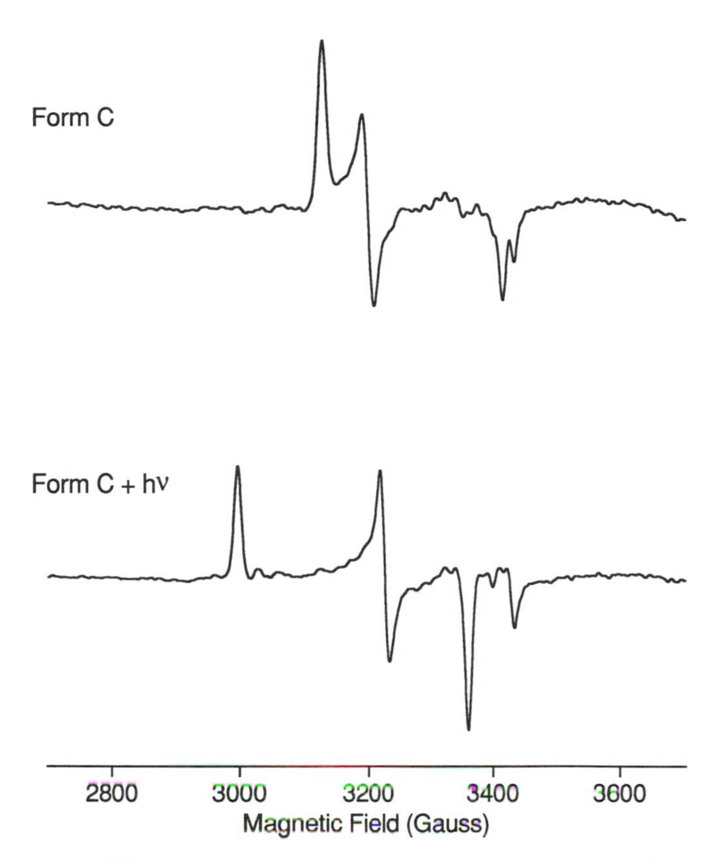

Figure 14. EPR spectra of Thiocapsa roseopersicina *hydrogenase, Ni–C (top) and Ni–L (bottom). Spectra were taken at 77 K, at a microwave frequency of 9.62 GHz, a microwave power of 20 mW, and a modulation amplitude of 4 G.*

observed. The resonance for the exchangeable proton that comes from dihydrogen is clearly shown in ^{2}H-ENDOR spectra obtained by preparing the sample in D_2O buffer (Figure 15). Exposure of this sample to light results in the conversion of the EPR spectrum from Ni–C to Ni–L and in the loss of the strongly coupled ^{2}H-ENDOR resonance. This resonance is restored upon annealing the sample, thus demonstrating that the conversion of Ni–C to Ni–L involves the photolysis of a proton derived from H_2 and that annealing the sample gives rise to the recombination of the proton with the active site.

If the photolysis involved dissociating a hydride (or H_2) ligand from the Ni site, changes in electron density, geometry, and structure that could be revealed by XAS would result. The Ni K-edge spectra taken on samples displaying EPR signals Ni–C and Ni–L are shown in Figure

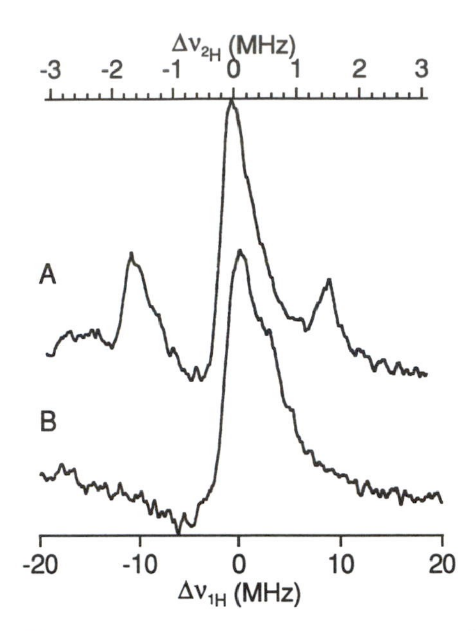

Figure 15. ^{2}H-ENDOR spectra of the solvent exchangeable protons associated with Thiocapsa roseopersicina *hydrogenase in form C (A) and its photoproduct (B).*

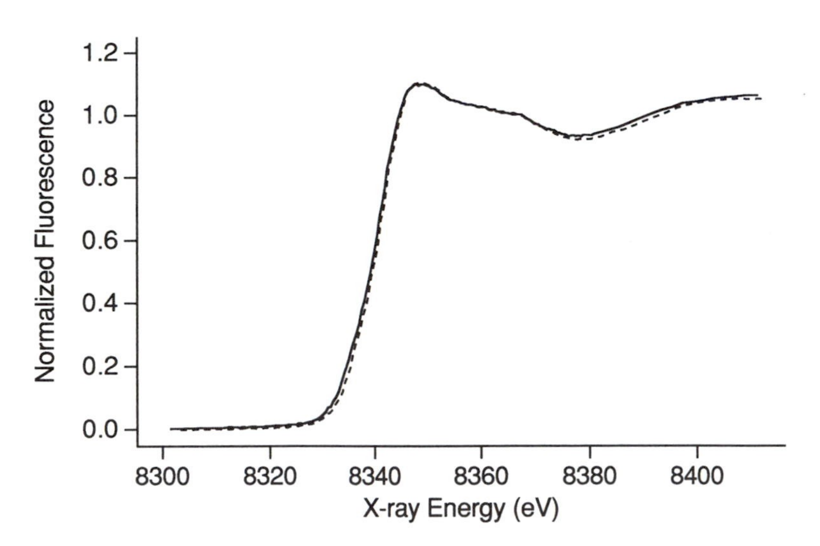

Figure 16. Ni K-edge XAS spectra obtained for Thiocapsa roseopersicina *hydrogenase form C (——) and its photoproduct (– – –). (Reproduced from reference 88. Copyright 1993 American Chemical Society.)*

16. These spectra show no change in the edge energy, pre-edge features that are sensitive to geometry, or XANES spectra (Table I) and suggest that the photodissociation of the hydrogenic proton(s) has no structural consequence for the Ni site. Similarly, the EXAFS spectra (Figure 17) reveal no evidence for a change in bond length for any of the nonhydrogen ligands of the Ni center (*88*).

Although it is possible that XAS is not sensitive to the structural changes that occur, the lack of *any* change in the XAS spectra suggests that Ni is not the H_2-binding site. The fact that ^{13}CO bound to the enzyme reveals a strong ^{13}C hyperfine interaction ($A_{x,y,z}$ = 85.3, 88.0, 90.3 MHz) (*26*) and that paramagnetic Ni hydrides also have large couplings to the H^- ligand, but H^- (or H_2) shows a much weaker coupling in Ni–C provides another argument against a strong bond to H^- or H_2 at the Ni site.

Other possible binding sites include Fe,S clusters, which are present in all H_2ases and are the likely binding sites for H_2 in Fe-only H_2ases (*1*), and sulfide or thiolate ligands. The possible presence of a Ni,Fe,S cluster in the active site in Fe,Ni H_2ases, as suggested by EPR (*2, 3, 108*) and EXAFS studies (*34*), could give rise to a mechanism by which

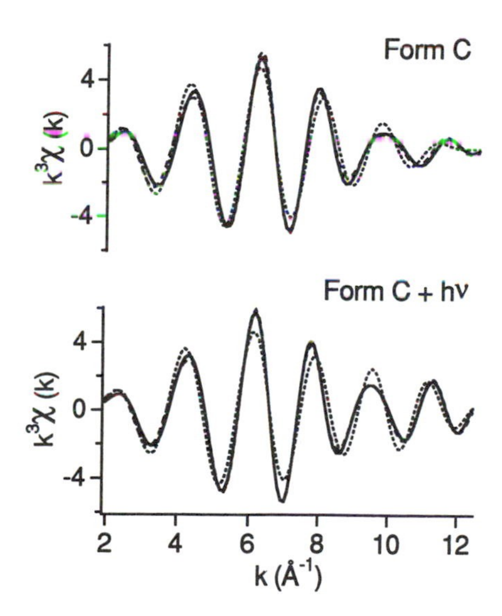

Figure 17. Ni K-edge Fourier-filtered EXAFS spectra from hydrogenase in form C (top) and its photoproduct (bottom). Two shell fits are shown as small dashed lines. Form C: (1–2)N,O 1.92 (2) Å + (3 ± 1)S at 2.22(2) Å; form C + light: (1–2)N,O at 1.92 (2) Å + (4 ± 1)S at 2.22(2) Å. Three-shell fits incorporating a long Ni–S interaction are shown as large dashed lines. Form C: (1–2)N,O at 1.92(2) Å + (3 ± 1)S at 2.22(2) Å, + (1)S at 2.75(2) Å; form C + light: (1–2)N,O at 1.94 (2) Å + (3 ± 1)S at 2.23(2) Å + (1)S at 2.70(2) Å.

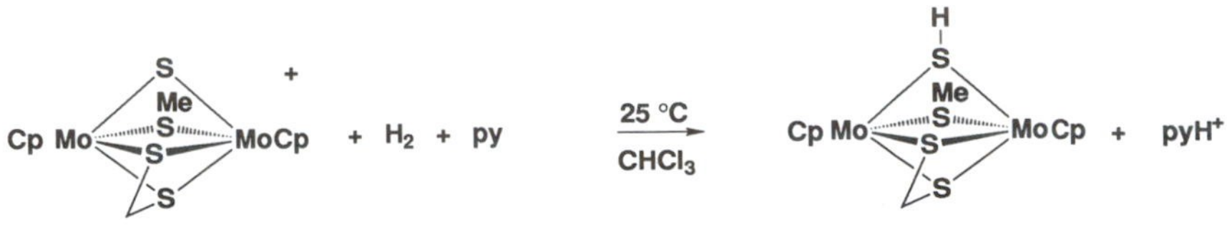

Scheme 3

weak coupling between the hydrogenic proton(s) bound to an Fe atom and the Ni center could occur. Another possibility is that activation of H_2 involves interaction with a sulfide ligand of Ni or an Fe,S cluster. It is not possible from XAS analysis to differentiate between thiolate and sulfide ligands; thus the possibility of a Ni sulfide ligand exists. Furthermore, heterolytic cleavage of H_2 (the reaction catalyzed by H_2ase) involving a bridging sulfide has been demonstrated to occur in one-model system (Scheme 3) (*109*). In this model system a base (pyridine) gets protonated, and the "hydridic" proton ends up converting a bridging sulfide ligand to a bridging hydrosulfide. Analogous chemistry could account for the presence of a strongly coupled solvent exchangeable proton in H_2ase (hydrosulfide) and a more weakly coupled solvent exchangeable proton corresponding to a protonated ligand, as revealed by the ENDOR studies.

Yet another possibility involves H_2 activation at thiolate ligands. This possibility is suggested by the chemistry of an Fe^{II} tetrathiolate complex (Scheme 4) (*110*), which evolves H_2 when H^+ is added to the system. Intermediates proposed for this reaction include thiol complexes derived from the protonation of the thiolate ligands. A role for a metal cluster in the catalysis is also suggested by the mechanism, which involves the formation of dimeric species in order to provide the two electrons necessary for the production of H_2.

Scheme 4

Summary and Conclusions

The redox role of the Ni center in hydrogenase was examined using a combination of biophysical and synthetic model approaches. The lack of a significant shift in the Ni K-edge energy or in the Ni-ligand bond lengths in hydrogenase, as observed by XAS on redox-poised enzyme samples, are inconsistent with a redox role for Ni. The bond lengths obtained from EXAFS analysis and the $S = 0$ determination for the Ni center from magnetization data that is available for an EPR silent form of hydrogenase from *D. baculatus* are both consistent with an oxidation state of II for the Ni center. Combined with the edge energy data, these results suggest that the oxidation state of Ni is II in all redox states of the enzyme. Although this conclusion is appealing in view of the redox chemistry of classical Ni(II) complexes, it does not explain the EPR spectra associated with the Ni site or provide any insight into the nature of the redox active species. It is clear from the observation of ^{61}Ni hyperfine splitting in the EPR spectra that the radical giving rise to the EPR signals observed at 77 K intimately involves the Ni center. However, we have argued that the EPR data are not unequivocal evidence for Ni-centered redox chemistry or the presence of Ni(III) or Ni(I) in the enzyme. These arguments are based on three facts:

1. S-centered radicals may exhibit *g*-values in the range of those observed in the enzyme and are therefore unreliable indicators of primarily metal vs. primarily ligand oxidation.
2. Oxidized model compounds featuring alkyl thiolate ligands exhibit small hyperfine couplings with the ^{61}Ni center and large hyperfine couplings with ligand methylene protons. These hyperfine couplings are indicative of substantial ligand oxidation and are similar to the hyperfine couplings observed for hydrogenase.
3. Studies of the redox chemistry of Ni complexes with alkyl thiolate ligands, like those expected in the biological site, suggest that the products of oxidations of such complexes will likely reflect S-oxidation.

If the changes that occur upon oxidation were largely localized on S (or delocalized over several S centers), this localization would provide one possible explanation for the observed insensitivity of the Ni site structure to the redox state of the enzyme. Alternatively, a redox site composed of protein substituents near the Ni site could also account for the observations.

The possible role of the Ni center as a hydrogen-binding site was also examined. ENDOR spectroscopy clearly shows that a proton from

H_2 binds to an atom that is in part responsible for the EPR spectrum of form C. Furthermore, studies of the photochemistry of form C using ENDOR spectroscopy demonstrate that the dissociation of this proton occurs during the photochemical process. This photolytic cleavage is reversed upon annealing the sample. Understanding the structure of the EPR active site in form C will be very important to understanding how the enzyme activates H_2. The assumption that Ni is the H_2 (or H^-) binding site in the enzyme is not firmly established and, given the highly delocalized nature of the species giving rise to EPR signal C and its associated ENDOR spectra, a number of alternative possibilities remain to be addressed.

The most striking result from the biophysical studies is the lack of any changes associated with the Ni site structure. This leads to a conclusion similar to one suggested for the Mo site in nitrogenase (*111*): that the role of the Ni in hydrogenase is to modify an existing active site, rather than to serve as an entirely new active center. The enzymology of the hydrogenases suggests that these modifications are associated with greater oxygen tolerance or specificity for hydrogen uptake.

Acknowledgments

This work was supported by National Institutes of Health Grant GM-38829 for Michael J. Maroney and National Science Foundation Grant MCB-9207974 for Brian M. Hoffman.

References

1. Adams, M. W. W. *Biochim. Biophys. Acta* **1990,** *1020,* 115–145.
2. Moura, J. J. G.; Teixeira, M.; Moura, I.; LeGall, J. In *The Bioinorganic Chemistry of Nickel;* Lancaster, J. R., Jr., Ed.; VCH: New York, 1988; pp 191–226.
3. Cammack, R.; Fernandez, V. M.; Schneider, K. In *The Bioinorganic Chemistry of Nickel;* Lancaster, J. R., Jr., Ed.; VCH: New York, 1988; pp 167–190.
4. Vignais, P. M.; Colbeau, A.; Willison, J. C.; Jouanneau, Y. *Adv. Microbiol. Physiol.* **1985,** *26,* 155–234.
5. Fauque, G.; Peck, H. D., Jr.; Moura, J. J. G.; Huynh, B. H.; Berlier, Y.; DerVartanian, D. V.; Teixeira, M.; Przybyla, A. E.; Lespinat, P. A.; Moura, I.; LeGall, J. *FEMS Microbiol. Rev.* **1988,** *54,* 299–344.
6. Ma, K.; Zirngibl, D.; Linder, D.; Stetter, K. O.; Thauer, R. K. *Arch. Microbiol.* **1991,** *156,* 43–48.
7. Zirngibl, C.; Hedderich, R.; Thauer, R. K. *FEBS Lett.* **1990,** *261,* 112–116.
8. Lancaster, J. R., Jr. *FEBS Lett.* **1980,** *115,* 285–288.
9. Graf, E. G.; Thauer, R. K. *FEBS Lett.* **1981,** *136,* 165–169.
10. Bryant, F. O.; Adams, M. W. W. *J. Biol. Chem.* **1989,** *264,* 507–509.

11. Hatchikian, E. C.; Fernandez, V. M.; Cammack, R. *FEMS Symp.* **1990,** *54,* 53–73.
12. Bastian, N. R.; Wink, D. A.; Wackett, L. P.; Livingston, D. J.; Jordan, L. M.; Fox, J.; Orme-Johnson, W. H.; Walsh, C. T. In *The Bioinorganic Chemistry of Nickel;* Lancaster, J. R., Jr., Ed.; VCH: New York, 1988; pp 227–247.
13. Gogotov, I. N. *Biochimie* **1986,** *68,* 181–187.
14. Ballantine, S. P.; Boxer, D. H. *J. Bacteriol.* **1985,** *163,* 454–459.
15. Seefeldt, L. C.; Arp, D. J. *Biochimie* **1986,** *68,* 25–34.
16. Hornhardt, S.; Schneider, K.; Friedrich, B.; Vogt, B.; Schlegel, H. G. *Eur. J. Biochem.* **1990,** *189,* 529–537.
17. Voordouw, G.; Menon, N. K.; LeGall, J.; Choi, E. S.; Peck, H. J.; Przybyla, A. E. *J. Bacteriol.* **1989,** *171,* 2894–2899.
18. He, S. H.; Teixeira, M.; LeGall, J.; Patil, D. S.; Moura, I.; Moura, J. J. G.; DerVartanian, D. V.; Huynh, B. H.; Peck, H. D., Jr. *J. Biol. Chem.* **1989,** *264,* 2678–2682.
19. Eidsness, M. K.; Scott, R. A.; Prickril, B. C.; DerVartanian, D. V.; Legall, J.; Moura, I.; Moura, J. J. G.; Peck, H. J. *Proc. Natl. Acad. Sci. U.S.A.* **1989,** *86,* 147–151.
20. Hsu, J. C.; Beilstein, M. A.; Whanger, P. D.; Evans, H. J. *Arch. Microbiol.* **1990,** *154,* 215–220.
21. Przybyla, A. E.; Robbins, J.; Menon, N.; Peck, H. D., Jr. *FEMS Microbiol. Rev.* **1992,** *88,* 109–135.
22. Moura, J. J. G.; Teixeira, M.; Xavier, A. V.; Moura, I.; LeGall, J. *J. Mol. Catal.* **1984,** *23,* 303.
23. Teixeira, M.; Moura, I.; Xavier, A. V.; Huynh, B. H.; DerVartanian, D. V.; Peck, H. D., Jr.; LeGall, J.; Moura, J. J. G. *J. Biol. Chem.* **1985,** *260,* 8942–8950.
24. van der Zwaan, J. W.; Albracht, S. P. J.; Fontijn, R. D.; Slater, E. C. *FEBS Lett.* **1985,** *179,* 271–277.
25. Cammack, R.; Fernandez, V. M.; Schneider, K. *Biochimie* **1986,** *68,* 85–91.
26. van der Zwaan, J. W.; Coremans, J. M. C. C.; Bouwens, E. C. M.; Albracht, S. P. J. *Biochim. Biophys. Acta* **1990,** *1041,* 101–110.
27. Fan, C.; Teixeira, M.; Moura, J.; Moura, I.; Huynh, B. H.; Le Gall, J.; Peck, H. D., Jr.; Hoffman, B. M. *J. Am. Chem. Soc.* **1991,** *113,* 20–24.
28. Coremans, J. M. C. C.; van der Zwaan, J. W.; Albracht, S. P. J. *Biochim. Biophys. Acta* **1992,** *1119,* 157–168.
29. Wang, C. P.; Franco, R.; Moura, J. J. G.; Moura, I.; Day, E. P. *J. Biol. Chem.* **1992,** *267,* 7378–7380.
30. Scott, R. A.; Czechowski, M.; DerVartanian, D. V.; LeGall, J.; Peck, H. D., Jr.; Moura, I. *Rev. Port. Quim.* **1985,** *27,* 67–70.
31. Scott, R. A.; Wallin, S. A.; Czechowski, M.; DerVartanian, D. V.; LeGall, J.; Peck, H. D., Jr.; Moura, I. *J. Am. Chem. Soc.* **1984,** *106,* 6864–6865.
32. Eidsness, M. K.; Sullivan, R. J.; Scott, R. A. In *The Bioinorganic Chemistry of Nickel;* Lancaster, J. R., Jr., Ed.; VCH: New York, 1988; pp 73–91.
33. Lindahl, P. A.; Kojima, N.; Hausinger, R. P.; Fox, J. A.; Teo, B. K.; Walsh, C. T.; Orme-Johnson, W. H. *J. Am. Chem. Soc.* **1984,** *106,* 3062–3064.
34. Maroney, M. J.; Colpas, G. J.; Bagyinka, C.; Baidya, N.; Mascharak, P. K. *J. Am. Chem. Soc.* **1991,** *113,* 3962–3972.

35. Bagyinka, C.; Whitehead, J. P.; Maroney, M. J. *J. Am. Chem. Soc.* **1993,** *115,* 3576–3585.
36. Albracht, S. P. J.; Kröger, A.; van der Zwaan, J. W.; Unden, G.; Böcher, R.; Mell, H.; Fontijn, R. D. *Biochim. Biophys. Acta* **1986,** *874,* 116–127.
37. Cammack, R.; Kovacs, K. L.; McCracken, J.; Peisach, J. *Eur. J. Biochem.* **1989,** *182,* 363–366.
38. Chapman, A.; Cammack, R.; Hatchikian, C. E.; McCracken, J.; Peisach, J. *FEBS Lett.* **1988,** *242,* 134–138.
39. Tan, S.-L.; Fox, J. A.; Kojima, N.; Walsh, C. T.; Orme-Johnson, W. H. *J. Am. Chem. Soc.* **1984,** *106,* 3064–3066.
40. Teixeira, M.; Moura, I.; Fauque, G.; Dervartanian, D. V.; Legall, J.; Peck, H. D., Jr.; Moura, J. J. G.; Huynh, B. H. *Eur. J. Biochem.* **1990,** *189,* 381–386.
41. Clark, K.; Penner-Hahn, J. E.; Whittaker, M. M.; Whittaker, J. W. *J. Am. Chem. Soc.* **1990,** *112,* 6433–6434.
42. Tan, G. O.; Ensign, S. A.; Ciurli, S.; Scott, M. J.; Hedman, B.; Holm, R. H.; Ludden, P. W.; Korszun, Z. R.; Stephens, P. J.; Hodgson, K. O. *Proc. Natl. Acad. Sci. U.S.A.* **1992,** *89,* 4427–4431.
43. Cramer, S. P.; Gray, H. B.; Rajagopalan, K. V. *J. Am. Chem. Soc.* **1979,** *101,* 2772–2774.
44. McDermott, A. E.; Yachandra, V. K.; Guiles, R. D.; Cole, J. L.; Dexheimer, S. L.; Britt, R. D.; Sauer, K.; Klein, M. P. *Biochemistry* **1988,** *27,* 4021–4031.
45. Goodin, D. B.; Yachandra, V. K.; Britt, R. D.; Sauer, K.; Klein, M. P. *Biochim. Biophys. Acta* **1984,** *767,* 209–216.
46. Wirt, M. D.; Sagi, I.; Chen, E.; Frisbie, S. M.; Lee, R.; Chance, M. R. *J. Am. Chem. Soc.* **1991,** *113,* 5299–5304.
47. Powers, L.; Blumberg, W. E.; Chance, B.; Barlow, C. H.; Leigh, J. S., Jr.; Smith, J.; Yonetani, T.; Vik, S.; Peisach, J. *Biochem. Biophys. Acta* **1979,** *546,* 520–538.
48. Cramer, S. P.; Eidsness, M. K.; Pan, W.-H.; Morton, T. A.; Ragsdale, S. W.; DerVartanian, D. V.; Ljungdahl, L. G.; Scott, R. A. *Inorg. Chem.* **1987,** *26,* 2477–2479.
49. Hu, V. W.; Chan, S. I.; Brown, G. S. *Proc. Natl. Acad. Sci. U.S.A.* **1977,** *74,* 3821–3825.
50. Peisach, J.; Powers, L.; Blumberg, W. E.; Chance, B. *Biophys. J.* **1982,** *38,* 277–285.
51. Tullius, T. D.; Kurtz, D. M., Jr.; Conradson, S. D.; Hodgson, K. O. *J. Am. Chem. Soc.* **1979,** *101,* 2776–2779.
52. Hamilton, G. A.; Adolf, P. K.; de Jersey, J.; Dubois, G. C.; Dyrkacz, G. R.; Libby, R. D. *J. Am. Chem. Soc.* **1978,** *100,* 1899–1912.
53. Ito, N.; Phillips, S. E. V.; Stevens, C.; Ogel, Z. B.; McPherson, M. J.; Keen, J. N.; Yadav, K. D. S.; Knowles, P. F. *Nature (London)* **1991,** *350,* 87–90.
54. Smith, T. A.; Penner-Hahn, J. E.; Berding, M. A.; Doniach, S.; Hodgson, K. O. *J. Am. Chem. Soc.* **1985,** *107,* 5945–5955.
55. Bair, R. A.; Goddard, W. A. *Phys. Rev. B* **1980,** *22,* 2767.
56. Kosugi, N.; Yokohama, T.; Asakura, K.; Kuroda, H. *Chem. Phys.* **1984,** *91,* 249.
57. Yokoyama, T.; Kosugi, N.; Kuroda, H. *Chem. Phys.* **1986,** *103,* 101.
58. Colpas, G. J.; Maroney, M. J.; Bagyinka, C.; Kumar, M.; Willis, W. S.; Suib, S. L.; Mascharak, P. K.; Baidya, N. *Inorg. Chem.* **1991,** *30,* 920–928.

59. Schulman, R. G.; Yafet, Y.; Eisenburger, P.; Blumberg, W. E. *Proc. Natl. Acad. Sci. U.S.A.* **1976,** *73,* 1384.
60. Hahn, J. E.; Scott, R. A.; Hodgson, K. O.; Doniach, S.; Desjardins, S. E.; Soloman, E. I. *Chem. Phys. Lett.* **1982,** *88,* 595.
61. Roe, A. L.; Schneider, D. J.; Mayer, R. J.; Pyrz, J. W.; Widom, J.; Que, L., Jr. *J. Am. Chem. Soc.* **1984,** *106,* 1676–1681.
62. Krüger, H.-J.; Holm, R. H. *J. Am. Chem. Soc.* **1990,** *112,* 2955–2963.
63. Huynh, B. H.; Patil, D. S.; Moura, I.; Teixeira, M.; Moura, J. J. G.; DerVartanian, D. V.; Czechowski, M. H.; Prickril, B. C.; Peck, H. D., Jr.; LeGall, J. *J. Biol. Chem.* **1987,** *262,* 795–800.
64. Teixeira, M.; Moura, I.; Xavier, A. V.; Moura, J. J. G.; LeGall, J.; DerVartanian, D. V.; Peck, H. D., Jr.; Huynh, B.-H. *J. Biol. Chem.* **1989,** *264,* 16435–16450.
65. Boussac, A.; Zimmermann, J.-L.; Rutherford, A. W.; Lavergne, J. *Nature (London)* **1990,** *347,* 303.
66. Browett, W. R.; Gasyna, Z.; Stillman, M. J. *J. Am. Chem. Soc.* **1988,** *110,* 3633–3640.
67. Schulz, C. E.; Rutter, R.; Sage, J. T.; G., D. P.; Hager, L. P. *Biochemistry* **1984,** *23,* 4743–4754.
68. Zimmer, M.; Schulte, G.; Luo, X. L.; Crabtree, R. H. *Angew. Chem. Int. Ed. Engl.* **1991,** *103,* 205–207.
69. Haines, R. I.; McAuley, A. *Coord. Chem. Rev.* **1981,** *39,* 77–119.
70. Nag, K.; Chakravorty, A. *Coord. Chem. Rev.* **1980,** *33,* 87–147.
71. Sugiura, Y.; Kuwahara, J.; Suzuki, T. *Biochem. Biophys. Res. Commun.* **1983,** *115,* 878–881.
72. Lappin, A. G.; Murray, C. K.; Margerum, D. W. *Inorg. Chem.* **1978,** *17,* 1630–1634.
73. Lappin, A. G.; McAuley, A. *Adv. Inorg. Chem.* **1988,** *32,* 241–294.
74. Krüger, H.-J.; Peng, G.; Holm, R. H. *Inorg. Chem.* **1991,** *30,* 734–742.
75. Baidya, N.; Olmstead, M. M.; Whitehead, J. P.; Bagyinka, C.; Maroney, M. J.; Mascharak, P. K. *Inorg. Chem.* **1992,** *31,* 3612–3619.
76. Baidya, N.; Olmstead, M. M.; Mascharak, P. K. *J. Am. Chem. Soc.* **1992,** *114,* 9666–9668.
77. Fox, S.; Wang, Y.; Silver, A.; Millar, M. *J. Am. Chem. Soc.* **1990,** *112,* 3218–3220.
78. Kumar, M.; Day, R. O.; Colpas, G. J.; Maroney, M. J. *J. Am. Chem. Soc.* **1989,** *111,* 5974–5976.
79. Mirza, S. A.; Pressler, M. A.; Kumar, M.; Day, R. O.; Maroney, M. J. *Inorg. Chem.* **1993,** *32,* 977–987.
80. Colpas, G. J.; Kumar, M.; Day, R. O.; Maroney, M. J. *Inorg. Chem.* **1990,** *29,* 4779–4788.
81. Franolic, J. D.; Wang, W. Y.; Millar, M. *J. Am. Chem. Soc.* **1992,** *114,* 6587–6588.
82. Handa, M.; Mikuriya, M.; Okawa, H.; Kida, S. *Chem. Lett.* **1988,** 1555.
83. Bradbury, J. R.; Masters, A. F.; C., M. A.; Brunette, A. A.; Bond, A. M.; Wedd, A. G. *J. Am. Chem. Soc.* **1981,** *103,* 1959–1974.
84. Magno, F.; Bontempelli, G.; Pillone, G. *J. Electroanal. Chem. Interfac. Electrochem.* **1971,** *30,* 375–383.
85. Krüger, H. J.; Holm, R. H. *Inorg. Chem.* **1989,** *28,* 1148–1155.
86. Gilbert, B. C. In *Structure and Reaction Mechanisms in Sulphur-Radical Chemistry Revealed by E.S.R. Spectroscopy;* Chatgilialoglu, C.; Asmus, K.-D., Eds.; Plenum: New York, 1990; pp 135–154.

87. Symons, M. C. R. *Chemical and Biochemical Aspects of Electron-Spin Resonance Spectroscopy;* John Wiley and Sons: New York, 1978.
88. Whitehead, J. P.; Gurbiel, R. J.; Bagyinka, C.; Hoffman, B. M.; Maroney, M. J. *J. Am. Chem. Soc.* **1993,** *115,* 5629–5635.
89. Kumar, M.; Colpas, G. J.; Day, R. O.; Maroney, M. J. *J. Am. Chem. Soc.* **1989,** *111,* 8323–8325.
90. Deutsch, E.; Root, M. J.; Nosco, D. L. *Adv. Inorg. Bioinorg. Mech.* **1982,** *1,* 269.
91. Adzamli, I. K.; Deutsch, E. *Inorg. Chem.* **1980,** *19,* 1366–1373.
92. Kriege, M.; Henkel, G. *Z. Naturforsch. B Chem. Sci.* **1987,** *42,* 1121–1128.
93. Tremel, W.; Kriege, M.; Krebs, B.; Henkel, G. *Inorg. Chem.* **1988,** *27,* 3886–3895.
94. Nicholson, J. R.; Christou, G.; Huffman, J. C.; Folting, K. *Polyhedron* **1987,** *6,* 863–870.
95. Chatgilialoglu, C.; Guerra, M. In *Alkanethiylperoxyl Radicals;* Chatgilialoglu, C.; Asmus, K.-D., Eds.; Plenum: New York, 1990; pp 31–36.
96. Chen, D.; Motekaitis, R. J.; Martell, A. E. *Inorg. Chem.* **1991,** *30,* 1396–1402.
97. Kimura, E.; Sakonaka, A.; Machida, R. *J. Am. Chem. Soc.* **1982,** *104,* 4255–4257.
98. Watanabe, Y.; Kuriki, N.; Ishiguro, K.; Sawaki, Y. *J. Am. Chem. Soc.* **1991,** *113,* 2677–2682.
99. Akasaka, T.; Haranaka, M.; Ando, W. *J. Am. Chem. Soc.* **1991,** *113,* 9898–9900.
100. Liang, J.-J.; Gu, C.-L.; Kacher, M. L.; Foote, C. S. *J. Am. Chem. Soc.* **1983,** *105,* 4717–4721.
101. Schrauzer, G. N.; Zhang, C.; Chadha, R. *Inorg. Chem.* **1990,** *29,* 4104–4107.
102. Farmer, P. J.; Solouki, T.; Mills, D. K.; Soma, T.; Russell, D. H.; Reibenspies, J. H.; Darensbourg, M. Y. *J. Am. Chem. Soc.* **1992,** *114,* 4601–4615.
103. Greenwood, N. N.; Earnshaw, A. In *Chemistry of the Elements;* Pergamon: New York, 1984; pp 911–913.
104. Morton, J. R.; Preston, K. F. *J. Chem. Phys.* **1984,** *81,* 5775–5778.
105. Symons, M. C. R.; Aly, M. M.; West, D. X. *Chem. Commun.* **1979,** 51–52.
106. Pappenhagen, T. L.; Margerum, D. W. *J. Am. Chem. Soc.* **1985,** *107,* 4576–4577.
107. Albeniz, A. C.; Heinekey, D. M.; Crabtree, R. H. *Inorg. Chem.* **1991,** *30,* 3632–3625.
108. Cammack, R.; Bagyinka, C.; Kovacs, K. L. *Eur. J. Biochem.* **1989,** *182,* 357–362.
109. Laurie, J. C. V.; Duncan, L.; Haltiwanger, R. C.; Weberg, R. T.; DuBois, M. R. *J. Am. Chem. Soc.* **1986,** *108,* 6234–6241.
110. Sellmann, D.; Geck, M.; Moll, M. *J. Am. Chem. Soc.* **1991,** *113,* 5259–5264.
111. Kim, J.; Rees, D. C. *Nature (London)* **1992,** *360,* 553–560.

RECEIVED for review July 19, 1993. ACCEPTED revised manuscript March 15, 1994.

3

Modeling Phenoxyl Radical Metalloenzyme Active Sites

David P. Goldberg and Stephen J. Lippard*

Department of Chemistry, Massachusetts Institute of Technology, Cambridge, MA 02139

A phenoxyl radical has been implicated in the functioning of three metalloenzymes, ribonucleotide reductase, galactose oxidase, and prostaglandin H synthase. A brief summary of these three systems is presented, focusing on the role of the metal–phenoxyl radical active site. Special emphasis has been given to the physical properties of the phenoxyl radical–containing metal active sites and to possible mechanisms for the generation and function of these active sites. Several ligand systems designed to provide a phenoxyl radical coordinated to a metal center are described. These systems include ligands containing sterically hindered phenol groups that allow for delocalization of the phenoxyl radical spin onto the metal center. We give a summary of progress in our own laboratory on modeling the active site of ribonucleotide reductase through the use of a new bidentate nitrogen donor ligand with a pendant phenol–phenoxyl radical arm. The synthesis and properties of a dinuclear iron compound containing this pendant phenoxyl radical ligand is described. This complex mimics some of the key physical properties of the diferric active site of ribonucleotide reductase.

RADICAL COFACTORS IN BIOLOGICAL SYSTEMS have become a subject of increasing interest in recent years (*1–3*). Tyrosine-based radicals, in particular, have now been identified in several enzymes (*4*). The tyrosine residue functions as a redox-active cofactor by interconverting between the oxidized phenoxyl radical and the normal phenol or phenolate states. More commonly known redox-active cofactors include transition metal ions, and a few enzymes use both tyrosine residues and metals as partners in effecting redox chemistry.

Mechanistic pathways that explore this theme have been proposed

*Corresponding author

0065–2393/95/0246–0061/$08.18/0

for three metalloenzymes: ribonucleotide reductase, galactose oxidase, and prostaglandin H synthase. In each of these cases mechanisms have been proposed that include one or two transition metals working in concert with a redox-active tyrosine group. Dioxygen and hydrogen peroxide have been implicated as agents responsible for generating these stable tyrosyl radicals, with a concomitant redox change at the metal site. We discuss these three systems and concentrate on the properties of the metal active site and the characterization of the tyrosyl radicals. Current mechanistic hypotheses about the generation and function of the tyrosyl radical and metal active sites found in these enzymes are presented. We focus on the metal-radical partnership as well as the roles of dioxygen and hydrogen peroxide in generating and maintaining the tyrosyl radical.

To model the oxo-bridged diiron(III) phenoxyl radical active site found in the R2 protein of *Escherichia coli* ribonucleotide reductase (RR), we designed and synthesized a new bidentate nitrogen donor ligand that harbors a pendant phenol moiety. Complexes of both the nonradical phenol and phenoxyl radical forms of this ligand are described, including a dinuclear iron compound that reproduces some of the key physical properties of the R2 protein active site. This work was carried out as part of a broader program in which first row transition metal complexes are synthesized as models for the active sites of a variety of metalloenzymes (5). These activities were expanded to include models for the active sites of enzymes containing both first row transition metal ions and phenoxyl radicals. Our ultimate goal is to prepare compounds that can serve as functional mimics for these sites. In particular, we want to understand better the mechanism of radical generation and the synergistic relationship between the metal ion and phenoxyl radical in the functioning biological systems.

Previous work by other groups has led to the synthesis of various ligands incorporating sterically hindered phenol substituents for the generation of phenoxyl radical metal complexes. These systems were designed to allow for delocalization of the phenoxyl radical spin from phenyl ring out onto the metal center. Usually, the putative phenoxyl radical complexes were generated in situ from the analogous phenolate complexes by addition of an appropriate oxidizing agent and were characterized mainly by electron paramagnetic resonance (EPR) spectroscopy. We describe several exemplary ligands used in this earlier work and summarize the synthesis and characterization of the phenol and phenoxyl radical complexes reported for these systems.

Phenoxyl Radical Metalloenzymes

Ribonucleotide Reductase. All living organisms depend upon one of three classes of ribonucleotide reductases in the biosynthesis of deoxyribonucleotides from ribonucleotides (*6*). One class contains a dinuclear iron center and stable tyrosyl radical essential for catalytic activity. The best studied ribonucleotide reductase in this class comes from *E. coli* and is composed of two homodimeric subunits R1 and R2. Substrate and allosteric effector molecules bind to the R1 subunit, which is believed to be the site of substrate reduction. The smaller R2 protein contains the diiron core and a proximal tyrosyl radical. This subunit can exist in four states: active R2, met R2, red R2, and apo R2. The active form contains an oxo-bridged diiron(III) core and a tyrosyl radical that is stable for several days at room temperature. The Fe(III)–O–Fe(III) center is maintained in the met R2 form, but the radical has been reduced to a normal tyrosine residue. This form of the enzyme has been characterized by X-ray crystallography at 2.2 Å resolution, and the structure of its diiron(III) site is shown in Figure 1 (*7, 8*).

In the met form, the two ferric ions are bridged by an oxo ligand and the carboxylate side chain of a glutamate residue. The Fe1 atom has a distorted square pyramidal geometry, alternately characterized as pseudo-octahedral if the aspartate residue is assigned as chelating bidentate, whereas Fe2 has nearly perfect octahedral symmetry. In the active form of the enzyme, Tyr 122 has been oxidized to a tyrosyl radical (*9*), which, as revealed by the X-ray structure of met R2, is located in a hydrophobic pocket with its hydroxyl group 5.3 Å away from Fe1. The pocket does not contain any oxidizable side chains near the radical site

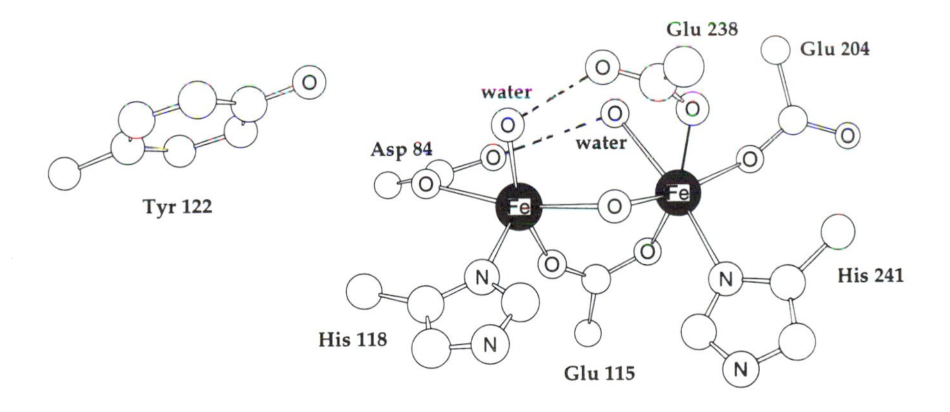

Figure 1. Structure of the R2 protein of E. coli *ribonucleotide reductase in the met form. (Adapted from reference 7, which reports the 2.2 Å crystal structure of the protein. Note that, in reference 7, Asp 84 is considered to be bidentate and chelating, but we prefer the monodentate, hydrogen-bonded representation depicted above based on an analysis of the Fe–O distances.)*

and protects Tyr 122 from solvent or other possible reductants. These features may account for the remarkable stability of the phenoxyl radical generated at Tyr 122.

Formation of the ribonucleotide reductase tyrosyl radical in vivo is believed to occur during the reaction of dioxygen with the reduced, diferrous form of R2. Experiments in vitro confirm that addition of O_2 to red R2 does lead to the immediate generation of the tyrosyl radical, together with the (μ-oxo)diiron(III) center. Recent studies using stopped-flow absorption, rapid freeze-quench EPR (*10*) and rapid freeze-quench Mössbauer spectroscopy (*11*) have detected the presence of two intermediates in the radical-generating reaction, either one of which could be responsible for producing the radical from Tyr 122. One intermediate is probably a peroxo-bridged diiron(III) species, whereas the other may be a dioxygen- or amino acid-derived radical bound to one or both ferric ions.

A mutant form of the reduced R2 protein, red R2 S211A, has been crystallographically characterized to 2.2 Å resolution (*12*), and structural features of its diiron(II) site are shown schematically in Figure 2. Cur-

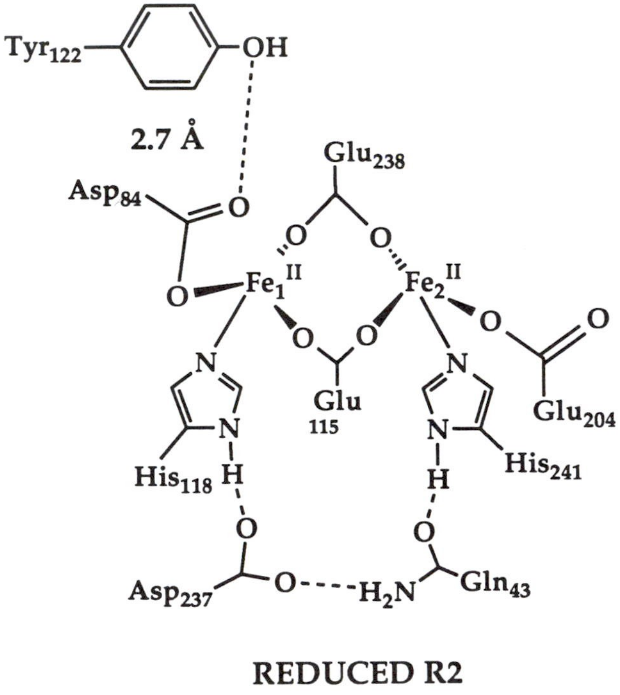

Figure 2. Structure of a S211A mutant of the R2 protein of E. coli ribonucleotide reductase in the reduced form. (Adapted from reference 11 reporting the 2.2 Å crystal structure.)

rently there are no X-ray data on wild-type (wt) red R2. The structures of wt met R2 and met R2 S211A are nearly identical, including details of the metal coordination sphere. This result suggests that the metal sites in wt red R2 and red R2 S211A are probably also the same. In the reduced protein, the bridging oxo group present in the met R2 structure is gone, and Glu 238 has changed from a terminal to a bridging ligand. This movement of Glu 238 is a good example of the ability of carboxylate residues to vary their binding modes in dinuclear iron cores and has been referred to as the carboxylate shift (13). No water ligands were located near the iron atoms in the red R2 S211A structure, and both Fe(II) ions display tetrahedral geometry.

The structural analysis of the red R2 protein revealed two features important to the proposed mechanisms for tyrosyl radical generation (12). The four-coordinate ferrous ions have available sites for binding dioxygen as either a terminal ligand to one of the iron atoms or as a bridging ligand. The latter binding mode permits a variety of possible structures. The largest change in the structure of red R2 S211A compared to the met form is a 2 Å movement in the position of the hydroxyl group of Tyr 209, a residue not shown in Figures 1 and 2. This motion opens a channel from the surface of the protein down to the diiron centers, which creates a possible pathway for O_2 to reach the iron site and nearby hydrophobic patch.

Reactions of the met R2 protein with H_2O_2 (*14, 15*), as well as O-atom donors such as peroxyacids and monoperoxophthalate (*15*) afford the tyrosyl radical in 25–35% yield. The reaction of H_2O_2 and met R2 supports a mechanism in which a diiron(III) peroxide intermediate is involved in the dioxygen-dependent tyrosyl radical generation from the reduced form of the protein. The single oxygen atom donor experiments suggest that a high-valent iron-oxo intermediate, such as an $\{Fe^{IV}O\}$ or $\{Fe^{V}O\}$ species, may also be competent to generate the tyrosyl radical in the R2 protein, possibly being formed in a subsequent step from the peroxide intermediate. Work by Stubbe and co-workers (*11*) revealing spectroscopic intermediates in the reaction of red R2 with dioxygen shows no evidence for the formation of such a high-valent iron-oxo species, however.

Galactose Oxidase. The enzyme galactose oxidase (*16–18*), secreted from the fungus *Dactylium dendroides*, is a 68-kDa monomeric polypeptide that effects the transformation of a wide variety of primary alcohols to aldehydes according to the generalized reaction given in equation 1. Dioxygen is the electron acceptor in this two-electron oxidation of alcoholic substrates. The catalytic reaction is believed to take place at or near a mononuclear copper active site. No other metal cofactors have been identified. The source of the second oxidizing equivalent necessary for converting alcohols to aldehydes remains undetermined, but a mounting body of evidence points to the presence of an organic radical cofactor working in conjunction with the active site Cu(II)–Cu(I) redox couple. Such a radical cofactor has been positively identified as originating from a tyrosine residue (*19*), which is modified by an unusual thioether linkage involving a nearby cysteine.

$$RCH_2OH + O_2 \rightarrow RCHO + H_2O_2 \quad (1)$$

The crystal structure of galactose oxidase from *D. dendroides* was determined to 1.7 Å resolution (*20*). The copper site is located near the solvent-accessible surface. An almost perfect square is formed by four ligands in the equatorial plane of the copper, two nitrogen donors from histidines and two oxygen donors from a monodentate acetate and a tyrosine residue. The acetate ligand comes from the buffer (pH 4.5) in which the native crystals were grown. Another data set (1.9 Å resolution) collected on native crystals grown in PIPES (1,4-piperazinediethanesulfonic acid) buffer (pH 7.0) revealed a water molecule, rather than acetate, at a distance of 2.8 Å from the cupric ion. This long $Cu \cdots OH_2$ distance suggests that the geometry is more tetrahedrally distorted at pH 7.0, the pH at which the enzyme is active. Such a distortion may lower the redox potential for the Cu(II)–Cu(I) couple proposed to participate in catalytic turnover. A fifth ligand, an oxygen from Tyr 495, occupies the axial position of the square-pyramidal Cu(II) atom, as shown schematically in Figure 3.

The X-ray structure also revealed that the tyrosine bonded in the equatorial plane, Tyr 272, has undergone a remarkable modification. A covalent bond has formed between a carbon atom ortho to the hydroxyl group and a cysteinyl sulfur atom from Cys 228. The cysteinyl carbon atom bonded to the sulfur atom lies in the plane of the tyrosine phenyl ring. This planarity suggests partial double-bond character in the C–S

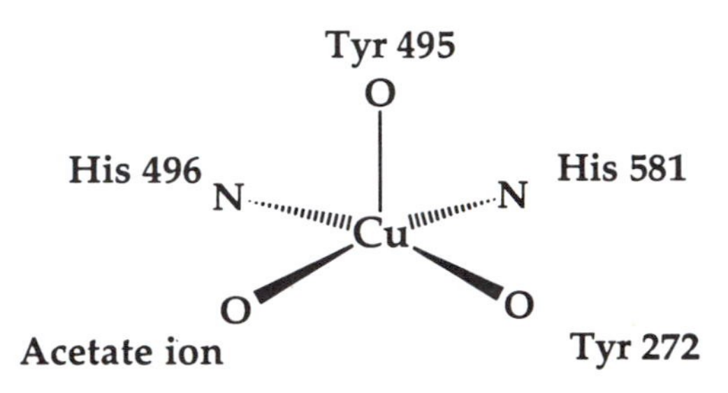

Figure 3. Sketch of the copper coordination in galactose oxidase, as determined by the 1.7 Å resolution crystal structure.

linkage. A further intriguing structural feature is the close contact made by this sulfur atom with a π-stacked tryptophan residue, the distance of S to all carbons in the six-membered ring being 3.84 Å. This stacking interaction is shown schematically in Figure 4. These geometric features suggest that the tyrosyl radical forms at Tyr 272 and is stabilized by distribution of spin density onto the cysteinyl sulfur atom and possibly over the entire π system of the proximal tryptophan group.

Galactose oxidase is EPR-silent, which suggests an antiferromagnetically coupled $S = 0$ ground state, consistent with the $S = 1/2$ tyrosyl radical being directly coordinated to the $S = 1/2$ Cu(II) in the active site. Such direct coupling is in accord with the assignment of Tyr 272 as the tyrosyl radical, because this residue is bonded to the Cu(II) ion. More detailed EPR and electron nuclear double resonance (ENDOR) studies of the apoenzyme and thioether-substituted model compounds provide additional support for Tyr 272 being the radical (*21*).

Further characterization of galactose oxidase by X-ray absorption spectroscopy has provided strong evidence for the assignment of a +2 oxidation state for the copper ion in both the reduced and oxidized forms (*22*). Such an assignment is consistent with a possible mechanism for alcohol oxidation by galactose oxidase (*23*). In this mechanism, shown in Figure 5, the alcohol first binds close to the Cu(II) ion and coordinated tyrosyl radical and then donates two electrons to give aldehyde and a Cu(I)–tyrosine species. The cuprous site is then reoxidized to a Cu(II)–tyrosyl radical species by dioxygen. As in ribonucleotide reductase, the active form of galactose oxidase, which contains the metal in the oxidized state and tyrosyl radical, is thought to be generated by dioxygen.

Prostaglandin (PG) H Synthase. The enzyme PGH synthase is a homodimeric protein consisting of subunits with an approximate molecular weight of 72 kDa and one Fe(III)-protoporphyrin IX (PPIXFe(III)) prosthetic group per subunit. This protein is responsible for the central reaction in the biosynthesis of prostaglandins and is selectively inhibited by antiinflammatory drugs such as aspirin and indo-

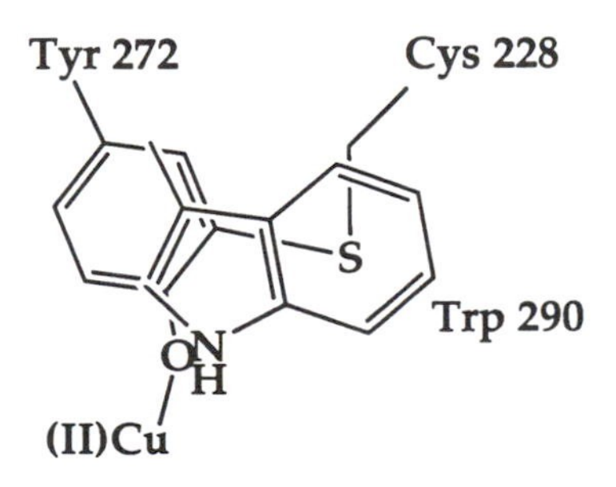

Figure 4. Drawing of the stacking interaction between Trp 290 and the thioether bond formed by Cys 228 with Tyr 272 in galactose oxidase as determined from the crystal structure.

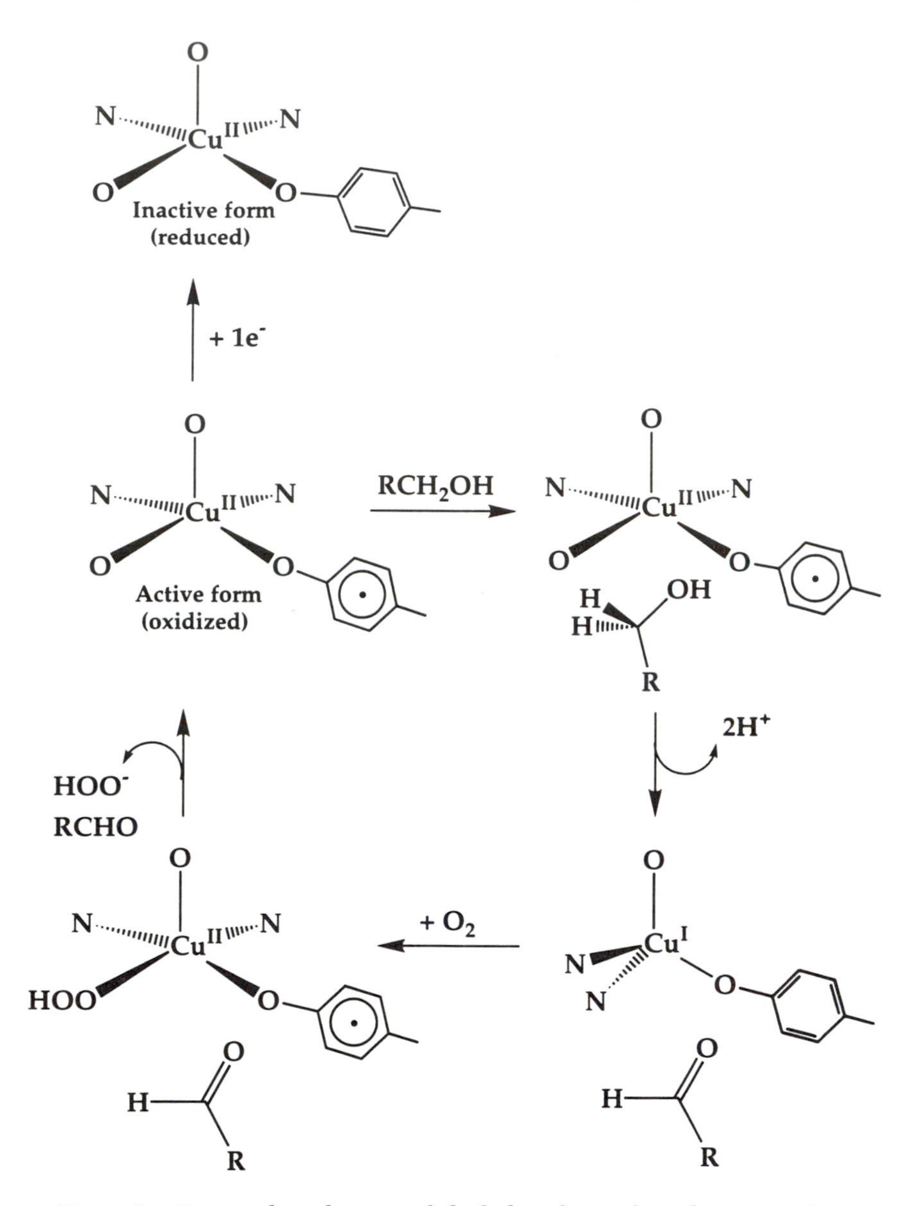

Figure 5. Proposed mechanism of alcohol oxidation for galactose oxidase.

methacin. Two independent reactions are effected by the synthase, a cyclooxygenase reaction in which arachidonic acid is converted into prostaglandin G_2 (PGG_2), and a peroxidase reaction in which PGG_2 is converted to prostaglandin H_2 (PGH_2) (*24*). A tyrosyl radical (*25*) has been implicated as the species responsible for the initial step in the cyclooxygenase activity, and a plausible mechanism involving its generation and role in the cyclooxygenase reaction is discussed in the next paragraph. In contrast to this mechanism, some evidence has been reported suggesting that tyrosyl radicals observed spectroscopically during the action of PGH synthase are not catalytically competent intermediates for the cyclooxygenase reaction.

Figure 6 presents a postulated mechanism for these two reactions. In the peroxidase reaction, a hydroperoxide substrate such as PGG_2 converts $\{PPIXFe(III)\}^+$ to $\{PPIXFe(V)O\}^{+4}$ (intermediate I) in a two-electron redox step to give PGH_2. Intermediate I then removes a hydrogen atom from a nearby tyrosine residue to give {PPIXFe(IV)O}-(TYR·) (intermediate II). The tyrosyl radical of intermediate II links

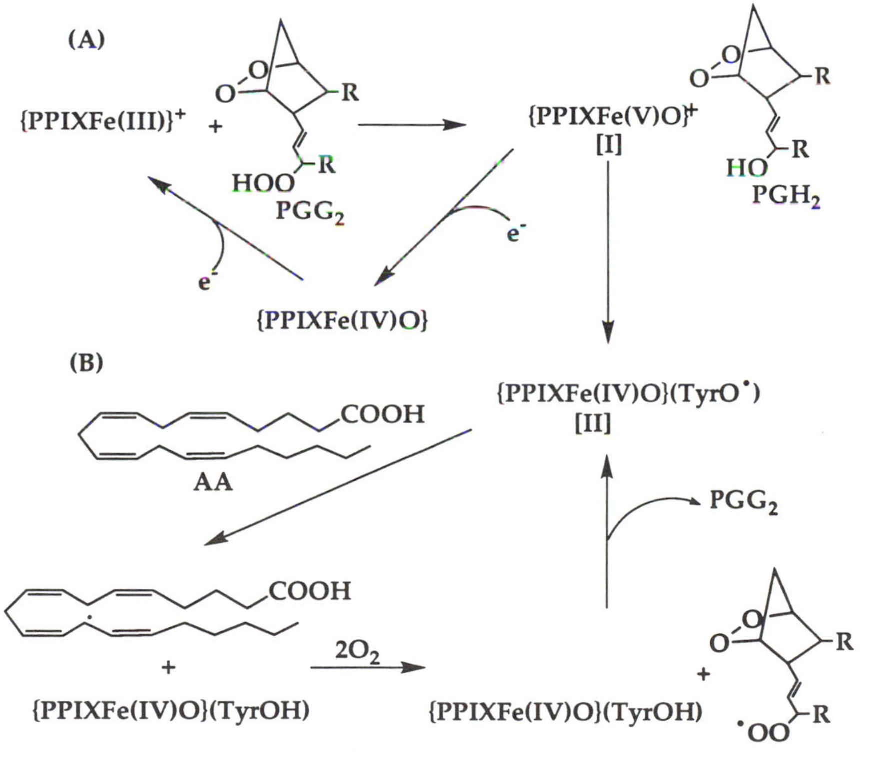

Figure 6. Proposed mechanism for the peroxidase (A) and the cyclooxygenase (B) reactions of prostaglandin H synthase.

the peroxidase and cyclooxygenase reactions. In the first step of the cyclooxygenase mechanism, the tyrosyl radical removes the 13-pro-S hydrogen atom from arachidonic acid, which then goes on to react with molecular oxygen to form PGG_2 (*26*).

Evidence from EPR spectroscopy supports the presence of a tyrosyl radical in the mechanism described in the preceding paragraph. A doublet signal appears in the EPR spectrum corresponding to the formation of a tyrosyl radical immediately after addition of arachidonic acid or PGG_2 to resting enzyme. This doublet signal decays within 1 min at −12 °C, with concomitant formation of a singlet that may originate from either the same or another tyrosyl radical having a different dihedral angle between the phenyl ring and the benzylic carbon CH_2 group (*27*).

When native PGH synthase was treated with tetranitromethane, three tyrosine residues, Tyr 355, Tyr 385, and Tyr 417, were modified (*28*). These three tyrosines are possibly important to cyclooxygenase activity. When the native enzyme was first incubated with indomethacin, a known cyclooxygenase inhibitor, none of these tyrosine residues was nitrated. In the same work, site-directed mutagenesis was used to replace each of these tyrosines with phenylalanine, and only the mutant Y385F lacked cyclooxygenase activity. This result suggests that Tyr 385 is located near the site of cyclooxygenase activity and is the source of the tyrosyl radical responsible for H atom abstraction from arachidonic acid.

Evidence suggests that the tyrosyl radicals giving rise to either the doublet or singlet signal are not kinetically competent intermediates in the cyclooxygenase pathway (*29*). In this study, addition of arachidonic acid to Fe–PGH synthase led to the formation of arachidonic acid metabolic products prior to the detection of either the doublet or singlet EPR signals assigned as tyrosyl radicals. When the Fe-PGH synthase was incubated with the peroxidase-reducing substrate phenol, formation of the tyrosyl radical signals was inhibited, but arachidonic acid metabolites were still produced. No EPR signals corresponding to tyrosyl radicals were detected for Mn–PGH synthase, but cyclooxygenase activity was still observed for the Mn-reconstituted enzyme. These data led the authors to conclude that the tyrosyl radicals detected by EPR spectroscopy were not catalytically competent intermediates for the cyclooxygenase reaction, but that tyrosyl radicals too short-lived to be observed on the EPR time scale may still be important in the cyclooxygenase mechanism.

Phenoxyl Radical Metal Complexes

A variety of phenoxyl radical metal complexes have been generated in situ by oxidation of the precursor phenol metal complex. Characterization of the resulting radical metal complexes has been mainly by EPR spectroscopy. The phenoxyl radical ligands were used in part as spin probes to interrogate the structure of the metal complexes. In all cases the ligands were designed to provide a pathway for delocalization of phenoxyl radical spin out of the phenyl ring. In several cases delocalization directly onto the metal center was reported to stabilize the phenoxyl radical compared to the corresponding uncomplexed radical ligand. Very few of these radical–metal complexes were isolated as solids, and little crystallographic data were reported for even the stable, nonradical phenol precursor complexes. Representative examples of these systems are discussed later.

Several sterically hindered phenol ligands were prepared by using Schiff-base systems. Condensation of 2-amino-4,6-di-*tert*-butylphenol with the appropriate *ortho*-hydroxy aromatic aldehyde or ketone afforded Schiff-base ligands of the kind shown in Figure 7A. The addition of Pd(II), Co(II), and VO^{2+} to these ligands formed 2:1 Schiff-base–M(II) complexes that could be isolated as solids. These compounds were treated with lead dioxide (PbO_2) to generate phenoxyl-radical complexes in situ, which were characterized by EPR spectroscopy. For example, in the case of [((*N*-3,5-di-*tert*-butyl-2-hydroxyphenyl)salicylaldiminato)$_2$Co(II)], an eight-line EPR signal was observed and interpreted as arising from hyperfine splitting of phenoxyl radical electron density delocalized onto the Co(II) ($I = 7/2$) nucleus. The authors reported that the free ligands, treated in the same manner with lead dioxide, gave only unstable species for which it was difficult to resolve an EPR signal and concluded that chelation to the metal stabilized the phenoxyl radical (*30, 31*).

A similar Schiff-base ligand, shown in Figure 7B, was allowed to react with Co(II) and Zn(II) salts to give phenolate complexes characterized by UV-visible (UV-vis) and IR spectroscopy. Treatment of these products with lead dioxide gave EPR signals suggesting Co(III) and Zn(II) phenoxyl radical complexes. The authors, relying on IR data, concluded that they were unable to synthesize the analogous Cu(II) Schiff-base complexes and claimed to have obtained products in which the ligand had rearranged to a quinolide state (*32*). A later report described the crystallographic characterization of mononuclear and dinuclear copper complexes in which the same Schiff-base ligand had been oxidized to give coupled quinone fragments. The authors suggested that dioxygen may have oxidized the coordinated Schiff-base ligand to form the coupled products (*33*).

Another interesting ligand containing a sterically hindered phenol group was designed as an analogue of *N*-phenylbenzohydroxamic acid, as shown in Figure 7C. This ligand provides two chemically distinct sites for potential radical formation upon oxidation, the phenol fragment for generation of a phenoxyl radical and the hydroxyamino group to afford

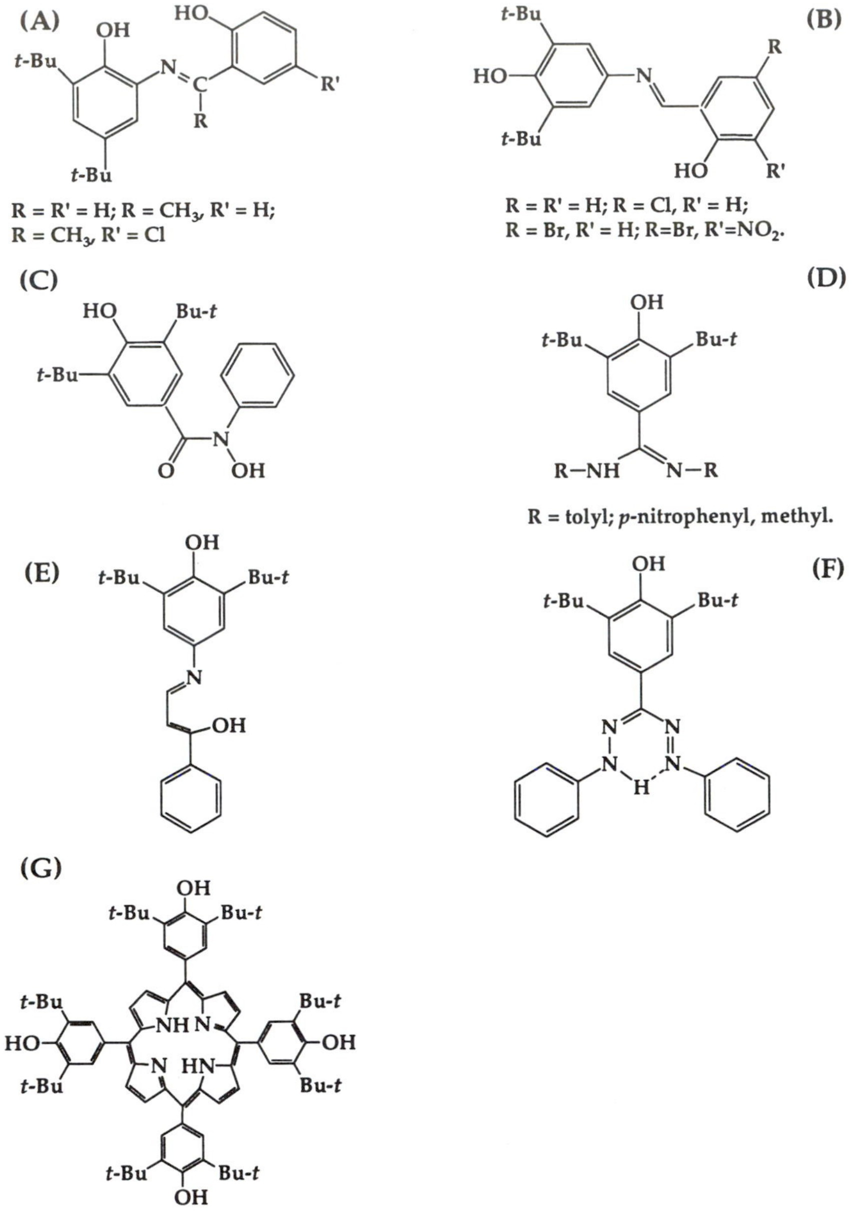

(A)

R = R' = H; R = CH_3, R' = H;
R = CH_3, R' = Cl

(B)

R = R' = H; R = Cl, R' = H;
R = Br, R' = H; R=Br, R'=NO_2.

(C)

(D)

R = tolyl; *p*-nitrophenyl, methyl.

(E)

(F)

(G)

Figure 7. Ligand systems containing sterically hindered phenol groups.

a nitroxyl radical. A palladium compound was prepared and characterized by elemental analysis. Its formula was consistent with two ligands bound to one metal, and IR data supported coordination by the carbonyl oxygen atom, most likely accompanied by chelation to the hydroxyamino oxygen atom. Oxidation of this complex with $K_3[Fe(CN)_6]$ gave rise to EPR spectra interpreted as both monoradical and diradical complexes, corresponding to one or both ligands being oxidized to phenoxyl radicals. A "dark-colored" crystalline material was isolated after oxidation and assigned as the diradical complex based on IR and EPR evidence. This diradical complex was reduced with dithionite to give the monoradical and diamagnetic starting material. Attempts were made to generate phenoxyl radical complexes by using this hydroxamic acid ligand and Zn(II), Co(II), Ni(II), Fe(II), and Fe(III), but in all these trials only nitroxide radical EPR signals were detected (*34*).

To study isomerization and tautomerism in a series of amidine-based ligands, a pendant, di-*tert*-butyl phenol group was attached to the central carbon atom of an amidine framework, as shown in Figure 7D. The phenoxyl radical, generated by oxidation with PbO_2, allowed for the study of processes, such as the 1,3-migration of the amidine H atom and other isomeric equilibria, by EPR spectroscopy. The Hg complex was formed by coordination of the di-*p*-tolyl amidine derivative to a phenylmercury group, and EPR measurements of the corresponding phenoxyl radical Hg complex suggested that 1,3-migration of the phenylmercury group is faster than the EPR time scale (*35, 36*). These amidine ligands provide a good example of the use of phenoxyl radicals as spin probes for intramolecular rearrangements.

The ligand 3-(3,5-di-*tert*-butyl-4-hydroxyanilino)-1-phenyl-2-propen-1-one, shown in Figure 7E, which has a N,O donor set similar to that of Schiff-base systems, also displayed prototropic tautomerism as seen in the amidine systems. The phenol prefers the keto-enamine form, but the imino-enol tautomer is favored for the phenoxyl radical. In this case, the phenylmercury derivative was synthesized by addition of PhHgOH to the phenoxyl radical generated in a previous oxidation step. Hyperfine splitting from ^{119}Hg (16.84% natural abundance) was observed in the EPR spectrum of the phenylmercury complex. Complexes of Cu(II), Pd(II), and Ni(II) were formed by addition of $Cu(OAc)_2$, $PdCl_2$, and $Ni(OAc)_2$, respectively, to the phenol form of the ligand. Data from IR spectroscopy and elemental analysis were consistent with formation of 2:1 complexes for these three metals. Oxidation of the Cu and Pd complexes by PbO_2 in tetrahydrofuran solutions gave rise to EPR signals suggesting that spin density for the phenoxyl radical was located in part on the metal center. In the case of Ni(II), oxidation by PbO_2 did not give any stable EPR signals (*37*).

An early report of an attempt to develop a biomimetic ligand with a stable pendant phenoxyl radical involved the synthesis of a diphenylformazan molecule shown in Figure 7F (*38*). The authors stated, "square planar coordination of a metal ion with four nitrogen atoms is characteristic of many enzyme systems," as their rationale for the design of this formazan-based ligand system. The complexes $[Ni(L{-}OH)_2]$ and $[Pd(L{-}OH)_2]$ were prepared and characterized by elemental analysis, electronic spectroscopy and mass spectrometry. Oxidation of benzene solutions of these complexes by aqueous $K_3Fe(CN)_6$ (pH 10) for 5–10 min led to the appearance of an EPR signal assigned to the monoradical $[M(L{-}OH)(L{-}O\cdot)]$ complexes $[M(L{-}O\cdot)_2]$. Longer periods of oxidation (30–40 min) gave rise to a broad singlet in the EPR spectrum attributed to the biradical complexes. Reduction of the complexes in discrete steps from the biradical to monoradical to bis(phenol) complexes was accomplished with sodium dithionite.

The isolation of $[M(L{-}O\cdot)_2]$ and $[M(L{-}OH)(L{-}O\cdot)]$ as dark green crystalline complexes was reported, although data to characterize these complexes were scanty. The compounds were found to be air-stable both in the solid state and in solution. The source of the high stability for these complexes was attributed to delocalization of the unpaired electron over the phenoxyl ring and chelate ring. The synthesis of the Co(II) and Cu(II) complexes of the formazan ligand were reported, although no preparative details or characterization data were given. The corresponding phenoxyl radical complexes of Co(II) or Cu(II) could neither be generated in situ nor isolated (*38*).

Another attempt at modeling metalloenzyme active sites involved the study of a tetra(4-hydroxy-3,4-di-*tert*-butylphenyl)porphyrin system, shown in Figure 7G. Once again $K_3Fe(CN)_6$ was employed as the agent of oxidation, and the formation of radicals was monitored by UV-vis and EPR spectroscopy. Initially, oxidation gave spectra consistent with a bisquinone structure, and further oxidation gave mono and biradical products according to the EPR data. The Zn(II) porphyrin complex was synthesized and subjected to the same oxidation procedure, affording the biradical Zn(II) complex that was isolated as dark green crystals (*39*).

In a later article, complexes of Ni(II), Cu(II), Pd(II), and VO^{2+} ions with the same tetra-substituted porphyrin were reported. Stepwise oxidation of these complexes gave products for which the authors proposed quinonoid, monoradical, and diradical structures. The most prolonged oxidations yielded the diradical products, which were isolated as dark purple crystals, relatively stable in air (*40*). The monoradical vanadyl complex was observed to be diamagnetic, suggesting antiferromagnetic coupling between the phenoxyl radical and unpaired electron on vanadium, whereas in the copper complex no such coupling was observed. More detailed studies of these systems seem warranted.

Model for the Ribonucleotide Reductase R2 Protein

To construct an oxo-, carboxylato-bridged diiron(III) model for the R2 active site, we designed a ligand that could coordinate to Fe(III) and provide a stable phenoxyl radical moiety at a distance from the metal as close as possible to the 5.3 Å distance observed in the crystal structure of the R2 protein. We wanted a ligand that provided a pendant phenoxyl radical in close proximity to the metal, but not coordinated directly to it through the phenolic oxygen atom. For an accurate R2 model, the ligand should not have a delocalization pathway of the radical spin directly onto the metal center, as do all of the ligands discussed in the preceding section.

The ligand that we designed and synthesized, 1,1-bis[2-(1-methylimidazolyl)]-1-(3,5-di-*tert*-butyl-4-oxylphenyl)ethane (BIDPhE), is shown in Figure 8 (*41*). It contains biomimetic imidazole donors with a pendant, sterically hindered phenoxyl radical arm. The quaternary carbon atoms in the ortho and para positions of the phenyl ring prevent coupling reactions that occur for unsubstituted phenoxyl radicals (*42*). The para quaternary center prevents delocalization of the radical spin onto the imidazoles or metal center. A related bidentate nitrogen donor, BIPhMe, had previously been shown to form (μ-oxo)bis(μ-carboxylato)diiron(III) complexes (*43*, *44*), and thus inspired our design of BIDPhE.

The phenoxyl radical ligand BIDPhE was generated by oxidation of the precursor phenol, BIDPhE–H, with $K_3Fe(CN)_6$. The radical proved to be quite stable and was isolated as a pure green solid. This material was used as the ligand to prepare the phenoxyl radical complex $[Zn(BIDPhE)Cl_2] \cdot MeOH$ from $ZnCl_2$. Binding of BIDPhE to Zn(II), a spectroscopically silent metal ion incapable of undergoing redox state changes under normal reaction conditions, allowed us to determine whether any changes in the physical properties of BIDPhE might arise simply following coordination of the imidazole donors. The use of Zn(II) also minimized the possibility that the metal center might reduce the BIDPhE.

The UV–vis and resonance Raman spectra of BIDPhE and $[Zn(BIDPhE)Cl_2]$ were nearly identical, suggesting that the pendant

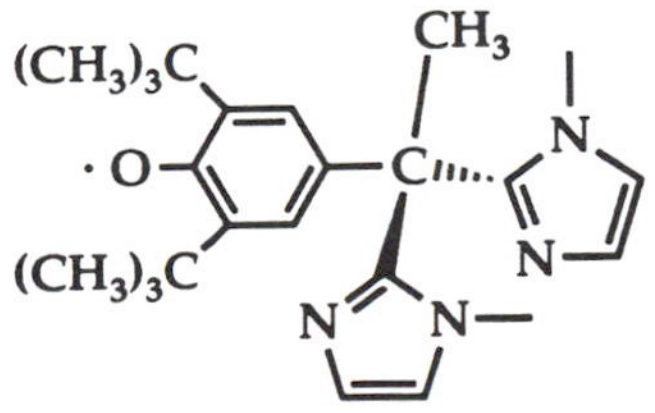

Figure 8. Drawing of BIDPhE.

radical in the Zn(II) complex was essentially unaffected by coordination through the imidazole rings. A crystallographic determination of the analogous phenol complex [Zn(BIDPhE–H)Cl_2] revealed **BIDPhE–H** bound in a bidentate manner through the imidazole nitrogen atoms to the tetrahedral Zn(II) ion, as expected. Comparison of the powder diffraction spectra of the **BIDPhE** and **BIDPhE-H** complexes showed them to be essentially identical, confirming that the radical ligand also coordinates through its imidazole donors, and not through the phenoxyl oxygen atom.

After the stability and preferred binding mode of **BIDPhE** had been established, efforts were made to prepare a complex of dinuclear iron(III) with oxo and carboxylato-bridges. Attempts to synthesize such a complex by well-established methods (*5, 44*) that used simple iron salts or basic iron carboxylates as starting materials proved unsuccessful. The remarkably stable solvento complex $[Fe_2O(XDK)(MeOH)_5(H_2O)](NO_3)_2$ (*45*) (**XDK** is xylenediamine bis(Kemp's triacid imide), shown in Figure 9) (*13, 46*), however, could be used as starting material to prepare $[Fe_2O(XDK)(BIDPhE)_2](NO_3)_2$, the first ($\mu$-oxo)bis($\mu$-carboxylato diiron(III)) phenoxyl radical complex. The solvento starting material allows facile substitution of bidentate nitrogen donors like **BIDPhE** for the terminal solvent molecules, without rearrangement or other reactions of its $\{Fe_2O(XDK)\}^{2+}$ core.

Addition of the solvento complex to two equivalents of **BIDPhE** in methylene chloride led to the isolation of $[Fe_2O(XDK)(BIDPhE)_2]$-$(NO_3)_2$ as a dark brown crystalline solid, the proposed structure for which is shown schematically in Figure 10. Although yet to be confirmed by X-ray crystallography, convincing evidence for this structure comes from elemental analysis, UV–vis, and resonance Raman spectroscopy. A characteristic band for the bent μ-oxo bridge appears at 524 cm^{-1} in the resonance Raman spectrum, as shown in Figure 11. A review of dinuclear ferric complexes with bent μ-oxo bridges,

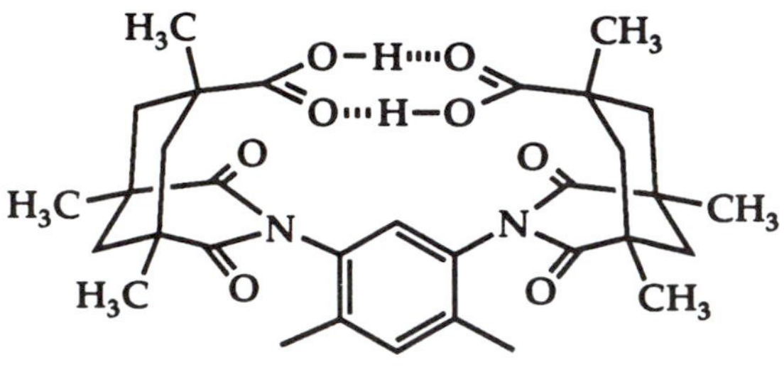

$XDKH_2$

Figure 9. Drawing of $XDKH_2$.

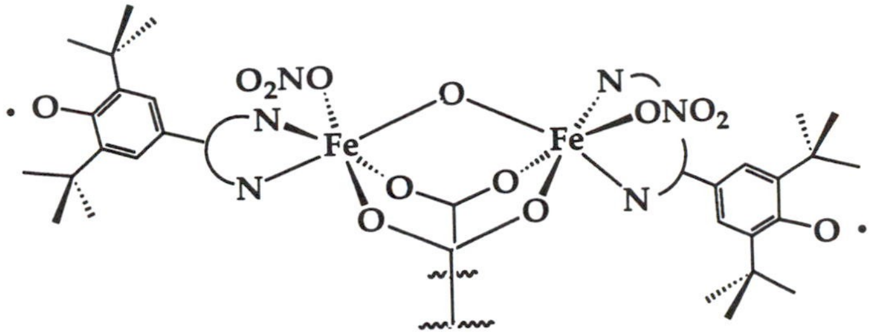

Figure 10. Schematic diagram of the proposed structure for $[Fe_2O(XDK)(BIDPhE)_2(NO_3)_2]$.

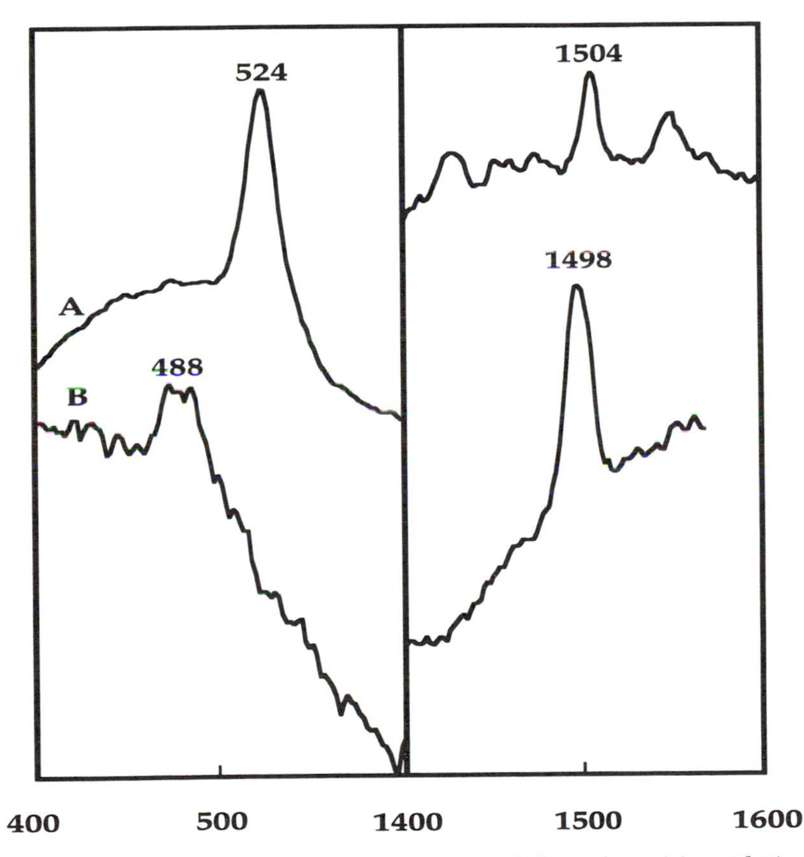

Figure 11. Resonance Raman spectrum of $[Fe_2O(XDK)(BIDPhE)_2(NO_3)_2 \cdot CH_2Cl_2$ *(A) dissolved in* CH_2Cl_2 *and wt ribonucleotide reductase R2 protein from* E. coli, *strain N6405/PSPS2, (B) dissolved in Tris buffer, pH 7.6, 5% glycerol. All spectra were recorded at room temperature with Kr ion laser excitation at 406.7 nm.*

including an analysis of their resonance Raman spectra, is available (*47*). This spectrum also shows a peak at 1504 cm^{-1} arising from the C–O stretch for BIDPhE (*48*). Included in Figure 11 is the resonance Raman spectrum of the native ribonucleotide reductase R2 protein, illustrating the similarity between the BIDPhE model complex and the active site of the R2 protein.

The magnetic interaction between the tyrosyl radical and the diiron(III) center in R2 has been probed by spin saturation-recovery pulsed EPR experiments (*49, 50*). A J value for the Fe_2(III) exchange coupling in R2 has been extracted by using these techniques. One immediate value of the BIDPhE model complex is its potential to serve as a small-molecule calibrant for such spin saturation–recovery measurements. The J value obtained from spin saturation-recovery for the Fe_2(III) exchange coupling in the protein can be compared to the strength of the coupling determined for the BIDPhE complex from superconducting quantum interference device (SQUID) measurements. Preliminary SQUID measurements reveal the BIDPhE model complex to exhibit antiferromagnetic coupling with a J value near $-120\ cm^{-1}$, assuming an isotropic exchange Hamiltonian, $H = -2JS_1 \cdot S_2$, typical for a (μ-oxo)(μ-carboxylato) Fe_2(III) complex (*51*). Pulsed EPR experiments on the BIDPhE model complex are in progress.

The BIDPhE model complex provides us with a unique opportunity to explore the reactivity of a phenoxyl radical–diiron(III) moiety with a variety of inhibitors of the R2 active site. Several studies have been carried out in which the R2 protein was treated with different radical inhibitors (*52–54*). The mechanism of inhibition, the oxidation products formed from the inhibitors and the possible concomitant reactivity of the diiron(III) center with these agents are all subjects of current investigation. The reactivity of the BIDPhE model complex may be more easily studied than that of R2, given the relative ease of handling a small molecule compared to a protein.

Summary and Conclusions

The generation, stability, and function of tyrosyl radicals in ribonucleotide reductase, PGH synthase, and galactose oxidase continue to be active areas of research. The difficulties encountered in preparing and handling these proteins, as well as in probing the physical properties and reactivity of their metal–phenoxyl radical active sites, make the preparation and investigation of stable phenoxyl radical metal model complexes an attractive goal.

The potential of the ligand systems shown in Figure 7 as physical and functional enzyme mimics has not been fully exploited. The porphyrin system described deserves further attention as a model for PGH synthase. Both the physical properties and possible reactivity of these porphyrin complexes with peroxide and arachidonic acid-related substrates warrant investigation. The copper chemistry reported for the Schiff-base system shown in Figure 7B is intriguing with respect to the galactose oxidase system, especially the reactivity toward dioxygen proposed for the Schiff-base Cu(II) complex. Isolation of a Cu(II) phenoxyl radical complex would allow one to test the hypothesis that such a system is capable of effecting the two-electron oxidation of alcohols to aldehydes. The Cu complex reported for the imino-enol ligand in Figure 7E could be a good candidate for study as a galactose oxidase model, if it could be isolated in pure form.

Much work remains to be done in developing functional mimics of the diiron center in ribonucleotide reductase, but the BIDPhE system affords some progress toward this goal by showing that it is at least possible to synthesize a diiron(III) phenoxyl radical complex. In a functional model of the R2 protein, one would like to be able to mimic the ability of the enzyme to generate the radical by addition of dioxygen to the reduced diferrous form. Work is under way to obtain the necessary Fe(II)/phenol starting materials with BIDPhE and related pendant phenol ligand systems. The generation of a diiron(III) phenoxyl radical complex by addition of dioxygen to a diferrous phenol precursor molecule would provide a system for a detailed study of the mechanism by which the iron atoms and dioxygen interact to generate a proximate phenoxyl radical. On the other hand, the failure of such diferrous complexes to generate phenoxyl radicals in the presence of O_2 could lead to a better understanding of the features used by the protein system to make such radical generation reactions possible.

Acknowledgments

This work was supported by a grant from the National Institute of General Medical Sciences. We thank R. H. Beer and J. G. Bentsen for useful discussions.

References

1. Stubbe, J. *Annu. Rev. Biochem.* **1989,** *58*, 257–285.
2. Ochiai, E.-I. *J. Chem. Ed.* **1993,** *70*, 128–133.
3. Pedersen, J. Z.; Finazzi-Agro, A. *FEBS Lett.* **1993,** *325*, 53–58.
4. Prince, R. C. *Trends Biochem. Sci.* **1988,** *13*, 152–154.
5. Lippard, S. J. *Angew. Chem. IEE* **1988,** *27*, 344–361.
6. Reichard, P. *Science (Washington, D.C.)* **1993,** *260*, 1773–1777.
7. Nordlund, P.; Eklund, H. *J. Mol. Biol.* **1993,** *232*, 123–164.
8. Nordlund, P.; Sjöberg, B.-M.; Eklund, H. *Nature (London)* **1990,** *345*, 593–598.

9. Larsson, A.; Sjöberg, B.-M. *EMBO J.* **1986,** *5,* 2037–2040.
10. Bollinger, J. M., Jr.; Edmondson, D. E.; Huynh, B. H.; Filley, J.; Norton, J. R.; Stubbe, J. *Science (Washington, D.C.)* **1991,** *253,* 292–298.
11. Bollinger, J. M., Jr.; Stubbe, J.; Huynh, B. H.; Edmondson, D. E. *J. Am. Chem. Soc.* **1991,** *113,* 6289–6291.
12. Åberg, A. Doctoral Dissertation, Stockholm University, 1993.
13. Rardin, R. L.; Tolman, W. B.; Lippard, S. J. *New J. Chem.* **1991,** *15,* 417–430.
14. Sahlin, M.; Sjöberg, B.-M.; Backes, G.; Loehr, T.; Sanders-Loehr, J. *Biochem. Biophys. Res. Commun.* **1990,** *167,* 813–818.
15. Fontecave, M.; Gerez, C.; Mohammed, A.; Jeunet, A. *Biochem. Biophys. Res. Commun.* **1990,** *168,* 659–664.
16. Hamilton, G. A. In *Copper Proteins;* Spiro, T. G., Ed.; John Wiley & Sons: New York, 1981; Vol. 3, pp 193–218.
17. Ettinger, M. J.; Kosman, D. J. In *Copper Proteins;* Spiro, T. G., Ed.; John Wiley & Sons: New York, 1981; Vol. 3; pp 219–261.
18. Avigad, G.; Amaral, D.; Asensio, C.; Horecker, B. L. *J. Biol. Chem.* **1962,** *237,* 2736–2743.
19. Whittaker, M. M.; Whittaker, J. W. *J. Biol. Chem.* **1990,** *265,* 9610–9613.
20. Ito, N.; Phillips, S. E. V.; Stevens, C.; Ogel, Z. B.; McPherson, M. J.; Keen, J. N.; Yadav, K. D. S.; Knowles, P. F. *Nature (London)* **1991,** *350,* 87–90.
21. Babcock, G. T.; El-Deeb, M. K.; Sandusky, P. O.; Whittaker, M. M.; Whittaker, J. W. *J. Am. Chem. Soc.* **1992,** *114,* 3727–3734.
22. Clark, K.; Penner-Hahn, J. E.; Whittaker, M. M.; Whittaker, J. W. *J. Am. Chem. Soc.* **1990,** *112,* 6433–6434.
23. Whittaker, M. M.; Whittaker, J. W. *J. Biol. Chem.* **1988,** *263,* 6074–6080.
24. Smith, W. L.; Marnett, L. J. *Biochim. Biophys. Acta* **1991,** *1083,* 1–17.
25. Smith, W. L.; Eling, T. E.; Kulmacz, R. J.; Marnett, L. J.; Tsai, A.-l. *Biochemistry* **1992,** *31,* 3–7.
26. Dietz, R.; Nastainczyk, W.; Ruf, H. H. *Eur. J. Biochem.* **1988,** *171,* 321–328.
27. Karthein, R.; Dietz, R.; Nastainczyk, W.; Ruf, H. H. *Eur. J. Biochem.* **1988,** *171,* 313–320.
28. Shimokawa, T.; Kulmacz, R. J.; DeWitt, D. L.; Smith, W. L. *J. Biol. Chem.* **1990,** *265,* 20073–20076.
29. Lassmann, G.; Odenwaller, R.; Curtis, J. F.; DeGray, J. A.; Mason, R. P.; Marnett, L. J.; Eling, T. E. *J. Biol. Chem.* **1991,** *266,* 20045–20055.
30. Kompan, O. E.; Ivakhnenko, E. P.; Lyubchenko, S. N.; Olekhnovich, L. P.; Yanovskii, A. I.; Struchkov, Yu. T. *J. Gen. Chem. USSR (Engl. Transl.)* **1990,** *60,* 1682–1690.
31. Ivakhnenko, E. P.; Lyubchenko, S. N.; Kogan, V. A.; Olekhnovich, L. P.; Prokof'ev, A. I. *J. Gen. Chem. USSR (Engl. Transl.)* **1986,** *56,* 765–768.
32. Medzhidov, A. A.; Kasumov, V. T.; Mamedov, K. S. *Sov. J. Coord. Chem. (Engl. Transl.)* **1981,** *7,* 28–32.
33. Medzhidov, A. A.; Kasumov, V. T.; Guseinova, M. K.; Mamedov, K. S.; Aliev, V. S. *Dokl. Akad. Nauk. SSSR* **1982,** *259,* 363–365.
34. Vovk, D. N.; Melezhik, A. V.; Pokhodenko, V. D. *J. Gen. Chem. USSR (Engl. Transl.)* **1982,** *52,* 1201–1206.
35. Ivakhnenko, E. P.; Shif, A. I.; Prokof'ev, A. I.; Olekhnovich, L. P.; Minkin, V. I. *J. Org. Chem. USSR (Engl. Transl.)* **1989,** *25,* 319–328.
36. Ivakhnenko, E. P.; Shif, A. I.; Olekhnovich, L. P.; Prokof'ev, A. I.; Minkin, V. I.; Kabachnik, M. I. *Dokl. Akad. Nauk SSSR* **1988,** *299,* 79–82.

37. Ivakhnenko, E. P.; Lyubchenko, S. N.; Kogan, V. A.; Olekhnovich, L. P.; Prokof'ev, A. I. *J. Gen. Chem. USSR (Engl. Transl.)* **1986,** *56,* 349–352.
38. Pokhodenko, V. D.; Melezhik, A. V.; Vovk, D. N. *Sov. J. Coord. Chem. (Engl. Transl.)* **1982,** *8,* 667–671.
39. Melezhik, A. V.; Pokhodenko, V. D. *J. Org. Chem. USSR (Engl. Transl.)* **1982,** *18,* 912–917.
40. Melezhik, A. V.; Vovk, D. N.; Pokhodenko, V. D. *J. Gen. Chem. USSR (Engl. Transl.)* **1983,** *53,* 1442–1447.
41. Goldberg, D. P.; Watton, S. P.; Masschelein, A.; Wimmer, L.; Lippard, S. J. *J. Am. Chem. Soc.* **1993,** *115,* 5346–5347.
42. Altwicker, E. R. *Chem. Rev.* **1967,** *67,* 475–531.
43. Tolman, W. B.; Liu, S.; Bentsen, J. G.; Lippard, S. J. *J. Am. Chem. Soc.* **1991,** *113,* 152–164.
44. Taft, K. L.; Masschelein, A.; Liu, S.; Lippard, S. J.; Garfinkel-Shweky, D.; Bino, A. *Inorg. Chim. Acta* **1992,** *198–200,* 627–631.
45. Watton, S. P.; Masschelein, A.; Rebek, J., Jr.; Lippard, S. J. *J. Am. Chem. Soc.* **1994,** *116,* 5196–5205.
46. Rebek, J., Jr.; Marshall, L.; Wolak, R.; Parris, K.; Killoran, M.; Askew, B.; Nemeth, D.; Islam, N. *J. Am. Chem. Soc.* **1985,** *107,* 7476–7481.
47. Sanders-Loehr, J.; Wheeler, W. D.; Shiemke, A. K.; Averill, B. A.; Loehr, T. M. *J. Am. Chem. Soc.* **1989,** *111,* 8084–8093.
48. Tripathi, G. N. R.; Schuler, R. H. *J. Chem. Phys.* **1984,** *81,* 113–121.
49. Sahlin, M.; Petersson, L.; Gräslund, A.; Ehrenberg, A.; Sjöberg, B.-M.; Thelander, L. *Biochemistry* **1987,** *26,* 5541–5548.
50. Hirsh, D. J.; Beck, W. F.; Lynch, J. B.; Que, L., Jr.; Brudvig, G. W. *J. Am. Chem. Soc.* **1992,** *114,* 7475–7481.
51. Que, L., Jr.; True, A. E. In *Progress in Inorganic Chemistry;* Lippard, S. J., Ed.; John Wiley and Sons: New York, 1990; Vol. 38; pp 97–200.
52. Atta, M.; Lamarche, N.; Battioni, J.-P.; Massie, B.; Langelier, Y.; Mansuy, D.; Fontecave, M. *Biochem. J.* **1993,** *290,* 807–810.
53. Lassmann, G.; Thelander, L.; Gräslund, A. *Biochem. Biophys. Res. Commun.* **1992,** *188,* 879–887.
54. Lam, K.; Fortier, D. G.; Thomson, J. B.; Sykes, A. G. *J. Chem. Soc., Chem. Commun.* **1990,** 658–660.

RECEIVED for review September 7, 1993. ACCEPTED revised manuscript January 12, 1994.

4

Reactivity Models for Dinuclear Iron Metalloenzymes

Oxygen Atom Transfer Catalysis and Dioxygen Activation

Adonis Stassinopoulos, Subhasish Mukerjee, and John P. Caradonna*

Department of Chemistry, Yale University, New Haven, CT 06511–8118

The reduced dinuclear iron sites found in hemerythrin, ribonucleotide reductase, purple acid phosphatase, and methane monooxygenase exhibit an extremely broad range of reactivity toward molecular oxygen ranging from reversible O_2 binding (transport activity) to irreversible O_2 reduction followed by oxidation of organic substrates (monooxygenase activity). Although synthetic complexes having similar electronic and structural features as these metalloenzyme active sites have been prepared and characterized, modeling the catalytic activity of these sites has proven difficult. Recent results examining the ability of synthetic diferrous complexes to perform catalytic chemistry are discussed. Some speculations on the mechanisms of these reactions are presented.

OCCASIONALLY, A SINGLE MOIETY may play such a central role in a particular area of chemistry that its name becomes synonymous with that field. The Fe–O(R)–Fe linkage is one such group. In addition to being found at the catalytic active sites of metalloproteins such as hemerythrin (Hr), ribonucleotide reductase (RNR), purple acid phosphatase (PAP), methane monooxygenase (MMO), the Fe–O(R)–Fe unit has intrigued the inorganic community, owing to its thermodynamic stability, magnetic behavior, and ability to exist in either the diferrous, mixed-valence, and diferric core oxidation states (*1*, *2*).

Although analysis of synthetic models has allowed a greater understanding of the structural, spectroscopic, and magnetic properties of

*Corresponding author

0065–2393/95/0246–0083/$11.24/0

this class of metalloproteins (*1, 2*), too little effort has been placed on investigating the reactivity properties of appropriately designed dinuclear ferrous systems. Despite the intrinsic difficulties of characterizing diferrous systems and the reactive intermediates formed by the interaction of these complexes with small molecules such as dioxygen, peroxides, peracids, and other oxygen atom donor molecules, this chemistry promises to bring rewards commensurate with the effort (*5–11*).

It is clear that fundamental studies investigating the modes of interaction of O_2 with diferrous complexes is essential for an understanding of the factors governing the chemistry of Hr, RNR, and MMO (*1, 2*). The reduced (μ-hydroxo)bis(μ-carboxylato)diiron core of Hr (Figure 1) consists of two distinct ferrous sites: the first is coordinatively saturated (FeN_3O_3) owing to the ligation of three histidine residues, whereas the second (FeN_2O_3), bound to only two other histidine ligands, is coordinatively unsaturated. The current model for Hr activity, based in part on crystallographic analyses of both the deoxy and oxy forms of the protein, holds that the binding of molecular oxygen to the five-coordinate ferrous center induces a coupled two-electron one-proton transfer from the diferrous core to O_2 to yield an η^1 peroxo diferric species that is stabilized by intracluster hydrogen bonding (Figure 2). The reduction

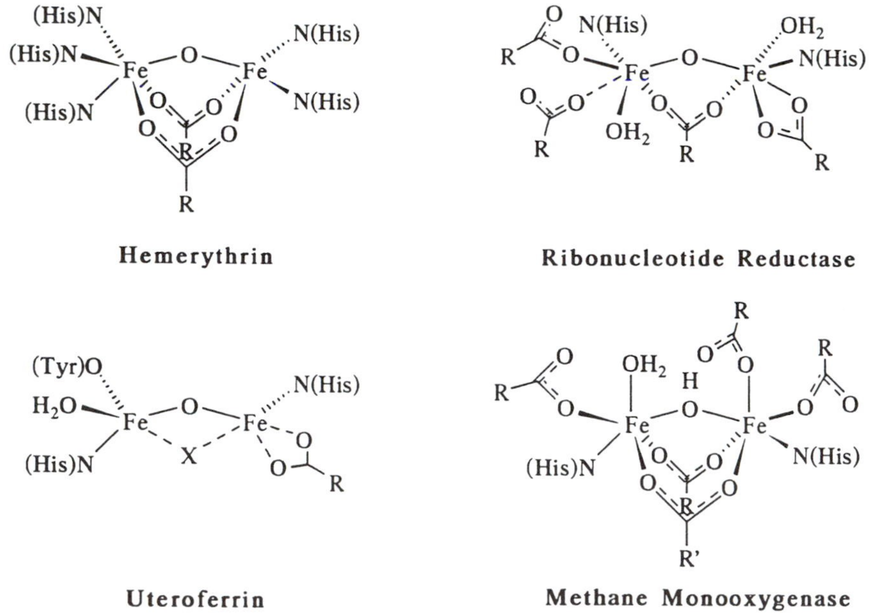

Hemerythrin **Ribonucleotide Reductase**

Uteroferrin **Methane Monooxygenase**

Figure 1. Comparison of the coordination spheres of the dinuclear iron sites in oxidized hemerythrin (Hr), ribonucleotide reductase (RNR), uteroferrin (Uf), and methane monooxygenase (MMO).

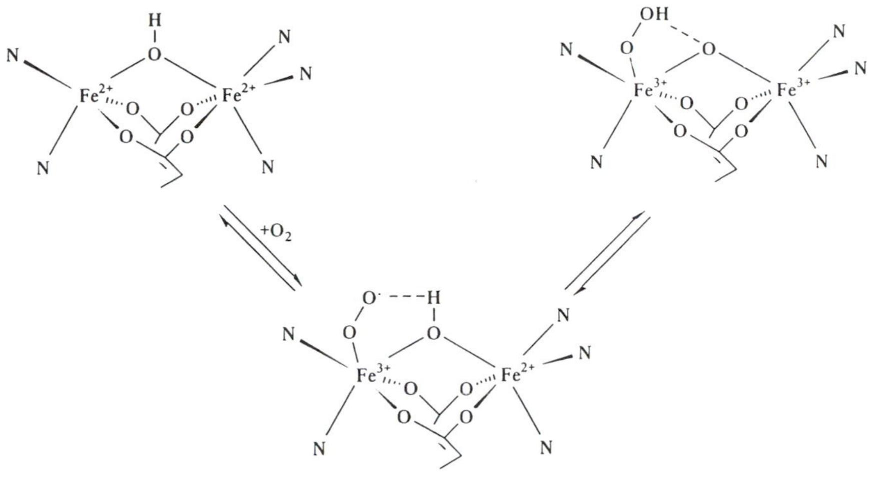

Figure 2. Proposed mechanism for the reversible dioxygen–hemerythrin interaction.

of O_2 is reversible, however, as required for the function of this transport protein.

The reduced form of RNR reacts with dioxygen to generate the μ-oxo diferric core (crystallographically defined (*3*)) and a tyrosyl radical necessary for the production of a reactive species responsible for the reduction of ribonucleotides. The intimate details of this dioxygen-based chemistry and the structure of the reduced enzyme are still unknown. Despite intensive spectroscopic characterization of the active site of MMO and the recent X-ray structural analysis of the hydroxylase component (*4*), even less is known concerning its mechanistic pathways responsible for the conversion of methane and other alkanes to their corresponding oxygenated products.

Although spectroscopic, mechanistic, and structural characterization of these metalloproteins is proceeding at a rapid pace (*2b*), the design, synthesis, and investigation of the properties of diiron(II) reactivity model systems has seriously lagged (*1, 2*). Models resembling the Hr sites have been synthesized and characterized but these complexes have structural and not functional similarities (*1, 2*). This chapter focuses on recent advances in the area of dinuclear non-heme iron metalloprotein reactivity model systems. Although several excellent papers have dealt with non-heme iron complexes as structural models (*1, 2*), the general aim of this review is to examine those iron-based systems that act as metalloenzyme reactivity models, a significantly more difficult chemical challenge. Owing to space limitations, a comprehensive discussion of small molecule catalysis is not possible. Instead, only representative examples of catalytic reactions mediated by well-characterized dinuclear iron complexes are presented.

A current challenge to the synthetic inorganic chemist is to identify through reactivity models those electronic and structural properties that control reversible versus irreversible electron donation to dioxygen (*5–11*). In this review, special emphasis will be placed on dinuclear non-heme iron complexes that catalyze oxidative transformations. We have not attempted to classify the oxidation reactions according to either the source of oxygen (dioxygen, peroxides, peracids, or other reagents such as iodosylbenzene, PhIO), target substrate (simple hydrocarbons such as alkanes, alkenes, and arenes), or proposed mechanism, because mechanistic studies have not been thoroughly performed for all systems and the potential for the existence of multiple reactive species that may readily interconvert during the reaction timecourse (*5*). For more detailed information, the reader is referred to the primary literature cited.

Oxidation Catalysts

Systematic examination of the catalytic properties of dimeric complexes was initiated shortly after the identification of dinuclear iron sites in metalloenzymes. The first report of a reactive dimeric system came from Tabushi et al. in 1980, who examined the catalytic chemistry of $[Fe^{3+}(salen)]_2O$, **1** (salen is *N,N'*-(salicylaldehydo)-1,2-ethylenediamine) (*12*). They reported interesting stereoselectivity in the oxidation of unsaturated hydrocarbons with molecular oxygen in the presence of mercaptoethanol or ascorbic acid and pyridine as a solvent ([**1**]$<<$[alkane]$<<$[2-mercaptoethanol]). With adamantane as substrate, they observed the formation of a mixture of (1- and 2-) adamantols and adamantanone (Table I) (*12*). Both the relative reactivity between tertiary and secondary carbons (maximum value is 1.05) and final yield ($\approx$12 turnovers per 12 hr) were dependent on the quantity of added 2-mercaptoethanol. Because autoxidation of adamantane gave a ratio of 3°/2° carbon oxidation of 0.18–0.42, the authors proposed two coexisting processes: autooxidation and alkane activation.

When, under identical conditions, ascorbic acid was used instead of mercaptoethanol, the reaction gave products with 3°/2° carbon reactivity of 0.28–0.42, suggestive of an autoxidation process (*12*). Furthermore, the kinetics of the reaction are biphasic for 2-mercaptoethanol and monophasic for ascorbic acid. These kinetics are consistent with the generation of a new catalytic system by the coordination of the thiol to the ferric center(s). For either reductant, bleaching of the complex was observed within minutes in the absence of substrate.

Although interesting reactivity is observed with **1**, its chemistry exhibits problems inherent to simple monobridged dinuclear systems.

Table I. Oxidation of Adamantane with Dioxygen Catalyzed by $Fe(Salen)_2O$, 1

Concentration of Reductant (M)	*Product Yield (%)*[a]			*Total TN*[b]
	1-Adamantol	*2-Adamantol*	*Adamantanone*	
2-Mercaptoethanol[c]				
1.3	67	162	51	2.8
0.64	71	163	36	2.7
0.32	65	106	59	2.3
0.16	110	92	158	3.6
0.064	129	56	135	3.2
0.032	60	8	42	1.1
Ascorbic acid[d]				
0.025	104	nd[e]	85	1.89
0.13	238	94	139	4.71
0.51	220	161	115	4.96

[a] Based on **1** used.

[b] TN is turnover number.

[c] [**1**]:[adamantane] = 1:474, [**1**] = 3.8×10^{-4} M, under 1 atm of oxygen, in 20 mL of pyridine. Reaction time is 4 h.

[d] Conditions same as in footnote *c* but [**1**] = 6.7×10^{-4} M.

[e] nd means not determined.

Many μ-oxo dimers are not stable under reducing conditions owing to instability of the μ-oxo bridge moiety in either the mixed-valence or diferrous states. For example, $[Fe(HBpz_3)]_2O(OAc)_2$ ($HBpz_3$ is hydrotris(pyrazolyl)borate) shows an electrochemical irreversible reduction wave even at very high scan rates (5 V/s) (*13*). The fact that the ascorbic acid does not show the same reactivity as 2-mercaptoethanol may be a consequence of differences in the redox potentials or the ability of the reductant to displace ligands and coordinate to the metal center.

Kitajima et al. (*14*) reported another example of a well-characterized dimeric system that was capable of catalyzing oxidation reactions. This group studied the catalytic activity of $[Fe_2O(HBpz_3)_2(OAc)_2]$, **2**, in CH_2Cl_2 in the presence of Zn dust and acetic acid. The structure of **2** was determined previously by Armstrong et al. (*13*). When crystallized from MeCN, **2** consists of two N_3O_3 octahedral centers bridged by one μ-oxo and two μ-acetato bridges. The equatorial planes of the two octahedra are defined by two bridging acetate oxygens and two pyrazolyl borate nitrogens, whereas the remaining nitrogen and the μ-oxo oxygen constitute the axial ligands of the octahedron. The average $Fe{-}O_{oxo}$ bond is 1.783 Å and the FeOFe unit is bent (124.6°). Careful magnetic susceptibility and Mössbauer spectroscopic studies showed that the ferric centers exhibit antiferromagnetic coupling ($J = -121$ cm^{-1}) (*13*). Electrochemical studies showed that the mixed-valence state of **2** is not stable, degrading to give the coordinatively saturated monomeric species $[Fe(HBpz_3)_2]^+$, **3** (*13*).

Catalytic reactions were performed in CH_2Cl_2 under an O_2 atmosphere; Zn was used as an electron source and acetic acid as a proton donor (*14*). Under these reaction conditions ([**2**]:[substrate] = 1:125), the production of adamantan-1-ol (248%), adamantan-2-ol (50%), and adamantan-2-one (108%) was observed. With cyclohexene as substrate, a mixture of cyclohexanol (54%), cyclohexanone (73%), and cyclohexene oxide (20%) was generated. In a similar experiment with cyclohexane, cyclohexanol (99%), and cyclohexanone (84%) were obtained. The product distribution is inconsistent with a free radical process; for adamantane, the 3°/2° carbon reactivity ratio is 2.2. Control experiments demonstrated that both Zn dust and acetic acid were necessary, whereas larger quantities of acetic acid quenched the reaction (Table II). This may be due to the acidolysis of the μ-oxo bond. Simple monomeric complexes such as FeClTPP (TPP is tetraphenylporphin), $Fe(acac)_3$ (acac is acetylacetonate), and $[Fe(HBpz_3)_2]^+$, **3**, were inactive as catalysts under identical conditions. Furthermore, $[Fe^{3+}(Salen)]_2O$, **1**, did not show any reactivity.

The pathway followed during these reactions appears different from that of the Gif system, because use of the same complexes under strict Gif conditions (pyridine–acetic acid–Zn powder) gives a different product ratio; the same reactivity and selectivity is obtained for all monomeric and dimeric complexes (*15–18*). The reactivity observed with **2** is reminiscent of the chemistry of the dinuclear ferrous active site found in MMO, although their mechanistic similarities must still be demonstrated (*19, 20*).

Kitajima et al. (*21*) also reported a more efficient catalyst through a serendipitous modification of complex **2**. They observed that an increase in the catalytic activity (by a factor of 1.5) occurred when hexafluoroacetyl acetone (hfacac) was used instead of acetic acid. The reaction between **2** and hfacac produced $[\{Fe(HBpz_3)(hfacac)\}_2O]$, **4** in 50–60% yield that was structurally characterized to reveal a ferric dimer with a single μ-oxo bridge (*21*). Each iron center is in a six-coordinate N_3O_3 environment. The two octahedral units are more distorted in this coordination environment than that of **2**; the FeOFe unit is bent with an angle of 169.4°. Each ferric center has one hfacac ligand bound in a bidentate mode.

When used as a catalyst in the presence of Zn dust and excess hfacac, **4** gave high turnover numbers in CH_2Cl_2 for a variety of substrates (Table III). Under these conditions ([**4**]:[substrate]:hfacac = 1:1106:287), adamantane gave admantan-1 and -2-ols (turnover numbers 46.6 and 1.7, respectively) and adamantan-2-one (trace), whereas cyclohexane

Table II. Hydrocarbon Oxidations Catalyzed by $[Fe_2O(HBpz_3)_2(OAc)_2]$, 2

		Product Yield (%)[a]			
Catalyst	*Substrate*	*1-Adamantol*	*2-Adamantol*	*Adamantanone*	*Total TN*[b]
2[c]	adamantane	50	108	248	4.06
2[d]	adamantane	0	0	0	no reaction
2[e]	adamantane	0	0	0	no reaction
$Fe(acac)_3$[f]	adamantane	trace	0	trace	not determined
$Fe(salen)_2O$[g]	adamantane			trace	not determined
Catalyst	*Substrate*	*Cyclohexanol*		*Cyclohexanone*	*TN*[b]
2	cyclohexane	99		84	1.83
Catalyst	*Substrate*	*Cyclohexenol*	*Cyclohexenone*	*Cyclohexene Epoxide*	*TN*[b,c]
2	cyclohexene	54	73	20	1.47

NOTE: Reactions were performed under 1 atm of O_2 at room temperature in 20 mL of CH_2Cl_2. 0.034 mmol of catalyst and 3.7 mmol of substrate were used, unless otherwise noted.

[a] Based on the complex.

[b] TN is turnover number; 1 turnover is equal to an equivalent of the catalyst concentration.

[c] 0.5 g of Zn and 0.05 mL of AcOH were used and 3.7 mmol of **2**.

[d] Acetic acid was omitted from the mixture and 0.03 mmol of **2** was used.

[e] Zn was omitted from the mixture and 0.03 mmol of **2** was used.

[f] 0.064 mmol of $Fe(acac)_3$ was used.

[g] 0.032 mmol of **2** was used.

Table III. Oxidation of Alkanes and Arenes by 4 Using Molecular Dioxygen

Substrate	*Products*[a]			*TN*[b]
Adamantane	1-adamantanol (96.5)	2-adamantanol (3.5)	2-adamantanone (trace)	48.3
Cyclohexane	cyclohexanol (95.75)	cyclohexanone (4.25)		4.7
Pentane	2-pentanol (100)			21.0
Benzene	phenol (100)			12.1
Toluene	*p*-anisole (70)	*o*-anisole (30)		9.0
Chlorobenzene	*p*-chlorophenol (76)	*o*-chlorophenol (24)		2.9

NOTE: Reaction conditions: 1 atm O_2; reagents in 20 mL of CH_2Cl_2 at 25 °C for 30 h. [4] = 7.5×10^{-5} M; [hfacac] = 0.0215 M; Zn powder, 0.5 g; [substrate] = 1.66 mmol.

[a] Amount of each product as a percentage of the total products is in parentheses.

[b] TN is turnover number, based on **4**.

and pentane gave predominately cyclohexanol and 2-pentanol, respectively. The most important transformations reported were the oxidations of benzene, toluene, and chlorobenzene to products of aromatic oxidation (Table III) that are not achieved by the Gif system. The addition of the spin trapping agent, *N-tert*-butyl-α-phenyl nitrone (BPN), indicated the production of hydroxyl radicals.

Although the [{Fe(HBpz$_3$)(hfacac)}$_2$O] system is quite intriguing, it is unclear whether ferrous decomposition products are responsible for the observed chemistry, particularly in light of the reported reactivity properties of several less well characterized mononuclear nonheme iron systems that are capable of hydroxylating aromatic compounds (*22–26*). The relationship between the chemistry of these iron-based systems, such as **4** and the Gif (and modified Gif) systems (*15–18*) is currently unclear.

Almost simultaneously, a collaborative effort modeling the reactivity of dinuclear sites of iron oxidative enzymes was reported by Vincent et al. (*27*). Their dinuclear model, $Fe_2O(OAc)_2Cl_2(bipy)_2$, **5** (bipy: 2,2′-bipyridine), synthesized by cleavage of tetrameric $[Fe_4O_2(OAc)_7(bipy)_2]^+$, **6**, (equation 1), was especially designed to have open or exchangeable coordination sites.

$$[Fe_4O_2(OAc)_7(bipy)_2]^+ + 2bipy + 4Cl^- \rightarrow 2Fe_2O(OAc)_2Cl_2(bipy)_2 + 3AcO^- \quad (1)$$

The structure of **5** consists of a diferric unit containing one μ-oxo and two μ-acetato bridges (Figure 3). Each octahedral iron center completes

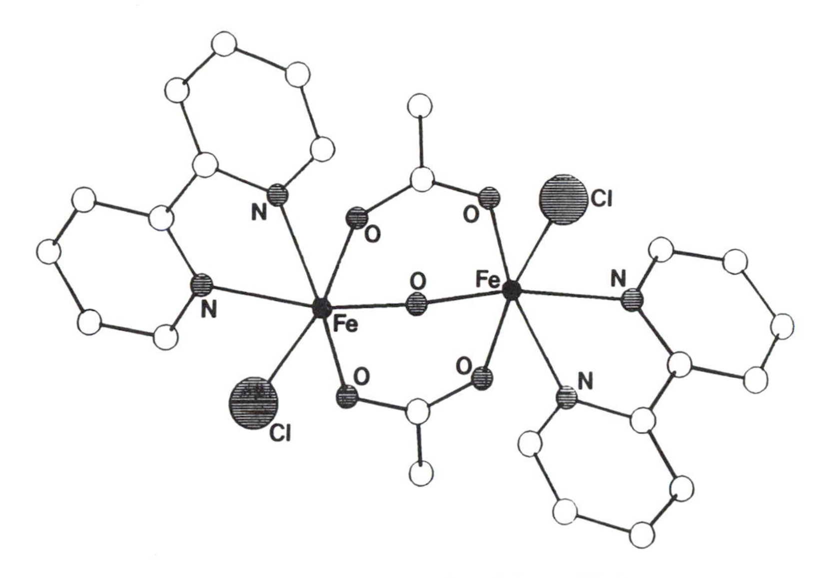

Figure 3. Crystal structure of $Fe_2O(OAc)_2Cl_2(bipy)_2$, **5**.

its coordination sphere with a molecule of bipy and a chloride anion. The FeOFe unit is bent with an angle of 123.9°. The Mössbauer spectrum shows parameters typical of a high-spin diferric, oxo-bridged system with two six-coordinate centers ($\delta = 0.37$ mm s^{-1}, $\Delta E_Q = 1.79$ mm s^{-1}), whereas magnetic susceptibility measurements show strong antiferromagnetic coupling between the ferric centers ($J = -132$ cm^{-1}). Quite interestingly, the cyclic voltammogram of this compound in MeCN indicates irreversible reduction waves in the 0- to −0.5-V range, consistent with instability of the structure during redox chemistry. Unfortunately, characterization of the decomposition products was not reported.

$Fe_2O(OAc)_2Cl_2(bipy)_2$ successfully hydroxylates C_6, C_3, and C_2 alkanes when *tert*-butyl hydrogen peroxide (TBHP) is used as the oxygen donor ([**5**]:[TBHP]:[substrate] = 1:150:1100); the observed reactivity is $C_6 > C_3 > C_2$ (Table IV). This work represents the first report of the oxidation of a small molecular weight alkane (ethane) by a characterized iron model compound. Reactions of this complex with Zn dust and acetic acid under 1 atm of dioxygen with cyclohexane gave rise to only cyclohexanone (turnover number: 2.5). The parent tetrameric compound, **6**, was reported to be a more efficient catalyst.

In view of the demonstrated ease in which the tetrameric cluster cleaves (equation 1) and the redox-induced decomposition of dimeric **5**, the integrity of the complexes at the end of the catalytic reaction and the nature of the iron species under a reducing environment are ambiguous. The possibility that mononuclear species are involved in the

Table IV. Oxidation of Hydrocarbons with $Fe_2O(OAc)_2Cl_2(bipy)_2$, 5, $[Fe_4O_2(OAc)_7(bipy)_2]^+$, 6, and $Fe_2O(OAc)(tmima)_2(ClO_4)_3$, 7, Using TBHP or H_2O_2 as Monooxygen Transfer Reagents

Substrate[a]	*Catalyst*	*Product (%)*[b]	*% TBHP Consumed*	*TN*[c]
Ethane	5	ethanol (<1)	95	1.2
Propane	5	2-propanol (8.8), 1-propanol (1.5)	82	13[d]
Cyclohexane	5	cyclohexanol (32), cyclohexanone (22)	90	72
Cyclohexane	5	cyclohexanol (32), cyclohexanone (22)		

Substrate[e]	*Catalyst*	*Product*	*OH/O Ratio*	*TN*[c]
Cyclohexane	5	cyclohexanol, cyclohexanone	0.8–1.1	~15
Cyclohexane	6	cyclohexanol, cyclohexanone	0.6–1.0	~7
Cyclohexane	7	cyclohexanol, cyclohexanone	0.9–1.0	~21

Substrate[f]	*Catalyst*	*Product (%)*[b]	OH/O Ratio	C_3/C_2[g]
Adamantane	5	R-1-ol (19), R-2-ol (9.1), R-2-one (7.2)	3.9	3.5
Adamantane	6	R-1-ol (8.7), R-2-ol (4.8), R-2-one (3.1)	4.4	3.3
Adamantane	7	R-1-ol (16.8), R-2-ol (8.5), R-2-one (6.4)	3.9	3.4

		Product (TN)[c]			
Substrate[h]	*Catalyst*	*PhCHO*	*$PhCH_2OH$*	*Cresols*	*Benzylic/Aromatic Activation Ratio*
Toluene	5	3.5	1.4	1.5	3.4
Toluene	6	1.9	0.6	0.6	4.2
Toluene	7	3.0	0.9	4.2	0.9

[a] Gas reactions were carried out in a Parr kinetic apparatus at pressures of 250 and 90 lb/in.^{2} at room temperature, in MeCN, with [TBPH]:[complex] = 150:1; [complex] = 0.7 mM. In cyclohexane, [TBPH]:[complex]:[substrate] = 150:1:1100. Reaction time is 3 days.

[b] Based on TBPH consumed. Percentage of starting material is in parentheses.

[c] TN is turnover number; oxidizing equivalents/mole of catalyst.

[d] Two-day reaction.

[e] $[H_2O_2]$:[complex]:[substrate] = 150:1:900, reaction time 6–18 h at room temperature, in MeCN.

[f] $[H_2O_2]$:[complex]:[substrate] = 150:1:20, reaction time 16 h at room temperature, in MeCN.

[g] Normalized per hydrogen.

[h] Same as footnote *f* but ratio = 150:1:300.

observed catalytic chemistry is supported by work from Leising et al. (*28*), who have shown that $[Fe(TPA)Cl_2]^+$ (TPA is tris(2-pyridylmethyl)amine) in the presence of TBHP is capable of catalyzing the oxidation of cyclohexane to cyclohexanol, cyclohexanone, chlorocyclohexane, and (*tert*-butylperoxy)cyclohexane. Mechanistic studies with $[Fe(TPA)Cl_2]^+$ suggested that heterolytic cleavage of the ROOH bond occurs leading to a high-valent iron–oxo intermediate that is capable of reacting with a variety of organic substrates (*28*). Thus, it is difficult to assess the mechanistic pathways followed by unstable dimer **5** during catalysis. Nonetheless, the reactivity of this dimer system suggests that the real structure of the active species in solution may have relevance to the catalytic center of MMO (*1*, *2*).

Fish et al. (*29*) followed these studies with an examination of the ability of **5**, **6** and $[Fe_2O(OAc)(tmima)_2](ClO_4)_3$, **7** (tmima is tris[(1-methylimidazol-2-yl)methyl]amine), to catalyze the oxidation of small molecule organics by using H_2O_2/O_2 as the oxidant. Under standard conditions ([complex]:$[H_2O_2]$:[substrate] = 1:150:900), moderate turnover numbers were found for **5–7** (Table IV); the oxidation products were the respective alcohol and ketone (OH/(O) ratio of 0.6:1.0). Control reactions, however, showed $Fe(ClO_4)_3{,}6H_2O$ to be a superior oxidation catalyst. Although both the alcohol and ketone were observed, the product ratio depended on initial reaction conditions (1.9–3.6). This dependence suggests a different mechanism for the two oxidations. Kinetic studies showed that cyclohexanol was not the precursor of the cyclohexanone; the true precursor was an active organic peroxo compound that was accumulating during the reaction. Titration of the active organic peroxo compound with iodine increased the observed cyclohexanone/cyclohexanol ratio. Finally, the oxidant was isolated and unequivocally identified as cyclohexyl hydroperoxide through ^1H NMR and comparison with an authentic sample.

The authors proposed that the peroxide is decomposed by the metal catalysts to the ketone and alcohol in a manner similar to that previously reported (*30–34*). This later system was also reactive toward adamantane (giving a high 3°/2° carbon activation ratio of ≈3.5) and other saturated alkanes. These catalysts also oxidize toluene at both the aliphatic and aromatic carbons (ratio: benzylic/aromatic = 3.4:0.9) (Table IV). Activation of the aromatic ring was attributed to the formation of hydroxyl radicals.

The free radical scavenger TBPH (TBPH: 2,4,6-tri-*tert*-butylphenol) quenched reactions catalyzed by **5–7**. Elimination of dioxygen had a similar effect. These results, in conjunction with the generation of cyclohexyl hydroperoxide, strongly suggest the existence of a radical chain process generating cyclohexyl radicals, which upon trapping by dioxygen, give the peroxide. The formation of this peroxo compound was attributed to either an iron superoxo Fe–OO· or a ferryl Fe=O species. A similar species must therefore be responsible for production of the oxygenated products for the other substrates as well (29).

Following a similar approach, Fontecave et al. (35) studied the catalytic activity of $[Fe_2OCl_6](NEt_4)_2$, **8**, $[Fe_2O(phen)_2(H_2O)_2](ClO_4)_4$, **9**, (phen is 1,10-phenanthroline) and $[\{Fe(ala)_2(H_2O)\}_3O](ClO_4)]_7$, **10** (ala is alanine). Initially, no activity was observed for the anaerobic oxidation of cyclohexane ([complex]:[TBHP]:[substrate] = 1:20:120) by **8** and **9**, with marginal activity reported for **10**. They observed, however, a significant increase in catalytic activity of **8**–**10** in the presence of excess imidazole. Cyclohexanone and cyclohexanol were observed in a 2:1 ratio (Table V). Under identical conditions, $FeCl_3$ and **8** gave the same product distribution. This observation, along with the well-established instability of **8** in MeCN, suggests that the active catalytic species is a monomeric complex. This conclusion is reinforced by data demonstrating that the catalytic activity of the reactive species is not affected by the presence of strong acid and that $FeCl_3$ gives the same results as the μ-oxo dimer. Although oxo-bridged species are known to be unstable under acidic conditions, no spectroscopic attempt was made to establish the nuclearity of the predominant or catalytically competent solution species. Another important unanswered question is whether the ketone arises from the oxidation of cyclohexanol in either the absence or presence of excess imidazole. The fact that the reaction is carried out under argon excludes the formation of the cyclohexyl hydroperoxide as an intermediate.

Table V. Oxidation of Cyclohexane by TBHP Catalyzed by $[Fe_2OCl_6](NEt_4)_2$, 8, $[Fe_2O(phen)_2(H_2O)_2](ClO_4)_4$, 9, and $[\{Fe(ala)_2(H_2O)\}_3O](ClO_4)_7$, 10

Catalyst[a]	*Cyclohexanol*	*Cyclohexanone*	*Total*[b]
8	0	0	0
8	9	18	45
8	0	0	0
$FeCl_3$	0	0	0
$FeCl_3$	8	21	50
9	13	12	37
9	12	18	48
10	5	3	11

NOTE: All values are yield in percent based on starting TBHP.

[a] [complex]:[imidazole]:[TBHP]:[substrate] = 1:50:20:120; [complex] = 15 μmol, in 2 mL of MeCn.

[b] A ratio of 1:1 and 1:2 is assumed for the alcohol and the ketone, respectively.

An interesting series of papers was published by the Sawyer group on catalytic oxidation of organic substrates with the use of simple ferrous coordination compounds (*36–40*). These studies constituted a continuation of earlier work designed to examine the reactivity properties of ferrous and ferric salts in anhydrous solvents (*38–40*). Their current work showed that $[Fe(PA)_2]$, **11** (PA is picolinic acid), [Fe(DPA)], **12** (DPA is 2,6-dicarboxylatopyridine), and the respective μ-oxo dimers catalyze the oxidation of alkanes, acetylenes, and arylolefins (by using H_2O_2 as oxidant) to give ketones, α-diketones, and aldehydes as the major products (Tables VIa–c), respectively (*36, 37, 40*). The reaction efficiencies were identical for **11**, **12**, and their μ-oxo dimers $(PA)_2FeOFe(PA)_2$, **13** and (DPA)FeOFe(DPA), **14**. The identical reactivities for the monomers and μ-oxo dimers indicate a facile interconversion between monomers and dimers under the reaction conditions. A 2:1 molar ratio of pyridine to acetic acid was found to maximize the catalytic activity; other solvents such as CH_3CN/pyridine or CH_3CN/acetic acid caused a decrease in reaction efficiency and a shift in product distribution toward the respective alcohol. The presence of pyridine was proposed to quench side reactions responsible for hydroxy radical formation. Interestingly, reactions performed in coordinating solvents such as DMF with either the di μ-hydroxy or the μ-oxo forms of **13** produced singlet O_2 that

Table VIa. Products and Conversion Efficiencies for Ketonization of Cyclohexane by H_2O_2 in Various Solvents, Catalyzed by $Fe(PA)_2$, 11, $(PA)_2FeOFe(PA)_2$, 13, $Fe(DPA)_2$, 12, and (DPA)FeOFe(DPA), 14

			Products[a]	
Iron Catalysts, 3.3 mM[b]	*Reaction Efficiency*[c] *(%, ±3)*	*Catalyst*[d] *Turnovers*	*Cyclohexanone (%, ±4)*	*Cyclohexanol (%, ±4)*
11	72	11	93	7
11[e]	72	6	95	5
11[f]	58	5	94	6
13[g]	72	11	93	7
12	73	12	>97	<3
14[g]	76	13	>97	<3

[a] Analyzed by capillary GC-MS.

[b] [Cyclohexane]:$[H_2O_2]$ = 10.41:1, $[H_2O_2]$ = 96 mM, pyridine/AcOH (2:1).

[c] 100% efficiency is one substrate per two moles of H_2O_2 added.

[d] Moles of substrate oxidized per mole of catalyst.

[e] 56 mM of H_2O_2 were used instead of 96 mM.

[f] Same as in footnote *e* and 101 mM of H_2O were added.

[g] [complex] = 1.7 mM.

Table VIb. Products and Conversion Efficiencies for the $Fe(PA)_2$, 11, Catalyzed (3.5 mM) Ketonization of Methylenic Carbon and the Dioxygenation of Acetylenes and Arylolefins by H_2O_2 (56 mM) in Pyridine/AcOH (2:1)

Substrate (1 M)	*Reaction Efficiency*[a] *(%, ±3)*	*Catalyst Turnovers*[b]	*Products*[c]
Cyclohexane	72	6	cyclohexanone (97), cyclohexanol (3)
n-Hexane	52	4	3-hexanone (53), 2-hexanone (46), 1-hexanol
Ethylbenzene	51	5	acetophenone (>96)
$PhCH_2Ph$ (0.6 M)	35	3	benzophenone (>96)
Toluene	9	<1	benzaldehyde (>96)
2-Methylbutane	32	3	3-methyl-2-butanone (>95), 2-methyl-1-butanol
Adamantane	32	3	2-adamantanone (43), 1-adamantanol (29), 1-pyridyladamantane (two isomers 10 and 18)
Cyclododecane	70	6	cyclododecanone (90), cyclododecanol (10)
Cyclohexene	59	5	2-cyclohexen-1-one (>95)
1,3-Cyclohexadiene	33	5	benzene (>95)
1,4-Cyclohexadiene	30 [70]	3 [11]	phenol (17), [PhH] (83)
Cyclohexanone	0		
Cyclohexanol	25	4	cyclohexanone (>95)
Diphenylacetylene (0.6 M)	40	3	PhC(O)C(O)Ph (>97)
cis-Stilbene	36	4	PhCH(O) (75), stilbene epoxide (25)
trans-Stilbene	48	4	PhCH(O) (63), stilbene epoxide (16), two others (21)

NOTE: Slow addition of H_2O_2.

[a]100% efficiency is one substrate per two moles of H_2O_2 added.

[b]Moles of substrate oxidized per mole of catalyst.

[c]Analyzed by capillary GC-MS. Percentage is in parentheses.

was proposed to come from the transition state complex responsible for the catalytic chemistry.

The reactivity of **11** was investigated in a stoichiometric reaction with H_2O_2 and PhSeSePh in the presence of a hydrocarbon substrate (2:2:1:100 mole ratio). These reagents react in pyridine/AcOH (1.8: 1) to give 2 equivalents of the phenylselenyl derivative of the hydrocarbon.

Table VIc. Ketonization of Methylenic Carbons and Dioxygenation of Aryl Olefins, Acetylenes, and Catechols via the $Fe^{2+}(DPAH)_2$, 16, Induced Activation of Dioxygen in 1.8:1 py/HOAc

Substrate	*Products (mM)*[a]	*Reaction Efficiency (%, ±3)*[b]
Cyclohexane (1 M)	cyclohexanone (4.4)	28
Ethylbenzene (1 M)	acetophenone (3.5)	22
Ethylbenzene (1 M) [+128 mM PhNHNHPh]	acetophenone (18.9)	
2-Methylbutane (1 M)	$Me_2CHC(O)Me$ (1.0)	6
2-Methylbutane (1 M) [+128 mM PhNHNHPh]	$Me_2CHC(O)Me$ (9.1)	
Cyclohexene	2-cyclohexen-1-one (1.2)	7
Diphenylacetylene (0.6 M)	PhC(O)C(O)Ph (2.2)	14
cis-Stilbene (1 M)	benzaldehyde (3.1)	10
1,2-$Ph(OH)_2$ (1 M)	HOC(O)CH=CHCH=CHC(O)OH (and its anhydride) (2.0)	13
PhCH(OH)C(O)Ph (0.3 M)	benzoic acid (5.2)	16
PhNHNHPh (100 mM)	PhN=NPh (100)	667[c]
$PhCH_2SH$ (128 mM)	$PhCH_2SSCH_2Ph$ (64)	800[c]
H_2S (128 mM)	S_8 (16.0)	800[c]

NOTE: $Fe^{2+}(DPAH)_2$, **16**, (32 mM); O_2 1 atm; 3.5 mL of solvent in a reaction cell with 6 mL of headspace.

[a]Analyzed by capillary GC-MS.

[b]100% efficiency is one substrate per two moles of H_2O_2 added.

[c]100% represents one substrate oxidation per $(DPAH)_2FeO(DPAH)_2$ reaction intermediate.

A Fenton-like mechanism was proposed to explain this product distribution. Hydroxyl radicals, formed by a reaction between the iron complex and H_2O_2, abstract protons from the substrate to form carbon radicals R· (equations 2–4) (*38*, *39*). These are subsequently trapped by the diphenylselenide to give a phenylselenyl derivative (equation 4). Increasing the ratio of H_2O_2 to **11** switches the reactivity from stoichiometric to catalytic (Scheme 1).

$$\underset{\mathbf{11}}{Fe(PA)_2} + HOOH \rightarrow (PA)_2Fe(OH) + OH\cdot \quad (2)$$

$$OH\cdot + RH \rightarrow R\cdot + H_2O \quad (3)$$

$$2R\cdot + PhSeSePh \rightarrow 2RSePh \quad (4)$$

The first step in the proposed mechanism for the **9**/H_2O_2/pyridine/acetic acid catalytic system is the reaction of **11** with H_2O_2 to form the μ-dihydroxy diferric species $[(PA)_2Fe(OH)_2]_2$, **13b**. This species is thought to be in equilibrium with the μ-oxo diferric species $[(PA)_2Fe]_2O$, **13**. Reaction of **13** with another molecule of H_2O_2 leads to the formation of the reactive species $(PA)_2FeO(OO)Fe(PA)_2$, **15**. Oxidation of the substrate by **15** leads to product formation and **13b**. When **15** reacts with excess H_2O_2, it produces singlet oxygen and **13**. Analogous reactions can explain the reaction chemistry of Fe(DPA), **12** through intermediates analogous to **11** and **12** (*36, 37*).

Evidence for species such as the dihydroxy-bridged dimers exists for ferric complexes of picolinate and 2,6 dicarboxylatopyridine. Complexes of the type $[(PA)_2Fe(OH)_2]_2$, **13b**, and $(H_2O)LFe(OH)_2FeL(OH_2)$, where L = 2,6-dicarboxylato(4-X)pyridine, (X = H, NMe_2, OH), have been structurally and spectroscopically characterized (*41–43*). The magnetic exchange interactions in these systems are characterized by weak antiferromagnetic coupling; values of $J = -8\ cm^{-1}$ were reported for **13b**.

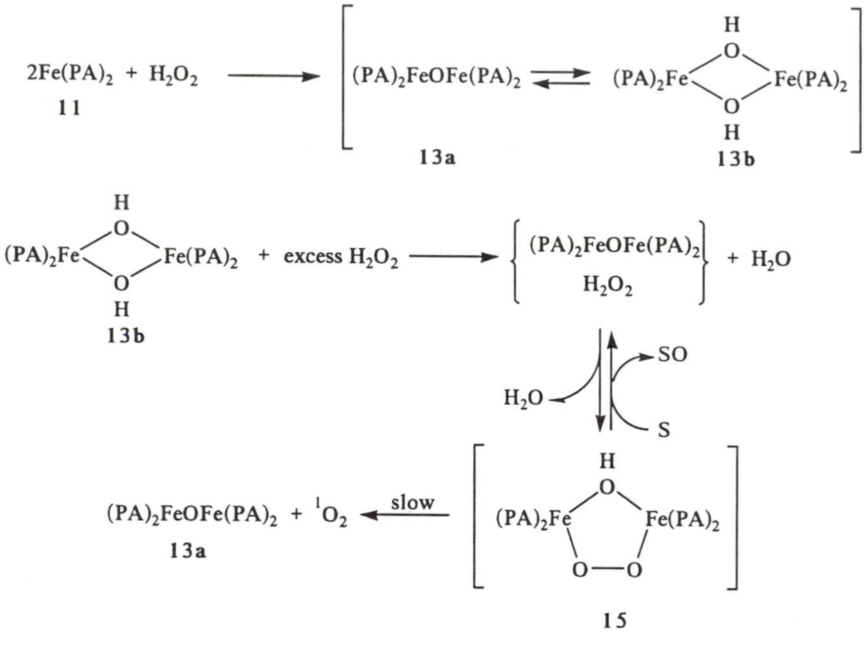

$2Fe(PA)_2 + H_2O_2$ (**11**) → [$(PA)_2FeOFe(PA)_2$ (**13a**) ⇌ $(PA)_2Fe(\mu\text{-}OH)_2Fe(PA)_2$ (**13b**)]

$(PA)_2Fe(\mu\text{-}OH)_2Fe(PA)_2$ (**13b**) + excess H_2O_2 → { $(PA)_2FeOFe(PA)_2$, H_2O_2 } + H_2O

{ $(PA)_2FeOFe(PA)_2$, H_2O_2 } ⇌ ($-H_2O$; S → SO) [$(PA)_2Fe(\mu\text{-}OH)(\mu\text{-}O\text{—}O)Fe(PA)_2$ (**15**)]

15 —slow→ $(PA)_2FeOFe(PA)_2$ (**13a**) + $^1O_2$

Scheme 1

The authors proposed the following mechanism to explain the unusual organic transformations. The μ-peroxo-μ-oxo complex, **15**, reacts in a biradical fashion with the protons on a methylenic carbon to generate a carbene and two hydroxy radicals that recombine to give a gem-diol. This diol will in turn collapse to the ketone product. The diketonization of acetylenes was explained in a similar fashion with the formation of a four-member ring intermediate (Scheme 2) (*36, 37*).

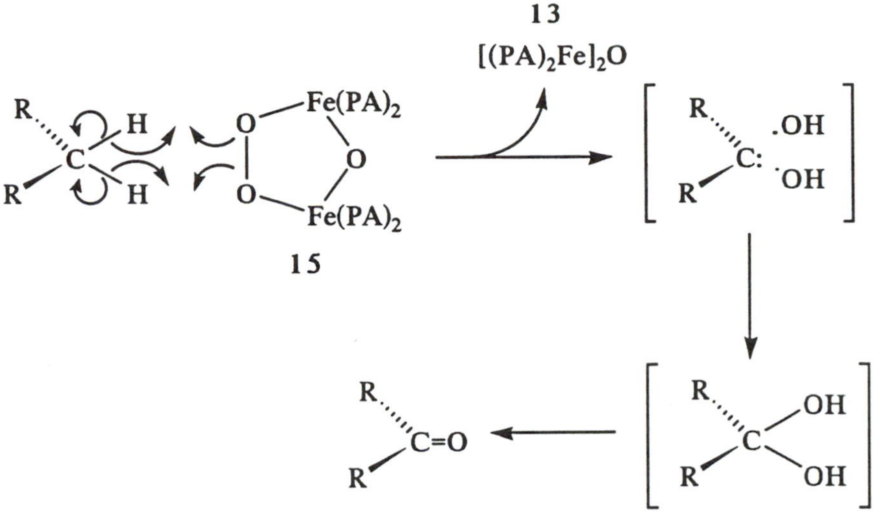

Scheme 2

The catalytic activity of $[Fe(DPAH)_2]^{2+}$, **16**, was investigated in the presence of oxygen and excess substrate by using pyridine/acetic acid (1.8:1) as solvent (*36, 37*). Although the reactivity profile exhibited by **16** was analogous to that observed for **11**, only stoichiometric product production was seen. The authors reported inactivation of the catalyst through formation of the μ-oxo dimer $(DPAH)_2FeOFe(DPAH)_2$, **17**. In the presence of a strong reducing reagent (e.g., PhNHNHPh, NH_2NH_2), the dimer is converted to the ferrous complex and the system recovers its catalytic ability (Scheme 3) (*36, 37*).

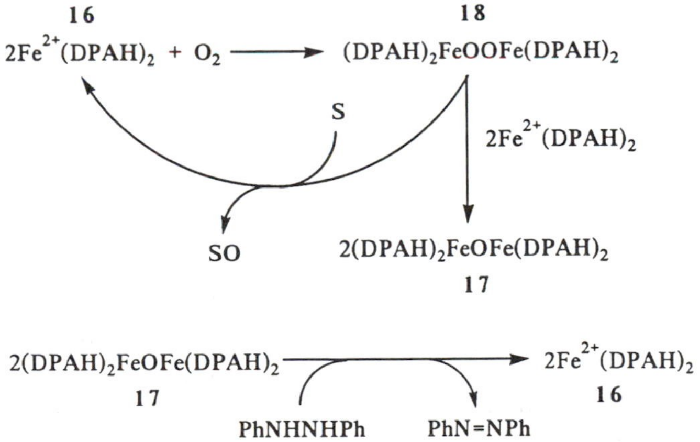

Scheme 3

In this system, **16** is proposed to react with dioxygen to give $(DPAH)_2FeOOFe(DPAH)_2$, **18**, which reacts with substrates to generate 2 mol of **16** and oxygenated products. In a significant side reaction, **16** can react with 2 mol of **18** to give 2 mol of the respective μ-oxo dimer **17**, which is catalytically inactive. Because **17** can be reduced to its monomeric ferrous precursor by a variety of reductants, this reaction can become catalytic, if performed in the presence of an appropriate reductant (*36, 37*).

Although the reactivity properties of this system are quite intriguing, there is currently insufficient characterization of the iron species to adequately assess the details of the proposed mechanism. The distribution of solution species and their nuclearity have yet to be established. Electrochemical measurements of the 1:1, 1:2, and 1:3 ferrous/PA mixtures confirms the rich solution chemistry expected for such a system. A multitude of species have been proposed based on electrochemical data, although definitive evidence is lacking. The introduction of either O_2 or H_2O_2 generates additional uncharacterized species. Furthermore, no details have been presented concerning the interconversion between **13** and **13b** during the catalytic production of singlet oxygen in DMF. These observations suggest that any proposed detailed mechanism is premature, because a variety of intermediate species and subsequently reaction pathways may be operational.

The reactivity and spectroscopic properties of diferrous complexes were also examined by Stassinopoulos and Caradonna (*44, 45*), who developed a system based on simple diamide ligands that are known to stabilize metals in high oxidation states. The diferrous complex $Fe_2(H_2Hbab)_2(N\text{-}MeIm)_2 \cdot MeOH$, **19**, ($H_4Hbab$ is 1,2-bis(2-hydroxybenzamido)benzene; *N*-MeIm is *N*-methylimidazole), obtained from reacting of the dianion of the ligand with *trans*-$Fe(N\text{-}MeIm)_2Cl_2(MeOH)_2$ in MeOH is a symmetric di-μ-phenoxy bridged dimer (Figure 4). Each ferrous center adopts a five-coordinate trigonal bipyramidal geometry. In addition to the two bridging phenolate oxygen atoms, the NO_4 coordination sphere about each iron consists of terminal phenolate and amide carbonyl oxygen atoms (from one of the ligands) and a nitrogen from a coordinated *N*-MeIm. The coordinatively unsaturated iron centers are 3.165 Å apart. Crystals grown from *N,N*-dimethylformamide (DMF)/ Et_2O solvent show that one of the *N*-MeIm ligands has been replaced by two coordinated DMF solvent molecules, resulting in an asymmetric five- and six-coordinate complex $Fe_2(H_2Hbab)_2(N\text{-}MeIm)(DMF)_2$, **20** (Figure 4). The unusual ligand arrangement observed in **19** is also present in **20**. Although the $Fe(1)NO_4$ site is structurally equivalent in **19** and **20**, the $Fe(2)O_6$ site shows octahedral geometry with two axial and two bridging phenoxy oxygens, an amide-carbonyl oxygen and two molecules

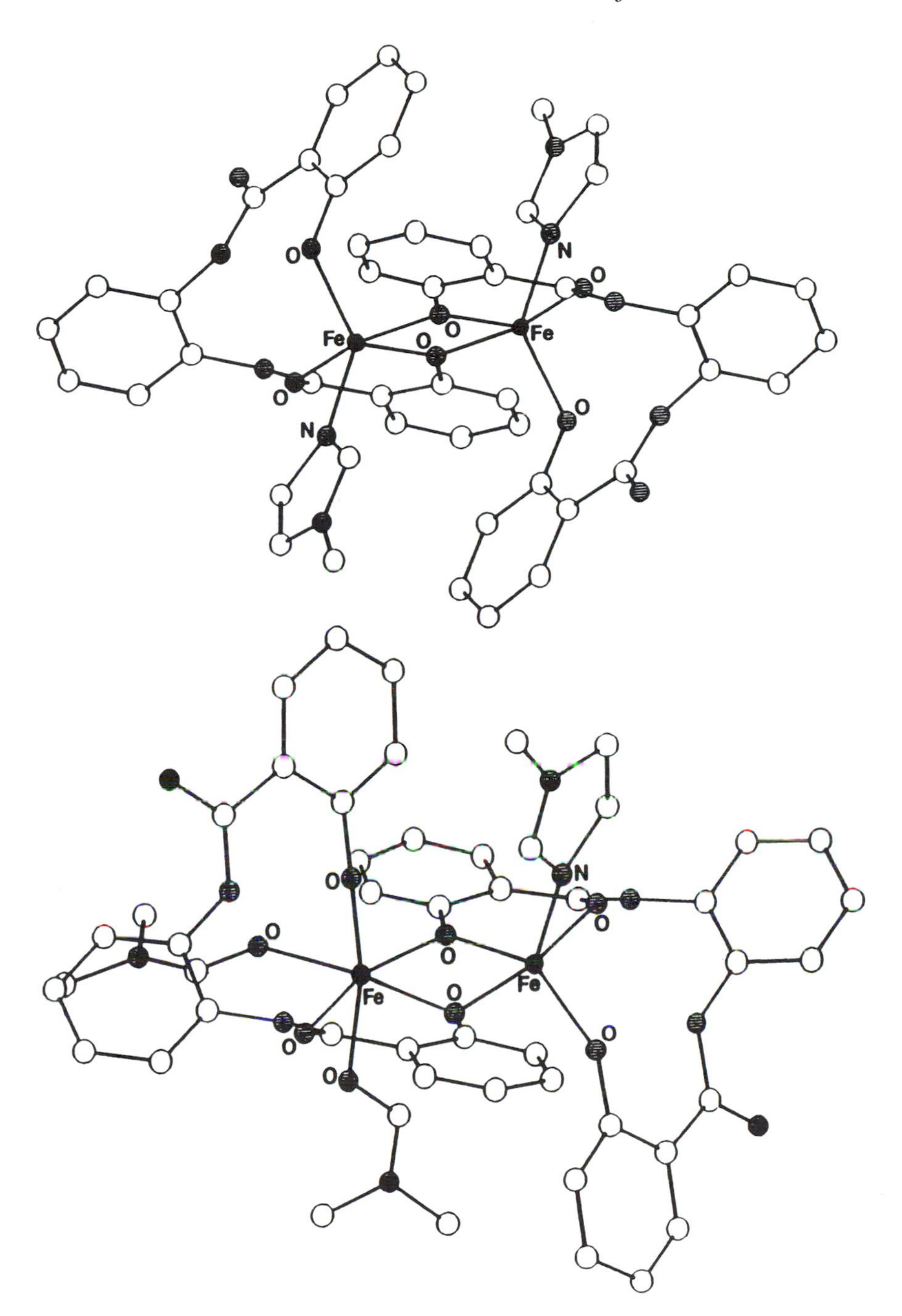

Figure 4. Crystal structures of the symmetric $Fe_2(H_2Hbab)_2$(N-*MeIm*)$_2$, **19**, *and the asymmetric five- and six-coordinate complex* $Fe_2(H_2Hbab)_2$(N-*MeIm*)(*DMF*)$_2$, **20**.

of DMF. Owing to the distortions associated with the coordination change, the two iron atoms are 3.19 Å apart.

Mössbauer studies show a single quadrupole doublet for crystalline **19**, (δ = 1.18 mm s^{-1}; ΔE_Q = 3.26 mm s^{-1}) consistent with the crystallographically equivalent five-coordinate sites of the ferrous dimer (*45*). Mössbauer investigation of microcrystalline compound **20**, however, shows two overlapping quadrupole doublets (δ_1 = 1.27 mm s^{-1}; ΔE_{Q1} = 3.35 mm s^{-1}; δ_2 = 1.30 mm s^{-1}; ΔE_{Q2} = 3.00 mm s^{-1}) that are expected for the two ferrous environments observed in the crystal structure. The Mössbauer spectrum of a frozen DMF solution of **19** has essentially the same features found for **20**, (δ_1 = 1.24 mm s^{-1}; ΔE_{Q1} = 3.08 mm s^{-1}; δ_2 = 1.29 mm s^{-1}; ΔE_{Q2} = 2.54 mm s^{-1}). This similarity suggests that the iron centers are in different environments in solution. Isothermal distillation techniques indicate that **20** maintains its dimeric structure in solution. Preliminary temperature-dependent magnetic susceptibility measurements of microcrystalline **19** indicate weak ferromagnetic interaction with $J \approx 2.5$ cm^{-1}. Low-temperature electron paramagnetic resonance (EPR) studies (5 K) also show the existence of a $g \approx 16$ integer spin EPR signal. Electrochemical studies of **20** in DMF show two quasireversible peaks (−500, −250 mV vs. saturated calomel electrode), each corresponding to a one-electron oxidation. The comproportionation constant, K_{com}, is 1.7×10^4, indicating a stable mixed-valence state. In addition to the [Fe^{2+}, Fe^{2+}] dimer, the [Fe^{2+}, Fe^{3+}], **21**, [Fe^{3+}, Fe^{3+}], **22** and di-μ-(OMe)-[Fe^{3+},Fe^{3+}], **23**, complexes were synthesized by either electrochemical or chemical methods and spectroscopically characterized. These core oxidation states can be readily interconverted by a variety of chemical methods.

The Caradonna group studied the ability of **19** and **20** to catalyze the decomposition of peracids in MeOH by using TBPH as trapping reagent (*44*). Both *meta*-chloroperbenzoic acid (*m*CPBA) and phenylperacetic acid (PPAA) were catalytically decomposed to yield 2,4,6-tri-*tert*-butylphenoxy radical, TBP·, and HCHO; no active oxygen was found at the end of the reaction. The mechanism of catalytic peracid decomposition (homolytic vs. heterolytic) was examined by using PPAA as substrate. Analysis of the PPAA decomposition products showed that **19** induced a heterolytic pathway; no products derived from the benzyl radical were detected. Under similar conditions, the decomposition of PPAA by **22** was shown to follow a homolytic mechanism.

Catalytic atom transfer reactions catalyzed by **19** were investigated by using PhIO as an oxygen atom donor in 1%DMF/CH_2Cl_2 (*44*). Oxidation of olefins, organic sulfides and sulfoxides was observed. Oxidation of cyclohexene gave the epoxide along with the allylic oxidation products of cyclohexenone and cyclohexenol (ratio: 1:2:4; ≈30 turnovers). Oxidation of phenyl methyl sulfide gave the respective sulfoxide and sulfone (sulfoxide–sulfone, 9:1); oxidation of phenyl methyl sulfoxide gave the sulfone. Analogous reactions for **22** and the μ-oxo complex $[Fe_2(H_2Hbab)_2(Solv)_2O]$, **24**, showed substoichiometric oxidation. Although no chlorinated products were observed in the catalytic reactions with **19** in CH_2Cl_2, bromide incorporation into the organic products was observed when CH_2Br_2 was used as solvent indicating the presence of organic radicals.

Upon completion of the catalytic reactions, the iron complex forms the insoluble μ-oxo dimer, **24** (or the μ-dimethoxy dimer, **23**, when MeOH is present). Under anaerobic conditions in the absence of substrates, PhIO quantitatively converts **19** to **24**; the use of $PhI^{18}O$ led to incorporation of ^{18}O in the μ-oxo bridge of the dimer ($\nu_{as}Fe–^{16}O–Fe$ = 837 cm^{-1}). The minor product, **23**, was crystallographically characterized and shown to contain two FeO_6 centers having octahedral geometries (Figure 5) (J. P. Caradonna et al., unpublished results). Each iron center is coordinated to one amide-carbonyl and phenolate oxygen of each ligand in addition to the two bridging methoxy oxygens. The mechanism by which **19** is converted to **23** is not yet established.

The addition of either PhIO, C_6F_5IO, or excess peracid to an anaerobic solution of **19** at low temperatures in the absence of substrate generates a transient species with a dark orange-red color (*44*). This color bleaches rapidly upon precipitation of the μ-oxo dimer or the addition of substrate. This species has a solution half-life of $\approx$1.5 h at −80 °C. Characterization of the intermediate is not yet reported.

To account for the reactivity of **19** toward oxygen atom donors, a catalytic cycle was proposed. In this mechanism, the diferrous complex reacts with an oxygen donor to give an adduct with **19** that can either act as the atom transfer species or collapse to a ferryl species $[Fe^{4+}{=}O]$ intermediate. The data currently available does not allow for a more detailed description, although spectroscopic characterization and determination of the kinetic competence of the observed intermediate will allow the differentiation of several possible pathways.

One of the first iron systems containing a binucleating ligand that was capable of catalyzing oxidation reactions was reported by Nishida et al. (*46*). This group synthesized a series of diferric complexes by using ligands in which two *N,N*-bis(2-benzimidazolyl-methyl)amine moieties are linked by a variable length chain ($-(CH_2)_4-$, L^4; $-(CH_2)_6-$, L^6; $-CH_2CH(OH)CH_2-$, L^3). Although no crystal structures were reported, it is clear from the magnetic and electrochemical properties of the complexes that ligand L^3 gave dinuclear complexes that contained magnetic exchange interactions between the two ferric sites ($[Fe_2L^3]Cl_5$, **25**, and $[Fe_2L^3](NO_3)_5$, **26**). The temperature-dependent magnetic moment of **26** indicated the existence of the expected μ-alkoxy-bridge (μ_{eff} = 4.10

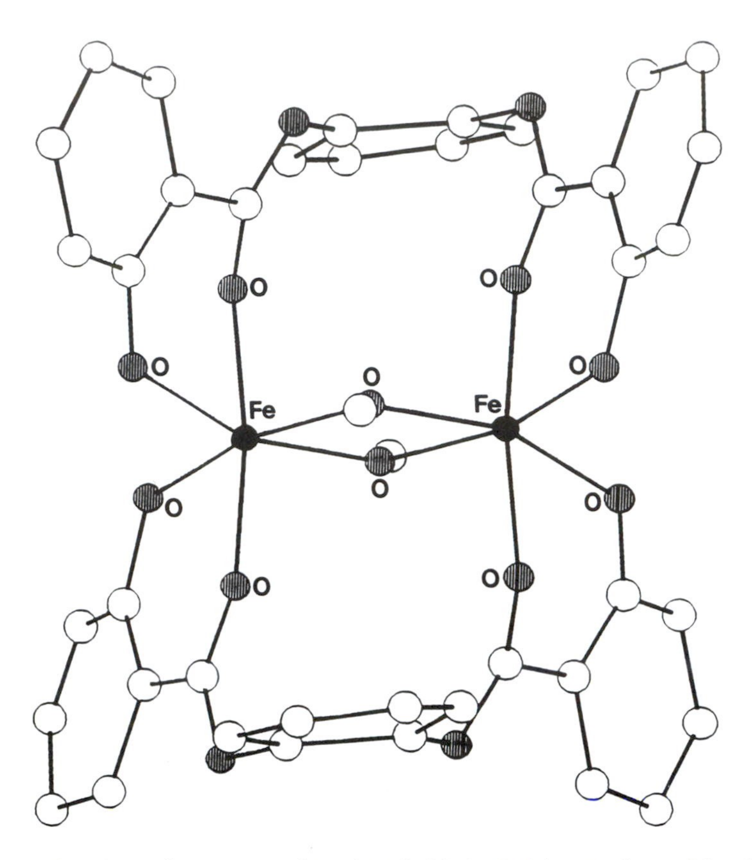

Figure 5. Crystal structure of $Fe_2(H_2Hbab)_2(\mu\text{-}OMe)_2$, **23**, *formed by the two-electron oxidation of* **19** *in the presence of MeOH.*

B_M at 299 K; 2.22 B_M at 87 K). Cyclic voltammetry data of **26** in DMF consisted of two quasireversible peaks separated by $\approx$550 mV. The fact that conductivity measurements reveal that **26** is a 2:1 electrolyte in MeOH suggests a coordinated nitrate ion.

Solutions of **26** were found to catalyze the oxidation of *N,N,N',N'*-tetramethyl-1,2-diaminobenzene (TMPD) to $TMPD^+$ in the presence of dioxygen. Control experiments with **25** showed lower catalytic activity, whereas the activity of the mononuclear complex $Fe(ibz)Cl_3$, **27**, (ibz is *N,N*-bis(2-benzimidazolyl-methyl)amine) was minimal. The authors proposed that the active species arose from the interaction of **26** with dioxygen. This interaction is not possible if the formulation of **26** as a diferric species is correct. However, if either dioxygen or **26** is reduced by a solution component or by TMPD, the redox catalysis can be explained.

$[Fe_2L^3](NO_3)^5$, **26,** binds catechol with a change in the visible spectrum from orange to dark green. This spectral change is not the same as that reported for coordination of phenol, suggesting a different coordination mode structure. Complex **26** also binds to H_2O_2, to form an adduct with λ_{max} = 600 nm (ϵ_M = 1500). This adduct was proposed to be a side-on peroxo adduct of the diferric center, but no other spectroscopic evidence was presented (*46*). Addition of 2,4-di-*tert*-butylphenol to the blue solution of the peroxide adduct caused bleaching within 2 h and resulted in oxidation of the phenol to the respective quinone 3,5-di-*tert*-butylquinone (5.2 turnovers). In the absence of the iron complex, phenol is not oxidized to catechol.

Brennan et al. (*47*) reformulated the structure of $[Fe_2L^3](NO_3)_5$, **26**, as $[Fe_2L^3(OH)(NO_3)_2](NO_3)_2$, based on ^1H NMR, extended X-ray absorption fine structure (EXAFS), X-ray diffraction, and conductivity measurements. In addition, they reported that two resonance-enhanced vibrations (ν_{Fe-O}: 476 cm^{-1}; ν_{O-O}: Fermi doublet, centered at 895 cm^{-1}) were observed during resonance Raman characterization of the peroxo adduct. These resonances were unaffected by D_2O but were observed to shift to 457 and 854 cm^{-1}, respectively, in $H_2{}^{18}O$. Mössbauer studies showed a single quadrupole doublet (δ = 0.54 mm s^{-1}, ΔE_Q = 0.84 mm s^{-1}) different from that of **26** (δ = 0.49 mm s^{-1}, ΔE_Q = 0.66 mm s^{-1}), indicating that both iron centers are similarly affected by the binding of peroxide. Proton NMR studies indicate an increase in the antiferromagnetic coupling upon adduct formation from $J = -20$ cm^{-1} for **26** to approximately -70 cm^{-1} for the peroxide adduct. Based on these data and conductivity measurements in CH_3CN indicating that the adduct is a 1:1 electrolyte, the peroxide adducty of **26** was formulated as $[Fe_2L^3(\mu\text{-}\eta^1,\eta^1\text{-}O_2)(NO_3)_2](NO_3)$ (*47*). The relevance of this peroxide complex and its chemistry to the putative intermediates in the oxygenation of fully reduced methane monooxygenase and ribinucleotide reductase are as yet firmly established.

Murch et al. (*48*) were the first to reported the synthesis of a dinuclear peroxide complex capable of olefin epoxidation and allylic oxidation. Their approach used the *N,N'*-(2-hydroxy-5-methyl-1,3-xylylene)bis(*N*-carboxy-methyl-glycine) ligand (L), or its 5-chloro variant (L′) to make the $(Me_4N)^+$ salts of the bis-μ-acetato-diferric complexes $[Fe_2L(OAc)_2]$, **28**, and $[Fe_2L'(OAc)_2]$, **29**. X-ray crystallographic analysis of **28** shows (Figure 6) the NO_5 coordination spheres about each ferric center and the positions of the μ-phenoxo and two μ-acetato bridges. A similar structure was inferred for **29**.

When H_2O_2 was added to a solution of **29**, a new visible band at 470 nm replaces the phenolate-to-Fe^{3+} charge-transfer band at 450 nm. Laser excitation (514.5 nm) of the 470-nm absorption enhances a Raman feature at 884 cm^{-1} (peroxide O–O stretch). This enhancement suggests

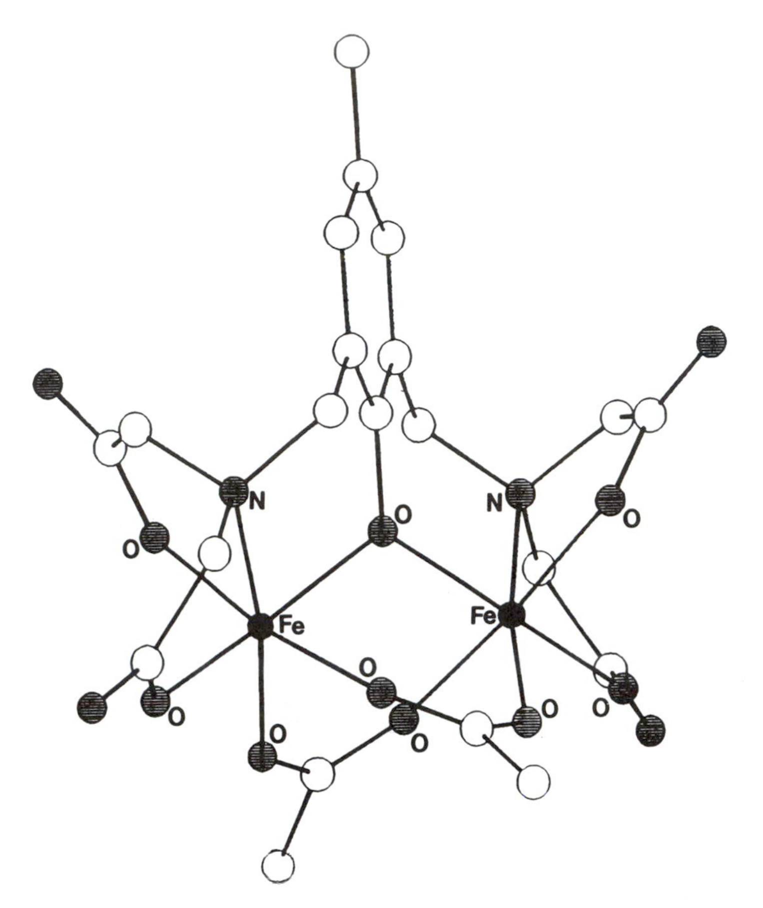

Figure 6. Crystal structure of $[Fe_2L(OAc)_2]$, **28**, *where L =* N,N′-*(2-hydroxy-5-methyl-1,3-xylylene)bis(*N-*carboxymethylglycine).*

the electronic absorption arises from a Fe^{3+}–peroxide charge-transfer transition. The ^1H NMR spectrum of the peroxide complex is consistent with a loss of the apparent C_2 symmetry of **29**, indicating a change in ligand arrangements. A bridging peroxo structure, similar to the one proposed for $[Co_2(BPMP)(OAc)O_2](ClO_4)_2$, (BPMP is 2,6-bis[bis(2-pyridylmethyl)aminomethyl]-4-methylphenol) was proposed for this adduct (*49*, *50*).

Complex **29** catalyzes the disproportionation of hydrogen peroxide to oxygen and water (*48*). In the absence of readily oxidizable substrates, it degrades. In the presence of olefins, however, it catalyzes the formation of epoxides. This catalysis was demonstrated for cyclohexene (1.6 turnovers), styrene (3.2 turnovers), and *cis*-stilbene (2.5 turnovers). The formation of the epoxides is not exclusive, because allylic oxidation products were also observed for cyclohexene (3-cyclohexenol, 0.9 turnovers; 3-cyclohexenone, 0.4 turnovers), whereas C–C bond cleavage products were observed for styrene (benzaldehyde, 1.5 turnovers). *cis*-Stilbene gave predominately the *trans*-epoxide (95%), indicating a nonconcerted process for the epoxidation process. Because the oxidation reaction is accompanied by destruction of the complex through autoxidation, it is not certain whether the side products are a result of the intrinsic reactivity of the active complex or a side reaction resulting from its decomposition products. Furthermore, it is uncertain whether the peroxy complex or species resulting from the decomposition of the peroxy adduct is responsible for the oxidation reactions.

Based on a comparison of **29** with an analogous mononuclear complex $Fe(Cl_2HDA)(H_2O)_2$, **30** [Cl_2HDA is *N*-(4,6-dichloro-2-hydroxybenzyl)-*N*-(carboxy-methyl)glycine], which does not bind peroxide or acetate, the authors suggested the requirement of a dimeric structure for the formation of peroxo adducts (*48*). Although **30** does not epoxidize olefins, it catalyzes the disproportionation of H_2O_2 and generates cation radicals of TBPH and *o*-dianisidine. This system, which requires a dinuclear ferric complex, is therefore significantly different from monomeric $Fe(acac)_3$, which catalyzes the epoxidation of olefins (*51, 52*).

An intriguing non-heme iron system was reported by Que and coworkers (52), who have published spectroscopic data that suggest the existence of a high-valent non-heme $[Fe{=}O]^{n+}$ intermediate in a system that is capable of catalytically oxidizing simple organic substrates. The complex $[Fe_2(TPA)_2O](ClO_4)_4$, **31**, synthesized by reacting simple ferrous salts with the TPA ligand in the absence of any coordinating ligands, is proposed to have a bent μ-oxo dimer structure based on ^1H NMR and Mössbauer spectroscopic parameters; this complex has not yet been crystallographically characterized (52).

In a manner analogous to monomeric Fe(TPA) complexes, **31** catalyzes the room temperature hydroxylation of cyclohexane by H_2O_2 or TBHP. During this reaction, a fleeting green color was observed, which prompted a low temperature investigation of the iron species in solution. This green species has a $t_{1/2} \approx 2$ h at -40 °C, allowing the possibility of spectroscopic investigations.

The visible spectrum of this intermediate consists of a band at $\lambda_{max} = 614$ nm. Excitation at 614 nm gives resonance Raman enhanced bands at 416 and 666 cm^{-1} that shift to 408 and 638 cm^{-1} upon addition of $H_2{}^{18}O$, indicating exchange with water. These bands are unaffected by the addition of D_2O. This behavior is consistent with an FeO stretch and the shift observed upon substitution agrees with the expected shift of 29 cm^{-1}. The second peak at 416 cm^{-1} was attributed to a metal–ligand vibration coupled to the iron–oxo stretch.

Additional EPR and Mössbauer spectroscopic characteristics of the intermediate were reported. The EPR spectrum consists of peaks at g values 3.95, 4.4, and 2.0, which were interpreted to arise from the ground Kramers doublet of a half-integral spin system. The intensity is 0.47 spin/Fe atom. The spectrum is reported to be similar to an $S = 3/2$ multiplet with zero field splitting ($D > 0$) and $E/D = 0.04$. Although the EPR experiment can only detect the transient paramagnetic species, the Mössbauer spectra obtained by reacting **31** with 5 equivalents of H_2O_2 is consistent with two major species, a diamagnetic dimer, **32**, (65%) characterized by an unresolved doublet with $\Delta E_{Q1} = 1.63$ mm s^{-1}, $\Delta E_{Q2} = 1.15$ mm s^{-1}, and $\delta_1 = \delta_2 = 0.44$ mm s^{-1}, and a transient paramagnetic species, **33**, (30%) characterized by a single quadrupole doublet with $\Delta E_Q = 0.42$ mm s^{-1} and $\delta = 0.07$ mm s^{-1}. These data, which suggest that **33** is oxidized relative to **32**, were interpreted to imply that **33** most likely contains a Fe^{4+} center; the chemical isomer shift of **33** is similar to authentic Fe^{4+} centers in both heme and non-heme environments (53). The initial model proposed to account for the spectroscopic properties of the new oxidation product involved a $S = 1$ ferryl center ($D > 15$ cm^{-1}, $E/D = 0.04$) that is ferromagnetically coupled to an $S = 1/2$ radical, with coupling strength of $J/D \approx 1.5$.

A mechanism to account for the formation of the ferryl species, was postulated. According to this mechanism (equations 5 and 6), one of the iron atoms in the μ-oxo dimer **31** is oxidized by two electrons.

$$[LFe^{3+}\text{–}OFe^{3+}L]^{4+} + H_2O_2 \rightarrow [LFeO]^{3+} + [LFeOH]^{2+} + OH^- \quad (5)$$

$$[LFeO]^{3+} + [LFeOH]^{2+} \rightarrow [L'Fe^{3+}OFe^{3+}L'] \quad (6)$$

Cleavage of the first species to the transient ferryl complex and a ferric hydroxo monomeric complex can explain the near 50% yield of the ferryl compound. The ferric monomer subsequently dimerizes to the paramagnetic dimer, **32**, responsible for the 65% of the Mössbauer signal. The dimer formed after the oxidation is proposed to be similar to the starting material, based on comparisons of its Mössbauer spectrum with that of **31**.

However, there is still substantial ambiguity concerning the nuclearity of **33** and the exact nature of the this high-valent iron species and the mechanism proposed to account for its formation should therefore be considered quite speculative. Future spectroscopic studies will undoubtedly shed new light on the nature of this interesting intermediate.

Reactions with Dioxygen. Despite the tremendous interest in the atom transfer chemistry of synthetic dinuclear iron centers, little is known about their interactions with dioxygen. This chemistry is of extreme interest owing to the demonstrated or postulated interactions of molecular oxygen with the diferrous sites of hemerythrin, methane monooxygenase, and ribonucleotide reductase (*1, 2*). Examination of the interaction of dioxygen with the five-coordinate (μ-hydroxo)bis(μ-carboxylato) diferrous core of hemerythrin (Figure 2) shows reversible reduction of O_2 to bound peroxide. This chemistry, which is observed for relatively nitrogen-rich coordination environments, is in vivid contrast to irreversible reduction of O_2 observed for the diferrous sites of methane monooxygenase and ribonucleotide reductase that have oxygen-rich coordination environments. Current efforts to study the interactions of O_2 with nonheme diferrous model complexes are just beginning to define the interesting chemistry that awaits. Although early studies demonstrated the possible quantitative aerial oxidation of diferrous complexes to μ-oxo diferric complexes, the intimate mechanistic details of these transformation were poorly defined (*54, 55*).

The reactivity of synthetic diferrous centers with molecular oxygen has been examined by several groups. $[Fe_2(Me_3TACN)_2(OH)(OAc)_2]^+$, **34** (1,4,7-trimethyl-1,4,7-triazacyclononane), reacts with dioxygen to give the corresponding μ-oxo diferric complex (*55*), whereas exposure of $[Fe_2(BPMP)(O_2Pr)_2]^-$, **35**, to O_2 results in oxidation of the diferrous core (*56, 57*); no dioxygen adducts were observed during the course of these reactions. These reduced complexes are coordinatively saturated and were thought to generate their oxidized species via autoxidation processes although the loss of a ligand in a pre-equilibrium step to afford vacant coordination sites cannot yet be unambiguously ruled out.

The synthesis and reactivity of an asymmetric diferrous complex containing an open coordination position was reported by Lippard and co-workers (*58, 59*). Reacting $Fe(O_2CH)_2 \cdot 2H_2O$ with bis(1-methylimidazol-2-yl)phenyl-methoxymethane (BIPhMe) in MeOH under anaerobic conditions afforded $[Fe_2(O_2CH)_4(BIPhMe)_2]$, **36**. This complex contains a novel core arrangement in that the ferrous centers are bridged by one monodentate and two bidentate formate ligands. The coordination sphere about the six-coordinate iron is completed by the two imidazoles of the bidentate BiPhMe and a fourth monodentate formate ligand, whereas the second iron center is five-coordinate and contains the second BiPhMe ligand. There is, however, a weak interaction between an oxygen atom of the monodentate bridging formate with the five-coordinate iron center (d_{Fe-O} = 2.74 Å).

Solid-state magnetic susceptibility data for **36** is consistent with little or no exchange coupling between the two ferrous centers (*58, 59*). The Mössbauer spectrum (zero field, 4.2 K) of **36** shows a broad asymmetric doublet consistent with two distinct sites (δ_1 = 1.26 mm s^{-1}, ΔE_Q = 2.56 mm s^{-1}; δ_2 = 1.25 mm s^{-1}, ΔE_Q = 3.30 mm s^{-1}). EPR spectrum (X-band, 7 K) of the diferrous system contains a broad feature at $g \approx 16$, indicative of an integer S = 4 spin state.

Interestingly, upon exposure to air, solutions of **36** give rise to $[Fe_2O(O_2CH)_4(BIPhMe)_2]$, **37**, which contains a (μ-oxo)bis(μ-carboxylato) diferric core (*58, 59*). Each iron center also contains one BIPhMe ligand and a monodentate formate ligand *cis* to the oxo bridge. The spectroscopic characteristics of **37** are analogous to other μ-oxo diferric complexes. Isotopic exchange studies demonstrated that the origin of the oxo group was dioxygen rather than adventitious water.

Lippard and co-workers (*58, 59*) proposed a route (Scheme 4) for the production of **37** from **36** based on the stoichiometry of iron complex to dioxygen. Binding of dioxygen to **36** might yield a reactive superoxo or peroxo adduct that might oxidatively react with solvent, ligand, or adventitious protons to yield **37** directly, using one mole of O_2 per mole of **36**.

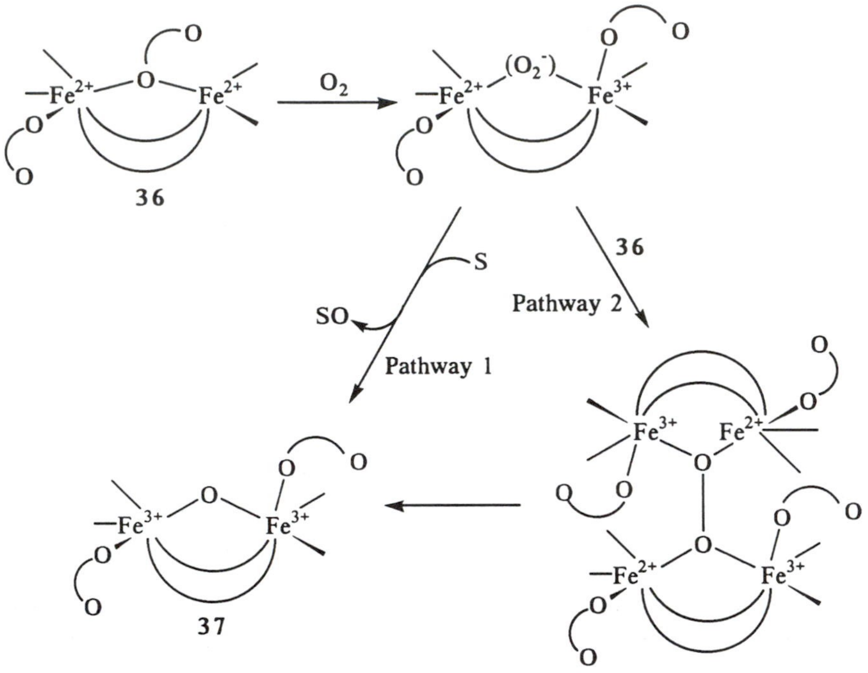

Scheme 4

A second pathway involves the formation of a mixed-valent $[Fe^{2+}, Fe^{3+}]$-superoxo species that could react with a second molecule of **36** to form a peroxo-bridged tetranuclear $[Fe_2^{2+}, Fe_2^{3+}]$ cluster. Homolytic cleavage of the peroxo O–O bond followed by electron transfer and rearrangement would yield **37**. These two pathways differ in their oxygen stoichiometry: pathway 1 has a ratio of O_2/reduced iron dimer of 1:1, whereas pathway 2 has a ratio of 1:2. Manometric measurements of O_2

uptake showed the conversion of **36** to **37** to occur with 0.6 mol of O_2 consumed per mole of **36**, suggesting that pathway 2 is being followed. The high yield of the conversion argues against any significant oxidative consumption of ligand as a possible source of electrons in pathway 1.

EPR spectra and power saturation studies of solutions of **36** exposed to air show the presence of a magnetically exchange-coupled mixed-valence species with $J = -31$ (2) cm^{-1} (*58*, *59*). The authors interpreted these results as being consistent with the formation of a dinuclear [Fe^{2+}, Fe^{3+}]-superoxo complex as proposed in the previously mentioned oxidation pathway. This would assume, however, that there is no coupling between the bound superoxide ion and the dinuclear center, which would give rise to an EPR silent complex. The inability of exogenous ligands such as halides, CO, HO^-, NO, and PPh_3 to coordinate to the five-coordinate ferrous site in **36** was presented as evidence that the dangling formate group protected the coordinatively unsaturated site and thereby precludes the direct binding of dioxygen to **36**. Furthermore, this indicates that some type of rearrangement of carboxylates about the core occurs upon oxidation. Interestingly, no PPh_3O was observed under these conditions, indicating either the inability of an intermediate species to act as an atom transfer species or the unreactivity or inaccessibility of the peroxo species owing to its mode of formation in homogeneous solutions. The authors could not rule out the possible role of mononuclear species resulting from bridge-breaking chemistry to account for their results. This system, although demonstrating the ability of O_2 to oxidized a reduced (μ-hydroxo)bis(μ-carboxylato)-diiron core, does so in a manner inconsistent with the chemistry of Hr or other dinuclear ferrous metalloenzyme active sites.

Que, Münck and co-workers (*60*) studied the reaction of [Fe_2(HPTB)(OBz)]$(BF_4)_2$, **38a** [HPTB: 1,3-bis[*N*,*N*-bis(2-benzimidazolylmethyl)amino]-2-hydroxypropane], its *N*-ethyl analogue **38b**, and its tetrakis(pyridine) analogue **38c**, with dioxygen. This system, based on **26** [which was initially synthesized by Sakurai et al. (*61*) and Nishida et al. (*46*)], form irreversible peroxo adducts. Structural analysis revealed two five-coordinate ferrous centers, bridged by μ-alkoxy and μ-benzoato groups. The iron centers in **38b** are approximately trigonal bipyramidal with amine nitrogen and benzoato oxygen atoms acting as axial donors (*60*). The Fe–Fe distance is 3.473 Å. The Mössbauer spectrum of **38b** consists of a single quadrupole split doublet (δ = 1.07 mm s^{-1} and ΔE_Q = 3.13 mm s^{-1}); magnetic susceptibility data indicated the iron centers are antiferromagnetically coupled ($J = -11$ cm^{-1}). When **38b** was reacted with 1 atm of dioxygen in dichloromethane (CH_2Cl_2), the solution color changed from light yellow to deep blue (λ_{max} = 588 nm). Manometric measurements showed the consumption of 1 mol of O_2 per mole of **38**. The O_2 adduct of **38a** and **38b** were stable indefinitely in CH_2Cl_2 at −60 °C but decomposed upon warming. The addition of polar aprotic solvents, however, stabilized the adducts, allowing them to persist at ambient temperatures for short periods of time. The O_2 adduct of **38c** was not observed at −80 °C in the absence of a polar aprotic solvent. Resonance Raman spectra in CH_3CN/MeOH with 575 nm excitation wavelength enhanced features at 476 and 900 cm^{-1}, which were assigned the as ν_{Fe-O} and ν_{O-O} stretches, respectively. The blue chromophore was assigned to a Fe^{3+}-peroxo charge-transfer band. The Mössbauer spectrum consisted of a single quadrupole split doublet with parameters (δ = 0.52 mm s^{-1}; ΔE_Q = 0.72 mm s^{-1}) indicating a Fe^{2+} to Fe^{3+} change in oxidation state and a more symmetric electron density around the iron centers. The observation of a single doublet for the spectrum of the dioxygen adduct confirms a symmetric structure formulation. These data are consistent with a μ-1,2-peroxo adduct, a conclusion supported by a strong antiferromagnetically coupling interaction of $J \approx -140$ cm^{-1}. The effects of carboxylate substitution on the electronic spectrum of the peroxide adduct of **38b** suggested that the carboxylate ligand remains coordinated in the adduct. On the basis of the data mentioned previously, a tribridged (μ-1,2-peroxo)(μ-carboxylato)(μ-alkoxo) diferric core was proposed (*48*, *60*).

Differences in reactivity were observed also for **38a–c** (*60*). Although the addition of triphenyl phosphine (Ph_3P) or 2,4-di-*tert*-butylphenol accelerates the decomposition of the peroxide adduct of **38c** giving substoichiometric (0.5–0.6 equiv) quantities of the corresponding $OPPh_3$ or biphenol, respectively, these reagents have little effect on the stability of the peroxo adduct of **38b** at −50 °C. These studies suggest that the pyridine ligands enhance the electrophilic character of the peroxo moiety bound to the diferric core. Furthermore, this collection of data indicates that **38a–c** are not able to hydroxylate or oxygenate substrates such as alkanes that are substantially poorer oxo transfer substrates.

Interestingly, the peroxide adducts of **38a–c**, formed from the addition of hydrogen peroxide with the diiron(III) complexes, showed slight differences in the absorption maxima wavelengths when compared with the corresponding peroxide complexes formed in the reaction of dioxygen with the diiron(II) complexes (*45*, *48*, *61*, *62*). These variations are thought to arise from variations in the solvent systems, as the use of aqueous hydrogen peroxide introduces protons that are not present in the dioxygen reactions typically performed in dry, aprotic organic solvents.

The synthesis, characterization and reaction chemistry of $[Fe_2(TPA)_2(OAc)_2]^{2+}$, **39**, also was reported (*63*). Crystallographic studies show that

39 contains an inversion center located in the center of a bis(μ-acetato)diferrous core. Each acetate binds both iron centers in a O,O′-mode using a syn lone pair [d_{Fe-O} = 1.998 (2) Å] of one oxygen atom and the anti lone pair [d_{Fe-O} = 2.145 (2) Å] of the other oxygen atom. The distorted octahedral environment of each ferrous center is completed by the four nitrogen atoms of the TPA ligand. The Mössbauer spectrum of **39** consists of a single quadrupole split doublet, suggestive of a anisotropic electronic environment (δ = 1.12 mm s−*1*; ΔE_Q = 3.33 mm s^{-1}; g = 0.26 mm s^{-1}) (*63*). Magnetic properties of **39** are consistent with weak antiferromagnetic coupling between the two ferrous centers mediated through the acetate bridges even at d_{Fe-Fe} of 4.288 (2) Å. EPR spectra of acetonitrile solutions of **39** show features (10% of the total iron) at g = 9.3, characteristic of an uncoupled S = 2 center, indicating that the acetate dimer structure breaks to give monomeric units. This conclusion is also supported by NMR data (*63*).

The reactivity of $[Fe_2(TPA)_2(OAc)_2]^{2+}$, **39**, toward dioxygen was also reported (*63*). The resulting product was consistent with the formulation $[Fe_2(TPA)_2O(OAc)_2]^{2+}$, **40**, and is thought to be similar to two structurally characterized dimers, $[Fe_2(TPA)_2O(OAc)]^{3+}$, **41**, an unsymmetric complex containing a (μ-oxo)(μ-carboxylato)diferric core (*64, 65*) and $[Fe_2(TPA)_2O(Cl)_2]^{2+}$, **42**, a centrosymmetric complex containing a linear oxo bridge (*65*).

The UV–vis spectrum of **40** is similar to the spectrum reported for **42** with absorptions at 320 nm (ϵ_M = 10,000), 360 nm (ϵ_M = 8,000), and 500 nm (sh), with no strong absorption bands in the 400–700 nm region typically observed for complexes containing bent Fe–O–Fe moieties (*66*). The assignment of **40** as a μ-oxo diferric complex with monodentate terminal acetates is consistent with NMR and IR data.

Oxygen uptake studies show that the conversion of **39** to **40** requires 0.6 (1) mol of O_2, indicating that four ferrous atoms are oxidized per consumed O_2 (*63*). This stoichiometry is similar to that reported for the autoxidation of ferrous porphyrins. The proposed mechanism, shown in Scheme 5, suggests that the autoxidation is initiated by the interaction of O_2 with monomeric units of **39** that are either coordinatively unsaturated or contain a labile ligand such as solvent. The reaction of a second monomeric unit with the transient superoxo ferric complex gives rise to a (μ-peroxo)diferric species similar to that reported for [Fe{HB(3,5-$iPr_2pz)_3$}(OBz)(CH_3CN)] (*67*). This species is then somehow reduced by two electrons from **39** to yield two molecules of **40**. No intermediates in the autoxidation of **39** have yet been detected even at low temperatures (−80 °C), unlike the porphyrin systems in which intermediate species such as [(Por)$Fe^{2+}O_2$], [(Por)Fe^{3+}–OO–Fe^{3+}(Por)], and [(Por)-Fe^{4+}=O] have been spectroscopically identified (*68*).

The lack of observable intermediates may reflect differences in the

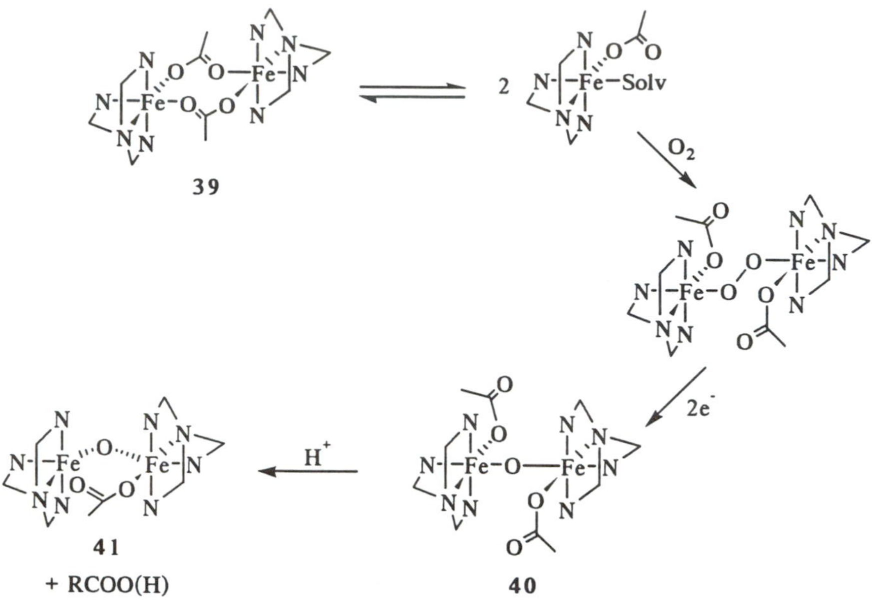

Scheme 5

steric properties of ligand systems in that sterically unencumbering ligands are not expected to stabilize the proposed μ-peroxo diferric species leading to rapid reduction by ferrous precursors.

The oxidized dimer, $[Fe_2(TPA)_2O(OAc)]^{3+}$, **41**, was shown to be an efficient catalyst for cyclohexane oxidation using *tert*-BuOOH as a source of oxygen (69). This catalyst reacts in CH_3CN to yield cyclohexanol (9 equiv), cyclohexanone (11 equiv), and (*tert*-butylperoxy)cyclohexane (16 equiv) in 0.25 h at ambient temperatures and pressures under an inert atmosphere. The catalyst is not degraded during the catalytic reaction as determined by spectroscopic measurements and the fact that it can maintain its turnover efficiency with subsequent additions of oxidant. Solvent effects on product distribution were significant; benzonitrile favored the hydroxylated products at the expense of (*tert*-butylperoxy)cyclohexane, whereas pyridine had the opposite effect. Addition of the two-electron oxidant trap, dimethyl sulfide, to the catalytic system completely suppressed the formation of cyclohexanol and cyclohexanone, but had no effect on the production of (*tert*-butylperoxy)cyclohexane. These and other studies suggested that cyclohexanol and cyclohexanone must arise from an oxidant different from that responsible for the formation of (*tert*-butylperoxy)cyclohexane. Thus, two modes of *tert*-BuOOH decomposition were postulated; a heterolytic pathway that generates a high-valent iron-oxo species that produces cyclohexanol and cyclohexanone and a homolytic pathway that generates *tert*-BuO· and *tert*-BuOO· radicals that induce the formation of (*tert*-butylperoxy)cyclohexane. The heterolytic decomposition of the alkyl peroxide is thought to be initiated by dissociation of the bridging anion from one iron center, thereby generating an open site for coordination of the alkyl hydroperoxide anion. Results from this study also indicated that the ability of the oxidant to abstract a hydrogen atom, as monitored by isotope effects of cyclohexane hydroxylation, is independent of the bridging anion but significantly modulated by the nature of the tripodal ligand.

The aforementioned dinuclear ferrous systems all exhibit irreversible oxygenation leading to peroxide formation. These studies have been complemented by Hayashi et al. (*70*), who reported reversible dioxygen binding by two classes of (μ-alkoxo)diferrous complexes, $[Fe_2(6\text{-Me-TPDP})(RCO_2)(H_2O)]^{2+}$ (**42**, R = CF_3; **43**, R = C_6H_5) based on the sterically demanding dinucleating ligand 6-Me-TPDP (*N,N,N′,N′*-tetrakis(2-(6-methylpyridyl)methyl)-1,3-diamino-propane-2-olate). Interestingly, the crystal structure of **43** shows two distinct iron coordination geometries. Octahedral Fe(1) binds to three N atoms from half of the symmetric ligand, two O atoms from the alkoxo and benzoate bridges and a water molecule that is located trans to the μ-alkoxo group. Fe(2) adopts a trigonal bipyramidal geometry consisting of the three N atoms from the ligand and two O atoms from the bridging groups. Although both **42** and **43** are stable toward dioxygen in the solid state, they rapidly react with O_2 to give a deep blue color in CH_2Cl_2 and CH_3CN even at room temperature. The chromophore bleaches within several minutes. However, reversible formation of a dioxygen adduct with either **42** or **43** was reported at −20 °C in CH_2Cl_2 and CH_3CN, giving rise to a species with λ_{max} = 618 nm. Bubbling Ar through the blue solution causes the complete disappearance of the band at 618 nm and a return of the spectrum of the diferrous complex. These spectral changes can be observed repeatedly.

The parent dinuclear complex, $[Fe_2(TPDP)(C_6H_5CO_2)]^{2+}]$, containing the TPDP ligand without the Me groups in the 6-position of the pyridyl rings, undergoes spontaneous irreversible oxidation of the ferrous centers by dioxygen with no observable oxygenated intermediate even at −40 °C. These observations suggest that the methyl groups on the pyridyl rings in TPDP decrease the basicity of the pyridyl nitrogen atoms, thereby weakening the interaction of the ligand toward the iron centers and destabilizing the peroxide adduct, allowing reversible oxygenation.

Preliminary resonance Raman spectroscopic characterization of the peroxide adduct of **42** showed two pairs of bands consistent with an

Table VII. Selected Properties of Reactive Diferrous Complexes

Complex	*μ-OR*	r_{Fe-Fe} (Å)	*FeOFe (deg)*	*Ligand Set*	δ *(mm s⁻¹)*	ΔE_Q *(mm s⁻¹)*	−J *(cm⁻¹)*	*Chemistry*[a]
19 $[Fe_2(H_2Hbab)_2(N\text{-}MeIm)_2]$	$(O\text{-aryl})_2$	3.165 (7)	98.9	NO_4	1.18	3.26	≈ −2.5	ao, e, po, ro[b]
20 $[Fe_2(H_2Hbab)_2(N\text{-}MeIm)(DMF)_2]$	$(O\text{-aryl})_2$	3.190 (4)	97.7	NO_4	1.27	3.35	<0	ao, e, po, ro[b]
			100.7	O_6	1.30	3.00		
34 $[Fe_2(Me_3TACN)_2(OH)(OAc)_2]^+$	OH	3.32 (1)	113.2	N_3O_3	1.16	2.83	13	ro
35 $[Fe_2(BPMP)(O_2Pr)_2]^-$	*O*-aryl	3.348 (1)	108.9	N_3O_3	1.20	2.72	<0	ro
36 $[Fe_2(O_2CH)_4(BIPhMe)_2]$	*O*-acyl	3.5736 (8)	113.0	N_2O_4	1.26	2.56	≈0	ro
38 $[Fe_2(N\text{-}Et\text{-}HPTB)(OBz)]^{2+}$	*O*-alkyl	3.473 (4)	124.0	N_3O_2	1.07	3.13	≈11	ro, po
39 $[Fe_2(TPA)_2(OAc)_2]^{2-}$	—	4.288 (2)	—	N_4O_2	1.12	3.33	≈ −1	ro
42 $[Fe_2(6\text{-}MeTPDP)(RCO_2)(H_2O)]^{2+}$	*O*-alkyl	3.684 (1)	131.2	N_3O_3	—	—	—	ro
				N_3O_2				

[a] ao, allylic oxidation of alkenes; e, epoxidation of alkenes; po, oxidation of phosphines to phosphine oxides; ro, reduces dioxygen.
[b] PhIO used as oxidant.

unsymmetrical coordination of the peroxide ligand. Bands at 918 and 891 cm^{-1} that shift to 889 and 857 cm^{-1} with $^{18}O_2$ were assigned as O–O stretching mode (ν_{O-O}, Fermi resonance between ν_{O-O} and $2\nu_{Fe-O}$) of coordinated peroxide, whereas the bands at 486 and 450 cm^{-1} that shift to 479 and 442 cm^{-1} with $^{18}O_2$ were assigned to the Fe–O stretching mode (ν_{Fe-O}). Analogous results were observed for **43**. Additional investigations of these complexes are underway.

Summary and Conclusions

The detection of the first non-heme ligand supported ferryl species is only the start of our understanding the chemistry of non-heme iron centers. From the examples reported in this review, it is clear that this area of bioinorganic chemistry is growing quickly. However, it is also evident that there are still many important problems to be solved. Although past efforts stressed structural or electronic active site models, the next set of goals involves the development of reactivity models that can shed light on the intrinsic reactivity properties of non-heme iron complexes. Table VII summarizes the reactivity properties of those diferrous complexes discussed in this chapter. Only **19** and **20** exhibit catalytic oxygen atom transfer (using PhIO as oxygen atom source); the reduction of O_2 to H_2O_2 is stoichiometric at best. Thus, another difficulty that must be overcome involves the reduction of the spent diferrous complexes from the diferric core oxidation state without irreversibly altering the nature of the iron complex by complexation with reductant. Many of the studies described herein avoided this problem by examining the interaction of diferric complexes with H_2O_2. Although this experimental design is a valid first approach, it is still essential to examine the process of O_2 interacting with reduced iron centers. Perhaps some of the most challenging directions involve modeling oxidative enzymes by stabilizing and characterizing the reactive intermediates generated during catalysis. Many of the catalytic systems described herein suffer from the problem that they degrade during the oxidative transformations. This degradation is a serious problem that must be overcome if the role of the ligand environment on reactivity chemistry is to be explored. This goal will undoubtedly be met by using a combination of rational ligand design and the use of low temperature spectroscopic techniques.

Acknowledgments

Preparation of this manuscript was supported in part by the Petroleum Research Fund, administered by the American Chemical Society, the Department of Energy, the Camille and Henry Dreyfus Foundation, and the Alfred P. Sloan Foundation.

References

1. (a) Lippard, S. J. *Angew. Chem. Int. Ed. Engl.* **1988,** *27,* 344. (b) Que, L., Jr.; Scarrow, R. C. In *Metal Clusters in Proteins;* Que, L., Jr., Ed.; ACS Symposium Series 372, American Chemical Society: Washington, DC, 1988; Chapter 8, p 152.
2. (a) Kurtz, D. M., Jr. *Chem. Rev.* **1990,** *90,* 585. (b) Que, L., Jr.; True, A. E. *Prog. Inorg. Chem.* Lippard, S. J., Ed.; Wiley: New York, **1990,** *38,* 97.
3. Nordlund, P.; Sjöberg, B.-M.; Eklund, H. *Nature (London)* **1990,** *345,* 593.
4. Rosenzweig, A. C.; Frederick, C. A.; Lippard, S. J.; Nordlund, P. *Nature (London)* **1993,** *366,* 537.
5. Martell, A. E.; Sawyer, D. T. *Oxygen Complexes and Oxygen Activation by Transition Metals;* Plenum Press: New York, 1987.
6. Spiro, T. G. *Metal Ion Activation of Dioxygen;* Wiley: New York, 1980.
7. Jones, R. D.; Summerville, D. A.; Basolo, F. *Chem. Rev.* **1979,** *79,* 139.
8. Niederhoffer, E. C.; Timmons, J. H.; Martell, A. E. *Chem. Rev.* **1984,** *84,* 137.
9. Smith, T. D.; Pilbrow, J. R. *Coord. Chem. Rev.* **1981,** *39,* 295.
10. Traylor, T. G.; Traylor, P. S. *Annu. Rev. Biophys. Bioeng.* **1982,** *11,* 105.
11. Sawyer, D. T. In *Oxygen Complexes and Oxygen Activation by Transition Metals;* Martell, A. E.; Sawyer, D. T., Eds.; Plenum Press: New York, 1987; pp 131–148.
12. Tabushi, I.; Nakajima, T.; Seto, K. *Tetrahedron Lett.* **1980,** *21,* 2565.
13. Armstrong, W. H.; Spool, A.; Papaefthymiou, G. C.; Frankel, R. B.; Lippard, S. J. *J. Am. Chem. Soc.* **1984,** *106,* 3653.
14. Kitajima, N.; Fukui, H.; Moro-oka, Y. *J. Chem. Soc. Chem. Commun.* **1988,** 485.
15. Barton, D. H. R.; Hay-Motherwell, R. S.; Motherwell, W. B. *J. Chem. Soc. Perkin Trans. 1* **1983,** 445.
16. Barton, D. H. R.; Boivin, J.; Gastiger, M.; Morzycki, J.; Hay-Motherwell, R. S.; Motherwell, W. B.; Ozbalik, N.; Schwartzentruber, K. M. *J. Chem. Soc. Perkin Trans. 1* **1986,** 947.
17. Barton, D. H. R.; Halley, F.; Ozbalik, N.; Young, E.; Balavoine, G.; Gref, A.; Boivin, J. *New J. Chem.* **1989,** *13,* 177.
18. Barton, D. H. R.; Csuhai, E.; Doller, D.; Balavoine, G. *J. Chem. Soc. Chem. Commun.* **1990,** 1787.
19. Green, J.; Dalton, H. *J. Biol. Chem.* **1989,** *264,* 17698.
20. Fox, B. G.; Borneman, J. G.; Wackett, L. P.; Lipscomb, J. D. *Biochemistry* **1990,** *29,* 6419.
21. Kitajima, N.; Ito, M.; Fukui, H.; Moro-oka, Y. *J. Chem. Soc. Chem. Commun.* **1991,** 102.
22. Udenfriend, S.; Clark, C. T.; Axelrod, J.; Brodie, B. B. *J. Biol. Chem.* **1954,** *206,* 731.
23. Nofre, C.; Cier, A.; Lefier, A. *Bull. Soc. Chim. Fr.* **1961,** 530.
24. Ullrich, V. Z. *Naturforsch. Teil B Chem. Sci.* **1969,** *24,* 699.
25. Lindsay Smith, J. L.; Shaw, B. A. J.; Foulkes, D. M.; Jeffrey, A. M.; Jerina, D. M. *J. Chem. Soc. Perkin Trans. 2* **1977,** 1583.
26. Kunai, A.; Hata, S.; Ito, S.; Sasaki, K. *J. Org. Chem.* **1986,** *51,* 3471.
27. Vincent, J. B.; Huffman, J. C.; Christou, G.; Li, Q.; Nanny, M. A.; Hendrickson, D. N.; Fong, R. H.; Fish, R. H. *J. Am. Chem. Soc.* **1988,** *110,* 6898.
28. Leising, R. A.; Norman, R. E.; Que, L., Jr. *Inorg. Chem.* **1990,** *29,* 2552.

29. Fish, R. H.; Konings, M. S.; Oberhausen, K. J.; Fong, R. H.; Yu, W. M.; Christou, G.; Vincent , J. B.; Coggin, D. K.; Buchanan, R. M. *Inorg. Chem.* **1991,** *30,* 3002.
30. Groves, J. T.; Van Der Puy, M. *J. Am. Chem. Soc.* **1976,** 98, 5290.
31. Faraj, M.; Hill, C. L. *J. Chem. Soc. Chem. Commun.* **1987,** 1487.
32. Lau, T.-C.; Che, C.-M.; Lee, W.-O.; Poon, C.-K. *J. Chem. Soc. Chem. Commun.* **1988,** 1406.
33. Mimoun, H.; Saussine, L.; Daire, E.; Postel, M.; Fischer, J.; Weiss, R. *J. Am. Chem. Soc.* **1983,** *105,* 3101.
34. Saussine, L.; Brazi, E.; Robine, A.; Mimoun, H.; Fischer, J.; Weiss, R. *J. Am. Chem. Soc.* **1985,** *107,* 3534.
35. Fontecave, M.; Roy, B.; Lambeaux, C. *J. Chem. Soc. Chem. Commun.* **1991,** 939.
36. Sheu, C.; Richert, S. A.; Cofré, P.; Ross, B., Jr.; Sobkowiak, A.; Sawyer, D. T.; Kanofsky, J. R. *J. Am. Chem. Soc.* **1990,** *112,* 1936.
37. Sheu, C.; Sobkowiak, A.; Jeon, S.; Sawyer, D. T. *J. Am. Chem. Soc.* **1990,** *112,* 879.
38. Cofré, P.; Richert, S. A.; Sobkowiak, A.; Sawyer, D. T. *Inorg. Chem.* **1990,** *29,* 2645.
39. Sheu, C.; Sobkowiak, A.; Zhang, L.; Ozbalik, N.; Barton, D. H. R.; Sawyer, D. T. *J. Am. Chem. Soc.* **1989,** *111,* 8030.
40. Sobkowiak, A.; Tung, H.-C.; Sawyer, D. T. *Prog. Inorg. Chem.* Lippard, S. J., Ed.; Wiley: New York, **1992,** *40,* 291.
41. Schugar, H. J.; Rossman, G. R.; Gray, H. B. *J. Am. Chem. Soc.* **1969,** *91,* 4564.
42. Thich, J. A.; Ou, C. C.; Powers, D.; Vasiliou, B.; Mastropaolo, D.; Potenza, J. A.; Schugar, H. J. *J. Am. Chem. Soc.* **1976,** *98,* 1425.
43. Ou, C. C.; Lalancette, R. A.; Potenza, J. A.; Schugar, H. J. *J. Am. Chem. Soc.* **1978,** *100,* 2053.
44. Stassinopoulos, A.; Caradonna, J. P. *J. Am. Chem. Soc.* **1990,** *112,* 7071.
45. Stassinopoulos, A.; Schulte, G.; Papaefthymiou, G. C.; Caradonna, J. P. *J. Am. Chem. Soc.* **1991,** *113,* 8686.
46. Nishida, Y.; Takeuchi, M.; Shimo, H.; Kida, S. *Inorg. Chim. Acta* **1984,** *96,* 115.
47. Brennan, B. A.; Chen, Q.; Juarez-Garcia, C.; True, A. E.; O'Conner, C. J.; Que, L., Jr. *Inorg. Chem.* **1991,** *30,* 1937.
48. Murch, B. P.; Bradley, F. C.; Que, L., Jr. *J. Am. Chem. Soc.* **1986,** *108,* 5027.
49. Suzuki, M.; Kanatomi, H.; Murase, I. *Chem. Lett.* **1981,** 1745.
50. Suzuki, M.; Ueda, I.; Kanatomi, H.; Murase, I. *Chem. Lett.* **1983,** 185.
51. (a) Tohma, M.; Tomita, T.; Kimura, M. *Tetrahedron Lett.* **1973,** 4359. (b) Yamamoto, T.; Kimura, M. *J. Chem. Soc. Chem. Commun.* **1977,** 948.
52. Leising, R. A.; Brennan, B. A.; Que, L., Jr.; Fox, B. G.; Münck, E. *J. Am. Chem. Soc.* **1991,** *113,* 3988.
53. (a) Schulz, C. E.; Rutter, R.; Sage, J. T.; Debrunner, P. G.; Hager, L. P. *Biochemistry* **1984,** *23,* 4743. (b) Collins, T. J.; Kostka, K. L.; Münck, E.; Uffelman, E. S. *J. Am. Chem. Soc.* **1990,** *112,* 5637.
54. Wieghardt, K.; Pohl, K.; Gebert, W. *Angew. Chem. Int. Ed. Engl.* **1983,** *22,* 727.
55. Hartman, J. R.; Rardin, R. L.; Chaudhuri, P.; Pohl, K.; Weighardt, K.; Nuber, B.; Weiss, J.; Papaefthymiou, G. C.; Frankel, R. B.; Lippard, S. J. *J. Am. Chem. Soc.* **1987,** *109,* 7387.
56. Borovik, A. S.; Que, L., Jr. *J. Am. Chem. Soc.* **1988,** *110,* 2345.

57. Borovik, A. S.; Hendrich, M. P.; Holman, T. R.; Münck, E.; Papaefthymiou, V.; Que, L., Jr. *J. Am. Chem. Soc.* **1990,** *112,* 6031.
58. Tolman, W. B.; Liu, S.; Bentsen, J. G.; Lippard, S. J. *J. Am. Chem. Soc.* **1991,** *113,* 152.
59. Rardin, R. L.; Poganiuch, P.; Bino, A.; Goldberg, D. P.; Tolman, W. B.; Liu, S.; Lippard, S. J. *J. Am. Chem. Soc.* **1992,** *114,* 5240.
60. (a) Menage, S.; Brennan, B. A.; Juarez-Garcia, C.; Münck, E.; Que, L., Jr. *J. Am. Chem. Soc.* **1990,** *112,* 6423. (b) Dong, Y.; Menage, S.; Brennan, B. A.; Elgren, T. E.; Jang, H. G.; Pearce, L. L.; Que, L., Jr. *J. Am. Chem. Soc.* **1993,** *115,* 1851.
61. Sakurai, T.; Kaji, H.; Nakahara, A. *Inorg. Chim. Acta* **1982,** *67,* 1.
62. (a) Nishida, Y.; Takeuchi, M. *Z. Naturforsch.* **1987,** *42b,* 52. (b) Nishida, Y.; Yoshizawa, K.; Takahashi. S.; Watanabe, I. *Z. Naturforsch.* **1992,** *47c,* 209.
63. Menage, S.; Zang, Y.; Hendrich, M. P.; Que, L., Jr. *J. Am. Chem. Soc.* **1992,** *114,* 7786.
64. Yan, S.; Cox, D. D.; Pearce, L. L.; Juarez-Garcia, C.; Que, L., Jr.; Zhang, J. H.; O'Connor, C. J. *Inorg. Chem.* **1989,** *28,* 2507.
65. Norman, R. E.; Yan, S.; Que, L., Jr.; Backes, G.; Ling, J.; Sanders-Loehr, J.; Zhang, J. H.; O'Connor, C. J. *J. Am. Chem. Soc.* **1990,** *112,* 1554.
66. Norman, R. E.; Holz, R. C.; Menage, S.; O'Connor, C. J.; Zhang, J. H.; Que, L., Jr. *Inorg. Chem.* **1990,** *29,* 4629.
67. Kitajima, N.; Fukui, H.; Moro-oka, Y.; Mizutani, Y.; Kitagawa, T. *J. Am. Chem. Soc.* **1990,** *112,* 6402.
68. Latos-Grazynski, L.; Cheng, R.-J.; LaMar, G. N.; Balch, A. L. *J. Am. Chem. Soc.* **1982,** *104,* 5992.
69. Leising, R. A.; Kim, J.; Perez, M. A.; Que, L., Jr. *J. Am. Chem. Soc.* **1993,** *115,* 9524.
70. Hayashi, Y.; Suzuki, M.; Uehara, A.; Mizutani, Y.; Kitagawa, T. *Chem. Lett.* **1992,** 91.

RECEIVED for review July 19, 1993. ACCEPTED revised manuscript September 16, 1994.

5

Electronic Structures of Active Sites in Copper Proteins and Their Contributions to Reactivity

Edward I. Solomon, Michael D. Lowery, David E. Root, and Brooke L. Hemming

Department of Chemistry, Stanford University, Stanford, CA 94305

Many active sites in copper proteins exhibit unique spectral features in comparison to small inorganic complexes. These spectral features are becoming well-understood and reflect unusual electronic structures that make key contributions to the reactivity of these sites in biology. In blue copper proteins, the unique spectral features reflect a ground-state, redox-active wave function that has high anisotropic covalency involving a cysteine ligand that activates this residue for rapid, directional, long-range electron transfer. For hemocyanin and tyrosinase, the characteristic spectral features reflect a peroxide–binuclear cupric bond that has very strong σ-donor and π-acceptor interactions that stabilize the oxy site with respect to loss of peroxide. In tyrosinase, this bonding mode activates peroxide for hydroxylation of phenolic substrates by making the peroxide less negative but with an unusually weak O–O bond. Finally, for the multicopper oxidases, magnetic circular dichroism and X-ray absorption studies first showed the presence of a trinuclear copper cluster site, which is the minimal structural unit required for the multielectron reduction of dioxygen to water. A peroxide level intermediate in this reduction has been obtained and is found to have strikingly different spectral features than those associated with bound peroxide in oxyhemocyanin and oxytyrosinase. This demonstrates a fundamentally different electronic and geometric structure for peroxide binding in the multicopper oxidases that promotes the further reduction of peroxide to water at the trinuclear copper cluster site.

MANY CLASSES OF ACTIVE SITES in copper proteins exhibit unique spectral features when compared to simple, high-symmetry transition-

0065–2393/95/0246–0121/$12.32/0

metal complexes. These active sites derive from the unusual geometric and electronic structures that can be imposed on the metal ion in a protein environment. It has been a general goal of our research to understand these electronic structures and to evaluate their contributions to the reactivities of these active sites in catalysis (*1–3*).

Our progress in four areas will be summarized (*4, 5*). First, if one is to understand the origin of these unique spectral features, one must understand the electronic structure of "normal" high-symmetry, transition-metal complexes. Square-planar cupric chloride has served as an electronic structural model complex and is now one of the best understood molecules in inorganic chemistry (*6*). Its spectral features and the electronic structure these reflect will be briefly described in the following section (*4*). Having provided this description of a "normal" copper site, the unique spectral features of the blue copper active site will then be addressed. An understanding of these features provides insight into ground and excited state contributions to the rapid rate of long-range electron transfer observed in this family of proteins. Next we focus on the coupled binuclear copper proteins, hemocyanin, and tyrosinase. These proteins have similar active sites that generate the same oxy intermediate involving peroxide bound to two copper(II) ions. The hemocyanins reversibly bind dioxygen and function as oxygen carriers in arthropods and molluscs, whereas the tyrosinases have highly accessible active sites that bind phenolic substrates and oxygenate them to *ortho*-diphenols. Their oxy sites exhibit unique excited state spectral features that reflect novel peroxide–copper bonding interactions that make a significant electronic contribution to the binding and activation of dioxygen by these sites. In the final section, spectroscopic studies of the multicopper oxidases, which include laccase, ascorbate oxidase, and ceruloplasmin, are summarized. These enzymes catalyze the four-electron reduction of dioxygen to water. Spectral studies have demonstrated that the multicopper oxidases contain a fundamentally different coupled binuclear copper site (called type 3) when compared with hemocyanin and tyrosinase. The type 3 site is part of a trinuclear copper cluster that plays the key role in the multielectron reduction of dioxygen by this important class of enzymes.

Normal Copper Complexes

Placing a cupric ion with its nine d electrons in an octahedral ligand field produces a ^2E_g ground state (Figure 1). This ground state is unstable to a Jahn–Teller distortion that lowers the symmetry and energy of the complex. The Jahn–Teller distortion normally observed is a tetragonal elongation along the z-axis and contraction in the equatorial x,y plane that ultimately results in a square-planar ligand environment as in D_{4h}-

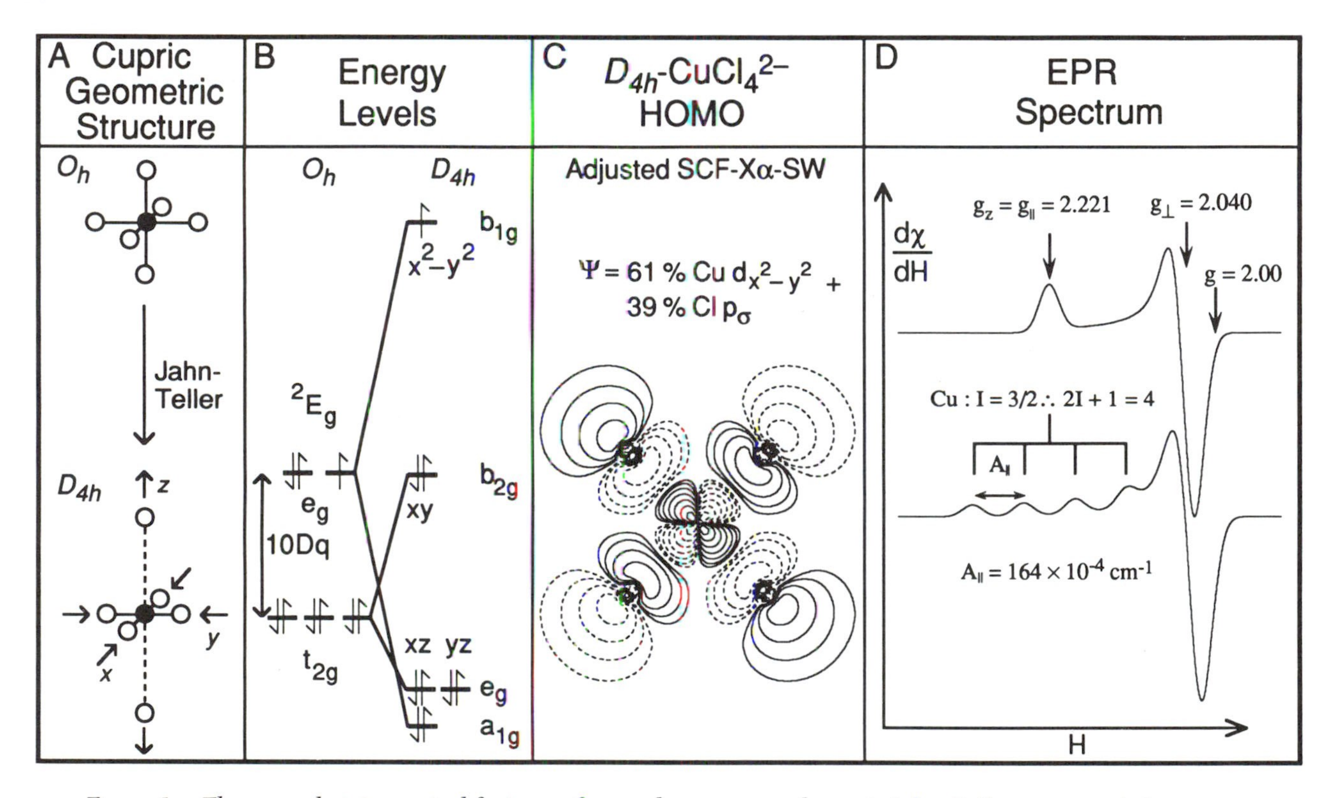

Figure 1. The ground-state spectral features of normal copper complexes. A: Jahn–Teller tetragonal elongation of an octahedral CuL_6 complex to the square planar limit. B: Energy level correlation diagram for the Jahn–Teller distortion depicted in A. C: SCF-Xα-SW wave function contour and charge decomposition for the HOMO of D_{4h}-$CuCl_4^{2-}$ (18, 19). *D: X-band EPR spectrum for tetragonal Cu(II) with D_{4h}-$CuCl_4^{2-}$ parameters.*

$CuCl_4^{2-}$ (Figure 1A). A key feature of ligand field theory (*7–11*) is that the *d* orbital splitting is very sensitive to the environment of the ligands around the metal center. The *d* orbital splitting experimentally determined using optical spectroscopy (*12*) for D_{4h}-$CuCl_4^{2-}$ is given in Figure 1B. The half-occupied $d_{x^2-y^2}$ orbital is at highest energy as it has the largest repulsive interaction with the ligands in the equatorial plane. A more complete description of this half-occupied ground state is provided by molecular orbital (MO) theory. In particular, self-consistent field-Xα-scattered wave (SCF-Xα-SW) calculations (*13–17*), adjusted to ground-state parameters (*18–21*) as will be described, are in good agreement with spectral data over many orders of magnitude in energy. These calculations (*18, 19*) generate a description of the ground state of D_{4h}-$CuCl_4^{2-}$ that has 61% Cu $d_{x^2-y^2}$ character with the remaining part of the wave function being delocalized equivalently into the four $p\sigma$ orbitals of the chloride ligands that are involved in antibonding interactions with the metal ion (Figure 1C). The unpaired electron in this wave function produces the electron paramagnetic resonance (EPR) spectrum shown in Figure 1D, in which $g_{\|}$ (corresponding to the magnetic field oriented along the *z*-axis of the complex) $> g_{\perp} > 2.00$, that is characteristic of this $d_{x^2-y^2}$ ground state. Additionally, copper has a nuclear spin (I_N = $^3/_2$) that couples to the electron spin to produce a four-line hyperfine splitting (*A*) of the EPR spectrum. Tetragonal cupric complexes generally have a large hyperfine splitting in the $g_{\|}$ region ($A_{\|} > 130 \times 10^{-4}$ cm^{-1}); that of D_{4h}-$CuCl_4^{2-}$ is 164×10^{-4} cm^{-1} (*22*).

With respect to excited states, optical excitation of electrons from the filled Cu *d* orbitals to the half-filled $d_{x^2-y^2}$ orbital (Figure 2A) produces Laporté-forbidden ligand-field transitions (*23*). These transitions are weak in the absorption spectrum with molar extinction coefficients (ϵ) of 30–50 M^{-1} cm^{-1} in the 12,000–16,000 cm^{-1} region (*12*) (Figure 2B, right). Observed at higher energy in the absorption spectrum are the Laporté-allowed ligand-to-metal charge-transfer transitions that are at least two orders of magnitude more intense than the ligand-field transitions (*24*) (Figure 2B, left). The energies and intensities of these charge-transfer (CT) transitions allow one to probe the specific bonding interactions of the ligand with the metal center (*24*). Chloride has three valence 3*p* orbitals that split into two sets on binding to copper (Figure 2C). The $p\sigma$ orbital is oriented along the Cl–Cu bond and is stabilized to higher binding energy due to strong overlap with the Cu^{2+} ion. The two chloride $p\pi$ orbitals are perpendicular to the Cl–Cu bond and hence are more weakly interacting with the metal and at lower binding energy.

The intensity associated with charge-transfer excitation of an electron from these filled ligand orbitals into the half-occupied Cu $d_{x^2-y^2}$ orbital also reflects metal–ligand bonding. Charge-transfer intensity is proportional to $(RS)^2$, where *S* is the overlap of the donor and acceptor

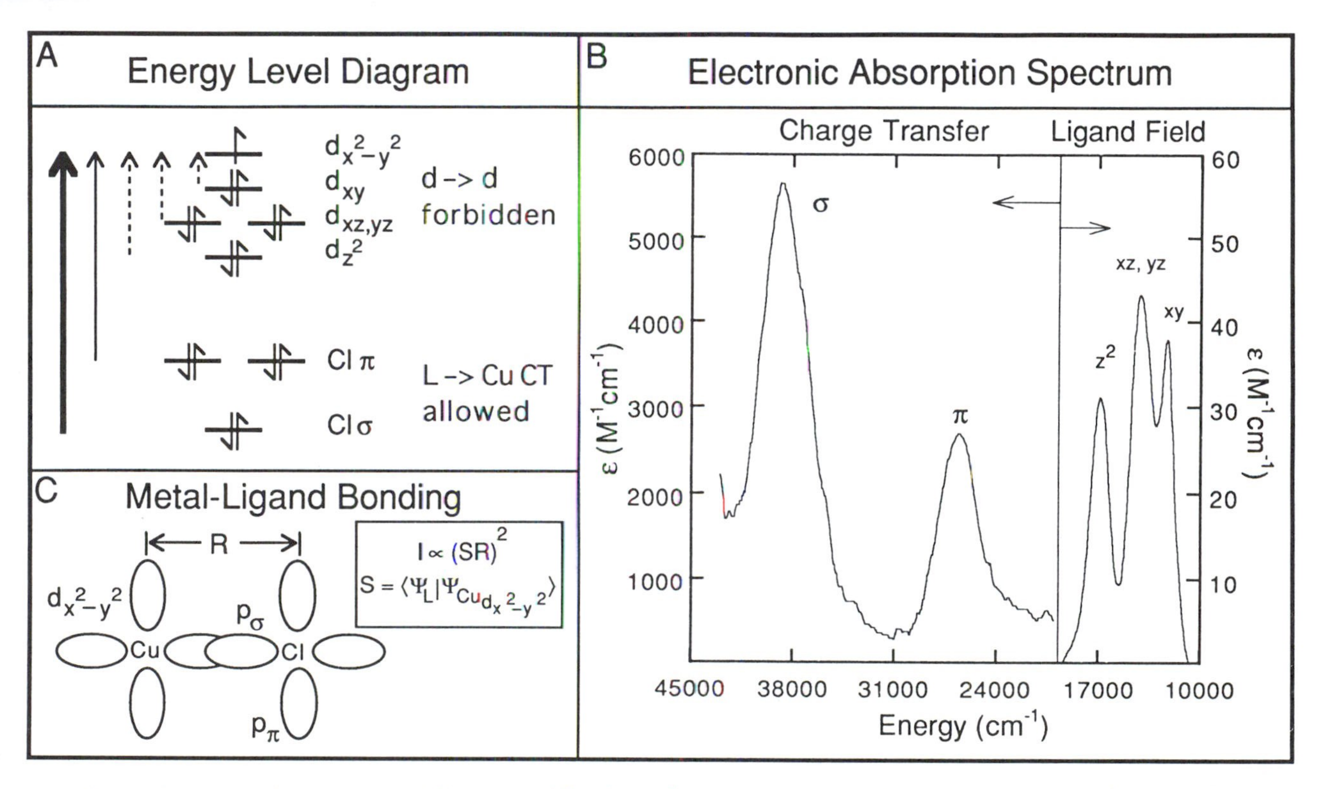

Figure 2. Excited-state spectral features of D_{4h}-$CuCl_4^{2-}$. *A: Energy level diagram showing the ligand-field* (d → d) *and charge-transfer (CT) optical transitions. The intensity of the transitions is approximated by the thickness of the arrow with the very weak ligand-field transitions represented as a dotted arrow. B: Electronic absorption spectrum for* D_{4h}-$CuCl_4^{2-}$ (12). *C: Schematic of the* σ *and* π *bonding modes between the Cu* $3d_{x^2-y^2}$ *and Cl 3p orbitals.*

orbitals involved in the charge-transfer transition and R is the metal–ligand bond length (25). Thus, the Cl $p\sigma \rightarrow$ Cu $d_{x^2-y^2}$ charge-transfer transition is at high energy and intense from large overlap, while the Cl $p\pi \rightarrow d_{x^2-y^2}$ charge transfer transition is at lower energy and weaker (Figure 2C). (Note that the ligand orbitals are actually linear combinations of the orbitals from the four chloride ligands. One combination of the Cl $p\sigma$ orbitals has e_u symmetry. The CT transition from this orbital is electric dipole allowed and responsible for the intense band at 36,000 cm^{-1}. The Cl $p\pi$ set also contains a linear combination having e_u symmetry. Configurational interaction with $e_u(p\sigma)$ contributes to the intensity of the $p\pi \rightarrow d_{x^2-y^2}$ band at 26,500 cm^{-1}). The key points to be emphasized here are that the charge-transfer transitions sensitively probe the ligand-metal bond and that for "normal" complexes one should observe a lower-energy weak π and higher-energy intense σ charge-transfer transition as is observed experimentally for $CuCl_4^{2-}$ in Figure 2B (left).

Blue Copper Proteins

As predicted by spectroscopy (26), the blue copper site has a structure very different from the normal tetragonal geometry of cupric complexes. The copper site in plastocyanin (27) has a distorted tetrahedral structure with a thiolate sulfur of cysteine (Cys) 84 bound with a short Cu–S bond length of 2.13 Å, a thioether sulfur of methionine (Met) 92 bound with a long Cu–S bond length of 2.90 Å, and two fairly normal histidine (His) N–Cu ligands (Figure 3A). This site has characteristic spectral features (*1–3*) that include an intense absorption band ($\epsilon \sim 3{,}000$–$5{,}000$ M^{-1} cm^{-1}) in the 600 nm ligand-field region (Figure 3B) and a small parallel hyperfine splitting ($A_{\parallel} \leq 70 \times 10^{-4}$ cm^{-1}) (Figure 3C). These unusual spectral features are now well-understood and help to define the ground-state wave function of the blue copper site. This is extremely important in that this is the half-occupied orbital that takes up and transfers the electron in the redox functioning of this center. A detailed experimental and theoretical description of the ground state provides fundamental insight into the active site contribution to the long-range electron-transfer reactivity exhibited by the blue copper proteins (*2, 3, 28*).

The EPR spectrum of the blue copper protein plastocyanin (Figure 3C) has $g_{\parallel} > g_{\perp} > 2.00$, and thus the copper site must have a $d_{x^2-y^2}$ ground state. First, we are interested in determining the orientation of the $d_{x^2-y^2}$ orbital relative to the distorted tetrahedral geometry observed in the protein crystal structure. Single crystal EPR spectroscopy allowed us to obtain this orientation and located the unique (i.e., z) direction in this distorted site (29). Plastocyanin crystallizes in an orthorhombic space group with four symmetry related molecules in the unit cell. The orientation of the plastocyanin copper sites in the unit cell are shown in

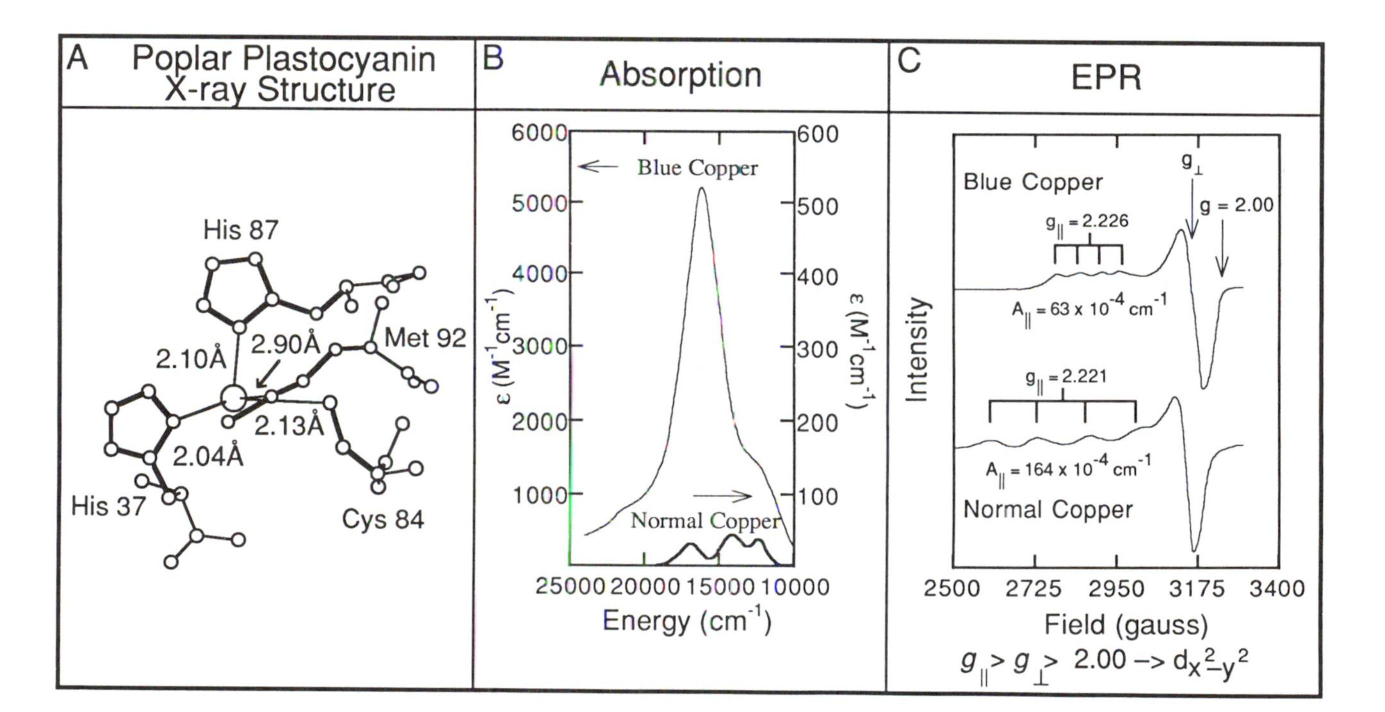

Figure 3. Blue copper proteins. A: X-ray structure of poplar plastocyanin (27). B: Absorption spectrum of plastocyanin and "normal" D_{4h}-$CuCl_4^{2-}$ *(ε scale expanded by 10). C: X-band EPR spectrum of plastocyanin and* D_{4h}-$CuCl_4^{2-}$.

Figure 4A (27). Figure 4B presents the EPR spectra obtained if one rotates the crystal around the *a* axis with the magnetic field in the *b/c* plane. The key feature to note in the figure is that one observes an $\sim g_{\|}$ EPR signal (with four parallel hyperfine components) with the field along the *c* axis and an $\sim g_{\perp}$ spectrum as the field is rotated perpendicular to this direction. Thus, $g_{\|}$ is nearly colinear with the *c* axis. Referencing to the four blue copper sites in the unit cell, each has its Met S–Cu bond approximately along the *c* axis. Therefore, $g_{\|}$ is approximately along the long Met S–Cu bond. A more quantitative fit of the EPR data shows that $g_{\|}$, which defines the *z*-axis of the site, is ~5° off the Met S–Cu bond and places the $d_{x^2-y^2}$ orbital perpendicular to this direction and within 15° of the plane defined by the thiolate S and two imidazole N ligands.

The next feature of the ground-state wave function to be discussed is the origin of the small parallel hyperfine splitting ($A_{\|} < 70 \times 10^{-4}$ cm^{-1}). Distorted tetrahedral cupric sites, for example D_{2d}-$CuCl_4^{2-}$, often exhibit small $A_{\|}$ values similar to the blue copper proteins and the mechanism for reducing the parallel hyperfine value has been thought to have a common origin. In D_{2d}-$CuCl_4^{2-}$, the small $A_{\|}$ value has been attributed to the effect of Cu 4p mixing into the $d_{x^2-y^2}$ orbital, which is allowed in lower-symmetry metal sites (30). In D_{2d} symmetry, the $4p_z$ orbital is allowed by group theory to mix into the $d_{x^2-y^2}$ orbital. The spin dipolar interaction of the $4p_z$ orbital with the copper nuclear spin opposes that of the electron spin in the $d_{x^2-y^2}$ orbital and reduces the $A_{\|}$ value (4*p* mixing is forbidden in D_{4h}-symmetry). Twelve percent $4p_z$ mixing is required to lower the $A_{\|}$ value to that value observed for D_{2d}-$CuCl_4^{2-}$ and plastocyanin (31, 32).

The nature of the 4*p* mixing into the $d_{x^2-y^2}$ orbital of plastocyanin is determined by the effective symmetry of the ligand field of the active site. A combination of low-temperature magnetic circular dichroism (LT-MCD) spectroscopy (33) and ligand field theory (29) has been used to obtain the splitting of the *d* orbitals of the blue copper site. Five nondegenerate ligand-field levels are observed indicating a rhombically distorted site (Figure 5B, left); however, the splitting is close to an axial limit. The *d* orbital splitting for the axial limit (Figure 5B, right) corresponds to a C_{3v} axially elongated tetrahedral structure in which the *z*-axis corresponds to the long thioether S–Cu bond (Figure 5C). However, in C_{3v} symmetry, the $d_{x^2-y^2}$ orbital is only allowed to mix with the $4p_x$, $4p_y$ levels and, in this case, the spin dipolar interactions would be complementary and increase the $A_{\|}$ value.

Thus we needed to determine experimentally the nature of the 4*p* mixing into the $d_{x^2-y^2}$ level. This determination was accomplished by going up about 10 orders of magnitude in photon energy and performing X-ray absorption spectroscopy (XAS) at the Cu-K edge (34). The 8979-eV pre-edge peak, corresponding to the Cu 1*s* → $3d_{x^2-y^2}$ transition, can

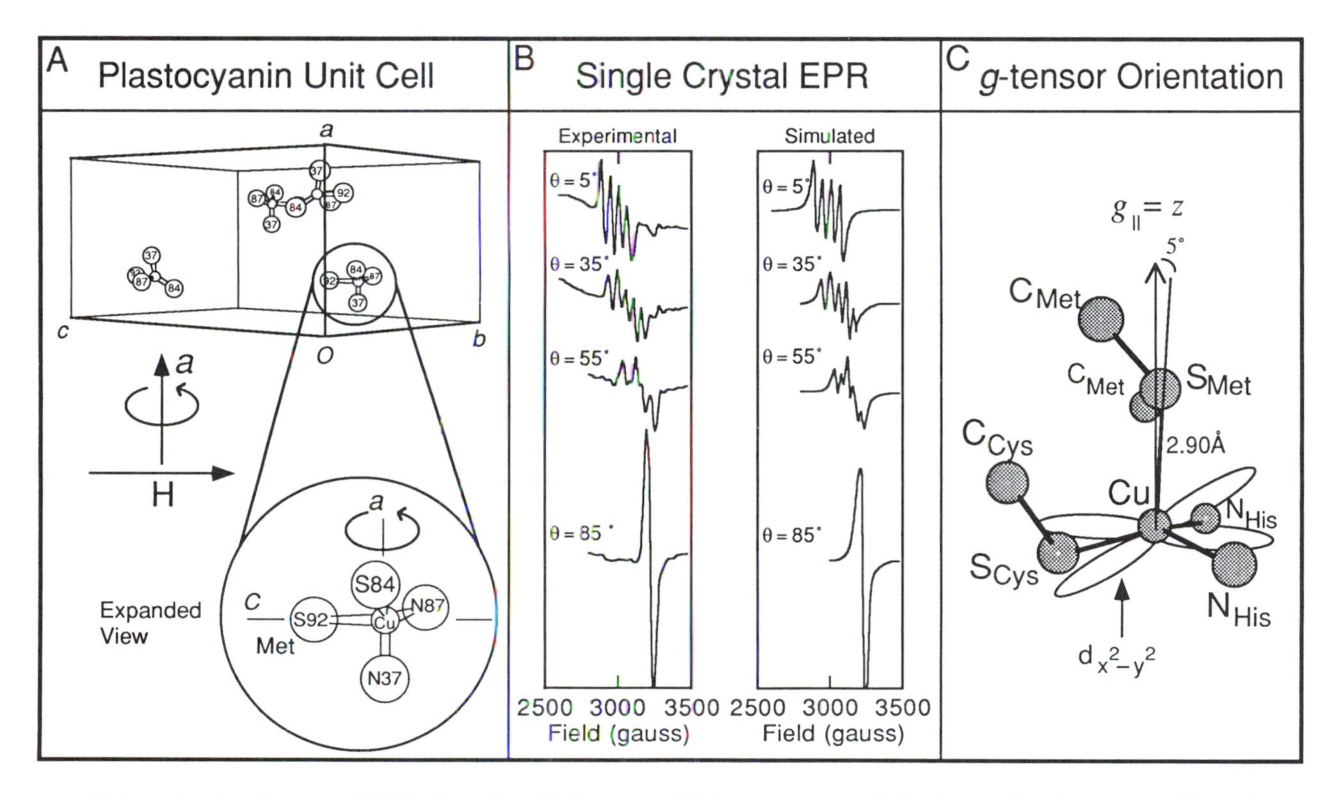

Figure 4. Single-crystal EPR of poplar plastocyanin (29); orientation of the $d_{x^2-y^2}$ *orbital. A: Unit cell and molecular orientation with respect to the applied magnetic field. B: EPR spectra and simulations for the crystal orientations shown. C: Orientation of the* $g_{\parallel}$ *direction and the* $d_{x^2-y^2}$ *orbital superimposed on the blue copper site.*

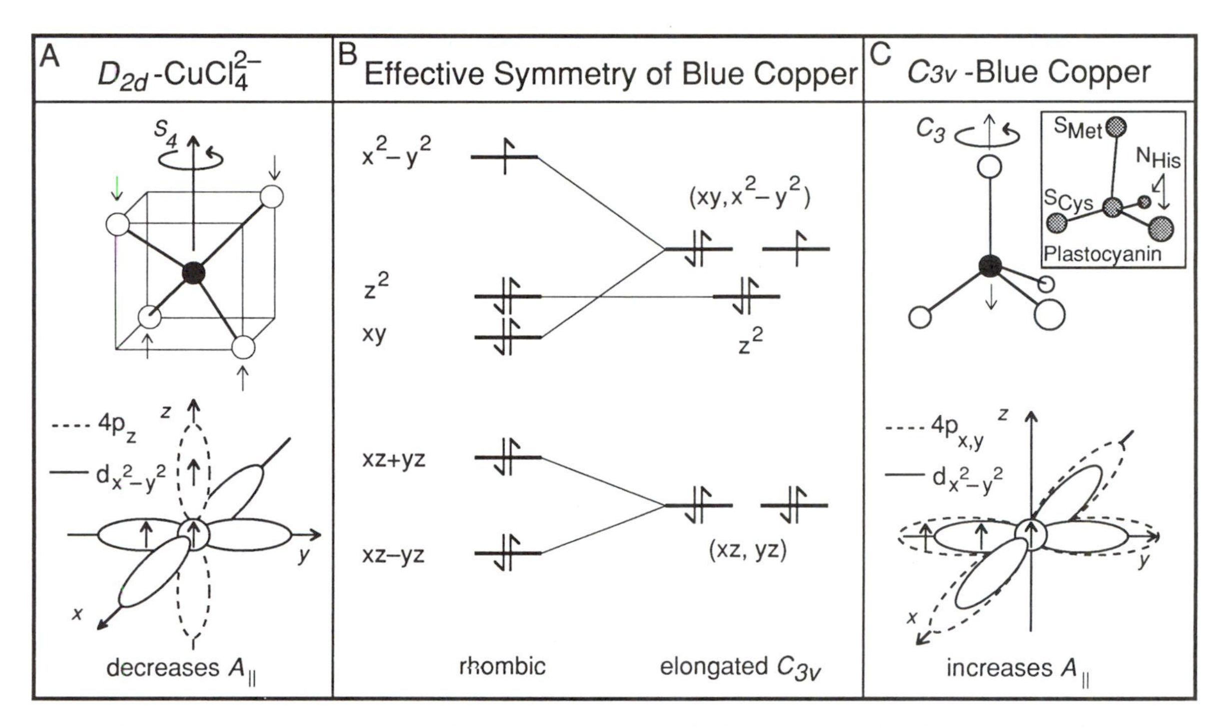

Figure 5. Origin of small $A_{||}$ values (34). A: 4p$_z$ mixing with the $d_{x^2-y^2}$ orbital in D_{2d} symmetry lowers $A_{||}$. B: Energy levels for blue copper site (left) and its axial C_{3v} limit (right). C: Effect of 4p$_{x,y}$ mixing with the $d_{x^2-y^2}$ orbital on $A_{||}$ for a C_{3v} distortion.

have no electric dipole intensity unless the copper site is distorted such that some Cu $4p$ orbital character mixes into the half-occupied $d_{x^2-y^2}$ level (Figure 6A). The intensity of the 8979-eV feature is much higher in the plastocyanin and D_{2d}-$CuCl_4^{2-}$ edges relative to the square planar D_{4h}-$CuCl_4^{2-}$ complex (*34*). This higher intensity indicates the presence of $4p$ mixing in the $d_{x^2-y^2}$ orbitals of plastocyanin and D_{4h}-$CuCl_4^{2-}$ (Figure 6B). It was then important to determine whether the $4p$ mixing in the blue copper site involves the p_z or p_x,p_y orbitals. This may be accomplished through analysis of the polarized single crystal X-ray absorption spectra of plastocyanin. The single crystal EPR data (*29*) determined the z-axis as being approximately colinear with the long thioether S–Cu bond. Polarized edge data were obtained (*35*) with the **E** vector of light parallel and perpendicular to this direction. No 8979-eV intensity was observed for $\mathbf{E}_{\parallel}\ z$, requiring that there is no $4p_z$ mixing, whereas all the 8979-eV intensity was observed in the $\mathbf{E}_{\parallel}\ x,y$ spectrum. Thus, only Cu $4p_z,p_y$ mixes into the $d_{x^2-y^2}$ orbital of plastocyanin and the small $A_{\parallel}$ value of the blue copper site cannot be due to the generally accepted mechanism of $4p_z$ mixing.

Having eliminated $4p_z$ mixing, we can focus on the alternative explanation for the small $A_{\parallel}$ value of the blue copper site that is covalent delocalization of the electron spin onto the ligands (*33, 36*). Covalency reduces the hyperfine interaction of the unpaired electron with the copper nuclear spin. We have approached the inclusion of covalency in the description of the ground state of the blue copper site through a quantitative consideration of its g values. Multifrequency EPR spectroscopy (*36*) gives the experimental g values for plastocyanin listed on the left in Table I. If the ground state only involves an unpaired electron in a $d_{x^2-y^2}$ orbital, there would only be a spin angular momentum contribution to the g values and thus the g values would be 2.00 and isotropic. Ligand field theory (LFT) (*7–11*) allows for the inclusion of some orbital angular momentum into the $d_{x^2-y^2}$ ground state through its spin-orbit mixing with ligand-field excited states. This orbital contribution leads to the devia-

Table I. Blue Copper Covalency → Quantitative Analysis of g Values

				SCF-Xα-SW d *Levels, CT Levels* $\lambda_{Cu}\,L \cdot S + \lambda_L\,L \cdot S$	
g	*Plastocyanin Experimental*	(x^2-y^2) *Spin Only*	*LFT* d *Orbitals* $+ \lambda\,L \cdot S$	*Normal Radii*	*Adjusted Radii*
g_x	2.047	2.00	2.125	2.046	2.059
g_y	2.059	2.00	2.196	2.067	2.076
g_z	2.226	2.00	2.479	2.159	2.226

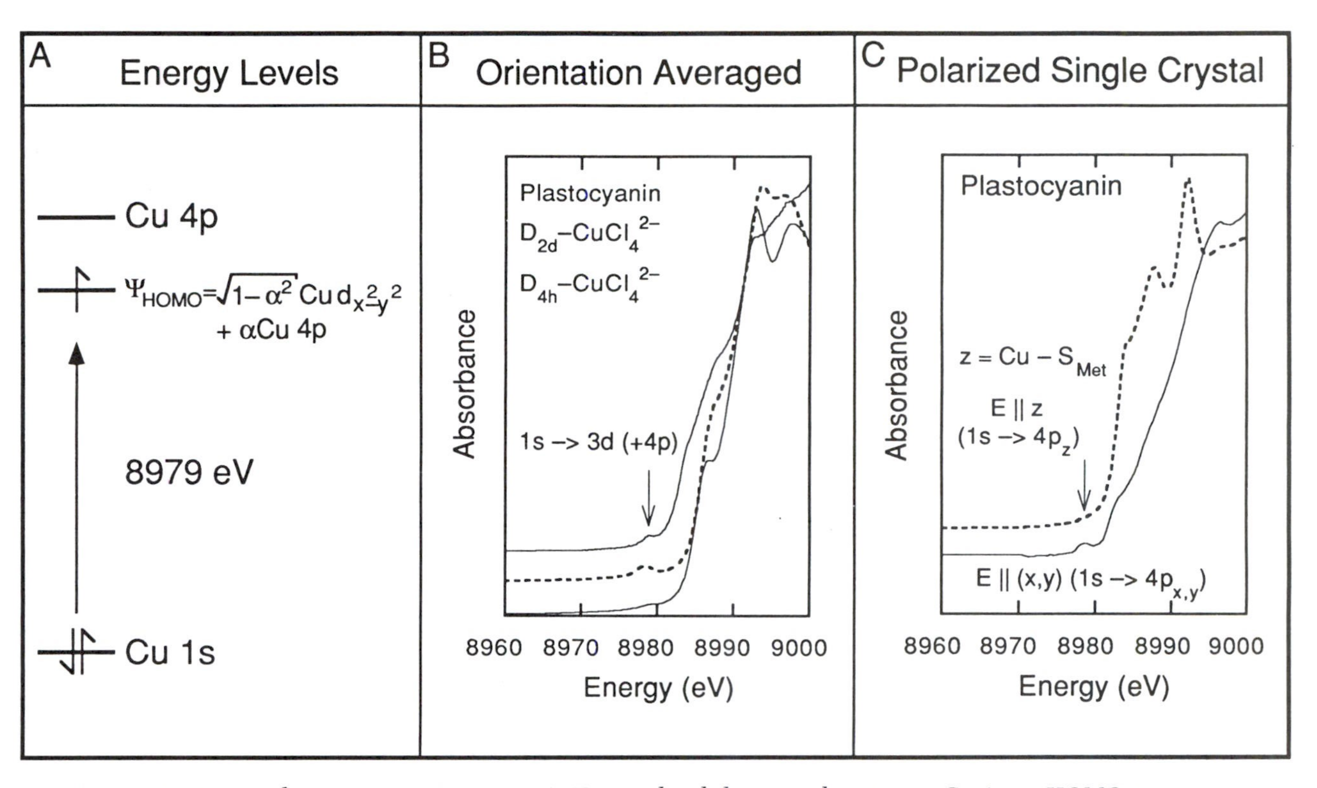

Figure 6. X-ray absorption spectroscopy. A: Energy level diagram depicting a Cu 1s → HOMO transition at ~8979 eV. B: Orientation averaged XAS spectra. C: Polarized single-crystal XAS spectra for poplar plastocyanin. (Data are from reference 35.)

tions of the g values from 2.00. A complete ligand-field calculation for plastocyanin gives the g values listed in the third column of Table I. Here, $g_{\parallel} > g_{\perp} > 2.00$, which is consistent with a spin-orbit mixed $d_{x^2-y^2}$ ground state. However, the calculated values are larger than the experimental values. This is due to the fact that the ligand field calculations use pure d orbitals that have too much orbital angular momentum. Covalent delocalization of the unpaired electron onto the ligands reduces the orbital contribution to the g values.

SCF-Xα-SW calculations (*33, 36*) were pursued to describe the bonding in the blue copper site. The wave functions obtained from these calculations were used to calculate the ground state g values and evaluate the covalent description generated by these calculations relative to experiment. The g value calculation included all the antibonding (d) and bonding (charge transfer) levels and included spin-orbit mixing from both the metal and the ligands. A Zeeman operator was applied to the spin-orbit corrected ground state making no assumption concerning the orientation of the principal axes. A $\mathbf{g}^2$ tensor was generated and diagonalized to obtain the principal component g values that can be compared to experiment (fourth column). It is observed that although the g values are reduced from those obtained from the ligand-field calculation due to the inclusion of covalency, they are closer to 2.00 than is obtained experimentally. Thus, the SCF-Xα-SW calculations are producing too covalent a description of the active site. There is one set of adjustable parameters in this calculation, which is the sphere sizes used in the scattered wave solutions. Those employed in this initial calculation are the standard spheres normally used that are defined by the Norman criteria (*37*). We systematically varied these spheres (increasing the metal sphere increases its electron density, lowers its effective nuclear charge, and reduces its interaction with the ligands), and iteratively repeated this g value protocol until the calculated values were in good agreement with experiment (*33, 36*). This approach provided the experimentally adjusted description of the ground state wave function of the blue copper site that is given in Figure 7A.

These Xα calculations provide a description of the ground state of the blue copper site that is highly covalent. The covalency is strongly anisotropic with delocalization predominantly into the S$p\pi$ orbital of the thiolate (Figure 7A). We have been able to experimentally test the key features of this ground state using a variety of spectroscopic methods. First, the high covalency can be probed by copper L-edge spectroscopy (*38*). The electric dipole intensity of the Cu $2p \rightarrow \Psi_{HOMO}$ (HOMO is the highest occupied molecular orbital) transition at 930 eV reflects the Cu $2p \rightarrow$ Cu $3d$ transition probability and probes the amount of Cu $d_{x^2-y^2}$ character in the ground-state wave function. From Figure 7C, it is ob-

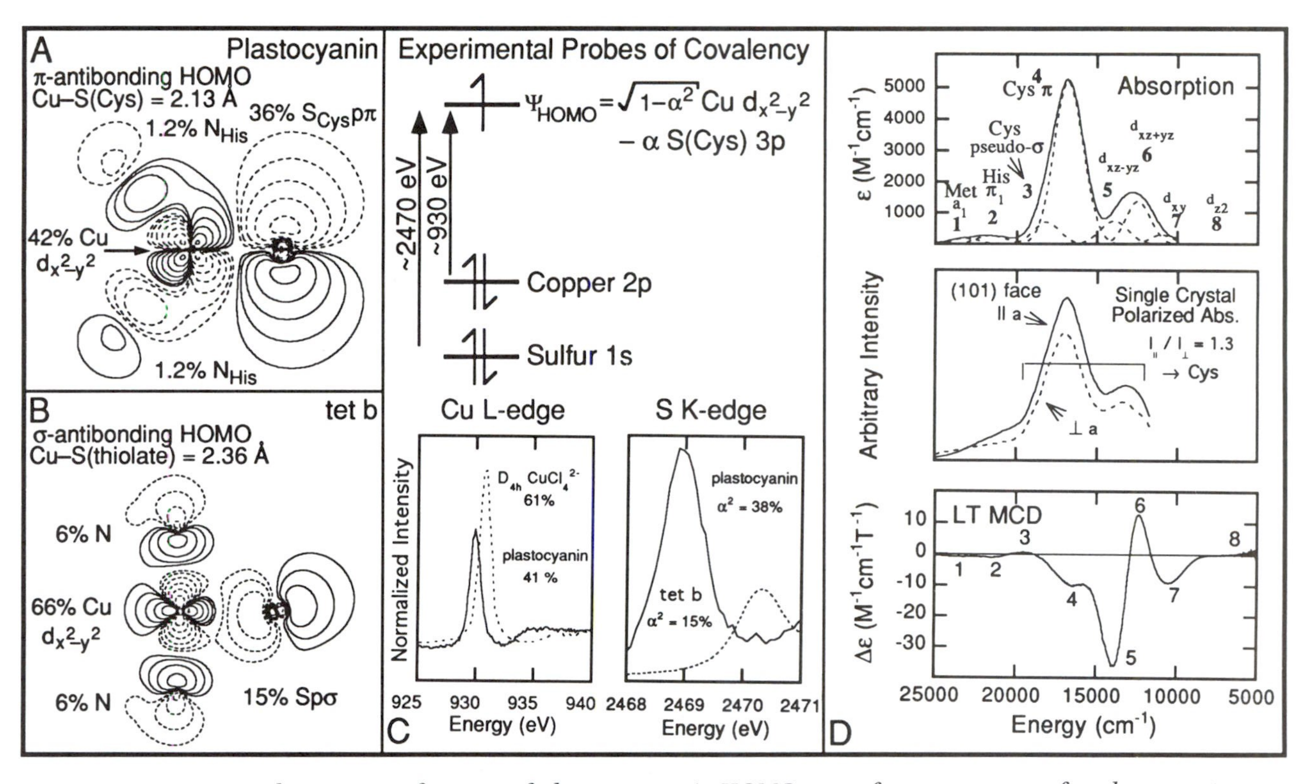

Figure 7. Ground-state wave function of plastocyanin. A: HOMO wave function contour for plastocyanin (28). B: HOMO wave function contour for the thiolate copper complex tet b *(34). C: Copper L-edge (38) and sulfur K-edge (34) spectra as probes of metal–ligand covalency. D: Absorption, single-crystal polarized absorption, and low-temperature MCD spectra of plastocyanin. The absorption spectrum has been Gaussian resolved into its component bands as in reference 33.*

served that the 930-eV peak in plastocyanin has 67% of the intensity of D_{4h}-$CuCl_4^{2-}$, which is known to have 61% Cu $d_{x^2-y^2}$ character (Figure 1C). Thus, the ground state of the blue copper site is estimated from experiment to have 41% Cu $d_{x^2-y^2}$ character. This is in good agreement with the adjusted SCF-Xα-SW calculations (42%). Second, the sulfur contribution to the HOMO can be studied using sulfur K-edge spectroscopy (*34*) in which the electric dipole intensity now reflects the S1*s* → S3*p* character in the HOMO. From Figure 7C, plastocyanin exhibits an intense sulfur pre-edge feature at 2469 eV. It has 2.6 times the intensity of the tet *b* model complex of Schugar (*39*), which contains a normal 2.36 Å copper-thiolate sulfur bond and has 15% sulfur *p* character in the ground state (Figure 7B). Thus, the blue copper site is also experimentally estimated to have 38% sulfur *p* character from the cysteine ligand, again in good agreement with the Xα calculations (36%).

The final feature of the ground-state wave function is elucidated through the assignment of the characteristic excited-state absorption spectral features of plastocyanin (Figure 7D). Although there are in fact eight bands required to fit a combination of absorption (Abs), circular dichroism (CD), and magnetic circular dichroism (MCD) spectra of the blue copper site (*33*), at low resolution the absorption spectrum was regarded originally as having a low-energy weak and higher-energy intense (i.e., the 600 nm, 16,000 cm^{-1}) band pattern (*1, 26*). Polarized single crystal spectral studies over this region (*29*) showed the same polarization ratio for both bands, which required that both bands be associated with the Cys S–Cu(II) bond. Thus, in parallel to the Cl^- → Cu(II) charge-transfer assignment presented earlier, these were assigned as low-energy weak π and higher-energy intense σ charge transfer transitions involving the thiolate sulfur. However, MCD spectroscopy showed that this assignment was not correct. All four of the low-energy bands (5–8 in Figure 7D) that comprise this region are weak in the absorption spectrum but quite intense in the low-temperature MCD spectrum (*33*). Because MCD *C*-term intensity for Cu(II) requires spin-orbit coupling, and hence *d* orbital character, this leads to the assignment of bands 5–8 as *d* → *d* transitions. Thus, the 600-nm band (*4*), that is intense in the absorption spectrum and weak in the low-temperature MCD spectrum, is the lowest-energy charge-transfer transition from the thiolate and must be the Cys $p\pi$ → Cu $d_{x^2-y^2}$ charge-transfer transition. The Cys $p\sigma$ → Cu $d_{x^2-y^2}$ is a weak band at higher energy. The key point is that for the blue copper site one has a low-energy intense π and higher energy weak σ charge–transfer transition to the Cu $d_{x^2-y^2}$ orbital. Inasmuch as charge-transfer intensity reflects orbital overlap, this overlap requires that the $d_{x^2-y^2}$ orbital have its lobes bisected by the Cys S–Cu bond (Figure 7A) and thus be involved in a strong π antibonding interaction with the thiolate as also obtained from the Xα calculations. The strong π interaction rotates the $d_{x^2-y^2}$ orbital by 45° relative to its usual orientation along the ligand–copper bond as, for example, in the tet *b* model complex (Figure 7B). This rotation of the $d_{x^2-y^2}$ orbital derives from the quite short blue copper Cys S–Cu bond length of 2.13 Å.

Thus the SCF-Xα-SW calculations are producing an accurate description of the ground state of the blue copper site and one can now correlate this with crystal structure information to obtain significant insight into function. In particular, the X-ray structure of ascorbate oxidase (*40*) shows that the cysteine ligand of a blue copper site in this multicopper oxidase is flanked on either side in the sequence by histidines that are ligands to two of the coppers in a trinuclear copper cluster site (discussed in the later section, *Multicopper Oxidases*) (*41*). This blue copper site transfers an electron rapidly in the reduction of O_2 at the trinuclear copper center. As can be seen from the Xα calculated wave function contour that we have superimposed on the crystal structure of the blue center in ascorbate oxidase (Figure 8), the ground-state wave function provides a highly anisotropic covalent pathway involving the cysteine sulfur. The covalency activates this residue for directional electron transfer. In addition, the low-energy, intense Cys $\pi \rightarrow$ Cu $d_{x^2-y^2}$ charge-transfer transition in the blue copper absorption spectrum provides an efficient hole superexchange pathway for rapid electron transfer between the blue and trinuclear copper cluster sites (*2*). Clearly, as shown in Figure 8, the unique electronic structure of the blue copper center reflects a ground-state wave function that plays a critical role in its functioning of rapid long-range electron transfer to a specific location in or on the protein.

Coupled Binuclear Copper Proteins

The binuclear copper proteins hemocyanin and tyrosinase reversibly bind dioxygen and in the case of tyrosinase activate it for hydroxylation of phenol to *ortho*-diphenol and further oxidation to *ortho*-quinone (Figure 9) (*42*). Both proteins have essentially the same oxy active sites (*42, 43*) that involve two Cu(II) ions (shown by X-ray absorption edge data (*44, 45*)) and bound peroxide (shown by the unusually low O–O stretching frequency of 750 cm^{-1} observed in the resonance Raman spectrum (*46–48*)). As will be summarized in this section, the unique vibrational and ground and excited state electronic spectral features of this oxy site are now understood. These generate a detailed description of the peroxide–copper bond that provides fundamental insight into the reversible binding and activation of dioxygen by this site.

The ground state of oxyhemocyanin exhibits no EPR signal. This results from a strong antiferromagnetic coupling of the two Cu(II) ions

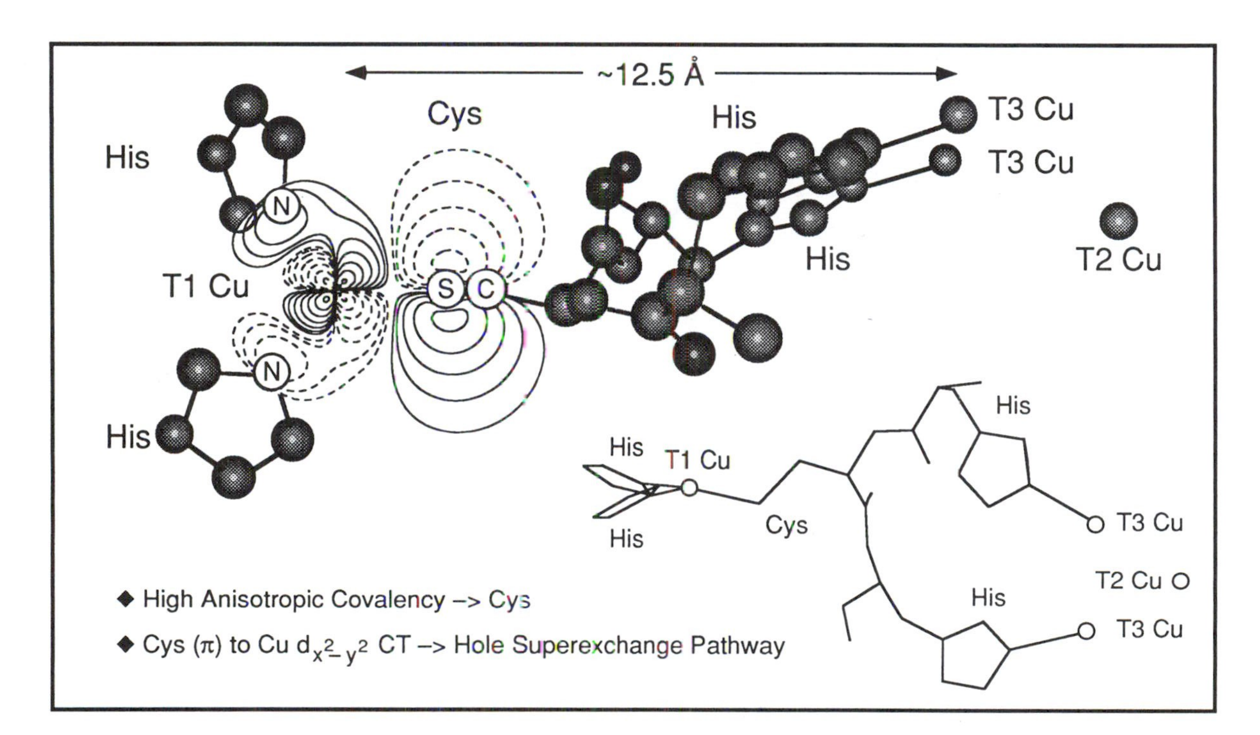

Figure 8. Proposed electron transfer pathway in blue copper proteins. The plastocyanin wave function contours have been superimposed on the blue copper (type 1) site in ascorbate oxidase (40). The contour shows the substantial electron delocalization onto the cysteine Spπ orbital that activates electron transfer to the trinuclear copper cluster at 12.5 Å from the blue copper site. This low-energy, intense Cys Sp → Cu charge-transfer transition provides an effective hole superexchange mechanism for rapid long-range electron transfer between these sites (2, 3, 28).

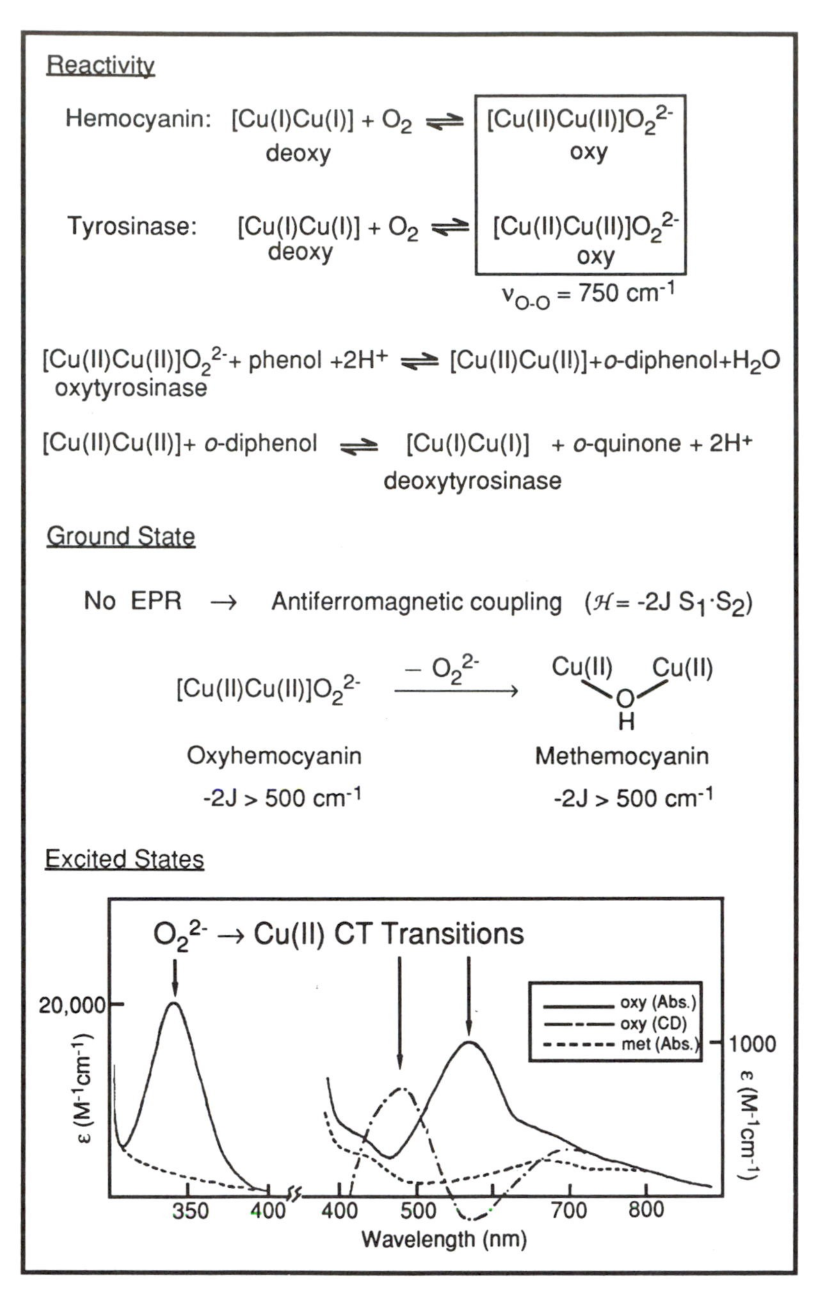

Figure 9. Coupled binuclear copper proteins; ground- and excited-state spectral features.

($-2J > 500\ cm^{-1}$), (*49, 50*) hence its classification as a *coupled* binuclear copper site (*51*). Displacement of the peroxide produces a met derivative that also has two Cu(II) ions that are strongly antiferromagnetically coupled ($-2J > 500\ cm^{-1}$) (Wilcox, D. E.; Westmoreland, T. D.; Sandusky, P. O.; Solomon, E. I., unpublished results). Thus, there must be an endogenous bridge present in the met derivative. The crystal structure of deoxyhemocyanin (*52*) shows no protein residues capable of bridging the copper ions in the vicinity of the binuclear copper site so this bridging ligand is likely to be hydroxide. With respect to the excited state spectroscopy, oxyhemocyanin exhibits a moderately intense band in the absorption spectrum at ~600 nm ($\epsilon \sim 1000\ M^{-1}\ cm^{-1}$) and an extremely intense band at ~350 nm ($\epsilon \sim 20{,}000\ M^{-1}\ cm^{-1}$). Displacement of peroxide on going to the met derivative (solid to dashed spectrum in Figure 9, bottom) eliminates these features as well as a band at 480 nm that is present in the CD but not the absorption spectrum (*53*). These three bands can be assigned as peroxide-to-copper charge transfer transitions and will be seen to provide a detailed probe of the peroxide–copper bond. We are particularly interested in (1) the fact that there are three charge transfer bands, (2) the selection rules associated with the presence of a band in the CD but not absorption spectrum, and (3) the high energy and intensity of the 350-nm band.

We first consider peroxide bound end-on to a single Cu (II) ion (Figure 10A). The valence orbitals of peroxide involved in bonding are the π^* set. These orbitals split into two nondegenerate levels (labeled π_σ^* and π_v^*) on bonding to the metal ion. The π_σ^* orbital is oriented along the Cu–O bond and has strong overlap with the $d_{x^2-y^2}$ orbital producing a higher-energy, intense charge-transfer transition. The peroxide π_v^* orbital is vertical to the Cu–O bond and weakly π interacting with the copper, producing a lower-energy, relatively weak transition. Thus, end-on peroxide bonding is dominated by the σ donor interaction of the O_2^{2-} π_σ^* orbital with the $d_{x^2-y^2}$ orbital. This predicted low-energy weak/high-energy intense charge-transfer spectrum is observed experimentally for the $[Cu_2(XYL\text{-}O\text{-})(O_2)]^+$ complex prepared by Karlin (*54*) that has O_2^{2-} end-on bound to a single Cu(II) ion (based on mixed isotope effects on its resonance Raman spectrum) (Figure 10B) (*55*). Note, however, that there are only two bands in the charge–transfer spectrum of this monomeric complex and that the π_σ^* transition is considerably lower in energy (500 nm) and weaker in intensity ($\epsilon \sim 5000\ M^{-1}\ cm^{-1}$) than the 350 nm O_2^{2-} charge-transfer band in oxyhemocyanin.

The fact that three peroxide-to-copper charge–transfer transitions are observed in oxyhemocyanin and oxytyrosinase led us to consider the spectral effects of bridging peroxide between two Cu(II) ions. A transition-dipole vector-coupling (TDVC) model was developed that predicts that each charge-transfer state in a Cu–peroxide monomer

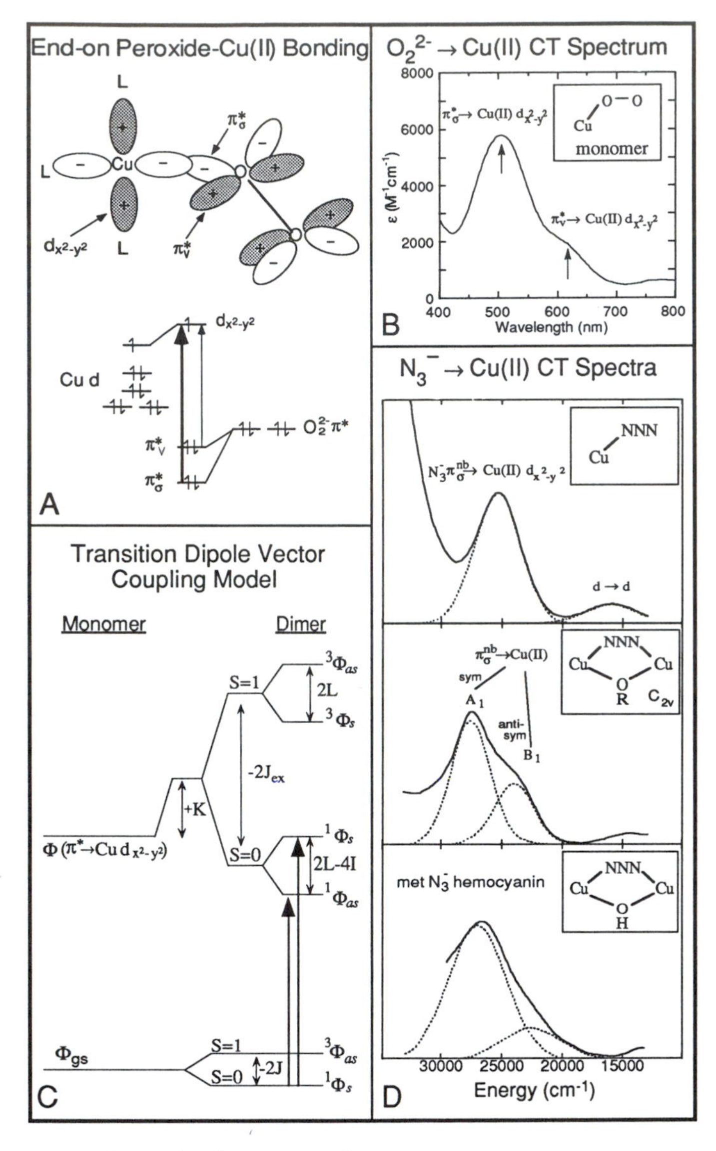

Figure 10. Peroxide charge-transfer transitions in copper monomers and dimers. A: Orbital interactions involved in end-on peroxide–copper bonding and predicted charge-transfer transitions (thickness of arrow indicates relative intensity). B: Charge-transfer absorption spectrum of peroxide bound to a single Cu(II) ion (Adapted from ref. 55). C: Ground-state and charge-transfer excited state splittings due to dimer interactions in a peroxide bridged copper dimer. K *is the coulomb dimer interaction,* J_{ex} *is the excited-state magnetic exchange, and* I *and* L *are the coulomb and exchange contributions to the excitation transfer between halves of the dimer, respectively. D: Azide-to-copper charge-transfer spectra of model complexes and met azide hemocyanin (Adapted from ref. 56).*

complex will split into four states in a dimer (Figure 10C) (*53, 56–58*). We have further developed a general model of excited state dimer interactions, the valence bond-configurational interaction (VBCI) model. This model reduces to the TDVC model in the in-state limit but the VBCI treatment gives a quantitative description of the dimer splittings in terms of parameters that can be evaluated using SCF-Xα molecular orbital calculations (*58–60*). First, there is a singlet–triplet antiferromagnetic splitting in the excited state just as there is in the ground state but considerably larger in magnitude (*58*). In addition, both the singlet and triplet states are split further into two states that correspond to symmetric and antisymmetric combinations of the $O_2^{2-} \rightarrow$ Cu(II) charge-transfer transition to each copper in the bridged dimer. As the antiferromagnetically coupled ground state is a singlet, only the two transitions to the singlet excited states should have absorption intensity. This predicted splitting into two bands is observed (*56*) in a series of azide model complexes prepared by Sorrell (*61*), Reed (*62, 63*), and Karlin (*64*) (Figure 10D). Azide bound to a single Cu(II) ion exhibits a $\pi_\sigma^{nb} \rightarrow$ Cu(II) charge-transfer transition that is analogous to the peroxide $\pi_\sigma^* \rightarrow$ Cu(II) charge-transfer transition. As predicted, bridging the azide in a *cis* μ-1,3 geometry between two Cu(II) ions results in a splitting of the monomer charge-transfer transition into two bands, the symmetric (A1 in the C_{2v} dimer symmetry) and antisymmetric (B1) components of the π_σ^* charge-transfer transition. Note in Figure 10D (bottom) that binding N_3^- to the met hemocyanin derivative produces the same A1/B1 charge-transfer intensity pattern indicating that azide also bridges in a *cis* μ-1,3 geometry in met hemocyanin (*56*).

The preceding discussion shows that the presence of more than two peroxide-to-copper CT transitions in oxyHc requires that this ligand bridge the copper centers. In 1989, Kitajima obtained the first crystal structure of a side-on bridging peroxide in transition-metal chemistry (*65, 66*) and, in 1992, Magnus determined that Limulus polyphemus oxyHc has the side-on bridging structure shown in Figure 11 (top) (*67*). The VBCI model can then be used to predict the energy splittings and symmetries and hence the selection rules for the peroxide-to-copper CT transitions in the effective D_{2h} symmetry of the side-on bridged site (the *trans* axial His ligands reduce the site symmetry to C_{2h} but have a small effect on the spectrum) (*59, 60*). Bridging the peroxide in the side-on structure results in a splitting of the π_v^* into two components. The low-energy component is electric dipole allowed (z) and should appear in the absorption spectrum, but with limited intensity inasmuch as it is a π_v^* charge-transfer transition. This can be associated with the 600-nm absorption band. The second component of the peroxide π_v^* charge-transfer transition is predicted to be only magnetic dipole allowed (R_y) and thus it should contribute to the CD but not absorption spectrum.

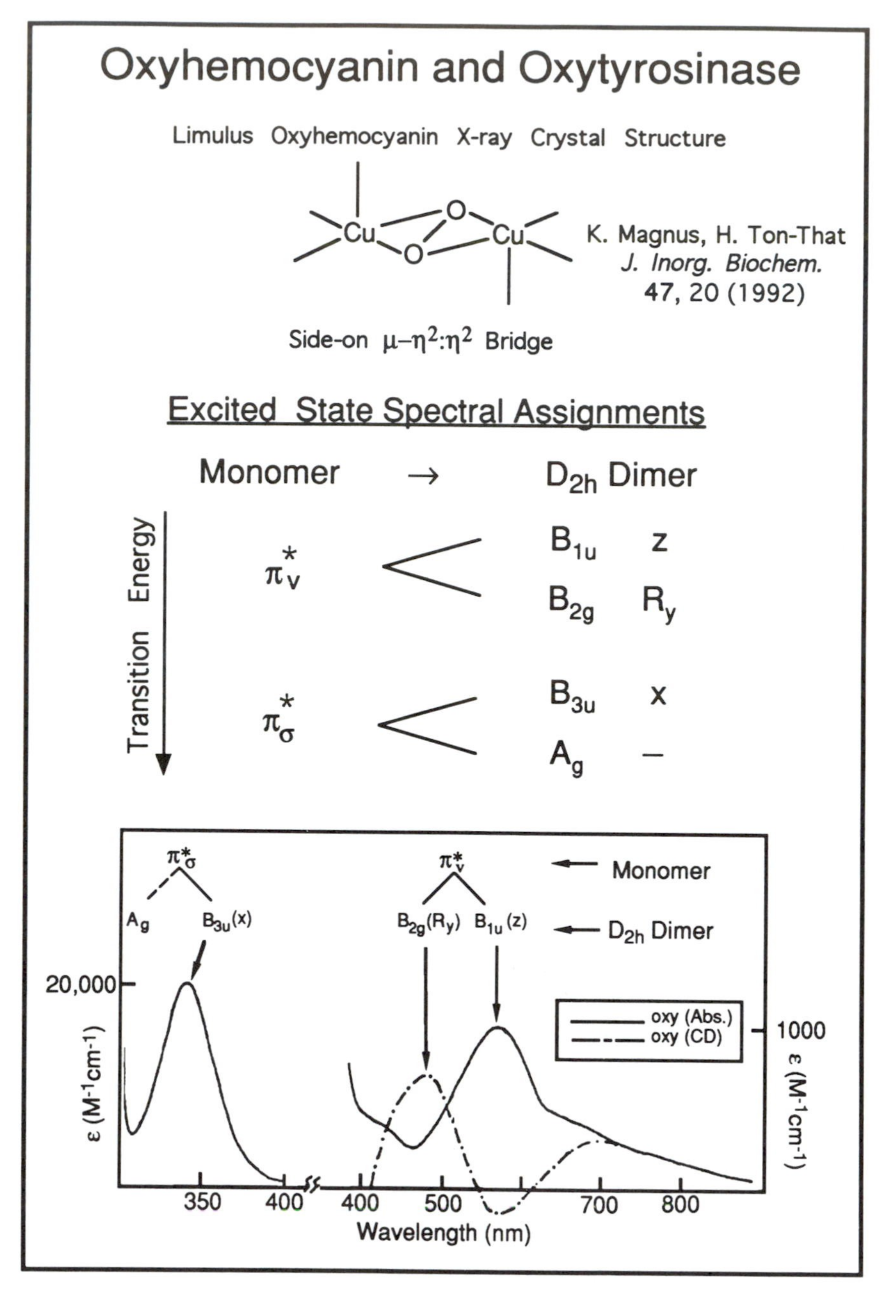

Figure 11. Excited-state spectral assignments for the D_{2h} μ-η^2:η^2 peroxo-copper unit in oxyhemocyanin and oxytyrosinase.

The 480-nm CD feature can be assigned to this transition. The π_σ^* also splits into two bands with the lower-energy component being electric dipole allowed and having the dominant absorption intensity. This can be associated with the 350-nm absorption band in oxyhemocyanin. Thus, the side-on bridging peroxide produces the three observed charge-transfer transitions with one being present in the CD but not the absorption spectrum. However, one must still account for the high intensity and energy of the 350-nm O_2^{2-} π_σ^* charge-transfer transition and the low vibrational frequency of the O–O stretch. Thus we proceed to evaluate quantitatively the electronic structure associated with the side-on bridging peroxide and compare this to the more commonly observed end-on bridging peroxide structure both theoretically and experimentally.

Broken-symmetry, spin-unrestricted SCF-Xα-SW MO calculations were performed to describe the electronic structures associated with both the end-on and side-on bridging peroxide geometric structures (*68, 69*). These calculations are appropriate for antiferromagnetically coupled dimers (*70, 71*). In Figure 12, we focus on the interaction of the highest occupied molecular orbital (HOMO) and the lowest unoccupied molecular orbital (LUMO), that are the symmetric and antisymmetric combinations of $d_{x^2-y^2}$ orbitals on each copper, with the valence orbitals of the peroxide. For the end-on bridged structure (Figure 12, left), the bonding is consistent with the qualitative description presented earlier. The peroxide π_σ^* orbital is stabilized through a bonding interaction with the LUMO on both coppers. Thus, in the end-on bridged geometry peroxide acts as a σ donor ligand with one bonding interaction with each of the two coppers. A very different bonding description is obtained for the side-on bridged peroxide. In this structure, the π_σ^* orbital is again stabilized by a σ donor interaction with the LUMO on both coppers. In the side-on structure the bonding/antibonding interaction of the π_σ^* orbital is larger than in the end-on structure because the peroxide now occupies two coordination positions on each of the two coppers. Thus, peroxide behaves as an extremely strong σ donor in the side-on structure. Furthermore, the side-on peroxide is predicted to have an additional bonding interaction with the $d_{x^2-y^2}$ orbitals on the coppers. This involves stabilization of the HOMO through its interaction with the high-energy unoccupied σ^* orbital on the peroxide. This additional bonding interaction shifts electron density from the copper ions onto the peroxide. Thus, peroxide also acts as a π acceptor ligand using this highly antibonding σ^* orbital.

It was of critical importance to evaluate experimentally this unusual electronic structure description for the side-on bridged peroxide and its relation to the spectral features of the μ-η^2:η^2 model complex and oxyhemocyanin. This evaluation was accomplished through a series of

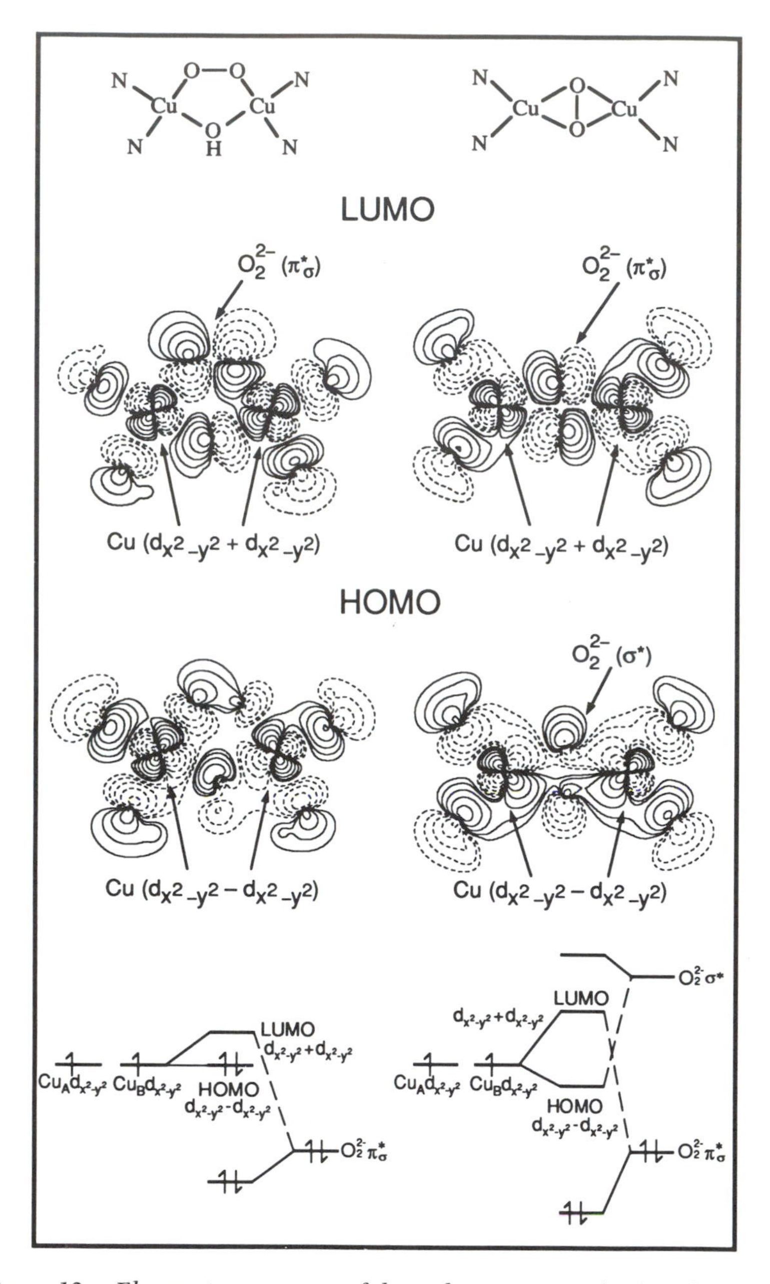

Figure 12. Electronic structures of the end-on cis-μ-*1,2* (C_{2v}) *and side-on* μ-η^2:η^2 (D_{2h}) *models of the oxyhemocyanin active site. Wave function contours of the HOMO and LUMO and energy level diagrams showing dominant orbital contributions.*

studies of the charge-transfer and vibrational spectral features of end-on (*72*) and side-on (*73*) bound peroxide–copper model complexes prepared by Karlin (*74*) and Kitajima (*65, 66*). The σ-donor ability of peroxide can be related to the intensity (and energy (*58*)) of the $\pi_\sigma^* \rightarrow$ Cu(II) charge-transfer transition (Figure 13). The idea is that as the wave function of the occupied O_2^{2-} π_σ^* orbital gains copper character, α (i.e., the coefficient for the amount of copper character in the wave function is α), its σ-donor interaction with the copper increases. This wave function gives an approximation for the ligand-to-metal charge-transfer intensity that (along with geometric factors) is proportional to α^2. Thus, the peroxide $\pi_\sigma^* \rightarrow$ Cu(II) charge-transfer intensity increases as its σ-donor interaction with the copper increases. If we normalize to the π_σ^* charge-transfer intensity of the end-on peroxide monomer complex shown in Figure 10B, the O_2^{2-} charge-transfer intensity of the *trans* μ-1,2 end-on bridged complex increases by a factor of two, consistent with peroxide binding to each of two Cu(II) ions. Although no *cis* model complex exists, our Xα calculations indicate that peroxide binding in this geometry should have a similar σ-donor interaction with the coppers as peroxide bridged in the *trans* complex. However, the side-on bridged μ-η^2:η^2 complex and oxyhemocyanin exhibit an extremely intense π_σ^* charge-transfer transition. The intensity of this transition quantitates to ~4 times the σ-donor interaction of peroxide bound to a single Cu(II) ion. This is consistent with the Xα calculations and the fact that in this geometry peroxide has four bonding interactions with the two Cu(II) ions. The extremely high intensity of the 350-nm band in oxyhemocyanin also quantitates to having ~4 σ-donor interactions with the binuclear copper site consistent with the side-on peroxide bridged structure of oxyhemocyanin.

One can probe the π-acceptor ability of the peroxide through a study of its intraligand stretching force constant (and hence O–O bond strength), which is obtained from a normal coordinate analysis (NCA) of vibrational spectra (Figure 13). One would expect this force constant to increase as the σ-donor interaction of the peroxide with the copper increases, because this increased interaction removes the electron density from a π-antibonding orbital on the peroxide that increases its intraligand bond strength. This increase in bond strength is observed experimentally by comparing the end-on monomer to the *trans* end-on dimer, in which the O–O vibrational frequency increases from 803 to 832 cm^{-1}. This increase in vibrational frequency is consistent with the *trans* end-on dimer having σ-donor interactions with two coppers (*72*). However, on going to the side-on peroxo bridging geometry, the O–O stretching frequency dramatically decreases in the μ-η^2:η^2 model complex and oxyhemocyanin, yet in this geometry the peroxide is the strongest σ donor, based on the high charge-transfer intensity associated with four

σ–Donor Ability: $O_2^{2-} \rightarrow$ Cu charge transfer transition intensity (and energy)

$$\Psi_{\pi_\sigma^*} = (1-\alpha^2)^{1/2}\,\pi_\sigma^* - \alpha\, d_{x^2-y^2}$$

$$I_{CT} \propto \alpha^2 (1-\alpha^2)(r \cos \varphi)^2$$

π-Acceptor Ability: $\nu_{O\text{-}O}\ (+\nu_{M\text{-}L}) \xrightarrow{NCA} k_{O\text{-}O}$

$k_{O\text{-}O}$ increases for σ-donor ($O_2^{2-}\ \pi^*$)

$k_{O\text{-}O}$ decreases for π-acceptor ($O_2^{2-}\ \sigma^*$)

	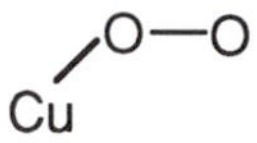 Cu–O–O (end-on) K. Karlin	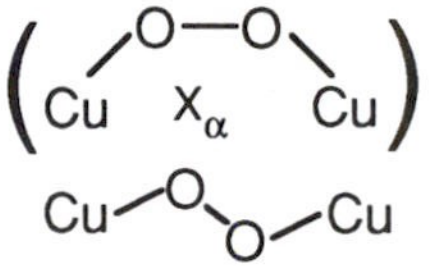 (Cu–O–O–Cu, X_α); Cu–O–O–Cu (trans) K. Karlin	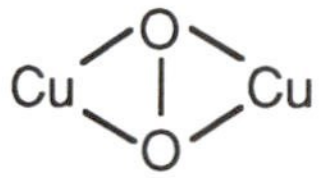 Cu(μ-O–O)Cu (side-on) N. Kitajima	oxyHc 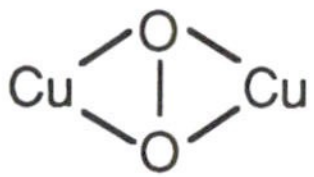 Cu(μ-O–O)Cu *L. Polyphemus*
σ-donor: relative α^2 (osc. str.)	1 (0.105)	1.9 (0.252)	3.7 (0.479)	3.7 (0.488)
π-acceptor: $k_{O\text{-}O}$/mdyne/Å ($\nu_{O\text{-}O}/cm^{-1}$)	2.9 (803)	3.1 (832)	2.4 (763) *i*-Pr (749)	(750)

Figure 13. Electronic structures of end-on and side-on peroxide bridged models of oxyhemocyanin. Comparison of experimentally determined peroxide σ-donor and π-acceptor abilities.

bonding interactions to the two Cu(II) ions. Although the mechanical coupling of the vibrations in this geometry is complicated, a normal coordinate analysis on the side-on bridged model complex gives a significantly lower O–O force constant indicating an extremely weak O–O bond (*73*). This is experimental evidence confirming the Xα calculated prediction that side-on bridged peroxide also participates in a π-acceptor interaction with the coppers that shifts some electron density into the strongly antibonding σ^* orbital on the peroxide.

With the novel electronic structure of the side-on bridged peroxy-binuclear cupric site of oxyhemocyanin and oxytyrosinase having been determined, it is now possible to evaluate electronic structure contributions to the functions of these protein active sites (*68, 69*) (Figure 14). The combination of strong σ-donor and π-acceptor interactions with the coppers leads to a very strong dioxygen–copper bond in the side-on structure that would contribute to reversible O_2 binding and, in particular, stabilize the bound peroxide with respect to decay to the inactive met site in hemocyanin. These bonding interactions further provide a major electronic structure contribution to the mechanism of dioxygen activation in oxytyrosinase. The strong σ-donor interaction with the coppers results in a less negative peroxide, promoting electrophilic attack on substrate while the π acceptor contribution to the bonding shifts a small amount of electron density into the peroxide σ^* orbital that leads to an extremely weak O–O bond activating it for cleavage. As shown in Figure 14, we found that a major difference between hemocyanin and tyrosinase is that the phenolic substrate can access and bind to the Cu center in tyrosinase. The substrate then contributes electron density into the LUMO, which is antibonding with respect to both the O–O and Cu–O bonds, and thus further initiates oxygen transfer in catalysis.

Multicopper Oxidases

The final section of this chapter focuses on the multicopper oxidases. The multicopper oxidases use at least four copper ions, which are grouped into three types based on their spectral properties, to couple four one-electron oxidations of substrate to the four-electron reduction of dioxygen to water (Figure 15). Two of the coppers form an antiferromagnetically coupled pair that is referred to as a type 3 center. The type 1 center is a blue copper site as discussed earlier, and the type 2 copper is "normal" in the sense of having a tetragonal Cu(II) EPR spectrum ($g_{\|} > g_{\perp} > 2.00$, $A_{\|} > 140 \times 10^{-4}$ cm^{-1}) and weak ligand field absorption features in the visible spectrum (*75*) as discussed earlier for normal copper complexes. Laccase is the simplest of the multicopper oxidases and contains one of each type of center for a total of four copper ions in the native enzyme (*1, 2, 76–80*). Understanding the reactivity of

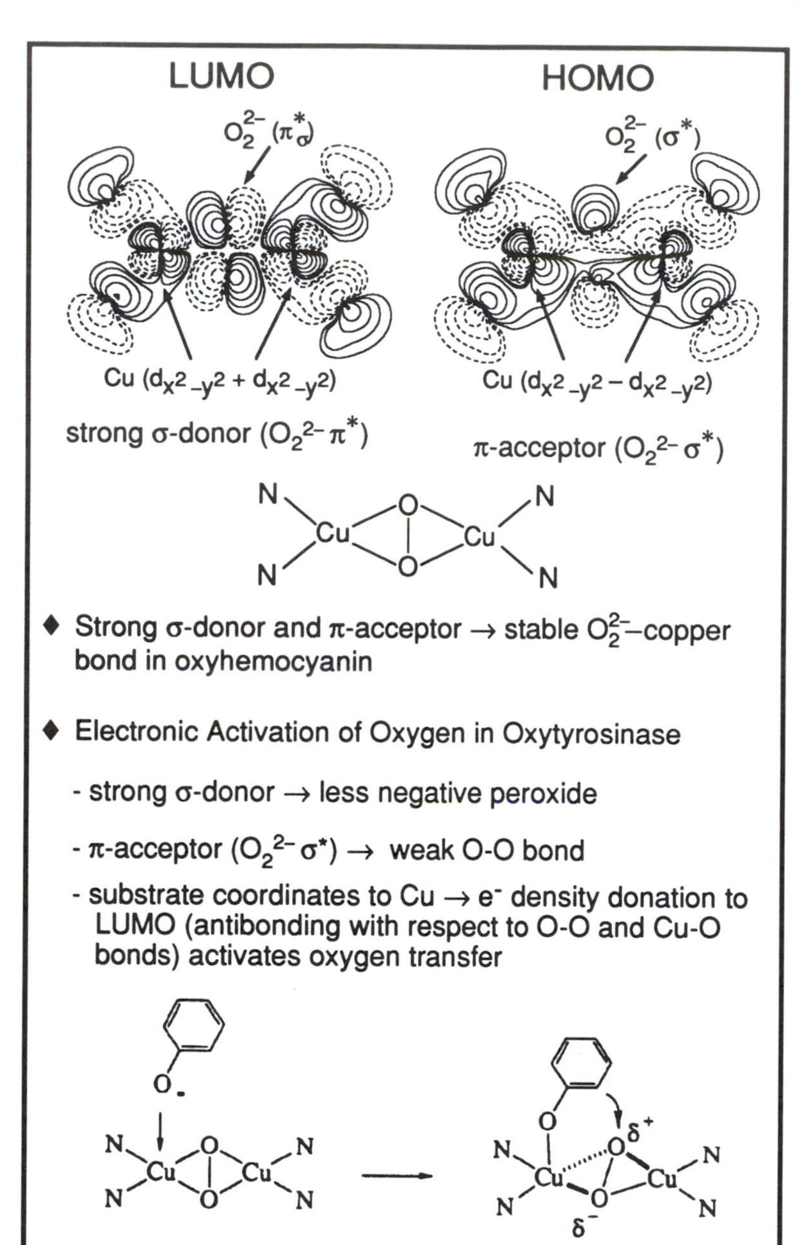

Figure 14. Electronic structural contributions to oxygen binding and activation in hemocyanin and tyrosinase.

Reactivity

$$4AH + O_2 \longrightarrow 4A + 2H_2O$$

Number of Centers

Multicopper Oxidases:	Type 1 (Blue)	Type 2 (Normal)	Type 3 (Coupled Binuclear)	Total Cu
Laccase	1	1	1	4
Ascorbate Oxidase	2	2	2	8
Ceruloplasmin	2	1	1 or 2	5 - 7
Laccase Derivatives:				
Type 2 Depleted (T2D)	1	--	1	3
Type 1 Hg Sub. (T1Hg)	Hg^{2+}	1	1	3

Figure 15. Multicopper oxidases: reactivity and stoichiometry.

this enzyme is a complex problem, and two derivatives have served to simplify this system. In type 2 depleted (T2D) laccase, the type 2 copper is reversibly removed leaving the type 1 and type 3 centers (*81*). The type 1 mercury substituted derivative (T1Hg) is formed by replacing the type 1 copper with the spectroscopic and redox innocent mercuric ion (*82*).

The goal of our research on the multicopper oxidases has been to determine the spectral features of the type 3 (and type 2) centers, to use these spectral features to define geometric and electronic structural differences relative to hemocyanin and tyrosinase, and to understand how these structural differences contribute to their variation in biological function. The hemocyanins and tyrosinases reversibly bind and activate dioxygen whereas the multicopper oxidases catalyze its four-electron reduction to water.

We start by defining the spectral features associated with each type of copper in native and T2D laccase (Figure 16). The EPR spectrum of the native enzyme contains contributions from two distinct cupric sites (Figure 16A); one with a large and a second with a small parallel hyperfine splitting. These have been assigned to the type 2 and type 1 copper sites, respectively. The EPR spectrum of the T2D derivative contains a single component, with a small hyperfine coupling, that is assigned to the type 1 copper center (*83*). The type 3 copper atoms are EPR nondetectable and, by analogy to hemocyanin and tyrosinase, can

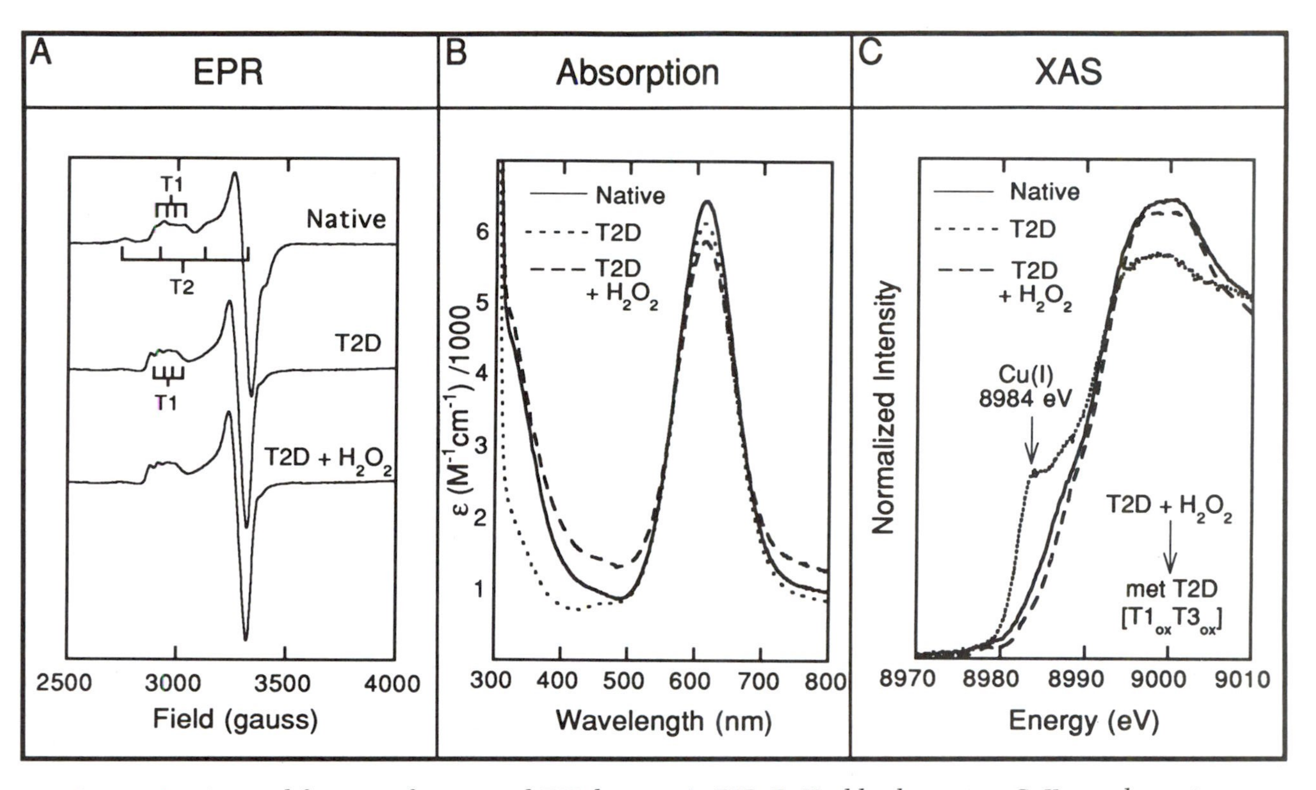

Figure 16. Spectral features of native and T2D laccase. A: EPR. B: Visible absorption. C: X-ray absorption spectra of native, T2D, and T2D laccase following reaction with hydrogen peroxide.

be considered to be a coupled binuclear copper site. The absorption spectrum of native and T2D laccase (Figure 16B) exhibits an intense thiolate S $\rightarrow$ Cu(II) charge-transfer transition at 600 nm, which is assigned to the type 1 copper center (*4*). The only spectral feature that has been associated with the type 3 center is the absorption band centered at 330 nm ($\epsilon \sim 3000\ M^{-1}\ cm^{-1}$). This band decreases in intensity with reduction of the protein by two electrons (at the same potential). The 330-nm spectral region is expected to contain histidine and hydroxide-to-type 3 Cu(II) charge-transfer transitions (*84*).

The assignment of the 330-nm absorption band in native laccase to the type 3 center, however, was complicated by the absence of a 330-nm band in the absorption spectrum of T2D laccase (which still contains a type 3 site) (Figure 16B). We discovered a key reaction that clarified the assignment of the 330-nm absorption band. Addition of peroxide to T2D laccase leads to the reappearance of the 330-nm band (*85*). This indicates that the type 3 site in T2D laccase was reduced (even when exposed to dioxygen) but that the stronger oxidant, peroxide, was capable of oxidizing the type 3 center. This finding was confirmed using X-ray absorption studies at the Cu K-edge (9000 eV) (*86*). The T2D derivative exhibits a peak at 8984 eV (Figure 16C) that is characteristic of Cu(I) in a three-coordinate site. The magnitude of the 8984-eV band could be quantitated using a normalized edge method that we have developed. We determined that the T2D derivative contained a fully reduced type 3 site. Addition of peroxide eliminated the 8984-eV peak, which indicates that peroxide fully oxidizes the type 3 center to form a met T3 site.

Having defined the T2D derivative, we could study the type 3 site in the absence of the type 2 copper and compare it to the coupled binuclear site in hemocyanin and tyrosinase (Figures 17 and 18). First, as demonstrated from the X-ray edges in Figure 16C, the fully reduced type 3 site is strikingly different from that of hemocyanin and tyrosinase inasmuch as it does not react with dioxygen (*85*, *86*). Peroxide does oxidize the site, and we can further compare this met type 3 center in laccase to met hemocyanin. As with hemocyanin, the met type 3 site in the multicopper oxidases is strongly antiferromagnetically coupled (*49*, *50*, *75*). This indicates the presence of an endogenous hydroxide bridge, which has been confirmed by X-ray crystallography (*87*).

One-electron reduction of met derivatives of hemocyanin and T2D laccase produces the mixed-valent half-met sites that exhibit dramatic differences (Figure 18). In particular, half-met hemocyanin has very unusual coordination chemistry with respect to exogenous ligand binding. For example, azide binds to this half-met active site with an equilibrium binding constant that is more than two orders of magnitude greater than that of azide binding to aqueous Cu(II), and this binding

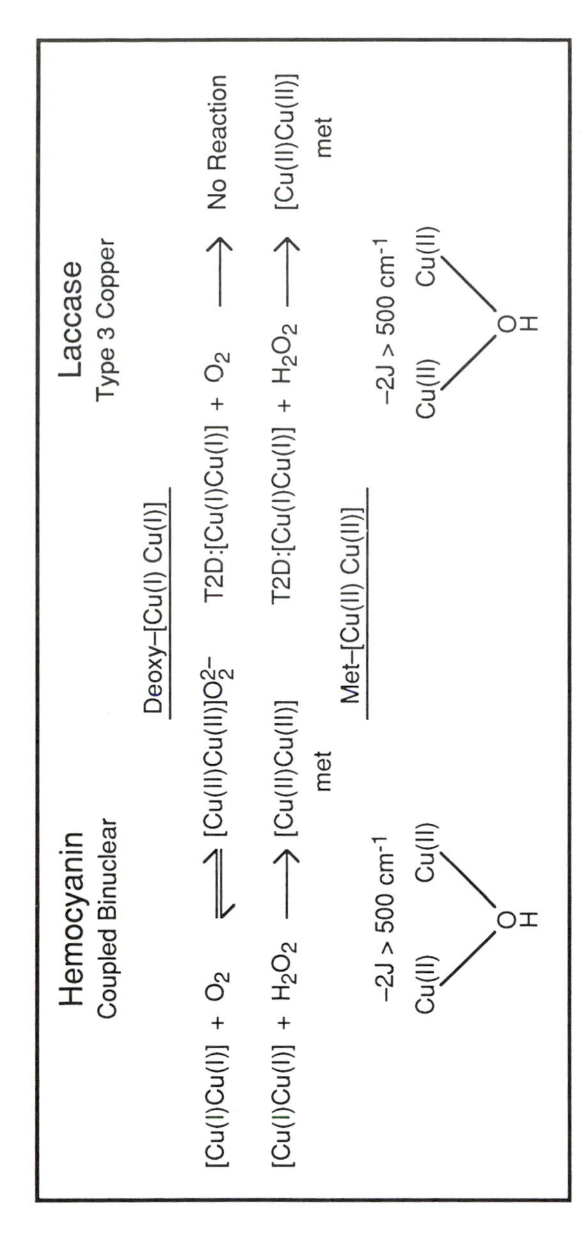

Figure 17. Comparison of the reactivity and magnetism of deoxy and met hemocyanin and the laccase type 3 copper site in the T2D derivative.

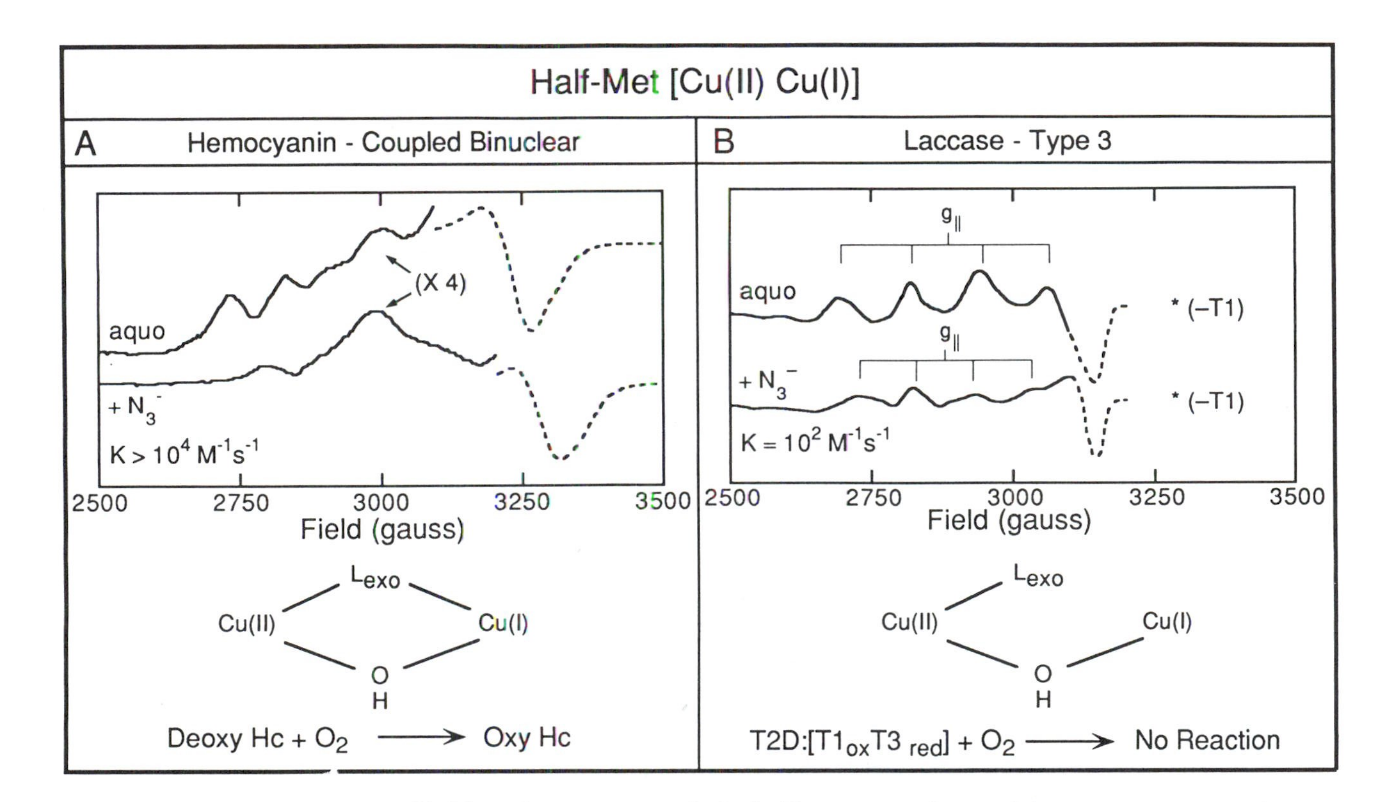

Figure 18. Comparison of half-met hemocyanin with the half-met type 3 (in T2D) laccase copper sites. A: EPR spectra and binding constants of exogenous azide binding. B: Spectroscopically effective structural models for exogenous ligand binding to the half-met derivatives and their relation to differences in dioxygen reactivity.

results in quite unusual mixed-valent spectral features (*88*). We have studied this unusual half-met hemocyanin chemistry and spectroscopy in some detail (*88*) and determined that these derive from the fact that exogenous ligands bridge between the Cu(II) and Cu(I) of this mixed-valent site (Figure 18A, bottom). Alternatively, the half-met type 3 site in T2D laccase exhibits normal Cu(II) EPR spectra for all exogenous ligand-bound forms and has an equilibrium binding constant consistent with aqueous Cu(II) chemistry, indicating that the exogenous ligands bind terminally to the Cu(II) of the half-met type 3 site (*84*). This difference in exogenous ligand binding modes (bridging vs. terminal) directly correlates with differences in O_2 reactivity of these binuclear copper sites as described above in that only the deoxyhemocyanin site reversibly binds dioxygen (Figure 18A).

The combination of the type 3 with the type 2 center does, of course, react with dioxygen in the native enzyme. This reaction led us to consider exogenous ligand interactions with both the type 3 and type 2 coppers in native laccase. An appropriate spectral method to study the interaction of exogenous ligands with each center is low-temperature MCD spectroscopy, which allows correlation of excited-state features with ground-state properties (*89–91*). In particular, the paramagnetic type 2 copper exhibits very different low-temperature MCD features relative to the antiferromagnetically coupled type 3 center (Figure 19). For the type 2 center, both the ground and excited states have $S = ½$ and split in a magnetic field. The selection rules for MCD spectroscopy predict that there should be two transitions to a given excited state that are of equal magnitude but of opposite sign. As the Zeeman splitting will be on the order of 10 cm^{-1} and absorption bands are on the order of a few thousand cm^{-1} broad, the positive and negative bands will mostly cancel and produce a broad, weak, derivative-shaped MCD signal known as an *A*-term. This is observed if both components of the ground state are equally populated. However, as one lowers the temperature the Boltzmann population of the higher-energy component is reduced, cancellation no longer occurs, and one observes intense, low-temperature MCD signals known as *C*-terms. These can be two to three orders of magnitude more intense than the high-temperature MCD signals.

For the type 3 center, the antiferromagnetic coupling leads to an $S = 0$ ground state that cannot split in a magnetic field. Thus, this site does not exhibit *C*-term intensity and the low-temperature MCD spectrum of native laccase will be dominated by the intense *C*-terms associated with paramagnetic copper centers (*89, 90*).

Low-temperature MCD spectroscopy was used to probe the effects of binding the exogenous ligand azide to native laccase (*89, 90*). Titration of the native enzyme with azide produces two $N_3^- \rightarrow$ Cu(II) charge-transfer transitions: one at 500 nm and a second more intense band at

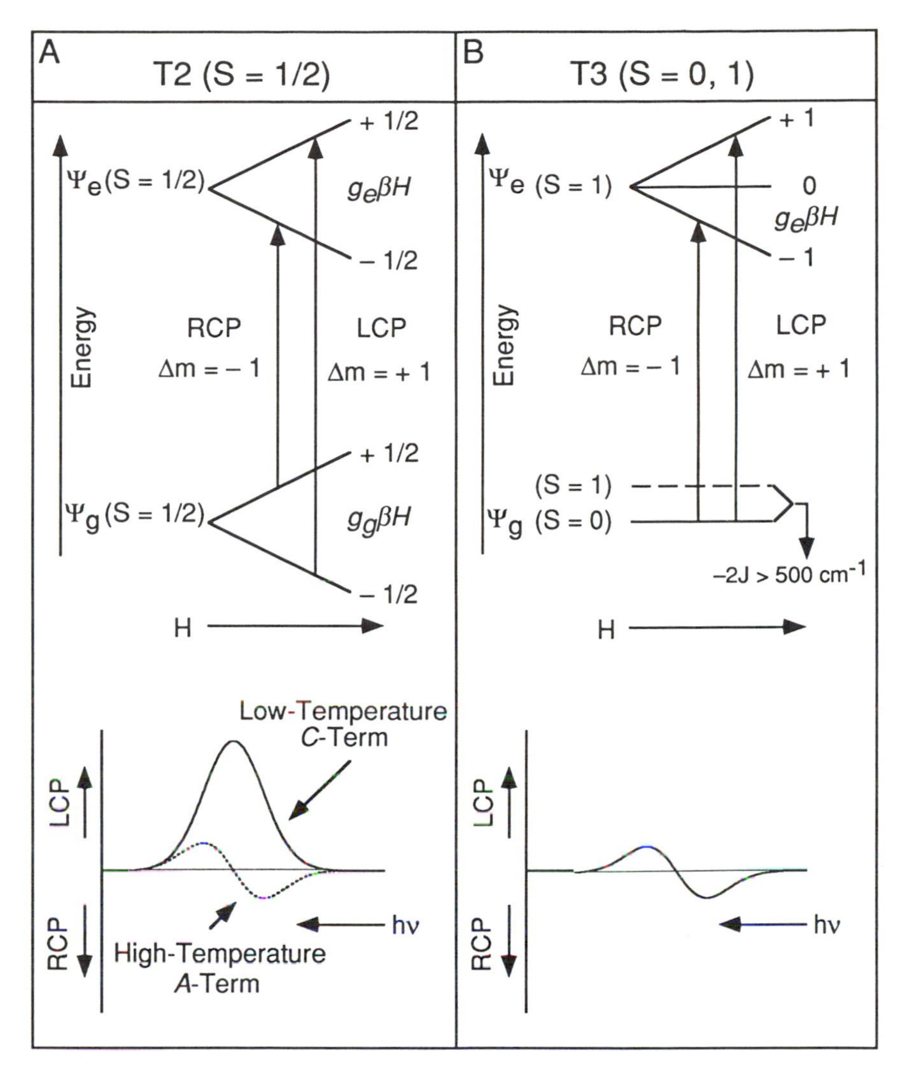

Figure 19. Model for the temperature dependence of the MCD bands of native laccase. A: Transitions and band profiles associated with type 2 copper. B: Transitions and band profiles associated with type 3 copper. Note the difference in temperature dependence of the MCD signal as described in the text.

400 nm (Figure 20A). The intensity of the 400-nm band as a function of azide concentration is plotted as a dashed line in Figure 20C. One can use low-temperature MCD to correlate these excited-state features to specific copper centers. The 500-nm absorption band has a negative low-temperature MCD signal associated with it at 485 nm that increases in magnitude with increasing azide concentration (Figure 20B). The

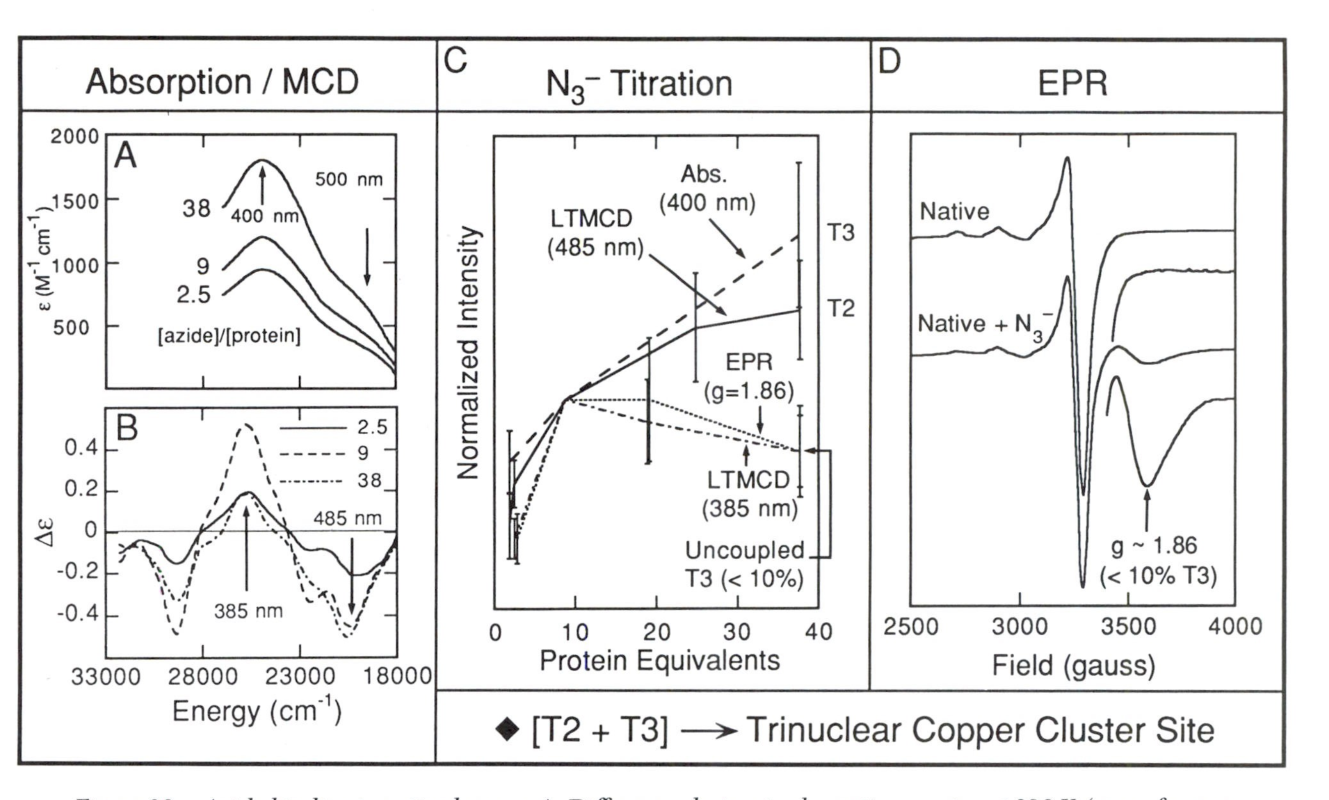

Figure 20. Azide binding to native laccase. A: Difference electronic absorption spectra at 298 K (see reference 89). B: Difference MCD spectra at 4.9 K (5 T). C: Changes in EPR, LTMCD, and absorption intensities plotted as a function of increasing azide concentration. D: EPR of laccase titrated with azide. Arrow indicates the new signal present at g = 1.86 and 8 K.

intensity of this MCD feature is plotted as a solid line in Figure 20C. The 500-nm absorption band has a corresponding low-temperature MCD signal; thus it must be associated with azide binding to the paramagnetic type 2 center. There is also an MCD signal in the region of the 400-nm absorption band; however, it does not exhibit the same behavior as the absorption intensity (Figure 20B). The 385-nm MCD signal first increases and then decreases in intensity with increasing azide concentration. Its magnitude is plotted as the dot–dash line in Figure 20C. Although the low-temperature MCD signal does not correlate with the 400-nm absorption band, it does correlate with an unusual $g = 1.86$ signal in the EPR spectrum (Figure 20D), which we have shown to be associated with <10% of the type 3 sites that become protonatively uncoupled (and hence paramagnetic) upon binding azide. Thus, the intense 400-nm absorption band has no low-temperature MCD signal associated with it, and it must correspond to azide bound to the MCD-silent coupled type 3 center.

The low-temperature MCD and absorption titration studies (Figure 10) have determined that azide binds to both the type 2 and type 3 centers with similar binding constants. A series of chemical perturbations and stoichiometry studies have shown that these effects are associated with the same azide. This demonstrates that one N_3^- bridges between the type 2 and type 3 centers in laccase. These and other results from MCD spectroscopy first defined the presence of a trinuclear copper cluster active site in biology (*89*). At higher azide concentration, a second azide binds to the trinuclear site in laccase. Messerschmidt et al. have determined from X-ray crystallography that a trinuclear copper cluster site is also present in ascorbate oxidase (*87*, *92*) and have obtained a crystal structure for a two-azide–bound derivative (*87*). It appears that some differences exist between the two-azide–bound laccase and ascorbate oxidase derivatives, and it will be important to spectroscopically correlate between these sites.

Having demonstrated that the type 3 center must be viewed as part of a trinuclear copper cluster, including the type 2 center, it was important to determine which coppers are required for the reactivity of the multicopper oxidases with dioxygen. We had already demonstrated using X-ray absorption edges (Figure 16C) that a reduced type 3 center in the presence of an oxidized type 1 center does not react with O_2 (*85*, *86*). We next looked at the reactivity of the fully reduced T2D [$T1_{red}$ $T3_{red}$] derivative with O_2. This had been generally viewed as the combination of copper centers in laccase required for dioxygen reactivity in the mechanistic proposals in the literature (*93*, *94*). From Figure 21A it is clear that the 8984-eV-reduced Cu K-edge peak does not change on exposure of fully reduced T2D laccase to O_2. This indicates that the type 2 center is required for dioxygen reactivity. Thus, the Cu K-edge

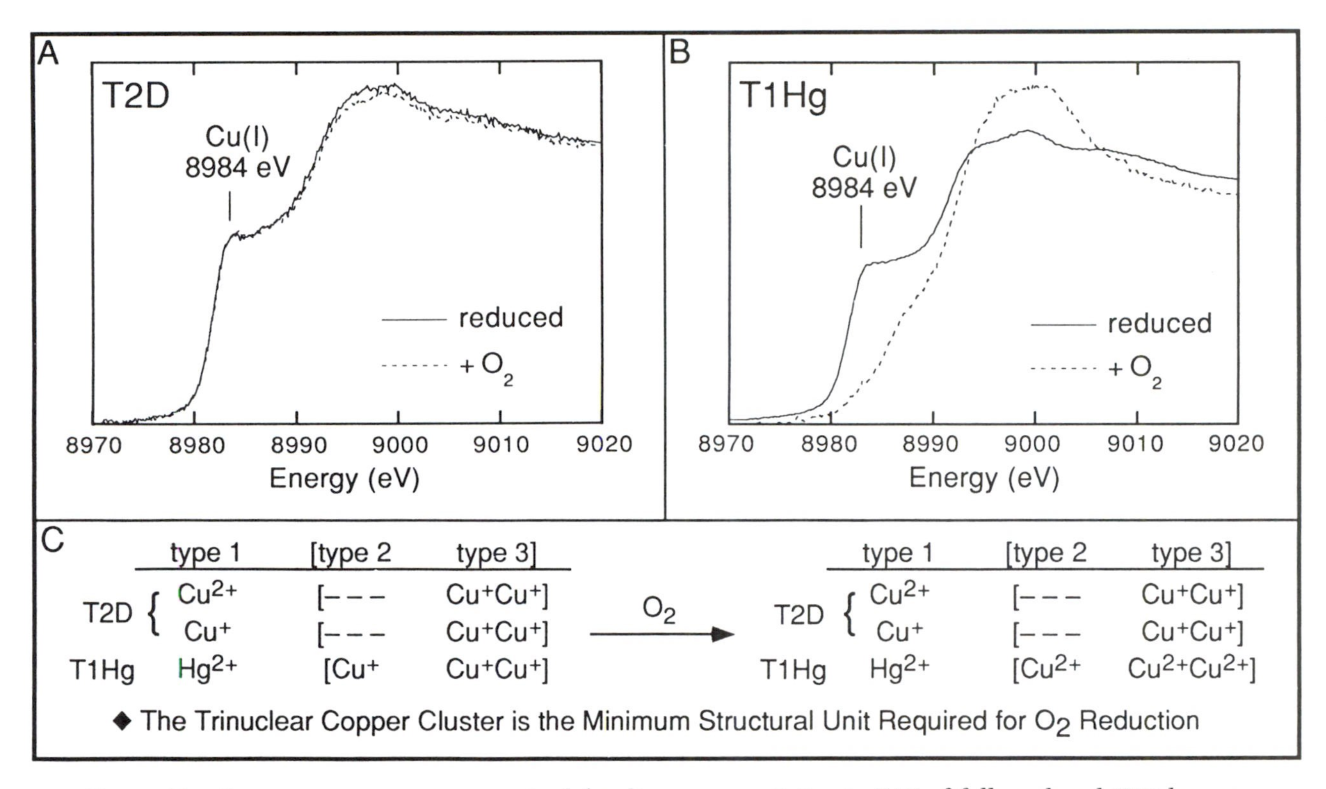

Figure 21. Laccase copper centers required for dioxygen reactivity. A: XAS of fully reduced T2D laccase and fully reduced T2D laccase following exposure to dioxygen. B: XAS of reduced T1Hg laccase and reduced T1Hg laccase following exposure to dioxygen. C: Summary of the reactivity of deoxy T2D, fully reduced T2D, and reduced T1Hg laccase with oxygen.

spectra of the T1Hg derivative, which contains a valid type 2/type 3 trinuclear copper cluster, was investigated. From Figure 21B, the fully reduced trinuclear copper cluster site rapidly reacts with O_2 eliminating the 8984-eV peak. Thus, the trinuclear copper cluster is the minimum structural unit required for O_2 reduction (95).

Because the mercuric ion in T1Hg laccase is redox inactive, this derivative has one less electron equivalent available for O_2 reduction than native laccase. This property enabled us to stabilize an oxygen intermediate in T1Hg laccase. A combination of low-temperature MCD and XAS has demonstrated that two coppers of the trinuclear cluster are oxidized in this intermediate (*41*). Thus, two electrons have been transferred to dioxygen and this species corresponds to a peroxide level intermediate that can be compared to the peroxo-binuclear cupric sites in oxyhemocyanin and oxytyrosinase. As is clear from Figure 22A, the peroxide intermediate in laccase has a strikingly different charge-transfer spectrum from that of oxyhemocyanin and oxytyrosinase. This requires a different geometric and electronic structure for this peroxy-trinuclear copper cluster site (*41*). Detailed spectral studies on this intermediate are presently underway (Shin, W.; Cole, J. L.; Root, D. E.; Solomon, E. I., unpublished results). However, at this point our data indicate that it corresponds to a hydroperoxide bound end-on to one of the coppers of an oxidized type 3 center and also likely bridging to a reduced type 2 copper center (Figure 22B). Messerschmidt et al. have since obtained a crystal structure of a low-affinity peroxide-bound adduct of ascorbate oxidase that is also described as having hydroperoxide end-on bound to one of the type 3 coppers. (This peroxide adduct has the three coppers of the trinuclear cluster unbridged and therefore uncoupled in contrast to the oxygen intermediate of T1Hg laccase.) The oxygen intermediate of T1Hg laccase indicates the mechanistic relevance of a end-on hydroperoxide-bound form of the protein. This difference in peroxide binding relative to hemocyanin and tyrosinase appears to play a key role in stabilizing the peroxide intermediate and promoting its irreversible further reduction to water at the trinuclear copper cluster site.

Summary

At this point the unique spectral features associated with the major classes of active sites in copper proteins are reasonably well understood and define active site electronic structures that provide significant insight into their reactivities in biology. For the blue copper sites, we determined that the unique spectral features derive from a ground-state wave function that has a high anisotropic covalency involving the thiolate ligand. This covalency provides a very efficient superexchange pathway

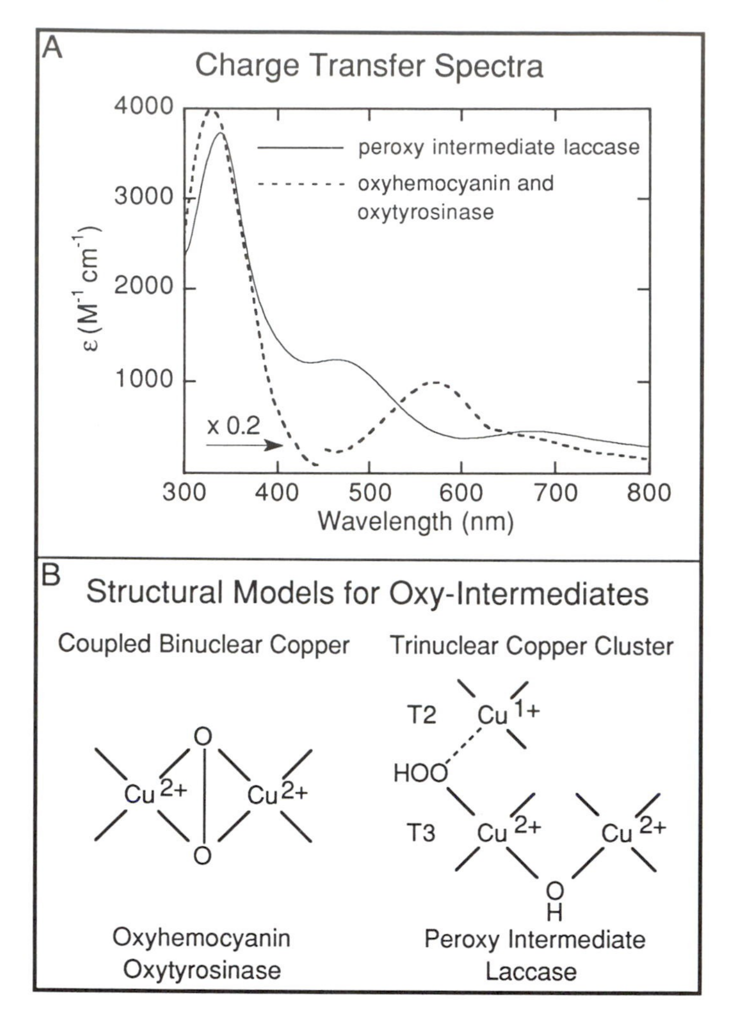

Figure 22. Comparison of oxygen intermediates. A: Electronic absorption spectra of the peroxy-intermediate in laccase versus oxyhemocyanin and oxytyrosinase. B: Proposed structural differences between peroxide binding in oxyhemocyanin and oxytyrosinase relative to the end-on bound hydroperoxide intermediate at the trinuclear copper cluster in laccase.

for long-range electron transfer. For the coupled binuclear copper active sites, we have seen that the unique spectral features of the oxy site correspond to a new bridging peroxide electronic structure that has very strong σ-donor and π-acceptor properties. These properties appear to make significant contributions to the reversible binding and activation of dioxygen by these active sites. In the multicopper oxidases, our spectral studies determined that the type 3 center is fundamentally different from the coupled binuclear copper site in hemocyanin and tyrosinase, that it is part of a trinuclear copper cluster, and that this trinuclear copper cluster is the structure required for O_2 reduction. We have now characterized a peroxide level intermediate at this trinuclear copper cluster site that is strikingly different from the peroxide bound in oxyhemocyanin and oxytyrosinase in that it is bound end-on as hydroperoxide. Our spectral studies presently underway should provide important insight into the geometric and electronic structure differences that are indicated by these spectral differences and their contribution to differences in biological function.

Acknowledgments

This research was supported by the National Science Foundation (CHE-9217628) for the blue copper studies and by the National Institutes of Health (DK-31450) for the coupled binuclear and multicopper oxidase studies. Edward I. Solomon expresses his sincere appreciation to all his students and collaborators who are listed as co-authors in the references for their commitment and contributions to this science.

References

1. Solomon, E. I.; Penfield, K. W.; Wilcox, D. E. *Struct. Bonding* **1983,** *53,* 1–57.
2. Solomon, E. I.; Baldwin, M. J.; Lowery, M. D. *Chem. Rev.* **1992,** *92,* 521–542.
3. Solomon, E. I.; Lowery, M. D. *Science (Washington, D.C.)* **1993,** *259,* 1575–1581.
4. Solomon, E. I.; Lowery, M. D. In *The Chemistry of Copper and Zinc Triads;* Welch, A. J.; Chapman, S. K., Eds.; The Royal Society of Chemistry: Cambridge, England, 1993; pp 12–29.
5. Solomon, E. I.; Hemming, B. L.; Root, D. E. In *Bioinorganic Chemistry of Copper;* Karlin K. D.; Tyeklár, Z., Eds.; Chapman & Hall: New York, 1992; pp 3–20.
6. Solomon, E. I. *Comments Inorg. Chem.* **1984,** *3,* 225–320.
7. Ballhausen, C. J. *Introduction to Ligand Field Theory;* McGraw-Hill: New York, 1962.
8. McClure, D. S. *Electronic Spectra of Molecules and Ions in Crystals;* Academic: New York, 1959.
9. Griffith, J. S. *The Theory of Transition Metal Ions;* Cambridge University Press: London, 1964.
10. Sugano, S.; Tanabe, Y.; Kamimura, H. *Multiplets of Transition Metal Ions in Crystals;* Academic: New York, 1970.
11. Figgis, B. N. *Introduction to Ligand Fields;* Interscience: New York, 1967.
12. Hitchman, M. A.; Cassidy, P. J. *Inorg. Chem.* **1979,** *18,* 1745–1754.
13. Johnson, K. H. *Adv. Quantum Chem.* **1973,** *7,* 143–185.
14. Johnson, K. H.; Norman, J. G., Jr.; Connolly, J. W. D. In *Computational Methods*

for Large Molecules and Localized States in Solids; Herman, F.; McLean, A. D.; Nesbet, R. K., Eds.; Plenum: New York, 1973; pp 161–201.
15. Connolly, J. W. D. In *Semiempirical Methods of Electronic Structure Calculations, Part A: Techniques;* Segal, G. A., Ed.; Plenum: New York, 1977.
16. Rosch, N. In *Electrons in Finite and Infinite Structures;* Phariseu, P.; Scheire, L., Eds.; Wiley: New York, 1977.
17. Slater, J. C. *The Calculation of Molecular Orbitals;* John Wiley & Sons: New York, 1979; p 104.
18. Gewirth, A. A.; Cohen, S. L.; Schugar, H. J.; Solomon, E. I. *Inorg. Chem.* **1987,** *26,* 1133–1146.
19. Solomon, E. I.; Gewirth, A. A.; Cohen, S. L. In *Understanding Molecular Properties;* Avery, J.; Dahl, J. P.; Hansen, A. E., Eds.; D. Reidel: Dordrecht, Netherlands, 1987; pp 27–68.
20. Didziulis, S. V.; Cohen, S. L.; Gewirth, A. A.; Solomon, E. I. *J. Am. Chem. Soc.* **1988,** *110,* 250–268.
21. Bencini, A.; Gatteschi, D. *J. Am. Chem. Soc.* **1983,** *105,* 5535–5541.
22. Chow, C.; Chang, K.; Willett, R. D. *J. Chem. Phys.* **1973,** *59,* 2629–2640.
23. Solomon, E. I.; Lowery, M. D.; LaCroix, L. B.; Root, D. E. In *Methods in Enzymology;* Riordan, J. F.; Vallee, B. L., Eds.; 1993; Vol. 226, Part C; pp 1–33.
24. Desjardins, S. R.; Penfield, K. W.; Cohen, S. L.; Musselman, R. L.; Solomon, E. I. *J. Am. Chem. Soc.* **1983,** *105,* 4590–4603.
25. Mulliken, R. S.; Rieke, C. A.; Orloff, D.; Orloff, H. *J. Chem. Phys.* **1949,** *17,* 1248–1267.
26. Solomon, E. I.; Hare, J. W.; Gray, H. B. *Proc. Natl. Acad. Sci. U.S.A.* **1976,** *73,* 1389–1392.
27. Guss, J. M.; Freeman, H. C. *J. Mol. Biol.* **1983,** *169,* 521–563.
28. Lowery, M. D.; Guckert, J. A.; Gebhard, M. S.; Solomon, E. I. *J. Am. Chem. Soc.* **1993,** *115,* 3012–3013.
29. Penfield, K. W.; Gay, R. R.; Himmelwright, R. S.; Eickman, N. C.; Norris, V. A.; Freeman, H. C.; Solomon, E. I. *J. Am. Chem. Soc.* **1981,** *103,* 4382–4388.
30. Bates, C. A.; Moore, W. S.; Standley, K. J.; Stevens, K. W. H. *Proc. Phys. Soc.* **1962,** *79,* 73.
31. Sharnoff, M. *J. Chem. Phys.* **1965,** *42,* 3383–3395.
32. Roberts, J. E.; Brown, T. G.; Hoffman, B. M.; Peisach, J. *J. Am. Chem. Soc.* **1980,** *102,* 825–829.
33. Gewirth, A. A.; Solomon, E. I. *J. Am. Chem. Soc.* **1988,** *110,* 3811–3819.
34. Shadle, S. E.; Penner-Hahn, J. E.; Schugar, H. J.; Hedman, B.; Hodgson, K. O.; Solomon, E. I. *J. Am. Chem. Soc.* **1993,** *115,* 767–776.
35. Scott, R. A.; Hahn, J. E.; Doniach, S.; Freeman, H. C.; Hodgson, K. O. *J. Am. Chem. Soc.* **1982,** *104,* 5364–5369.
36. Penfield, K. W.; Gewirth, A. A.; Solomon, E. I. *J. Am. Chem. Soc.* **1985,** *107,* 4519–4529.
37. Norman, J. G. J. *Mol. Phys.* **1976,** *31,* 1191–1198.
38. George, S. J.; Lowery, M. D.; Solomon, E. I.; Cramer, S. P. *J. Am. Chem. Soc.* **1993,** *115,* 2968–2969.
39. Hughey, J. L., IV; Fawcett, T. G.; Rudich, S. M.; Lalancette, R. A.; Potenza, J. A.; Schugar, H. J. *J. Am. Chem. Soc.* **1979,** *101,* 2617–2623.
40. Messerschmidt, A.; Ladenstein, R.; Huber, R.; Bolognesi, M.; Avigliano, L.; Petruzzelli, R.; Rossi, A.; Finazzi-Agro, A. *J. Mol. Biol.* **1992,** *224,* 179–205.

41. Cole, J. L.; Ballou, D. P.; Solomon, E. I. *J. Am. Chem. Soc.* **1991,** *113,* 8544–8546.
42. Jolly, R. L., Jr.; Evans, L. H.; Makino, N.; Mason, H. S. *J. Biol. Chem.* **1974,** *249,* 335.
43. Himmelwright, R. S.; Eickman, N. C.; LuBien, C. D.; Lerch, K.; Solomon, E. I. *J. Am. Chem. Soc.* **1980,** *102,* 7339–7344.
44. Woolery, G. L.; Powers, L.; Winkler, M.; Solomon, E. I.; Spiro, T. G. *J. Am. Chem. Soc.* **1984,** *106,* 86–92.
45. Woolery, G. L.; Powers, L.; Winkler, M.; Solomon, E. I.; Lerch, K.; Spiro, T. G. *Biochim. Biophys. Acta* **1984,** *788,* 155–161.
46. Freedman, T. B.; Loehr, J. S.; Loehr, T. M. *J. Am. Chem. Soc.* **1976,** *98,* 2809–2815.
47. Larrabee, J. A.; Spiro, T. G. *J. Am. Chem. Soc.* **1980,** *102,* 4217–4223.
48. Eickman, N. C.; Solomon, E. I.; Larrabee, J. A.; Spiro, T. G.; Lerch, K. *J. Am. Chem. Soc.* **1978,** *100,* 6529–6531.
49. Solomon, E. I.; Dooley, D. M.; Wang, R.-H.; Gray, H. B.; Cerdonio, M.; Mongo, F.; Romani, G. L. *J. Am. Chem. Soc.* **1976,** *98,* 1029–1031.
50. Dooley, D. M.; Scott, R. A.; Ellinghaus, J.; Solomon, E. I.; Gray, H. B. *Proc. Natl. Acad. Sci. U.S.A.* **1978,** *75,* 3019–3022.
51. Fee, J. A.; Malkin, R. M.; Malmström, B. G.; Vänngård, T. *J. Biol. Chem.* **1969,** *88,* 4200–4207.
52. Volbeda, A.; Hol, W. G. J. *J. Mol. Biol.* **1989,** *209,* 249–279.
53. Eickman, N. C.; Himmelwright, R. S.; Solomon, E. I. *Proc. Natl. Acad. Sci. U.S.A.* **1979,** *76,* 2094–2098.
54. Karlin, K. D.; Cruse, R. W.; Gultneh, Y.; Farooq, A.; Hayes, J. C.; Zubieta, J. *J. Am. Chem. Soc.* **1987,** *109,* 2668–2679.
55. Pate, J. E.; Cruse, R. W.; Karlin, K. D.; Solomon, E. I. *J. Am. Chem. Soc.* **1987,** *109,* 2624–2630.
56. Pate, J. E.; Ross, P. K.; Thamann, T. J.; Reed, C. A.; Karlin, K. D.; Sorrell, T. N.; Solomon, E. I. *J. Am. Chem. Soc.* **1989,** *111,* 5198–5209.
57. Ross, P. K.; Allendorf, M. D.; Solomon, E. I. *J. Am. Chem. Soc.* **1989,** *111,* 4009–4021.
58. Tuczek, F.; Solomon, E. I. *Inorg. Chem.* **1993,** *32,* 2850–2862.
59. Solomon, E. I.; Tuczek, F.; Root, D. E.; Brown, C. A. *Chem. Rev.* **1994,** *92,* 827–856.
60. Tuczek, F.; Solomon, E. I. *J. Am. Chem. Soc.* **1994,** *116,* 6916–6924.
61. Sorrell, T. N.; O'Connor, C. J.; Anderson, O. P.; Reibenspies, J. H. *J. Am. Chem. Soc.* **1985,** *107,* 4199–4206.
62. McKee, V.; Dagdigian, J. V.; Bau, R.; Reed, C. A. *J. Am. Chem. Soc.* **1981,** *103,* 7000–7001.
63. McKee, V.; Zvagulis, M.; Dagdigian, J. V.; Patch, M. G.; Reed, C. A. *J. Am. Chem. Soc.* **1984,** *106,* 4765–4772.
64. Karlin, K. D.; Cohen, B. I.; Hayes, J. C.; Farooq, A.; Zubieta, J. *Inorg. Chem.* **1987**, *26,* 147–153.
65. Kitajima, N.; Fujisawa, K.; Moro-oka, Y.; Toriumi, K. *J. Am. Chem. Soc.* **1989,** *111,* 8975–8976.
66. Kitajima, N.; Fujisawa, K.; Fujimoto, C.; Moro-oka, Y.; Hashimoto, S.; Kitagawa, T.; Toriumi, K.; Tatsumi, K.; Nakamura, A. *J. Am. Chem. Soc.* **1992,** *114,* 1277–1291.
67. Magnus, K.; Ton-That, H. *J. Inorg. Biochem.* **1992,** *47,* 20.
68. Ross, P. K.; Solomon, E. I. *J. Am. Chem. Soc.* **1990,** *112,* 5871–5872.
69. Ross, P. K.; Solomon, E. I. *J. Am. Chem. Soc.* **1991,** *113,* 3246–3259.

70. Noodleman, L.; Norman, J. G., Jr. *J. Chem. Phys.* **1979,** *70,* 4903.
71. Noodleman, L. *J. Chem. Phys.* **1981,** *74,* 5737–5743.
72. Baldwin, M. J.; Ross, P. K.; Pate, J. E.; Tyeklar, Z.; Karlin, K. D.; Solomon, E. I. *J. Am. Chem. Soc.* **1991,** *113,* 8671–8679.
73. Baldwin, M. J.; Root, D. E.; Pate, J. E.; Fujisawa, K.; Kitajima, N.; Solomon, E. I. *J. Am. Chem. Soc.* **1992,** *114,* 10421–10431.
74. Jacobson, R. R.; Tyeklar, Z.; Farooq, A.; Karlin, K. D.; Liu, S.; Zubieta, J. *J. Am. Chem. Soc.* **1988,** *110,* 3690–3692.
75. Cole, J. L.; Clark, P. A.; Solomon, E. I. *J. Am. Chem. Soc.* **1990,** *112,* 9534–9548.
76. Malmström, B. G. In *New Trends in Bio-inorganic Chemistry;* Williams, R. J. P.; DaSilva, R. J. R. F., Eds.; Academic: London, 1978; pp 59–78.
77. Fee, J. A. *Struct. Bonding* **1975,** *23,* 1–60.
78. Reinhammar, B. In *Copper Proteins and Copper Enzymes;* Lontie, R., Ed.; CRC Press: Boca Raton, 1984; Vol. III; pp 1–31.
79. Finazzi-Agro, A. *Life Chem. Rep.* **1987,** *5,* 199–209.
80. Ryden, L. In *Copper Proteins and Copper Enzymes;* R. Lontie, Ed.; CRC Press: Boca Raton, 1984; Vol. III.
81. Graziani, M. T.; Morpurgo, L.; Rotilio, G.; Mondovi, B. *FEBS Lett.* **1976,** *70,* 87–90.
82. Morie-Bebel, M. M.; Morris, M. C.; Menzie, J. L.; McMillin, D. R. *J. Am. Chem. Soc.* **1984,** *106,* 3677–3678.
83. Reinhammar, B. R. M. *Biochim. Biophys. Acta* **1972,** *275,* 245–259.
84. Spira-Solomon, D. J.; Solomon, E. I. *J. Am. Chem. Soc.* **1987,** *109,* 6421–6432.
85. LuBien, C. D.; Winkler, M. E.; Thamann, T. J.; Scott, R. A.; Co, M. S.; Hodgson, K. O.; Solomon, E. I. *J. Am. Chem. Soc.* **1981,** *103,* 7014–7016.
86. Kau, L.-S.; Spira-Solomon, D. J.; Penner-Hahn, J. E.; Hodgson, K. O.; Solomon, E. I. *J. Am. Chem. Soc.* **1987,** *109,* 6433–6442.
87. Messerschmidt, A. *Adv. Inorg. Chem.* **1993,** *40,* 121–185.
88. Westmoreland, T. D.; Wilcox, D. E.; Baldwin, M. J.; Mims, W. B.; Solomon, E. I. *J. Am. Chem. Soc.* **1989,** *111,* 6106–6123.
89. Allendorf, M. D.; Spira, D. J.; Solomon, E. I. *Proc. Natl. Acad. Sci. U.S.A.* **1985,** *82,* 3063–3067.
90. Spira-Solomon, D. J.; Allendorf, M. D.; Solomon, E. I. *J. Am. Chem. Soc.* **1986,** *108,* 5318–5328.
91. Stephens, P. J. *Adv. Chem. Phys.* **1976,** *35,* 197.
92. Messerschmidt, A.; Rossi, A.; Ladenstein, R.; Huber, R.; Bolognesi, M.; Guiseppina, G.; Marchesini, A.; Petruzzelli, R.; Finazzi-Agro, A. *J. Mol. Biol.* **1989,** *206,* 513–529.
93. Reinhammar, B.; Oda, Y. *J. Inorg. Biochem.* **1979,** *11,* 115–127.
94. Farver, O.; Goldberg, M.; Pecht, I. *Eur. J. Biochem.* **1980,** *104,* 71–77.
95. Cole, J. L.; Tan, G. O.; Yang, E. K.; Hodgson, K. O.; Solomon, E. I. *J. Am. Chem. Soc.* **1990,** *112,* 2243–2249.

RECEIVED for review July 19, 1993. ACCEPTED revised manuscript May 10, 1994.

6

Biomimetic Copper–Dioxygen Chemistry

Reversible O_2-Binding and Mechanistic Insights into Cu(I)/O_2-Mediated Arene Hydroxylation and Amide Hydrolysis

Narasimha N. Murthy and Kenneth D. Karlin[1]

Department of Chemistry, Remsen Hall, The Johns Hopkins University, Charles and 34th Streets, Baltimore, MD 21218–2685

Copper(I) complexes that bind dioxygen (O_2) reversibly can also effect either the biomimetic hydroxylation of arenes, or the hydrolysis of an unactivated amide such as dimethylformamide (DMF). Several classes of Cu_2O_2 (i.e., peroxo-dicopper(II)) complexes have been generated; an X-ray structure of a trans μ-1,2 peroxo complex, the kinetics and thermodynamics of formation, and spectroscopic characterizations have been achieved. One dinucleating ligand system affords μ-η^2:η^2 peroxo dicopper(II) species, subsequently attacking an arene substrate that is part of the ligand framework; this attack occurs via an electrophilic mechanism, accompanied by an "NIH shift." A different reaction pathway for hydrolysis of DMF occurs when using a dinucleating ligand that permits adjacent coordination of hydroxide and amide substrate. Possible mechanisms and the biological relevance of hydrolysis reactions mediated by dinuclear metal complexes are discussed.

THE COPPER(II) AND COPPER(I) IONS undergo facile redox interconversions for which the standard reduction potential is highly dependent on the nature of the ligands and coordination geometries observed (*1*). Thus, copper ion is a useful electron transfer or oxidation catalyst in the presence of dioxygen (O_2) (*2–4*). These properties have been put to advantage by nature, where copper-containing proteins (*5–11*) exist as electron

[1] Corresponding author

0065–2393/95/0246–0165/$09.62/0

transfer agents (*10, 11*), O_2-carriers (hemocyanin (Hc)) (Table I), oxygenases involved in O-atom incorporation into biological substrates (Table I), oxidases effecting dehydrogenation reactions while reducing O_2 to water (*12–14*), and even nitrogen-oxide reductases (*10, 15*). Much of our own effort has been concerned with O_2-reactions with synthetic copper(I) complexes, with particular interest in establishing the basic coordination chemistry involved in $Cu(I)_n/O_2$ interactions and reactivity patterns, in particular $Cu_n–O_2$ ($n = 1,2$) reversible binding, structure, and spectroscopy, and reactivity of these or derived species with substrates (*6, 7*). In this chapter, we describe a number of these investigations, bearing on models for hemocyanin and copper monooxygenases such as tyrosinase. As an outcome of these studies, we discovered a dicopper complex system effecting the hydrolysis of an unactivated amide under mild conditions. The dicopper complex system may be relevant to reactivity of an emerging area of metallobiochemistry, which involves peptidase or phosphatase hydrolytic enzymes containing di- or trinuclear metal ion active-site centers.

Hemocyanin and Reversible O_2-Binding in Model Compounds

Although there is a considerable body of data and understanding of heme-iron O_2-carriers and cytochrome P-450 monooxygenases and their model compound chemistry (*24–26*), much less is known about nonheme iron (*27*) and copper proteins (*5–23*) involved in O_2 metabolism (Table I) (*28*). Much of the attention in inorganic functional modeling of copper proteins involved in O_2 use has focused on hemocyanin (*5–10, 16, 17*), because of its intriguing dinuclear copper center that turns from colorless to intense blue upon conversion from its deoxy to oxy form (Figure 1). Invertebrate molluscan and arthropodal hemocyanins (Hcs) are very large (MW $4.5–90 \times 10^5$) O_2-transporting proteins. They consist of highly cooperative multisubunits, with molluscan hemocyanins containing 10 or 20 subunits, in which the functional unit has a molecular weight ~ 55,000. Arthropodal Hcs are hexamers or multihexamers with larger subunits (~75,000). Although differences exist between both classes of proteins, a variety of data indicate a closely related active site structure and binding mode.

Reduced hemocyanin (i.e., deoxy-Hc) is colorless, indicative of a $3d^{10}$ copper(I) formulation. Crystal structures of two deoxy forms are now available, and an investigation of the horseshoe crab *Limulus II* protein indicates that the two Cu(I) ions are 4.6 Å apart, each found in a trigonal-planar coordination environment with $Cu–N_{His} \sim 2.0$ Å (*17*). There is no bridging ligand in this form, and cooperative effects of O_2 binding are probably initiated and transmitted as a result of movement

Table I. Iron and Copper O_2-Carriers and Monooxygenases

Function	*Heme Iron*	*Nonheme Iron*	*Copper*
Dioxygen Transport			
$xM^{n+} + O_2 \rightleftarrows M^{(n+1)} + x(O_2^{x-})$	hemoglobin[a] myoglobin[a]	hemerythrin[a] myohemerythrin	hemocyanin[a] (*16, 17*)
Monooxygenases			
$XH_2 + O_2 \rightarrow XO + H_2O$ or $X + O_2 + DH_2 \rightarrow XO + H_2O + D$	cytochrome P-450 monooxygenases[a] secondary amine monooxygenase	*soluble*-methane monooxygenase[a] pteridine-dependent hydroxylases	tyrosinase (*9*) dopamine β-hydroxylase (*18, 19*) phenylalanine hydroxylase (*20*) peptidylglycine α-amidating monooxygenase (*21, 22*)
(D = e^- Donor)	nitric oxide synthase		*particulate*-methane monooxygenase (*23*)

[a] X-ray structure is available. References are given for the copper proteins only.

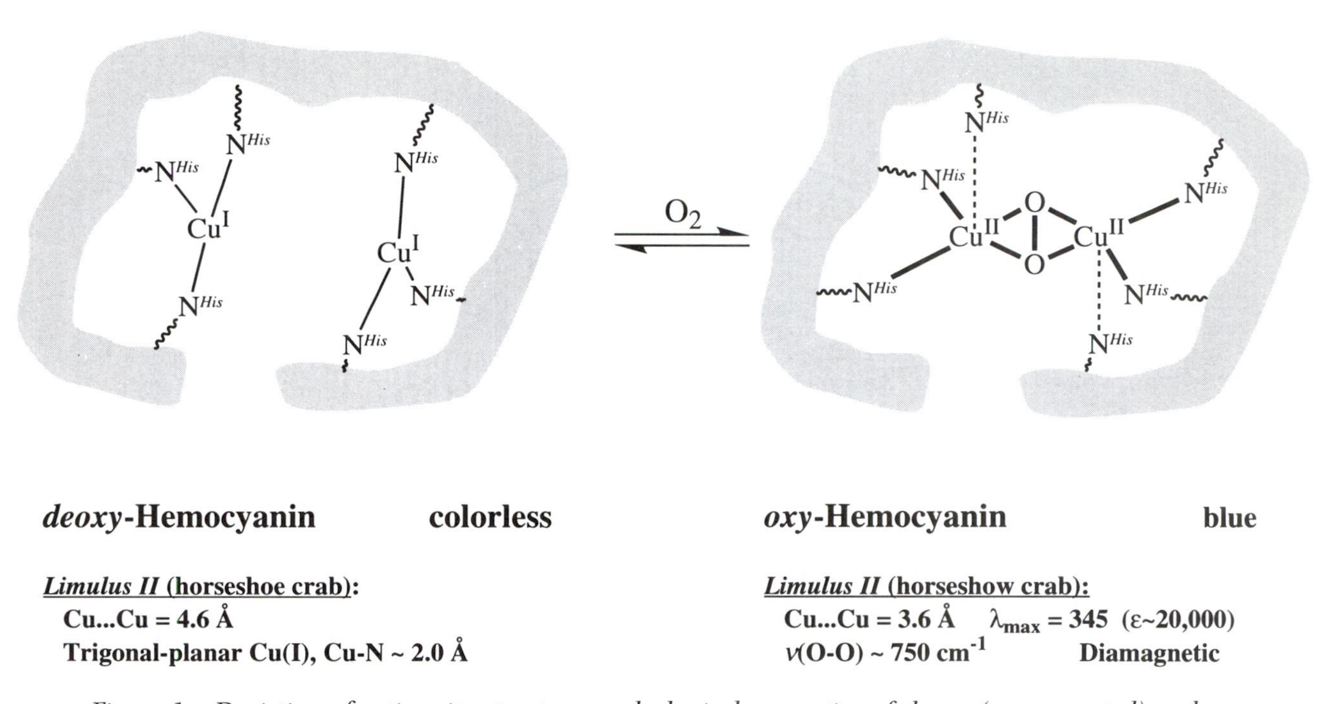

Figure 1. Depiction of active site structures and physical properties of deoxy (unoxygenated) and oxy (oxygenated) hemocyanin.

of histidine residue(s) and copper ion upon O_2 binding. Resonance Raman studies of the blue oxy-Hc form show that the dioxygen is bound as peroxide with $\nu_{O-O} \sim 750\ cm^{-1}$ (9). This dioxygen binding indicates an oxidative addition of O_2 to give a peroxodicopper(II) oxy-form. This oxy site is electron paramagnetic resonance (EPR) silent and diamagnetic, reflecting strong magnetic coupling between the two Cu(II) centers. An X-ray structure (*16*) has revealed a side-on μ-η^2:η^2-peroxo ligation with Cu· · ·Cu = 3.6 Å. The electronic spectrum of oxy-Hc is distinctive and dominated by $O_2^{2-} \rightarrow$ Cu(II) ligand-to-metal charge transfer transitions at λ_{max} of 345 (ϵ = 20,000 $M^{-1}\ cm^{-1}$) and 570 (ϵ = 1000 $M^{-1}\ cm^{-1}$) nm, with an additional circular dichroic feature at 485 nm (9). Kitajima and co-workers have described a structurally characterized model compound {Cu[HB(3,5-*i*Pr$_2$pz)$_3$]}$_2$(O_2), [HB(3,5-*i*Pr$_2$pz)$_3$ = hydrotris(3,5-diisopropyl-pyrazolyl)borate anion], with essentially identical peroxo-dicopper(II) ligation and physical properties (*29*) (Scheme 1).

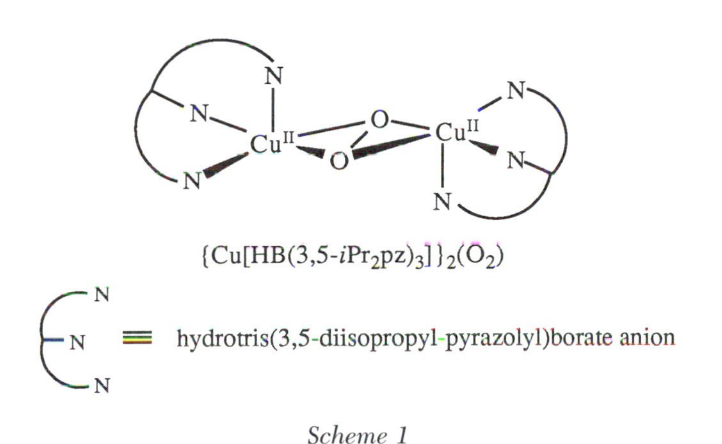

Scheme 1

A *trans*-μ-1,2-Peroxo Dicopper(II) Complex. Our own efforts have resulted in the structural and spectroscopic characterization of five types of copper–dioxygen complexes (*6*), distinguished on the basis of the ligands used for their synthesis and on their distinctive structures or physical properties. Thus, the manner in which hemocyanin binds O_2 is not the only one possible, and it is of considerable interest to deduce the structures, along with associated spectroscopy and reactivity of a variety of types. Dioxygen can bind to dinuclear transition metals in a variety of structural modes, shown in Figure 2. As mentioned, mode **C** is present in oxy-Hc and Kitajima's model complex (Scheme 1), whereas we have structural and spectroscopic evidence for types **A** (*30–32*), **B** (*33–35*), and **F** (*36–38*) for peroxo O_2^{2-} binding, and mode **D** (*39, 40*) in the case of hydroperoxo (OOH^-) complexes.

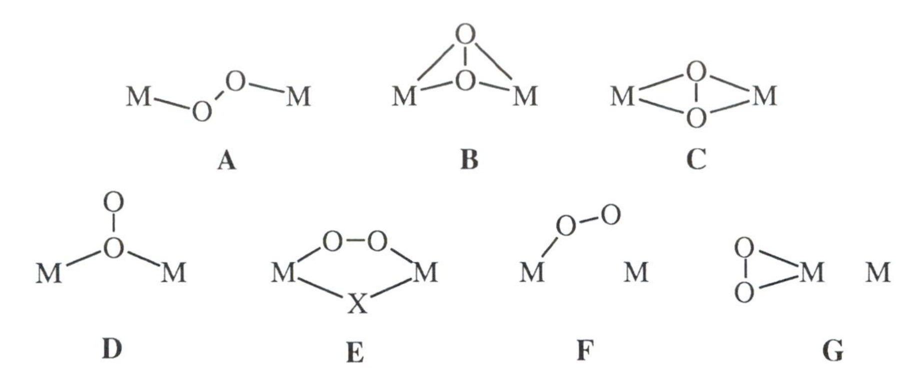

Figure 2. Possible O_2 (peroxide) binding modes to a dinuclear metal center.

The most well-characterized copper–dioxygen complex studied in our laboratories is $[\{(TMPA)Cu\}_2(O_2)]^{2+}$ (**3**, TMPA is tris[2-pyridylmethyl]amine), formed by reaction of mononuclear Cu(I) complex $[(TMPA)Cu(RCN)]^+$ (**1**) when reacted with O_2 at −80 °C (Figure 3) (*30*). Dioxygen binds strongly at low temperature in dichloromethane or propionitrile solvents to form an intense purple solution with λ_{max} = 440 (ϵ = 2000 M^{-1} cm^{-1}), 525 (11,500), 590 (sh, 7600) and a *d–d* band at 1035 (180) nm. Although the binding of O_2 is strong at low temperature,

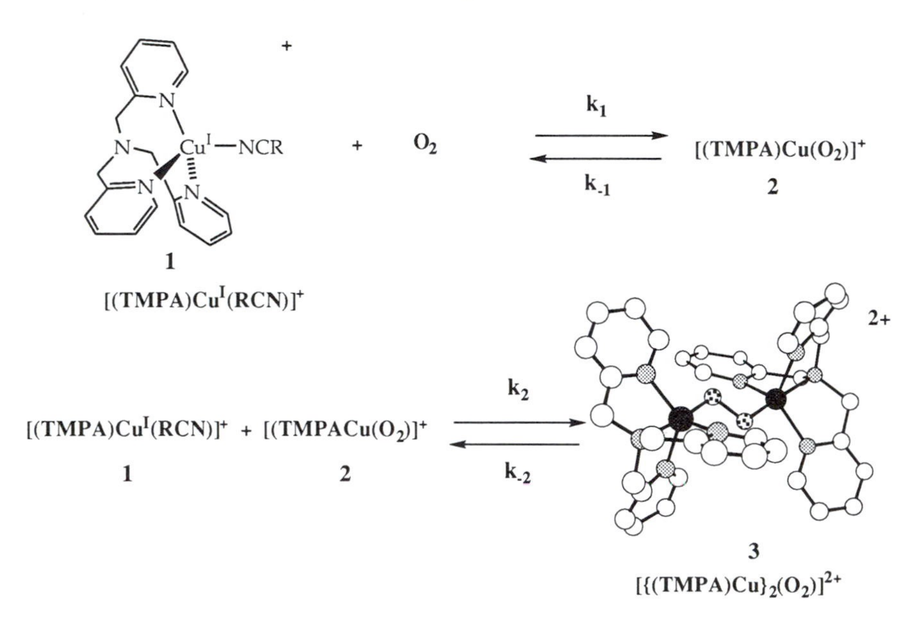

Figure 3. Scheme indicating the kinetics scheme for formation of $[\{(TMPA)Cu\}_2(O_2)]^{2+}$ (3) and Chem 3-D depiction of the X-ray structure of the product with trans-*μ-1,2-(O_2^{2-})-dicopper(II) coordination.*

it is reversible, as demonstrated by vacuum cycling experiments, i.e., application of vacuum while subjecting the solution to mild warming. Dioxygen can also be displaced from **3** by reaction with CO or PPh_3 to give the adducts $[(TMPA)Cu(CO)]^+$ and $[(TMPA)Cu(PPh_3)]^+$, respectively. X-ray data obtained for crystals of $[\{(TMPA)Cu\}_2(O_2)](PF_6)_2 \cdot 5Et_2O$ at −90 °C reveals a *trans*-μ-1,2-(O_2^{2-})-dicopper(II) coordinaton in **3** (*30, 31*). The Cu ions are pentacoordinate with distorted trigonal bipyramidal geometry and the peroxo oxygen atoms occupy the axial sites. The Cu–Cu′ distance is 4.359 Å and the O–O′ bond length is 1.432 Å. Resonance Raman studies showed an intraperoxide stretch (832 cm^{-1}) and a Cu–O stretch (561 cm^{-1}) (*32*).

A detailed kinetic study has been carried out for the reaction of $[(TMPA)Cu(RCN)]^+$ (**1**) with O_2, in collaboration with A. D. Zuberbühler (of the University of Basel) (Figure 3) (*41, 42*). The investigation revealed the intermediacy of an initially formed 1:1 adduct $[(TMPA)Cu(O_2)]^+$ (**2**), with λ_{max} = 410 nm, ϵ = 4000 M^{-1} cm^{-1}. The initial O_2 binding event occurs with an on-rate $k_1 = 2 \times 10^4$ M^{-1} s^{-1} and $k_1/k_{-1} = K_1 = 1.9 \times 10^3$ M^{-1} at 183 K, whereas values extrapolated to 25 °C are $k_1 \sim 10^8$ M^{-1} s^{-1} and $K_{form} \sim 0.3$ M^{-1}. These data represent the first of their kind for a primary 1:1 copper–dioxygen interaction. In comparison with a variety of cobalt or iron species that bind O_2, the formation of **2** is faster than rates seen for most LCo(II) 1:1 oxygenation reactions, where $k \sim 10^3$–10^6 M^{-1} s^{-1} (25 °C). For heme-proteins or porphyrin-Fe(II) model complexes, the O_2 on-rates ($k \sim 10^6$–10^9) are similar to that seen for formation of **2**. However, the off-rates for iron species appear to be much smaller, giving rise to large K_{eq} values in the range of 10^4–10^6 at 20 °C. This study and others reveal that the main source of the room temperature instability of Cu_2–O_2 complexes is their highly unfavorable $\Delta S°$ of formation.

N_n and Xylyl Dinuclear Complexes. Copper–dioxygen complexes (e.g., peroxo-dicopper(II) species) with a binding mode very similar to oxy-Hc, i.e., a side-on μ-η^2:η^2-peroxo ligation that appears to be bent (Figure 2, mode **B**) are formed in complexes $[Cu_2(L)(O_2)]^{2+}$, where L represents a dinucleating ligand having two PY2 (PY2 is bis[2-(2-pyridyl)ethyl]amine) units connected to a hydrocarbon linker, i.e., L = N_n (*33–35*), N_3OR (*34*) or R-XYL-X (*43–45*) (Scheme 2). The xylyl-containing systems not only bind O_2, but undergo a subsequent ligand hydroxylation reaction.

For tricoordinate complexes $[Cu_2(N_n)]^{2+}$ (**4a**) or tetracoordinate nitrile adducts $[Cu_2(N_n)(CH_3CN)_2]^{2+}$ (**4b**), reversible binding can be demonstrated (*33, 35*). These react with O_2 at −80 °C in CH_2Cl_2 solution, producing deep brown or purple species $[Cu_2(N_n)(O_2)]^{2+}$ (**5**) (*33, 35*). As found for our other copper–dioxygen complexes, reversible O_2 or

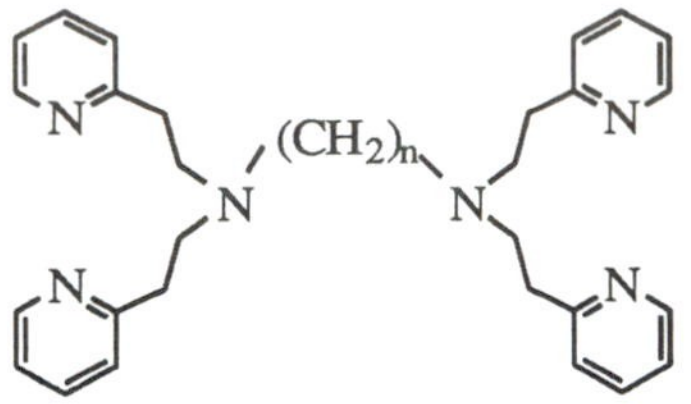

Nn (n = 3-5)

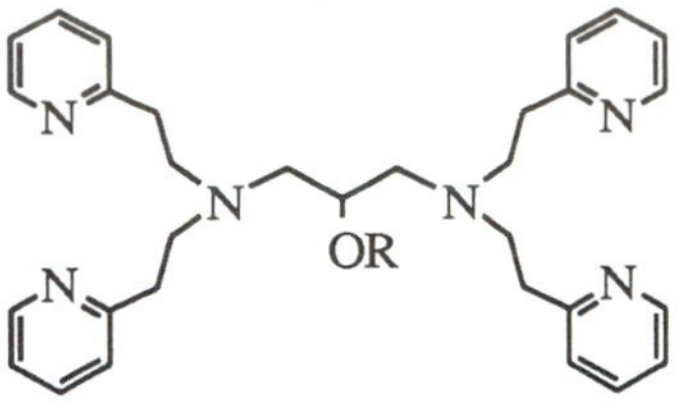

N3OR (R = $C(O)C_6H_4$-*p*-C_6H_5) or $C(O)(CH{=}CH)C_6H_5$)

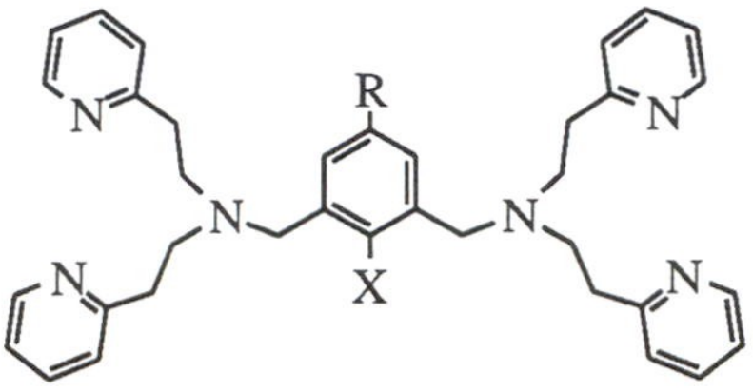

R-XYL-X (X = H, F, CH_3)

Scheme 2

CO binding can be followed spectrophotometrically; application of a vacuum while heating the complexes briefly removes the bound gaseous ligand, regenerating the colorless or light yellow-brown dicopper(I) precursor complexes **4a**. Release of intact O_2 from **5** has also been demonstrated. Carbon monoxide binding to **4a** is stronger than O_2, because CO can displace O_2 from **5** to give $[Cu_2(N_n)(CO)_2]^{2+}$ (**4c**). Thus, the relative binding strength of CO versus O_2 to reduced copper ion parallels that observed for heme-proteins and porphyrin-iron(II) complexes.

$[Cu_2(N_n)(O_2)]^{2+}$ (**5**) possess striking UV–visible (UV–vis) properties, with multiple and strong charge-transfer absorptions. The position and relative intensities of these bands vary with the length of the polymethylene unit connecting the two PY2 donor groups (*35*), reflecting subtle changes in the mode of O_2 binding. The characteristic 350–360 nm band with ϵ = 16,000–21,000 M^{-1} cm^{-1} dominates; the presence of this distinctive intense absorption in part provides indications for the possible close structural relationship of $[Cu_2(N_n)(O_2)]^{2+}$ (**5**) to the Cu_2O_2 oxy-Hc chromophore, with its 345 nm (ϵ = 20,000) feature.

We have yet to obtain vibrational data for complexes $[Cu_2(N_n)(O_2)]^{2+}$ (**5**), but a variety of other evidence is consistent with their peroxo-dicopper(II) formulation. These O_2-adducts (**5**) possess low-energy weak *d–d* absorptions, diagnostic of Cu(II) and not Cu(I), which has a filled shell d^{10} electronic configuration. X-ray absorption near-edge structure (**XANES**) measurements carried out on the N_3 and N_4 derivatives confirm the Cu(II) oxidation state (*34*), and extended X-ray absorption fine structure (EXAFS) spectroscopic data also allow the determination of the Cu· · ·Cu distances, which vary between 3.3 and 3.4 Å, depending on *n*. Further analysis of the outer shell multiple scattering effects due to axial versus equatorial pyridine ligands led to the proposed μ-η^2:η^2-peroxo coordination for $[Cu_2(N_n)(O_2)]^{2+}$ (**5**) (Scheme 3), which is mode **B** (Figure 2) shown in a previous paragraph. The "bent butterfly" structure is presumably caused by ligand constraints. Species **5** are EPR silent, have normal ^{1}H NMR spectroscopic properties and exhibit solution diamagnetism (*35*); thus, they appear to provide another class of compounds in which a single peroxo ligand bridges and strongly antiferromagnetically couples two Cu(II) ions. The proposed unusual bent side-on peroxo binding proposed in **5** is also justified in light of X-ray structures of (i) the μ-η^2:η^2-peroxo dicopper(II) complex structurally characterized by Kitajima and co-workers (Scheme 1) (*29*) and (ii) a structurally characterized bent μ-η^2:η^2-peroxo divanadium(V) complex (*46*).

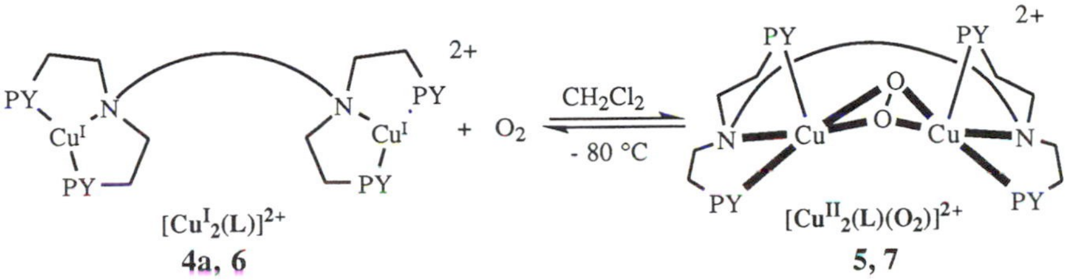

$[Cu^I_2(L)]^{2+}$ **4a, 6** + O_2 $\xrightleftharpoons[-80\ °C]{CH_2Cl_2}$ $[Cu^{II}_2(L)(O_2)]^{2+}$ **5, 7**

Scheme 3

A Functional Model for Copper Monooxygenases and the "NIH Shift" Mechanism

Dicopper(I) complexes with a R–XYL–H xylyl dinucleating ligand (Scheme 2) not only react with O_2 reversibly, but undergo a further hydroxylation reaction. Thus, reaction of **6** with O_2 leads to **8**, from which the free phenol R–XYL–OH can be isolated, completing the copper-mediated hydroxylation of the arene (Figure 4) (*43–45*). Both the precursor three-coordinate dicopper(I) complex $[Cu_2(R–XYL–H)]^{2+}$ (**6**, R = H; Cu· · ·Cu = 8.9 Å) and the hydroxylated product $[Cu_2(R–XYL–O–)(OH)]^{2+}$ (**8**, R = H; Cu· · ·Cu = 3.1 Å), with a phenoxo and hydroxo doubly bridged dicopper(II) coordination have been characterized by X-ray crystallography (*47*). The reaction of **6** with O_2 in DMF or CH_2Cl_2 provides nearly quantitative yields of $[Cu_2(XYL–O–)(OH)]^{2+}$ (**8**) and isotopic-labeling experiments using $^{18}O_2$ revealed that the source of oxygen atoms in **8** is dioxygen (*47*). Along with the observed reaction stoichiometry (Cu:O_2 = 2:1, manometry), the conversion **6** → **8** is reminiscent of the action of the copper monooxygenase tyrosinase. Here, a

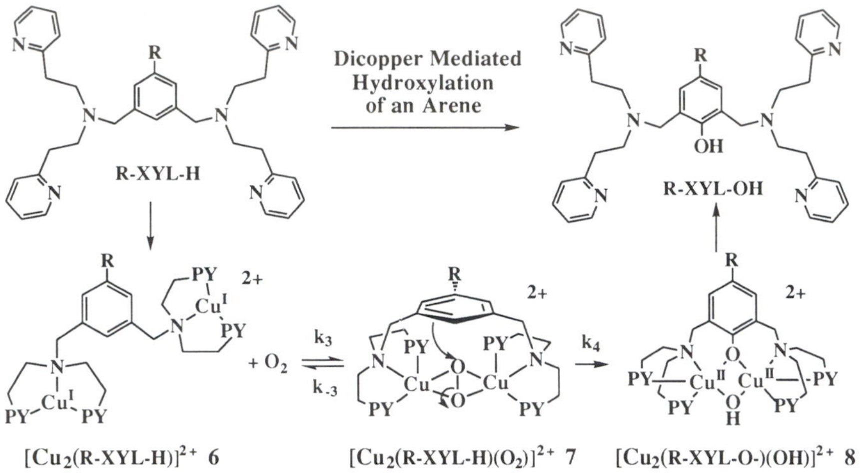

Figure 4. A copper monooxygenase model system, involving the dicopper-mediated hydroxylation of an arene.

dinuclear copper active center spectroscopically similar to that seen in hemocyanin effects an arene hydroxylation, converting phenols to *o*-catechols, which are subsequently oxidized to *o*-quinones (Scheme 4). Thus, key features also seen in our model system are (i) reversible binding of O_2 to the dinuclear copper(I) center and (ii) "appropriate" placement of the substrate for the subsequent O-atom transfer reaction. Thus, mechanistic insights were sought, because this biomimetic reaction system represents one in which dioxygen is incorporated into an unactivated (aromatic) C–H bond, under very mild conditions, i.e., in a very rapid reaction occurring at room temperature with 1 atm external O_2 pressure.

Low-temperature stopped-flow kinetic-spectroscopic studies with A. D. Zuberbühler (*43, 44*), provided the first evidence of reversible (i.e., k_3/k_{-3}, Figure 4) O_2 binding to $[Cu_2(R–XYL–H)]^{2+}$ **6** (R = H) in a $Cu:O_2$ = 2:1 stoichiometry (Figure 4). Multiwavelength data analyses ($\lambda > 360$ nm) revealed distinctive features attributable to intermediate species $[Cu_2(R–XYL–H)(O_2)]^{2+}$ (**7**), possessing a strong band in the 435–440 nm ($\epsilon \sim 3000–5000$) range. To illustrate the magnitude of the kinetics and thermodynamic parameters, $k_3 = 410\ M^{-1}\ s^{-1}$ and $k_3/k_{-3} = K_3 = 2.9 \times 10^7\ M^{-1}$ at 183 K for **6** (R = H), with $\Delta H° = -62 \pm 1$ kJ mol^{-1} and $\Delta S° = 196 \pm 6$ J K^{-1} mol^{-1} (*44*). Corroborative experimental information comes from bench-top studies of certain of these synthetic analogues. Thus, for the R = NO_2, F, and CN derivatives, the hydroxylation process is slowed to the point that the $[Cu_2(R–XYL–H)(O_2)]^{2+}$ (**7**) intermediates are stabilized at −80 °C and observable by usual low-temperature UV–vis spectroscopic methods. The similarity of spectral fea-

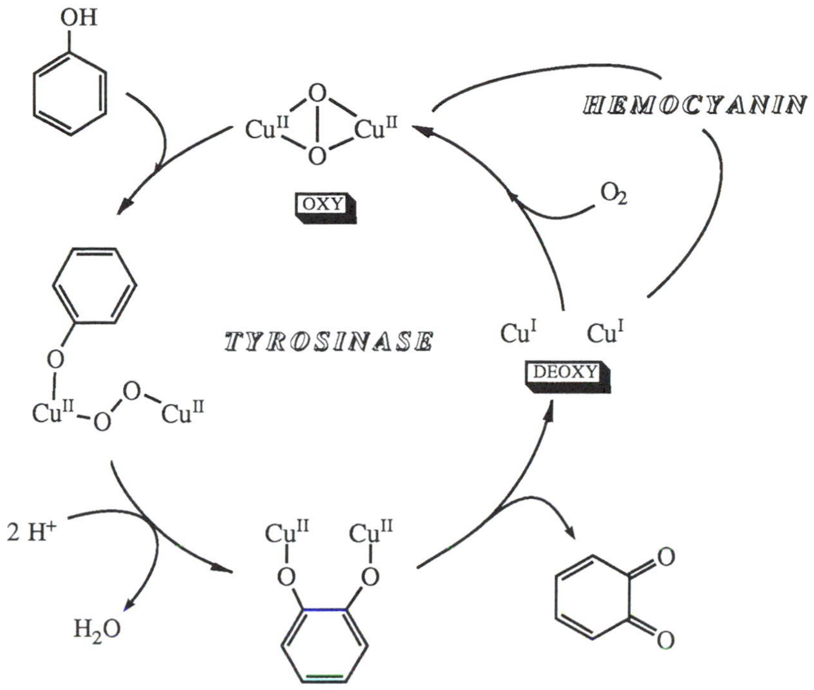

Scheme 4

tures of **7** with $[Cu_2(N_n)(O_2)]^{2+}$ (**5**), along with their closely related ligand structure, suggests that these xylyl derivatives also have the μ-η^2:η^2-peroxo coordination to two Cu(II) ions, i.e., Scheme 3.

The kinetic studies also indicated that the initial reversible O_2 binding to $[Cu_2(R–XYL–X)]^{2+}$ (**6**) is followed by an irreversible hydroxylation step (k_4, Figure 4). There is no measurable effect upon k_4 when X is deuterium. This finding is consistent with electrophilic attack on the arene substrate π-system, precluding rate-determining C–H bond cleavage. Supporting this view, we also find an increase in $\Delta H^{\neq}$ of the hydroxylation step (k_4) with electron-withdrawing ability of R, when studying the oxygenation of complexes $[Cu_2(R–XYL–H)]^{2+}$ (**6**, R = *t*-Bu, F, H, and NO_2).

The notion of an electrophilic attack mediated by the peroxo group in $[Cu_2(R–XYL–H)(O_2)]^{2+}$ (**7**) is also in accord with studies on reactivity comparisons of three classes of peroxo-dicopper(II) complexes, including $[\{(TMPA)Cu\}_2(O_2)]^{2+}$ (**3**) and $[Cu_2(N_n)(O_2)]^{2+}$ (**5**, n = 4) (Figure 5) (*48*). We found that the μ-η^2:η^2-O_2^{2-}) group in $[Cu_2(N_4)(O_2)]^{2+}$ (**5**) behaves as a nonbasic or electrophilic peroxo ligand, in contrast to the basic or nucleophilic behavior of the peroxo group in **3**, which possesses "end-on" coordination. For example, in reactions with H^+, CO_2, and PPh_3,

Basic/Nucleophilic Peroxide

Cu–O–O–Cu

3

(a) $\xrightarrow{PPh_3}$ O_2 + LCu(I)-PPh_3 complex

(b) $\xrightarrow{H^+}$ $\{L_nCu_2^{II}(OOH)\}$ $\xrightarrow{H^+}$ H_2O_2

(c) $\xrightarrow{RC(O)^+}$ $\{L_nCu_2^{II}O_2C(O)R\}$

(d) $\xrightarrow{CO_2}$ $\{L_nCu_2^{II}O_2CO\}$ $\longrightarrow$ $\{L_nCu_2^{II}CO_3\}$

(e) $\xrightarrow{PhOH}$ PhO^- + $\{L_nCu_2^{II}(OOH)\}$

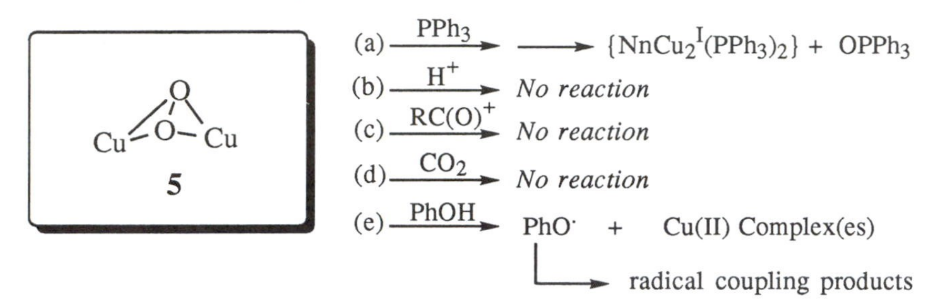

Figure 5. Summary of reactivity comparisons of end-on vs. side-on bound peroxo-dicopper(II) complexes (48).

$[Cu_2(N_4)(O_2)]^{2+}$ (**5**) does not readily protonate, it is unreactive toward CO_2 and it slowly oxygenates PPh_3, whereas **3** readily gives H_2O_2, carbonates, or liberates O_2, respectively.

A rather important mechanistic insight comes from experiments using $[Cu_2(R–XYL–Me)]^{2+}$ (**6**, X = Me), in which a methyl group was placed at the 2-position of the ligand (*45*). Instead of causing methyl hydroxylation or blocking ring attack, 2-hydroxylation occurs and the methyl group undergoes a 1,2-migration. When **6** (X = Me) is reacted with dioxygen in CH_2Cl_2 and the resulting solution is worked up for its organic products, phenol **9**, Me-PY2, PY2 and formaldehyde (detected as a Nash adduct) are isolated or detected in excellent yield and with good material balance (Figure 6). An isotope-labeling experiment using $^{18}O_2$ also established that the source of oxygen in **9** was dioxygen and confirmation of the regiochemistry in this product came from a crystallographic study of a dicopper(II) complex containing the phenolate derived from **9**.

The process observed is reminiscent of the "NIH shift", observed previously in iron hydroxylases, in which a reactive iron-oxy species (with an as yet undetermined identity) is an electrophile, attacking an arene substrate. This results in hydroxylation-induced migrations, due to the formation of carbonium ion intermediates and retention of heavier

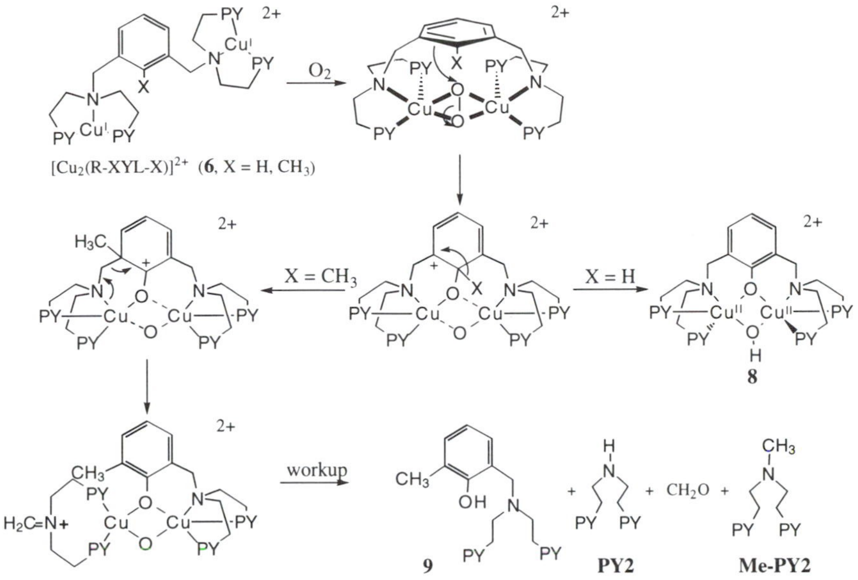

*Figure 6. Scheme showing the nature of products (i.e., **9**, PY2, formaldehyde and Me-PY2) obtained during the oxygenation of methyl-substituted xylyl ligand complexes $[Cu_2(R\text{–}XYL\text{–}Me)]^{2+}$ (**6**, X = Me). The proposed mechanism of copper mediated arene hydroxylation and "NIH shift" (1,2-migration) reactions is also outlined.*

substituents in preference to –H, during rearomatization. This comparison has lead us to suggest a similar reaction pathway, an "NIH shift" in copper chemistry. Thus, a detailed proposed mechanism for the hydroxylation of dicopper(I) complexes $[Cu_2(R\text{–}XYL\text{–}X)]^{2+}$ (**6**), can be outlined (Figure 6). $[Cu_2(R\text{–}XYL\text{–}Me)]^{2+}$ (**6**, X = H or Me) react with O_2, initially forming a Cu_2O_2 adduct suggested to have a μ-η^2:η^2-peroxo structure. This Cu_2O_2 adduct is capable of acting as an electrophile, and attacks the xylyl ligand π system (consistent with lack of 2-deuterium isotope rate effect upon k_4), which is located in a favorable proximity. In fact, a molecular model of $[Cu_2(R\text{–}XYL\text{–}X)(O_2)]^{2+}$ (**7**) suggests the O–O vector is well-aligned with and close to the p-π orbital of the arene carbon that is attacked, possibly an important factor in oxygen atom transfer reactions, as discussed by Sorrell (*49*). For **7** (X = H), direct loss of H^+ from the cationic intermediate produces rearomatized $[Cu_2(R\text{–}XYL\text{–}O\text{–})(OH)]^{2+}$ (**8**). For either X = H or Me, a 1,2-migration is likely, but loss of CH_3^+ is unlikely for X = CH_3, and rearomatization occurs with "assistance" of the amine nitrogen lone pair. This rearomatization leads to loss of an iminium ion in a retro-Mannich reaction. Under the experimental conditions employed, hydrolysis produces PY2 and the formaldehyde observed, and some reduction of the iminium salt can lead to a small amount of MePY2 observed.

Other researchers (*50–55*) have examined alternate xylyl systems similar to $[Cu_2(R\text{–}XYL\text{–}H)]^{2+}$ (**6**), using chelating groups other than PY2, seeing analogous hydroxylation reactions depending upon the exact nature of the dinucleating ligand. Sorrell (*49*) has studied closely related complexes in which 1-pyrazolyl or 2-imidazolyl donor groups fully or partially replace the 2-pyridyl ligands in XYL–H. Hydroxylation does not occur upon oxygenation of these dicopper(I) complexes and all react via four-electron reduction of O_2 to give bis(μ-hydroxo)copper(II) dimers. Also, if $–CH_2PY$ instead of $–CH_2CH_2PY$ (PY is 2-pyridyl) arms are used in the xylyl dinucleating ligands, only irreversible oxidation and no ligand hydroxylation takes place (K. D. Karlin and co-workers, unpublished results).

These and other observations (*51–55*) suggest that the tendency towards hydroxylation in these kinds of chemical systems is very sensitive to electronic effects, as well as copper chelation and peroxide proximity and orientation toward xylyl substrate. This view is supported by observations involving an unsymmetrical system, $[Cu_2(UN)]^{2+}$ (*56*), an analogue of $[Cu_2(XYL)]^{2+}$ (**6**)

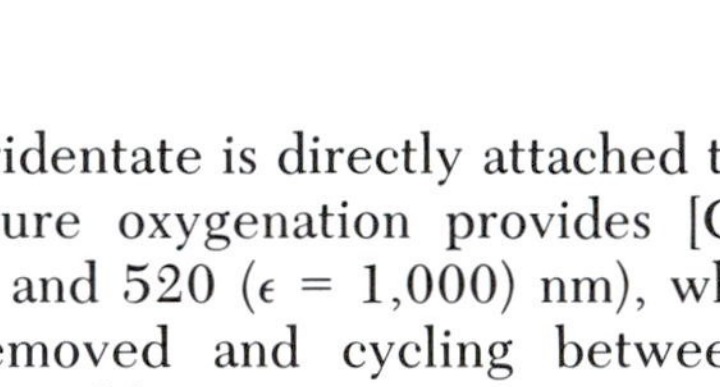

$[Cu_2^I(UN)]^{2+}$

in which one PY2 tridentate is directly attached to the central phenyl ring. Low-temperature oxygenation provides $[Cu_2(UN)(O_2)]^{2+}$ (λ_{max} = 360 (ϵ = 11,000) and 520 (ϵ = 1,000) nm), which is so stable that the O_2 can be removed and cycling between $[Cu_2(UN)]^{2+}$ and $[Cu_2(UN)(O_2)]^{2+}$ is possible. However, warming to room temperature gives hydroxylated product $[Cu_2(UN\text{–}O\text{–})(OH)]^{2+}$, with a structure analogous to that observed for $[Cu_2(XYL\text{–}O\text{–})(OH)]^{2+}$ (**8**) (Figure 4). The actual rate of conversion of $[Cu_2(UN)(O_2)]^{2+}$ to $[Cu_2(UN\text{–}O\text{–})(OH)]^{2+}$ has not yet been measured, but the process is clearly much slower than the hydroxylation in the parent xylyl system. The implication is that the unsymmetrical ligand alters the orientation and electronics of the peroxo ligand in the $[Cu_2(UN)(O_2)]^{2+}$ intermediate, disfavoring attack on the arene.

Thus, the xylyl system described in the previous paragraphs is very much like that of an enzyme active site; here, the active Cu_2O_2 intermediate is formed with ideal juxtaposition to the arene substrate.

Copper–Dioxygen Chemistry and Amide Hydrolysis

Because of the highly interesting O_2 reactivity with dicopper(I) of ligands like N_n, XYL, and UN, we have also expended some efforts with other modified analogues, including PD, based on a 1,3-phenylenediamine nucleus (PY is 2-pyridyl).

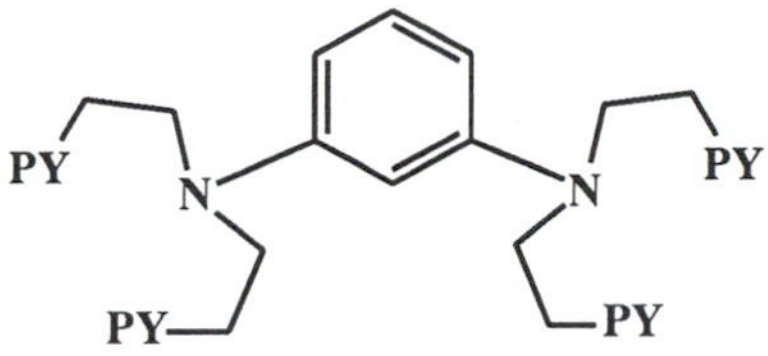

PD 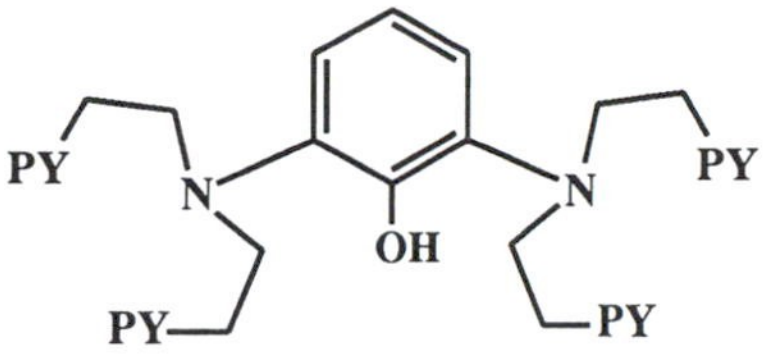 **PD-OH**

Although a hydroxylation reaction occurs upon addition of O_2 to a dicopper(I) complex of PD, there are marked differences in the chemistry, leading us to discover that the copper–dioxygen chemistry in this system is capable of effecting the hydrolysis of an unactivated amide under mild conditions. Through this chemistry, we can also isolate the hydroxylated ligand PD–OH and its anion forms a dinuclear Cu(II) complex possessing coordination and ligating properties well suited for amide hydrolysis under hydrolytic (i.e., nonoxidative) conditions (*57*).

A Room-Temperature Stable O_2 Adduct? Thus, oxygenation of $[Cu_2(PD)]^{2+}$ (**10**) in acetonitrile at room temperature generates a dark purple species with characteristic absorption at λ_{max} = 558 nm (ϵ = 3300 M^{-1} cm^{-1}), undoubtedly assignable as a charge-transfer transition. A corresponding stable purple solid can be isolated, formulated as $[Cu_2(PD)(O_2)]^{2+}$ (**11**), based on elemental analysis, mass spectrometric data, solution conductance and the observed Cu:O_2 = 2:1 (manometry in CH_3CN) stoichiometry of its formation (Figure 7) (*58*). Acetonitrile appears to be important for its solution stability, in which it appears to be essentially diamagnetic, because it exhibits a sharp, normal looking ^{1}H-NMR spectrum and has a magnetic moment 0.32 μ_B/Cu (room temperature) in CD_3CN (Evans method). As an isolated solid, it does not contain CH_3CN, and the magnetic coupling is weaker, i.e., 1.30 μ_B/Cu (room temperature). This material does possess an O_2-oxidizing equivalent, as seen from its reaction chemistry (*see* the next section); the original dinucleating ligand remains intact in **11**, based on the ability to recover unhydroxylated PD (81%, isolated) by extraction using

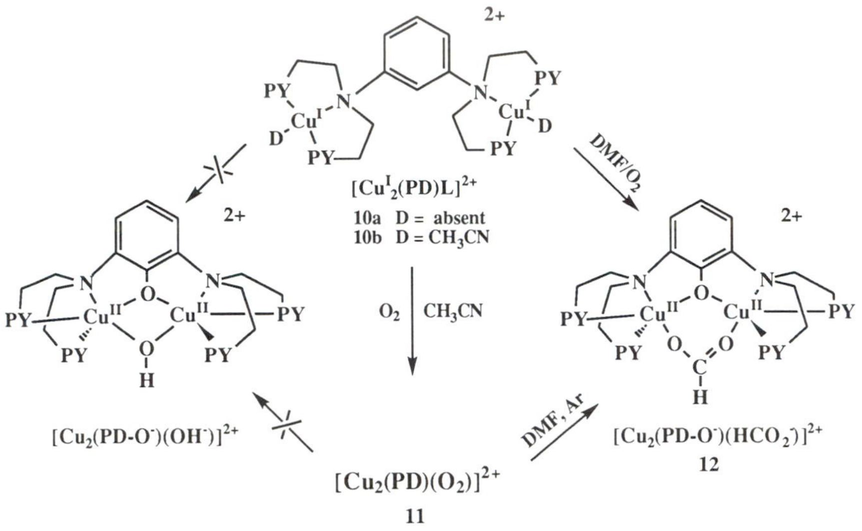

Figure 7. Copper(I)–O_2 chemistry involving the dinucleating PD ligand.

NH_4OH(aq). Thus, the spectroscopic properties and substrate reactivity (e.g., with H^+ or PPh_3) of **11** are very different from other Cu_2O_2 (e.g., peroxo-dicopper(II)) species such as $[\{(TMPA)Cu\}_2(O_2)]^{2+}$ (**3**) and $[Cu_2(N_n)(O_2)]^{2+}$ (**5**). We do not as yet have any clear ideas concerning the structure of **11**; one speculation is that the molecule is best described as an $Cu(I)_2$–O_2 species exhibiting a Cu → O_2 metal-to-ligand charge-transfer (MLCT) absorption [i.e., not peroxo-dicopper(II)] and that the O_2 molecule may be copper-bound but also closely associated with the phenyl ring of the PD ligand.

$Cu(I)_2$–O_2 Hydroxylation Reaction Accompanied by DMF Hydrolysis (57). Either by reaction of $[Cu_2(PD)(O_2)]^{2+}$ (**11**) with dimethylformamide (DMF) under argon at room temperature, or by direct addition of O_2 to a DMF solution of $[Cu_2(PD)]^{2+}$ (**10a**), a rapid change to green occurs. We originally expected that a hydroxo-bridged and phenoxo-bridged dicopper(II) complex analogous to **8** (i.e., $[Cu_2(PD–O^-)(OH^-)]^{2+}$) might be isolated, but it was not the case. Instead, the phenoxo-briged and formato-bridged species $[Cu_2(PD–O^-)(HCO_2^-)]^{2+}$ (**12**) is obtained in >70 % isolated yield (Figure 7). Interestingly, the acetate-bridged complex $[Cu_2(PD–O^-)(CH_3CO_2^-)]^{2+}$ is produced in a corresponding reaction carried out in *N,N*-dimethylacetamide. The structure of **12** has been confirmed in a X-ray crystallographic study (Figure 8).

Thus, this reaction has resulted in both the hydrolysis of DMF to give formate, as well as the hydroxylation of the PD ligand, to give the

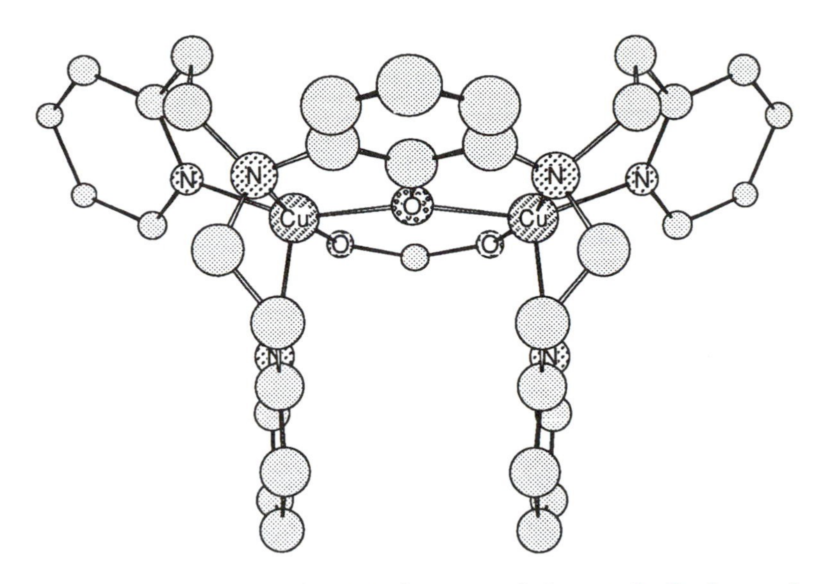

Figure 8. Perspective Chem3D drawing of formato-bridged complex $[Cu_2(PD\text{-}O^-)(HCO_2^-)]^{2+}$ *(**12**). Cu···Cu′ = 3.65 Å and* $Cu\text{-}O_{phenolato}\text{-}Cu'$ *= 135.3°.*

resulting phenolate. Amide hydrolysis is also indicated by the observed production of diethylamine (GC-MS analysis), when $[Cu_2(PD)]^{2+}$ (**10a**) is reacted with O_2 in *N,N*-diethylformamide. These transformations resemble the monooxygenase reactions described for the XYL system (i.e., **6** → **8**, *see* previous paragraphs), in which one atom of O_2 is incorporated into the arene substrate, whereas the other ended up as a bridging hydroxide ligand. In the present case, *no* corresponding μ-hydroxo complex (OH^- derived from O_2; cf. Figure 4) is produced, because an additional hydrolysis of DMF has occurred. Because formato complex $[Cu_2(PD\text{–}O^-)(HCO_2^-)]^{2+}$ (**12**) forms directly from $[Cu_2(PD)(O_2)]^{2+}$ (**11**) under Ar, the O-atom in the phenol PD–OH and one atom in the formato ligand in **12** are suggested to be derived from O_2. This indeed is the case, because mass spectral evidence indicates full incorporation of 18-O in free isolated PD–OH after reaction of $[Cu_2(PD)]^{2+}$ (**10a**) with $^{18}O_2$. Strong supporting evidence is also obtained by electrospray ionization or fast-atom-bombardment (FAB) mass spectrometric analysis carried out directly on the metal complex product **12**. Thus,

$$[Cu_2(PD)]^{2+}\ (\mathbf{10a}) + {}^{18}O_2 + HC(O)NMe_2 \rightarrow [Cu_2(PD\text{–}{}^{18}O^-)(HCO^{18}O^-)]^{2+}\ (\mathbf{12}) + HNMe_2$$

The possible course of reaction involving both arene hydroxylation and DMF hydrolysis is of considerable interest. (i) One can envision a DMF-induced hydroxylation of the arene in a process somewhat analogous to the hydroxylation reaction **6** → **8**, producing a species formulated as $[Cu_2(PD{-}O^-)(OH^-)]^{2+}$, however, not having the structure shown in Figure 7; production of a hydroxo ligand bound to only one Cu(II) ion might then attack DMF, producing **12**. (ii) Alternatively, the O_2 (peroxo?) group in $[Cu_2(PD)(O_2)]^{2+}$ (**11**) might directly react with DMF, and a subsequent intermediate could hydroxylate the ring. These possibilities are discussed in subsequent paragraphs, in light of the additional structural chemistry and hydrolysis reactivity discovered for dicopper(II) complexes with the $PD{-}O^-$ ligand.

DMF Hydrolysis in a Dinuclear Complex Containing Adjacent Exogenous Ligands. The structure of $[Cu_2(PD{-}O^-)(HCO_2{}^-)]^{2+}$ (**12**) suggests that the $PD{-}O^-$ dinucleating ligand may not be suitable for exogenous ligand μ-1,1-bridging (e.g., OH^-), but stabilizes 1,3-bridging interactions (e.g., O,O′-carboxylato). In fact using a dinucleating ligand similar to $PD{-}O^-$ (but possessing pyrazolyl rather than pyridyl donors), Sorrell and co-workers (59) showed that μ-1,3-bridging was preferred for acetate and azide ($N_3{}^-$), whereas μ-1,1-bridging occurs for the corresponding pyrazolyl-containing xylyl ligand analogous to $XYL{-}O^-$. Thus the $PD{-}O^-$ dinucleating ligand, by "pinning back" the Cu(II) ions via binding to the 1,3-phenylenediamine N-donors and bridging to the phenolato O-atom, enforces Cu···Cu distances > 3.5 Å, unsuitable for μ-1,1-bridging (Scheme 5: X = OH^-, halide$^-$, or $N_3{}^-$; X–Y = X is $N_3{}^-$ or $RCO_2{}^-$; X1 and X2 are neutral or anionic ligands). These distances suggest that the $[Cu_2(PD{-}O^-)]^{n+}$ framework could facilitate adjacent coordination of two ligands, perhaps a terminal hydroxo ligand and substrate (i.e., DMF), poised for intramolecular reaction and hydrolysis (57).

This notion is supported by a structural analysis of $[Cu_2(PD{-}O^-)(OMe^-)_2]^{1+}$ (**13**), which is produced by diphenylhydrazine reduction

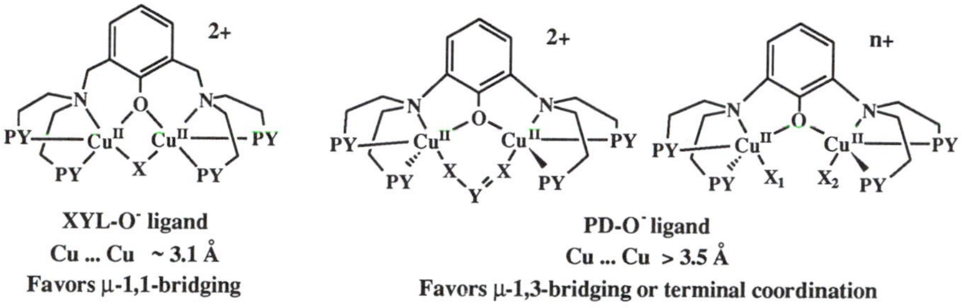

XYL-O⁻ ligand
Cu ... Cu ~ 3.1 Å
Favors μ-1,1-bridging

PD-O⁻ ligand
Cu ... Cu > 3.5 Å
Favors μ-1,3-bridging or terminal coordination

Scheme 5

of $[Cu_2(PD–O^-)(HCO_2^-)]^{2+}$ (**12**), followed by O_2 reoxidation of the resulting dicopper(I) species in methanol. In **13** (Figure 9), two adjacent terminal methoxide ligands are coordinated to the copper(II) ions of the dinuclear unit (*57*).

With structure $[Cu_2(PD–O^-)(OMe^-)_2]^{1+}$ (**13**), having adjacent terminal ligands, we surmised that this compound might well be suitable in promoting the hydrolysis of DMF, because we should be able to generate a copper(II) coordinated hydroxide ligand and adjacent labile site for DMF binding, i.e., a (OH^-)-Cu · · · Cu-(L) (L = H_2O or DMF) species. We tried to optimize such conditions by reacting **13** with DMF (23 °C) in the presence of one equivalent of $HClO_{4(aq)}$. Indeed, facile amide hydrolysis again occurs and $[Cu_2(PD–O^-)(HCO_2^-)]^{2+}$ (**12**) is generated in good yield (>60% isolated) (Figure 10). As followed by UV–vis spectrophotometric measurements of the disappearence of **13** or appearence of **12**, preliminary kinetic measurements indicate a pseudo-first-order process with k_{obs} = 0.3 h^{-1}. The suggested reaction intermediate for this process is $[Cu_2(PD–O^-)(OH^-)(S)]^{2+}$ (S = H_2O, MeOH, or DMF substrate), as depicted in Figure 10.

As related to the hydrolysis reaction promoted by dicopper complex **13** or $[Cu_2(PD–O^-)(OH^-)(DMF)]^{2+}$ (Figure 10), one needs to consider

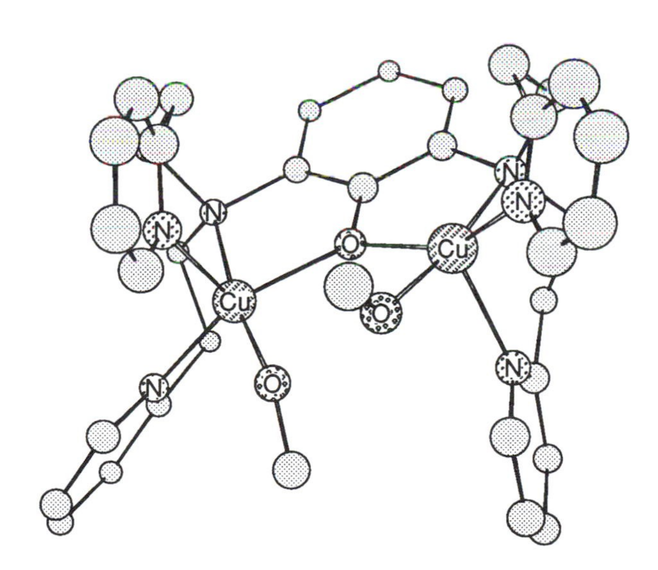

*Figure 9. Perspective Chem3D drawing of $[Cu_2(PD–O^-)(OMe^-)_2]^{1+}$ (**13**), with two adjacent terminally bound MeO^- ligands. Cu· · ·Cu′ = 3.74 Å and Cu–$O_{phenolato}$-Cu′ =137.4°.*

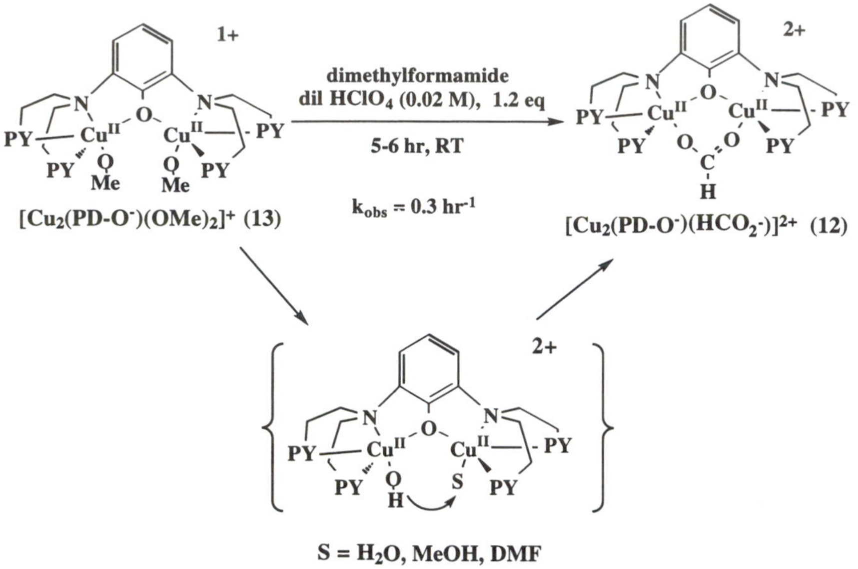

$[Cu_2(PD\text{-}O^-)(OMe)_2]^+$ (13) → dimethylformamide, dil $HClO_4$ (0.02 M), 1.2 eq; 5-6 hr, RT; $k_{obs} = 0.3\ hr^{-1}$ → $[Cu_2(PD\text{-}O^-)(HCO_2^-)]^{2+}$ (12)

S = H_2O, MeOH, DMF

*Figure 10. Complex **13** with two adjacent terminally bound ligands catalyzes the hydrolysis of dimethylformamide, giving formato-bridged complex **12**.*

the possibility that only one of the two metal ions is involved; we consider this unlikely. Support for this supposition comes from studies on the following mononuclear analogue, which has been crystallographically characterized (57):

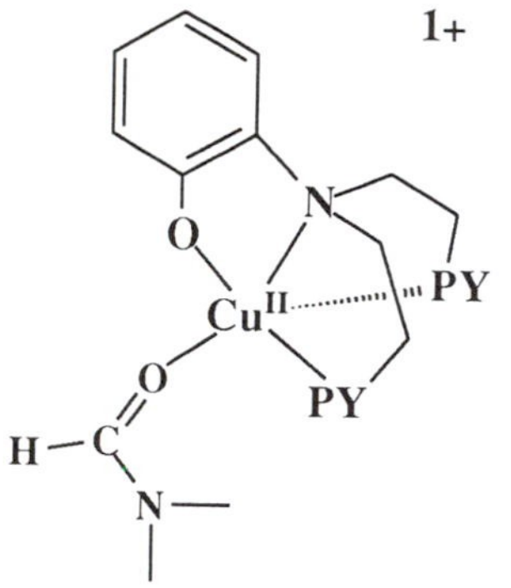

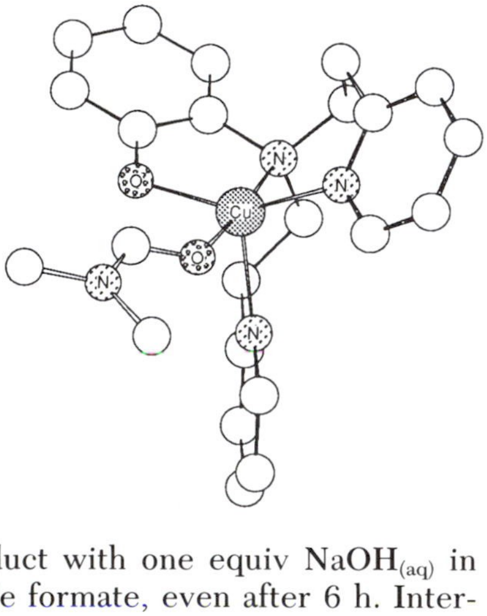

Reaction of this O-bonded DMF-adduct with one equiv $NaOH_{(aq)}$ in DMF solution produced no detectable formate, even after 6 h. Interestingly, a computer-generated model of $[Cu_2(PD{-}O^-)(OH^-)(DMF)]^{2+}$ (Figure 10) using the coordinates for the Cu(II)–DMF moiety observed in the mononuclear analogue indicated that the coordinated hydroxo oxygen atom could be within 2.5 Å of the DMF carbonyl carbon and supported the view that nucleophilic attack may be facilitated in this dinuclear complex.

Possible Hydrolytic Mechanisms: Mononuclear Metal Centers. There is considerable recent interest in the hydrolysis of amides by metal complexes (*60–63*), in large part due to their occurrence in biological systems; a typical peptide amide bond has a half-life of 7 y at pH 7 and 25 °C (*64*). Many of the well-known metalloproteins involved in hydrolytic processes contain mononuclear zinc active sites (*65–67*), and there have been numerous model systems studied, in particular those involving zinc(II) or cobalt(III) ions (*61*). Chin (*61*) summarized (Scheme 6) some of the basic ideas involving such mononuclear hydrolysis of amides, involving either (a) Lewis acid activation of substrate, facilitating solution attack by hydroxide, (b) a metal-hydroxide mechanism whereby binding of water to a Lewis acidic metal lowers its pK_a giving a bound hydroxide, which attacks substrate, or (c) a combination mechanism that is likely in metal complexes with two adjacent *cis* labile sites; here, the metal-bound hydroxide effects an intramolecular attack upon the coordinated amide substrate.

Dinuclear Hydrolytic Catalysis. Recent biochemical and protein crystallographic studies have revealed that di- or trinuclear metal ion centers (i.e., with Zn, Mg, Mn, Fe, Co, or Ni) effect peptidase or phosphatase (i.e., amide or phosphate ester hydrolysis) reactions in a variety of systems, Table II. Protein systems that carry out amide hy-

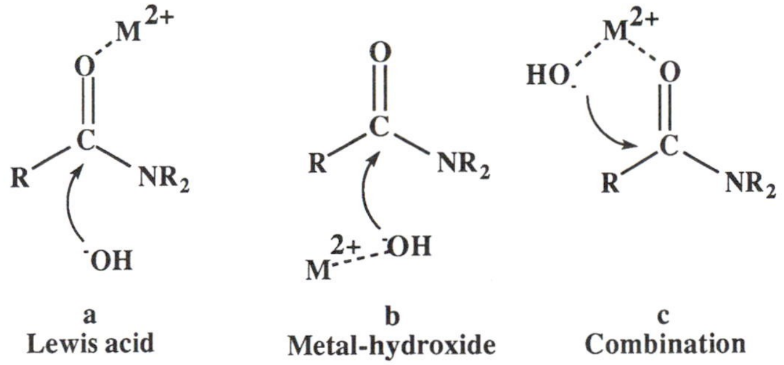

a Lewis acid — b Metal-hydroxide — c Combination

Scheme 6

drolysis reactions include aminopeptidases (*68*), which generally possess exchangeable metal-binding sites. X-ray crystallographic studies have revealed the active site structure in two cases. Bovine lens leucine aminopeptidase (*69*) possesses a di-zinc active site (Zn···Zn = 3.15 Å) structure, and in cobalt-dependent methionine aminopeptidase from *Escherichia coli* (*70*), the two metals are 2.9 Å apart (at 2.4 Å resolution).

There have been a few reports of "first generation" coordination complex structural models for the phosphatase enzyme active sites (*81,82*), whereas there are some examples of ester hydrolysis reactions involving dinuclear metal complexes (*83–85*). Kim and Wycoff (*74*) as well as Beese and Steitz (*80*) have both published somewhat detailed discussions of "two-metal ion" mechanisms, in connection with enzymes involved in phosphate ester hydrolysis. Compared to fairly simple chemical model systems, the protein active site mechanistic situation is rather more complex, because side-chain residues near the active site are undoubtedly involved in the catalysis, i.e, via acid–base or hydrogen-bonding interactions that either facilitate substrate binding, hydroxide nucleophilic attack, or stabilization of transition state(s). Nevertheless, a simple and very likely role of the Lewis-acidic metal ion center is to

Table II. Peptidases and Phosphatases with Polymetal Active Sites

Metal and Active Site	*Enzyme and Function*	*Ref.*
Peptidases		
2Zn	leucine aminopeptidase	69[a]
2Co	methionine aminopeptidase (*E. coli*)	70[a]
2Mn	arginase (L-arginine → L-ornithine + urea)	71
2Mn	enolase (hydratase)	72
2Ni	urease (urea → carbonic acid + ammonia)	73
Phosphatases		
3Zn (2Zn, 1Mg)	alkaline phosphatase (phosphomonoesterase)	74[a]
2Zn	phosphotriesterase (*Pseudomonas diminuta*)	75
3Zn	P1 nuclease (endonuclease)	76[a]
3Zn	phospholipase C, *Bacillus cereus*	77[a]
2Fe (Fe, Zn)	purple acid phosphatases	78
2Mn, $2M^{2+}$	ribonuclease H domain of HIV-1 reverse transcriptase (phosphodiesterase)	79[a]
2(Mg^{2+}, Mn^{2+}, or Zn^{2+})	*E. coli* DNA polymerase I (Klenow fragment) 3′,5′-exonuclease activity)	80[a]

[a] Protein X-ray structure is available.

facilitate proton loss from water, generating a hydroxide or $M–(OH)^-$ nucleophile. Lipscomb and co-workers (*86*) favor such a "general-base" mechanism for the di-zinc containing leucine aminopeptidase, because X-ray structural studies show that there are no active-site residues present that could themselves directly participate as nucleophile for carbonyl carbon attack.

Thus, it may be useful to diagram a number of possibilities for dinuclear metal ion catalysis with respect to phosphate ester or amide hydrolysis (Figure 11). Part a shows that substrate binding may be aided by ligation to one or both metals although ligation to both metals seems unlikely for a *neutral* amide. The other metal could then deliver the hydroxide nucleophile, or alternatively, the hydroxide nucleophile may come directly from solvent water (*85*). The transition state may be stabilized by metal-coordination of the developing negative charge on the O-atom of the substrate group. Part b of Figure 11 shows that reaction of a bound substrate (to one or both metals) may occur via a bridging $M–(OH^-)$-M group. In either of these cases, there is an underlying presumption that binding of water to a Lewis-acidic metal or metals favors deprotonation and generation of the hydroxide nucleophile. Part c of

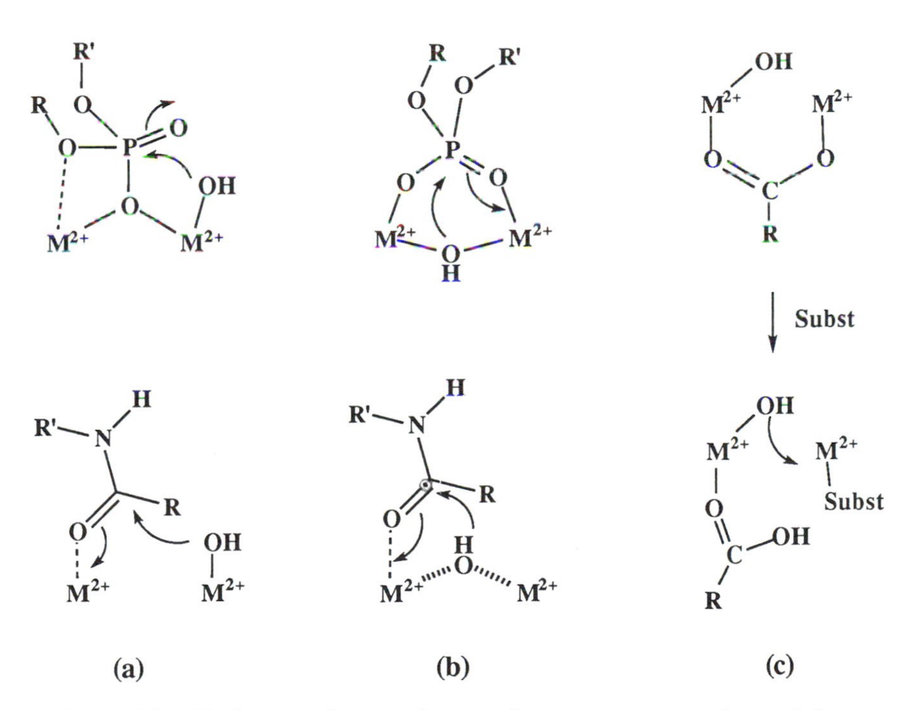

Figure 11. Mechanistic features that may be important in catalysis of phosphate ester or amide hydrolysis by a dinuclear metal center. See text for further discussion. (Adapted from references 74 and 80).

Figure 11 shows that based on the X-ray structures available (Table II), protein side-chain carboxylate (e.g., Asp or Glu) bridging of two metal ions is rather common in these hydrolytic metalloenzymes. One can speculate upon another mechanistic possibility, in which substrate binding may be accompanied by a "carboxylate shift" (*87*) or displacement (known to occur in the redox-active di-iron enzyme ribonucleotide reductase (*88*)), placing a RCO_2^-H or RCO_2 group as potential acid or base next to the substrate; its role might be as a proton donor to the $RC(O^-)(OH)$–NHR′ tetrahedral intermediate generated after hydroxide attack on substrate, thus facilitating product release.

In those situations in which three metal ions are found at the active site of phosphatases (*74, 76–78*), possible mechanisms become more complicated and numerous. The third metal may bind and be present in the active site in a structural role or as an enzyme activation step. For the catalytic event, two metal ions may be involved in substrate binding, while the third delivers the hydroxide nucleophile; other combinations are obviously possible.

Mechanistic Aspects of the DMF Hydrolyses Affected by 11 or 13. The mechanistic concepts discussed in previous paragraphs would seem to apply directly to the conversion of $[Cu_2(PD–O^-)(OMe^-)_2]^{1+}$ (**13**) to $[Cu_2(PD–O^-)(HCO_2^-)]^{2+}$ (**12**), as already alluded to in Figure 10. The PD–O$^-$ ligand is capable of supporting ligation by two adjacent terminal ligands. Thus, addition of $H_{(aq)}^+$ to **13** in DMF would release methanol, converting **13** to a species such as $[Cu_2(PD–O^-)(OH^-)(H_2O)]^{2+}$, with labile neutral water ligand. Facile exchange of the coordinated water by DMF substrate would give intermediate $[Cu_2(PD–O^-)(OH^-)(DMF)]^{2+}$ (Figure 12), well-suited for intramolecular attack by hydroxide on the carbonyl carbon, with stabilization of the tetrahedral intermediate product aided by Cu(II) binding. Release of dimethylamine gives formate-bridged product **12**.

The conversion of $[Cu_2(PD)(O_2)]^{2+}$ (**11**) to $[Cu_2(PD–O^-)(HCO_2^-)]^{2+}$ (**12**) is a considerably more interesting process, because hydrolysis is accompanied by PD ring hydroxylation. Qualitatively, the observed reaction of **11** with DMF is immediate, i.e., much faster than that observed in the purely hydrolytic process, $[Cu_2(PD–O^-)(OMe^-)_2]^{1+}$ (**13**) → $[Cu_2(PD–O^-)(HCO_2^-)]^{2+}$ (**12**). One possible explanation for this observation could involve a different reaction pathway, in which the O_2 (peroxo) ligand in **11** acts as a strong α-nucleophile. Thus, we speculate that **11** could react with DMF in such a manner, producing a peroxo-amidate intermediate (Figure 12). When this intermediate is formed in close proximity to the PD phenyl ring via Cu coordination, it could have oxidative capabilities like that of a peracid, effecting the hydroxylation reaction observed, leading to the formato product **12**. There are a num-

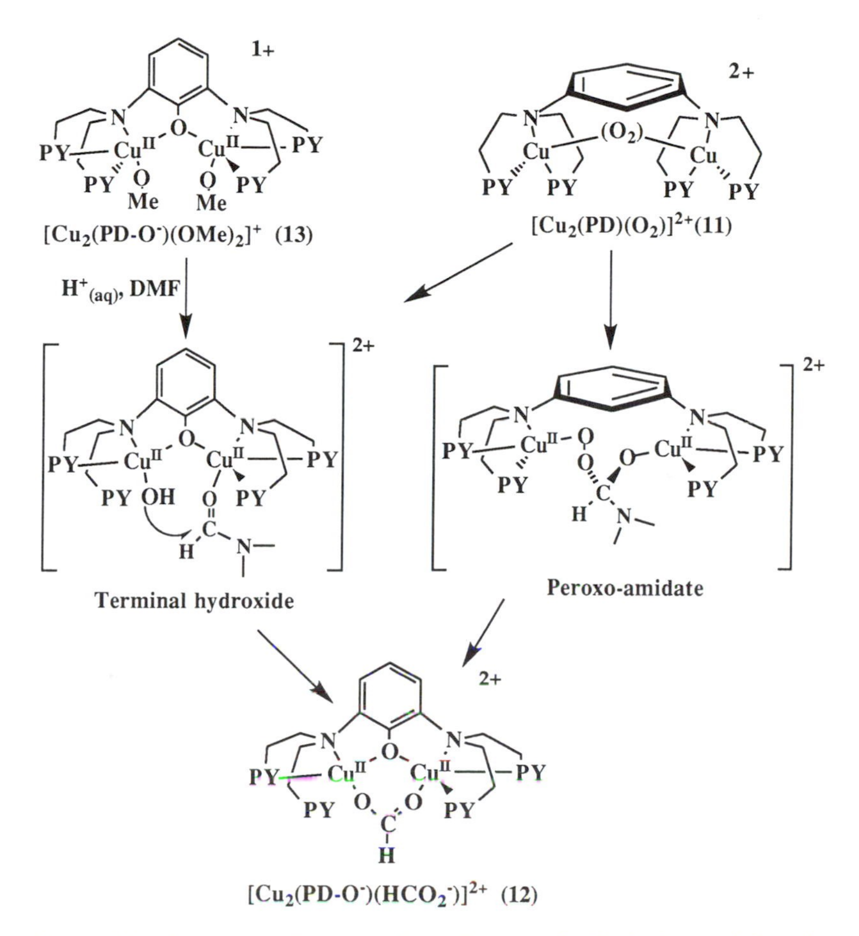

*Figure 12. Summary of proposed mechanisms for hydrolysis of dimethylformamide starting with either **11** or **13**.*

ber of recent examples in which a metal–peroxide group generated in proximity to a biological substrate has been shown to act as a nucleophile (*89–91*). For example, Robinson (*89*) as well as Coon (*90*) and their coworkers have provided evidence that a (porphyrin)iron–peroxide moiety can attack an exocyclic aldehyde, generating formaldehyde. Rana and Meares (*91*) have also demonstrated highly specific peptide amide hydrolysis by using hydrogen peroxide (and ascorbate) with an iron–EDTA molecule tethered to a protein.

Thus, the dicopper complexes with the PD or PD–O^- ligand provide very novel chemistry, clearly exhibiting cases of dinuclear metal-promoted hydrolytic processes. The range of possible substrates for this complex has yet to be studied and one wonders if similar chemistry can be observed for zinc or other metal analogues. A variety of mechanistic investigations will be required (e.g., kinetics and pH dependencies) to provide further insight into the hydrolytic chemistry and corroboration for these proposed reaction pathways.

Summary

Our interests in copper–dioxygen coordination chemistry relevant to copper proteins has led us to design and develop a variety of synthetically derived polydentate ligands, containing one or two metal ions. Reversible binding of O_2 by mono- or dinuclear copper(I) complexes has been demonstrated for several ligand systems, leading to the structural and spectroscopic characterization of several types of Cu_2O_2 (peroxo dicopper(II)) complexes. The dinucleating XYL, UN, and PD ligands involve chemistry in which Cu_2O_2 intermediates effect novel C–H activation reactions, i.e., the hydroxylation of an arene that is part of the dinucleating ligand framework. Such reactions are relevant to copper monooxygenases such as tyrosinase, and a variety of investigations involving copper complexes of XYL, plus derivatives or analogues have provided mechanistic insights, involving electrophilic attack of the Cu_2O_2 moiety upon the arene substrate. An emerging area of metallobiochemistry is the occurrence of peptidase or phosphatase enzymes possessing di- or trinuclear metal ion-containing active sites. The chemistry of copper complexes with the dinucleating PD and PD–O^- ligands not only involves $Cu(I)_2/O_2$ reactivity, but hydrolysis of an unactivated amide is also observed. The reaction chemistry and metal-complex structures observed in the system provide insights into the possible mechanisms.

Acknowledgment

We thank the National Institutes of Health (GM 28962) for support of the research described herein.

References

1. Karlin, K. D.; Gultneh, Y. *Prog. Inorg. Chem.* **1987,** *35,* 219–327.
2. *Dioxygen Activation and Homogeneous Catalytic Oxidation;* Simándi, L. I., Ed.; Elsevier: Amsterdam, Netherlands, 1991; Vol. 66.
3. Sheldon, R. A.; Kochi, J. M. *Metal-Catalyzed Oxidations of Organic Compounds;* Academic: New York, 1981.
4. Gampp, H.; Zuberbühler, A. D. In *Metal Ions in Biological Systems;* Sigel, H., Ed.; Marcel Dekker: New York, 1981; Vol. 12, pp 133–189.
5. Solomon, E. I.; Lowery, M. D. *Science (Washington, D.C.)* **1993,** *259,* 1575–1581.
6. Tyeklár, Z.; Karlin, K. D. In *Bioinorganic Chemistry of Copper;* Karlin, K. D., Tyeklár, Z., Eds.; Chapman & Hall: New York, 1993; pp 277–291.
7. Karlin, K. D.; Tyeklár, Z.; Zuberbühler, A. D. In *Bioinorganic Catalysis;* Reedijk, J., Ed.; Marcel Dekker: New York, 1993; Chapter 9, pp 261–315.
8. Kitajima, N.; Moro-oka, Y. *Chem. Rev.* **1994,** *94,* 737–757.

9. Solomon, E. I.; Baldwin, M. J.; Lowery, M. D. *Chem. Rev.* **1992,** *92,* 521–542 and references cited therein.
10. Adman, E. T. *Adv. Protein Chem.* **1991,** *42,* 145–197.
11. Chapman, S. K. *Perspect. Bioinorg. Chem.* **1991,** *1,* 95–140.
12. Messerschmidt, A. In *Bioinorganic Chemistry of Copper;* Karlin, K. D.; Tyeklár, Z., Eds.; Chapman & Hall: New York, 1993; pp 471–484.
13. Messerschmidt, A. *Biochem. Soc. Trans.* **1992,** *20,* 364–368.
14. See Fee, J. A.; Antholine, W. E.; Fan, C.; Gurbiel, R. J.; Surerus, K.; Werst, M.; Hoffman, B. M. In *Bioinorganic Chemistry of Copper;* Karlin, K. D.; Tyeklár, Z., Eds.; Chapman & Hall: New York, 1993; pp 485–500.
15. Tolman, W. B. Chapter 7 in this volume.
16. Magnus, K. A.; Hazes, B.; Ton-That, H.; Bonaventura, C.; Bonaventura, J.; Hol, W. G. *J. Proteins: Struct. Funct. Genet.* **1994,** *19,* 302–309.
17. Hazes, B.; Magnus, K. A.; Bonaventura, C.; Bonaventura, J.; Dauter, Z.; Kalk, K.; Hol, W. G. J. *Protein Sci.* **1993,** *2,* 597–619.
18. Klinman, J. P.; Berry, J. A.; Tian, G. In *Bioinorganic Chemistry of Copper;* Karlin, K. D., Tyeklár, Z., Eds.; Chapman & Hall: New York, 1993; pp 151–163.
19. Stewart, L. C.; Klinman, J. P. *Annu. Rev. Biochem.* **1988,** *57,* 551–592.
20. Pember, S. O.; Johnson, K. A.; Villafranca, J. J.; Benkovic, S. J. *Biochemistry* **1989,** *28,* 2124–2130.
21. Merkler, D. J.; Kulathila, R.; Consalvo, A. P.; Young, S. D.; Ash, D. E. *Biochemistry* **1992,** *31,* 7282–7288.
22. Eipper, B. A.; Milgram, S. L.; Husten, E. J.; Yun, H.-Y.; Mains, R. E. *Protein Sci.* **1993,** *2,* 489–497.
23. Chan, S. I.; Nguyen, H.-H. T.; Shiemke, A. K.; Lidstrom, M. E. In *Bioinorganic Chemistry of Copper;* Karlin, K. D., Tyeklár, Z., Eds.; Chapman & Hall: New York, 1993; pp 184–195.
24. Dawson, J. H. *Science (Washington, D.C.)* **1988,** *240,* 433–439.
25. Traylor, T. G. *Pure Appl. Chem.* **1991,** *63,* 265–274.
26. Watanabe, Y.; Groves, J. T. In *Mechanisms of Catalysis;* Sigman, D. S., Ed.; Academic: San Diego, CA, 1992; Vol. XX, pp 405–452.
27. Feig, A. L.; Lippard, S. J. *Chem. Rev.* **1994,** *94,* 759–805.
28. Karlin, K. D., *Science (Washington, D.C.)* **1993,** *261,* 701–708.
29. Kitajima, N.; Fujisawa, K.; Fujimoto, C.; Moro-oka, Y.; Hashimoto, S.; Kitagawa, T.; Toriumi, K.; Tatsumi, K.; Nakamura, A. *J. Am. Chem. Soc.* **1992,** *114,* 1277–1291.
30. Tyeklár, Z.; Jacobson, R. R.; Wei, N.; Murthy, N. N.; Zubieta, J.; Karlin, K. D. *J. Am. Chem. Soc.* **1993,** *115,* 2677–2689.
31. Jacobson, R. R.; Tyeklár, Z.; Farooq, A.; Karlin, K. D.; Liu, S.; Zubieta, J. *J. Am. Chem. Soc.* **1988,** *110,* 3690–3692.
32. Baldwin, M. J.; Ross, P. K.; Pate, J. E.; Tyeklár, Z.; Karlin, K. D.; Solomon, E. I. *J. Am. Chem. Soc.* **1991,** *113,* 8671–8679.
33. Karlin, K. D.; Haka, M. S.; Cruse, R. W.; Meyer, G. J.; Farooq, A.; Gultneh, Y.; Hayes, J. C.; Zubieta, J. *J. Am. Chem. Soc.* **1988,** *110,* 1196–1207.
34. Blackburn, N. J.; Strange, R. W.; Farooq, A.; Haka, M. S.; Karlin, K. D. *J. Am. Chem. Soc.* **1988,** *110,* 4263–4272.
35. Karlin, K. D.; Tyeklár, Z.; Farooq, A.; Haka, M. S.; Ghosh, P.; Cruse, R. W.; Gultneh, Y.; Hayes, J. C.; Zubieta, J. *Inorg. Chem.* **1992,** *31,* 1436–1451.
36. Karlin, K. D.; Cruse, R. W.; Gultneh, Y.; Farooq, A.; Hayes, J. C.; Zubieta, J. *J. Am. Chem. Soc.* **1987,** *109,* 2668–2679.

37. Pate, J. E.; Cruse, R. W.; Karlin, K. D.; Solomon, E. I. *J. Am. Chem. Soc.* **1987,** *109,* 2624–2630.
38. Blackburn, N. J.; Strange, R. W.; Cruse, R. W.; Karlin, K. D. *J. Am. Chem. Soc.* **1987,** *109,* 1235–1237.
39. Karlin, K. D.; Ghosh, P.; Cruse, R. W.; Farooq, A.; Gultneh, Y.; Jacobson, R. R.; Blackburn, N. J.; Strange, R. W.; Zubieta, J. *J. Am. Chem. Soc.* **1988,** *110,* 6769–6780.
40. Mahroof-Tahir, M.; Murthy, N. N.; Karlin, K. D.; Blackburn, N. J.; Shaikh, S. N.; Zubieta, J. *Inorg. Chem.* **1992,** *31,* 3001–3003.
41. Karlin, K. D.; Wei, N.; Jung, B.; Kaderli, S.; Zuberbühler, A. D. *J. Am. Chem. Soc.* **1991,** *113,* 5868–5870.
42. Karlin, K. D.; Wei, N.; Jung, B.; Kaderli, S.; Niklaus, P.; Zuberbühler, A. D. *J. Am. Chem. Soc.* **1993,** *115,* 9506–9514.
43. Cruse, R. W.; Kaderli, S.; Karlin, K. D.; Zuberbühler, A. D. *J. Am. Chem. Soc.* **1988,** *110,* 6882–6883.
44. Zuberbühler, A. D. In *Bioinorganic Chemistry of Copper;* Karlin, K. D.; Tyeklár, Z., Eds.; Chapman & Hall: New York, 1993; pp 264–276.
45. Nasir, M. S.; Cohen, B. I.; Karlin, K. D. *J. Am. Chem. Soc.* **1992,** *114,* 2482–2494.
46. Lapshin, A. E.; Smolin, Y. I.; Shepelev, Y. F.; Schwendt, P.; Byepesova, D. *Acta Crystallogr. Sect. C: Cryst. Struct. Commun.* **1990,** *46,* 1753–1755.
47. Karlin, K. D.; Hayes, J. C.; Gultneh, Y.; Cruse, R. W.; McKown, J. W.; Hutchinson, J. P.; Zubieta, J. *J. Am. Chem. Soc.* **1984,** *106,* 2121–2128.
48. Paul, P. P.; Tyeklár, Z.; Jacobson, R. R.; Karlin, K. D. *J. Am. Chem. Soc.* **1991,** *113,* 5322–5332.
49. Sorrell, T. N. *Tetrahedron* **1989,** *45,* 3–68.
50. Sorrell, T. N.; Vankai, V. A.; Garrity, M. L. *Inorg. Chem.* **1991,** *30,* 207–210.
51. Casella, L.; Gullotti, M.; Pallanza, G.; Rigoni, L. *J. Am. Chem. Soc.* **1988,** *110,* 4221–4227.
52. Casella, L.; Gullotti, M.; Bartosek, M.; Pallanza, G.; Laurenti, E. *J. Chem. Soc. Chem. Commun.* **1991,** 1235–1237.
53. Gelling, O. J.; an Bolhuis, F.; Meetsma, A.; Feringa, B. L. *J. Chem. Soc. Chem. Commun.* 1988, 552–554.
54. Gelling, O. J.; Feringa, B. L. *J. Am. Chem. Soc.* **1990,** *112,* 7599–7604.
55. Menif, R.; Martell, A. E.; Squattrito, P. J.; Clearfield, A. *Inorg. Chem.* **1990,** *29,* 4723–4729.
56. Nasir, M. S.; Karlin, K. D.; McGowty, D.; Zubieta, J. *J. Am. Chem. Soc.* **1991,** *113,* 698–700.
57. Murthy, N. N.; Mahroof-Tahir, M.; Karlin, K. D. *J. Am. Chem. Soc.* **1993,** *115,* 10404–10405.
58. Mahroof-Tahir, M. Ph.D. Dissertation, Johns Hopkins University, 1992.
59. Sorrell, T. N.; O'Connor, C. J.; Anderson, O. P.; Reibenspies, J. H. *J. Am. Chem. Soc.* **1985,** *107,* 4199–4206.
60. Fife, T. H. *Perspect. Bioinorg. Chem.* **1991,** *1,* 43–93.
61. Chin, J. *Acc. Chem. Res.* **1991,** *24,* 145–152.
62. Hay, R. W. In *Reactions of Coordinated Ligands;* Braterman, P. S., Ed.; Plenum: New York, 1989; Vol. 2, pp 223–364.
63. Sayre, L. M. *J. Am. Chem. Soc.* **1986,** *108,* 1632–1635.
64. Kahne, D.; Still, W. C. *J. Am. Chem. Soc.* **1988,** *110,* 7529–7534.
65. Coleman, J. E. *Annu. Rev. Biochem.* **1992,** *61,* 897–946.
66. Christianson, D. W. *Adv. Protein Chem.* **1991,** *42,* 281–355.

67. Vallee, B. L.; Auld, D. S. *Biochemistry* **1990,** *29*, 5647–5659.
68. Taylor, A. *FASEB J.* **1993,** *7*, 290–298.
69. Burley, S. K.; David, P. R.; Taylor, A.; Lipscomb, W. N. *Proc. Natl. Acad. Sci. U.S.A.* **1990,** *87*, 6878–6882.
70. Roderick, S. L.; Matthews, B. W. *Biochemistry* **1993,** *32*, 3907–3912.
71. Reczkowski, R. S.; Ash, D. E. *J. Am. Chem. Soc.* **1992,** *114*, 10992–10994.
72. Poyner, R. R.; Reed, G. H. *Biochemistry* **1992,** *31*, 7166–7173.
73. Jabri, E.; Carr, M. B.; Hausinger, R. P.; Karplus, P. A. *Science (Washington, D.C.)* **1995,** *268*, 998–1004.
74. Kim, E. E.; Wyckoff, H. W. *J. Mol. Biol.* **1991,** *218*, 449–464.
75. Chae, M. Y.; Omburo, G. A.; Lindahl, P. A.; Raushel, F. M. *J. Am. Chem. Soc.* **1993,** *115*, 12173–12174.
76. Volbeda, A.; Lahm, A.; Sakiuama, F.; Suck, D. *EMBO J.* **1991,** *10*, 1607–1618.
77. Hough, E.; Hansen, L. K.; Birknes, B.; Jynge, K.; Hansen, S.; Hordvik, A.; Little, C.; Dodson, E.; Derewenda, Z. *Nature (London)* **1989,** *338*, 357–360.
78. Vincent, J. B.; Crowder, M. W.; Averill, B. A. *Trends Biochem. Sci.* **1992,** *17*, 105–110.
79. Davies, J. F.; Hostomska, Z.; Hostomsky, Z.; Jordan, S. R.; Matthews, D. A. *Science (Washington, D.C.)* **1991,** *252*, 88–95.
80. Beese, L. S.; Steitz, T. A. *EMBO J.* **1991,** *10*, 25–33.
81. Chaudhuri, P.; Stockheim, C.; Wieghardt, K.; Deck, W.; Gregorzik, R.; Vahrenkamp, H.; Nuber, B.; Weiss, J. *Inorg. Chem.* **1992,** *31*, 1451–1457.
82. Uhlenbrock, S.; Krebs, B. *Angew. Chem. Int. Ed. Engl.* **1992,** *31*, 1647–1648.
83. Hikichi, S.; Tanaka, M.; Moro-oka, Y.; Kitajima, N. *J. Chem. Soc. Chem. Commun.* **1992,** 814–815.
84. Vance, D. H.; Czarnik, A. W. *J. Am. Chem. Soc.* **1993,** *115*, 12165–12166.
85. Wall, M.; Hynes, R. C.; Chin, J. *Angew. Chem. Int. Ed. Engl.* **1993,** *32*, 1633–1635.
86. Kim, H.; Lipscomb, W. N. *Biochemistry* **1993,** *32*, 8465–8478; Burley, S. K.; David, P. R.; Lipscomb, W. N. *Proc. Natl. Acad. Sci. U.S.A.* **1991,** *88*, 6916–6920.
87. Rardin, R. L.; Tolman, W. B.; Lippard, S. J. *New J. Chem.* **1991,** *15*, 417–430.
88. Atta, M.; Nordlund, P.; Aberg, A.; Eklund, H.; Fontecave, M. *J. Biol. Chem.* **1992,** *267*, 20682–20688.
89. Cole, P. A.; Bean, J. M.; Robinson, C. H. *Proc. Natl. Acad. Sci. U.S.A.* **1990,** *87*, 2999–3003 and references cited therein.
90. Vaz, A. D. N.; Roberts, E. S.; Coon, M. J. *J. Am. Chem. Soc.* **1991,** *113*, 5886–5887.
91. Rana, T. M.; Meares, C. F. *J. Am. Chem. Soc.* **1991,** *113*, 1859–1861.

RECEIVED for review July 19, 1993. ACCEPTED revised manuscript January 7, 1994.

7

Synthetic Modeling of the Interactions of Nitrogen Oxides with Copper Proteins

Copper Nitrosyl Complexes Relevant to Putative Denitrification Intermediates

William B. Tolman

University of Minnesota, 207 Pleasant Street S.E., Minneapolis, MN 55455

This chapter focuses on the chemistry of biomimetic copper nitrosyl complexes relevant to the NO-copper interactions in proteins that are central players in dissimilatory nitrogen oxide reduction (denitrification). The current state of knowledge of NO-copper interactions in nitrite reductase, a key denitrifying enzyme, is briefly surveyed; the syntheses, structures, and reactivity of copper nitrosyl model complexes prepared to date are presented; and the insight these model studies provide into the mechanisms of denitrification and the structures of other copper protein nitrosyl intermediates are discussed. Emphasis is placed on analysis of the geometric features, electronic structures, and biomimetic reactivity with NO or NO_2^- *of the only structurally characterized copper nitrosyls, a dicopper(II) complex bridged by* NO^- *and a mononuclear tris(pyrazolyl)hydroborate complex having a Cu(I)-NO formulation.*

THE CRITICAL ROLE OF METALLOPROTEINS in mediating interconversions of nitrogen oxides (N_xO_y) within the global nitrogen cycle is now well-established (*1–3*). In particular, copper-containing proteins from anaerobic bacteria function as catalysts for the dissimilatory reduction of nitrite (NO_2^-) and nitrous oxide (N_2O) to gaseous nitric oxide (NO, N_2O, or N_2). Dissimilatory nitrogen oxide reduction (denitrification) is a respiratory process with intriguing analogies to the dioxygen consumption performed by most organisms, but which differs from the assimilatory

0065–2393/95/0246–0195/$08.54/0

reduction process by which oxidized nitrogen compounds are reduced to ammonia that is subsequently used as a building block for the synthesis of nitrogen-containing biomolecules. Impetus for the study of denitrification is provided by the numerous environmental consequences (*4*) of the uptake or production of the simple inorganic nitrogen compounds involved in the process: NO_3^- and NO_2^-, important components of fertilizers and agents responsible for lake eutrophication; N_2O, a greenhouse gas that has been implicated in ozone depletion; and NO, an important pollutant and biological effector molecule. Developing an understanding of the mechanisms by which the copper-containing enzymes involved in denitrification reduce nitrogen oxides is thus an important research objective.

Significant information on the chemistry of nitrogen oxide copper–protein adducts can be uncovered by examining in detail the structures and reactivity of appropriate active site model complexes. This chapter will focus on the chemistry of biomimetic copper nitrosyl complexes relevant to the NO–copper interactions in proteins that are central players in dissimilatory nitrogen oxide reduction. Additional reasons for emphasizing copper nitrosyls include their presumed formation in experiments using NO as an O_2 analogue and probe of dioxygen-activating copper proteins (*5–13*) and the recent identification of NO as a pervasive biomolecule that has metal sites within proteins as a major target (*14–16*). After the current state of knowledge of NO–copper interactions in nitrite reductase (NiR), a central denitrifying enzyme, is briefly surveyed, the syntheses, structures, and reactivity of copper nitrosyl model complexes prepared to date will be presented, and the insight these model studies provide into the mechanisms of denitrification and the structures of other copper protein nitrosyl intermediates will be discussed.

The Copper Nitrite Reductase Nitrosyl Intermediate

Several copper-containing NiRs have been identified, but the most extensive structural and mechanistic studies have focused on the enzyme from *Achromobacter cycloclastes* (*17–25*). A 2.3-Å resolution X-ray crystal structure for this NiR in its oxidized form at pH 5.2 has been reported (*17*), and a representation of the active site is shown in Figure 1. Each monomer in the trimeric protein contains two copper ions, one of which (Cu-1) is ligated to a cysteine, a methionine, and two histidine residues in a geometry similar to that of "type 1" copper centers in proteins such as plastocyanin (*26*). The second "type 2" copper ion in NiR (Cu-2) is only 12.5-Å distant from the first and is bound to three histidine imidazoles (two from one monomer, the third from an associated subunit) and a fourth small ligand in an unusual tetrahedral arrangement. The

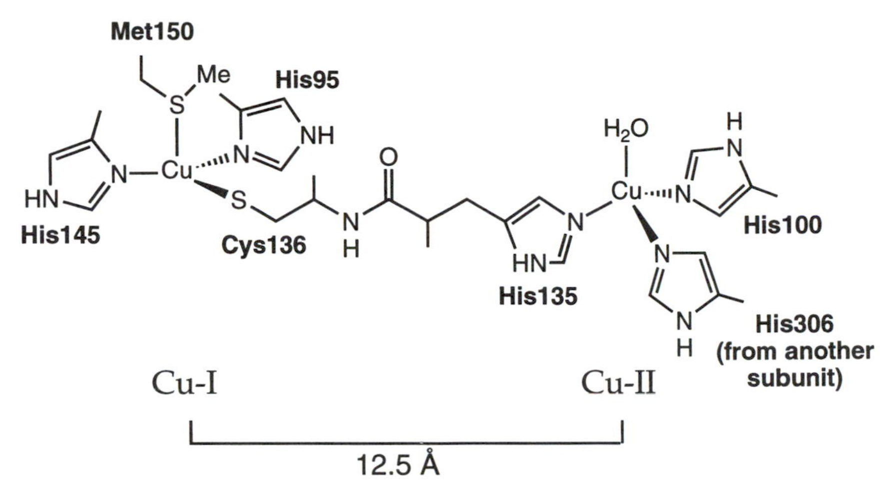

Figure 1. Schematic representation of the copper sites in nitrite reductase from Achromobacter cycloclastes *(17).*

copper ions are connected to each other by a dipeptide bridge, Cys-136–His-135. The fourth small ligand on Cu-2 is presumably a water molecule, making Cu-2 a good candidate for the site of substrate binding and reduction.

Preliminary observation of additional electron density at this fourth coordination position of Cu-2 upon soaking crystals with NO_2^- is consistent with this idea. Thus, from the structural data it would appear that Cu-1 is a type 1 center that functions to transfer electrons to the catalytic Cu-2 ion (*See* Note Added in Proof). It has been suggested, largely on the basis of electronic structural considerations (*27, 28*), that the Cys-136–His-135 link between Cu-1 and Cu-2 is a possible conduit for electron transfer between the two sites. An analogous dipeptide bridge between the type 1 center and the catalytic tricopper cluster in ascorbate oxidase (*29, 30*) may function similarly. Indeed, other close similarities between protein domains in ascorbate oxidase and NiR have been noted (*17*).

The idea that NO_2^- binding and reduction occurs at Cu-2 in NiR from *A. cycloclastes* is also supported by recent mechanistic experiments (*24*). In examinations of enzyme preparations having maximum Cu-1 content but depleted Cu-2 sites, specific activity (production of NO from NO_2^-) was found to be directly proportional to the Cu-2 content. This result argues against a previous suggestion (*25*) that Cu-1 is the site of catalysis in the enzyme and implies that the low specific activities of other copper NiRs that are reported to contain only type 1 centers (*31, 32*) may be due to the fact that type 2 centers are required in these systems, too (*33*). Thus, it has been proposed that small amounts of type 2 sites may actually be responsible for the observed activity in the purported type 1-only NiRs. Further indications of the critical substrate binding role of the Cu-2 site include spectroscopic data acquired for the NiR from *Alcaligenes xylosoxidans* (*33, 34*), reports of preliminary electron paramagnetic resonance (EPR) spectroscopy data that show changes only in the *A. cycloclastes* NiR Cu-2 signal upon treatment of solutions of the enzyme with excess NO_2^- (*24*), and electrochemical experiments that compare the apo, Cu-2-depleted, and fully metallated enzymes (*35*).

Several pieces of evidence, albeit mostly indirect, implicate a nitrosyl adduct to Cu-2 as a key reaction intermediate in the generation of NO or N_2O from NO_2^- (*21–25*). Thus, all proposed mechanisms of nitrite reduction have in common an initial, reversible dehydration of a coordinated NO_2^- to yield a copper nitrosyl, written as $Cu^+–NO^+$ to emphasize the presumed electrophilic character of the bound NO ligand and the need for an electron from copper during catalysis (*21–25*) (Scheme 1). The formulation of the copper nitrosyl is also based on analogies drawn to more fully characterized nitrosyls of iron-containing enzymes and model complexes (*36, 37*). Certainly, formation of the principal enzyme reaction product, free NO, during catalysis is readily explained by invoking ligand loss and Cu(II) generation from such a copper nitrosyl. Initial evidence for this species was provided by ^{15}N-labeled nitrite- and azide-trapping experiments performed with cell-free extracts from *A. cycloclastes*, which were interpreted to indicate partitioning between pathways of attack by the labeled substrates on the electrophilic $Cu^+–NO^+$ unit (*21*). Incorporation of ^{18}O from $H_2{}^{18}O$ into product N_2O when NO was introduced as substrate in whole cells has also been cited as supporting evidence for the nitrosyl intermediate (*22*), because it is known that such electrophilic $M–NO^+$ species undergo

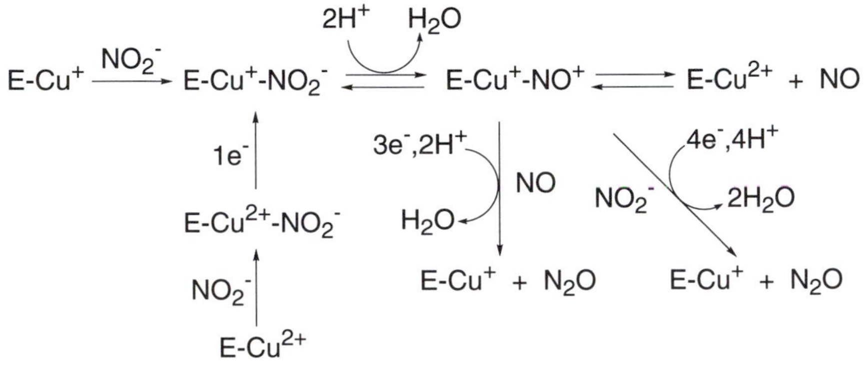

$$E\text{-}Cu^+ \xrightarrow{NO_2^-} E\text{-}Cu^+\text{-}NO_2^- \underset{}{\overset{2H^+ \;\; H_2O}{\rightleftharpoons}} E\text{-}Cu^+\text{-}NO^+ \rightleftharpoons E\text{-}Cu^{2+} + NO$$

$$E\text{-}Cu^{2+} \xrightarrow{NO_2^-} E\text{-}Cu^{2+}\text{-}NO_2^- \xrightarrow{1e^-} E\text{-}Cu^+\text{-}NO_2^-$$

$$E\text{-}Cu^+\text{-}NO^+ \xrightarrow[3e^-,2H^+ \;\to\; H_2O]{NO} E\text{-}Cu^+ + N_2O$$

$$E\text{-}Cu^+\text{-}NO^+ \xrightarrow[4e^-,4H^+ \;\to\; 2H_2O]{NO_2^-} E\text{-}Cu^+ + N_2O$$

Scheme 1

rapid and reversible hydration–dehydration processes that result in exchange of oxygen isotopes (*38–40*). In addition, a $Cu^{+}–NO^{+}$ species has been invoked to explain the observed incorporation of N atoms from both NO and NO_2^- into product N_2O (*23*) (Scheme 1). The reversible bleaching of the type 1 copper EPR and UV–vis signatures in NiR upon addition of NO has also been cited as evidence for the generation of a $Cu^{+}–NO^{+}$ intermediate during catalysis, although it is probable that the enzyme used in these experiments lacked the Cu-2 site subsequently shown to be linked to high specific activity (*25*). In sum, circumstantial evidence for the $Cu^{+}–NO^{+}$ denitrification intermediate exists. However, it has never been observed directly, and no spectroscopic evidence has been reported to either confirm or dispute its electronic structural formulation.

Studies of NO binding to other copper proteins have been equally inconclusive, and the geometric and electronic structures of the putative adducts that have been reported remain obscure. In attempts to produce protein–CuNO complexes, examples from each of the major classes of copper-containing enzymes have been treated with NO, including type 1 [e.g., azurin (*5*)], type 2 [e.g., superoxide dismutase (*5*)], type 3 [e.g., hemocyanin (*6–9*)], and others [e.g., galactose oxidase (*41, 42*), cytochrome c oxidase (*10–12*), laccase (*13*), and ascorbate oxidase (*5*)]. Evidence cited in these studies in support of successful generation of copper nitrosyls have included the observation of bleaching of EPR and UV-vis spectroscopic features upon treatment of oxidized type 1 centers with NO and restoration of these signals upon purging of protein solutions with N_2 or upon photolysis (*5, 25*), and the observation of an unusual EPR signal with a feature at $g < 2$ with hyperfine coupling to the nitrosyl nitrogen upon treatment of reduced laccase with NO (*13*). In some instances, NO binds but not to the copper center (*41, 42*). More commonly, complex redox chemistry ensues when copper proteins are exposed to NO, and much work is still needed to differentiate between NO binding, charge-transfer complex formation, and electron-transfer processes in these reactions and to determine the mechanisms of formation of other nitrogen oxides (NO_2^-, N_2O, etc.) that have been observed (*5–13*).

Copper Nitrosyl Complexes

The need for structural, spectroscopic, and reactivity benchmarks with which to compare and help interpret data acquired for all the copper protein systems, NiR in particular, has led to an interest in preparing copper nitrosyl model complexes that may be more readily characterized in detail. Such complexes are rare and, until recently, none had been structurally characterized by X-ray crystallography. Solids having the empirical formula CuX_2NO (X = Cl or Br) have been isolated from deep purple solutions formed upon exposure of cupric halides to NO in anhydrous alcohol or acetonitrile (*43*, *44*). The primary evidence for the binding of NO to copper in these species was provided by manometric NO-uptake experiments and the observation of bands between 1830 and 1870 cm^{-1} in the infrared spectra of the solids. No additional structural information has been reported. These compounds have been used as reagents for nitrosation of several classes of organic molecules, suggesting that the nitrosyl group is electrophilic (*45*). Interestingly, the complexes react with N_3^- to form N_2, N_2O, and the cuprous halide, a reaction relevant to that which is postulated to occur when azide is used to trap the putative electrophilic Cu^+–NO^+ NiR intermediate (*21*).

Weak and reversible binding of NO to the tetragonal Cu(II) complexes CuL_2 (L = dithiocarbamato, dithiophosphato, or 8-mercaptoquinolinato) has also been reported on the basis of observed broadening of EPR signals upon exposure of the compounds to NO (*46*). Regeneration of the original sharp spectra was observed upon subsequent purging with N_2 or Ar. It was suggested that the EPR signal broadening resulted from rapid exchange of bound with unbound NO but, again, no other structural information (including infrared spectra in this instance) were reported that substantiated the hypothesis of NO adduct formation.

The first copper nitrosyl complex to be structurally characterized, **2**, was prepared by treatment of a dicopper(I) compound, **1**, with NO+ (*47*) (Scheme 2). It was also synthesized from the dicopper(I) precursor, *n*-$Bu_4N(NO_2)$, and two equivalents of $HPF_6 \cdot Et_2O$ in a reaction that models the proposed initial nitrite dehydration step (Scheme 1) carried out by copper NiR (*48*). An X-ray crystallographic study revealed a symmetrically bound nitrosyl ligand in **2** with bond distances Cu–N = 2.036 (10) Å and N–O = 1.176 (1) Å. (The numbers in parenthe-

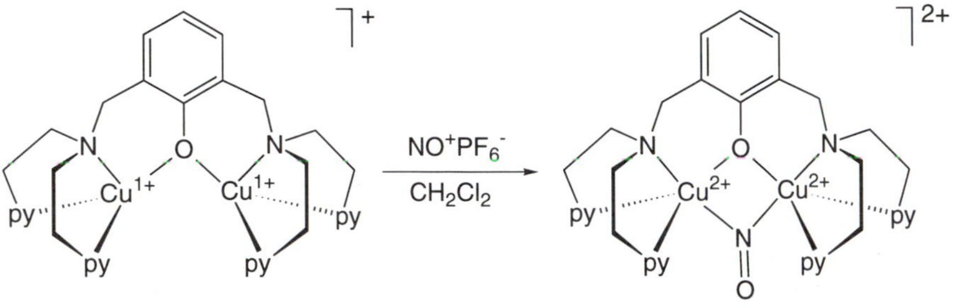

N O N Cu^{1+} Cu^{1+} py py py py $]^+$ **1**

$NO^+PF_6^-$ / CH_2Cl_2 →

N O N Cu^{2+} Cu^{2+} N O py py py py $]^{2+}$ **2**

Scheme 2

ses represent standard deviations.) The reported ν(NO) band at 1536 cm^{-1} is consistent with the bridged nitrosyl structure. Formulation of **2** as a Cu(II)–(NO^-)–Cu(II) complex, resulting from transfer of two electrons from the dicopper(I) precursor to NO^+, was based on the observation of square pyramidal copper geometries typical for Cu(II) ions and a feature in the electronic absorption spectrum at 730 nm ($\varepsilon \sim 500$) attributable to a Cu(II) dd transition. The complex has $\mu_{rt} = 0.59\ \mu_B$/Cu and is EPR-silent, presumably because the Cu(II) sites are strongly antiferromagnetically coupled.

Dinuclear **2** and its dicopper(I) precursor have been shown to mediate the formation of N_2O from NO_2^- or NO (*48*). For example, treatment of **2** with NO_2^- produced 1 mol N_2O, 0.5 mol O_2, and the oxo-bridged dicopper(II) complex **3** (Scheme 3). Formation of N_2O and coproduct **3** was also induced by treating dicopper(I) complex **1** with NO (Scheme 3). In similar reactions using related copper(I) starting materials, unstable colored intermediates were observed at low temperature, which were formulated as antiferromagnetically coupled dicopper(II) complexes containing two bridging nitrosyl (NO^-) ligands on the basis of manometric NO-uptake measurements, an observed low ν(NO) in the IR spectrum (1460 cm^{-1}), and the lack of an EPR signal.

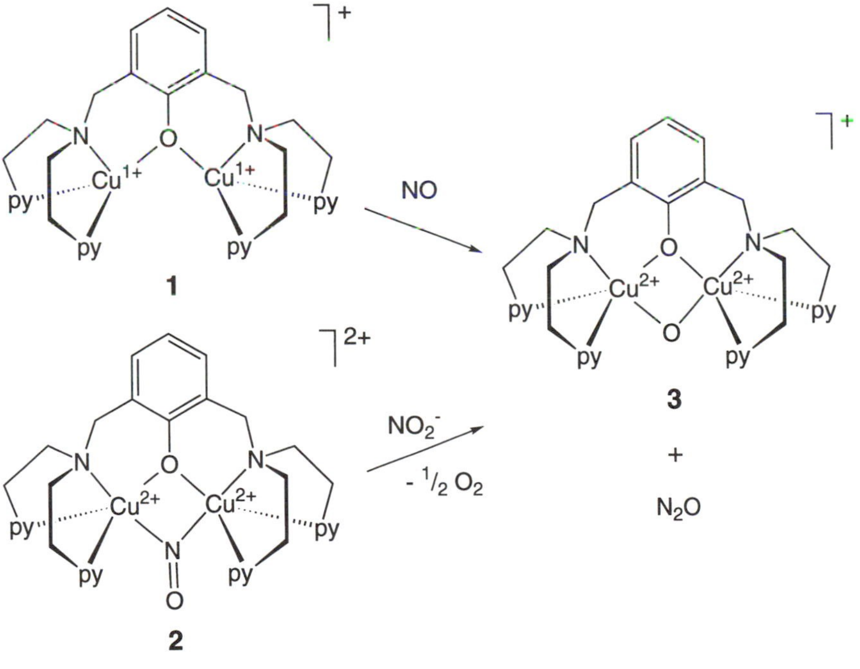

Scheme 3

It was suggested that a dibridged dinitrosyl structure for the unstable intermediates might facilitate the N–N coupling reaction, an hypothesis analogous to the postulated necessity for a cis disposition of two nitrosyls in mononuclear transition metal complexes that decompose to N_2O or N_2 (*49*). Both copper-promoted N_2O generation reactions (**2** plus NO_2^- and **1** and analogues plus NO) model proposed NiR N_2O production pathways insofar as N–N bond formation is induced via the intermediacy of a copper nitrosyl compound. Although a type 1 center 12.5-Å distant from the catalytic site in NiR probably supplies the needed electrons for nitrite reduction, however, the second much closer copper ion in the synthetic system that participates in nitrosyl complex formation and mediates the oxygen atom transfer reactions is absent in the enzyme. Thus, the relevance of this modeling chemistry to NiR structure and function is somewhat limited.

Motivated by the lack of well-characterized mononuclear models for copper NiR active site complexes, we have begun to explore the interactions of simple nitrogen oxides with molecules containing a single copper ion (*50–53*)(*See* Note Added in Proof). Taking our cue from previous successful preparations of reactive bioinorganic model complexes (i.e., *see* references 54 and 55), we have used sterically hindered 3,5-substituted tris(pyrazolyl)hydroborates ($Tp^{RR'}$; R = H, R′ = *t*-Bu; R = R′ = Ph; R = R′ = Me) as supporting ligands (*56*). The array of pyrazolyl N-donors in these ligands closely approximates the histidine imidazolyl groups coordinated to Cu(II) in NiR from *A. cycloclastes*. Inhibition of undesired polynuclear complex formation and increased kinetic stability of the targeted CuNO unit were anticipated to result from the presence of the bulky substituents attached to the tris(pyrazolyl) frame. Our initial synthetic strategy has involved treating suitable Cu(I) precursors with NO, because we reasoned that a mononuclear Cu(II)–(NO^-) species would result via a redox process analogous to, but involving one less electron than, that involved in the known reaction of NO^+ with the dicopper(I) compound **1** to generate the Cu(II)–(NO^-)–Cu(II) unit in **2** (*47*).

An array of new (*57*) and previously prepared (*58*) $Tp^{RR'}$Cu(I) starting materials were synthesized by mixing $MTp^{RR'}$ (M = K^+ or Tl^+) with CuCl in organic solvent (Scheme 4). X-ray crystal structure determinations revealed that all of the complexes, with the exception of **7**, are dimers bridged by two $Tp^{RR'}$ ligands in the solid state (*57, 58*) (Figures 2–4). For the dimers with R = R′ = Me or Ph (**4** and **5**, respectively), two pyrazolyl rings of each ligand are bonded to one Cu(I) ion and the third is coordinated to the other metal center, resulting in planar 3-coordinate Cu(I) geometries in the complexes (Figure 2). In solution, both molecules exhibit deceptively simple and sharp ^1H NMR spectra that do not broaden appreciably upon cooling, which is consistent with the existence

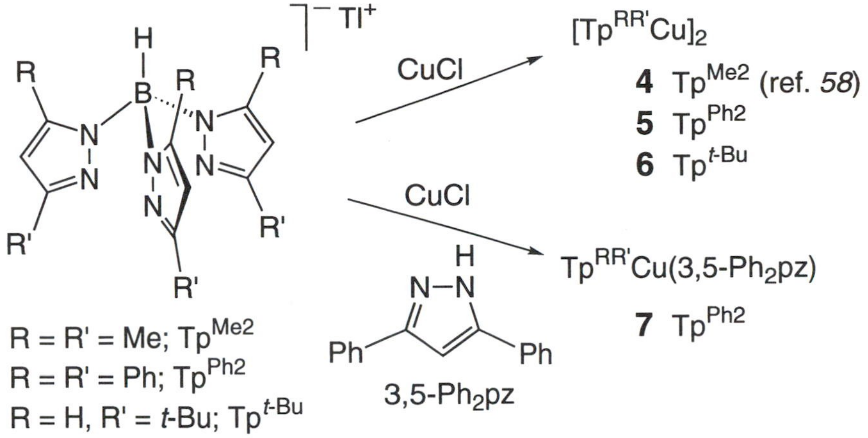

CuCl → $[Tp^{RR'}Cu]_2$

4 Tp^{Me2} (ref. *58*)
5 Tp^{Ph2}
6 $Tp^{t\text{-}Bu}$

CuCl, 3,5-Ph_2pz → $Tp^{RR'}Cu(3,5\text{-}Ph_2pz)$

7 Tp^{Ph2}

R = R' = Me; Tp^{Me2}
R = R' = Ph; Tp^{Ph2}
R = H, R' = *t*-Bu; $Tp^{t\text{-}Bu}$

Scheme 4

of rapid pyrazolyl exchange processes. In contrast, the dimer formed when R = H and R′ = *t*-Bu (**6**) contains linear two-coordinate Cu(I) ions and one uncoordinated pyrazolyl group on each bridging ligand (Figure 3). Fluxional processes are slowed in this complex due to the increased steric bulk of the pyrazolyl 3-substituent, allowing calculation of activation parameters for apparent intramolecular pyrazole exchange via line shape analysis of variable temperature ^{1}H NMR spectra [$\Delta H^{\ddagger} = 11.7$ (5) kcal mol^{-1} and $\Delta S^{\ddagger} = -9$ (2) eu]. Finally, mixing MTp^{Ph2} and 3,5-diphenylpyrazole (3,5-Ph_2pz) with CuCl yielded the distorted tetrahedral monomer **7** (Figure 4). This molecule is also highly fluxional in solution, with exchange of bound with added unbound pyrazole being particularly facile by ^{1}H NMR spectroscopy.

Dimer **6** and monomer **7** (dimer **5** as well, but more slowly) dissolved in organic solvent rapidly reacted with NO to form deep red or orange solutions, respectively (*52, 53*). The solutions exhibited strong IR absorptions at 1712 cm^{-1} ($Tp^{t\text{-}Bu}$) and 1720 cm^{-1} (Tp^{Ph2}) that were assigned as ν(NO) bands on the basis of their appropriate shifts to lower energy upon isotopic substitution (^{15}NO). From these data alone it was evident that copper nitrosyl adducts had formed. These results also indicated that the Tp^{Ph2} ligand causes the copper ion to be more electron-deficient than that bound to $Tp^{t\text{-}Bu}$, a conclusion corroborated by the carbonyl-stretching frequencies of the corresponding complexes $Tp^{RR'}$Cu(CO) [ν(CO) for $Tp^{t\text{-}Bu}$ = 2069 cm^{-1} (*57*); ν(CO) for Tp^{Ph2} = 2086 cm^{-1} (*59*)].

The respective red and orange colors of the NO-saturated solutions of the Cu(I) complexes result from electronic absorption features at 498 nm ($Tp^{t\text{-}Bu}$) and 478 nm (Tp^{Ph2}), respectively (Figure 5). Placing the

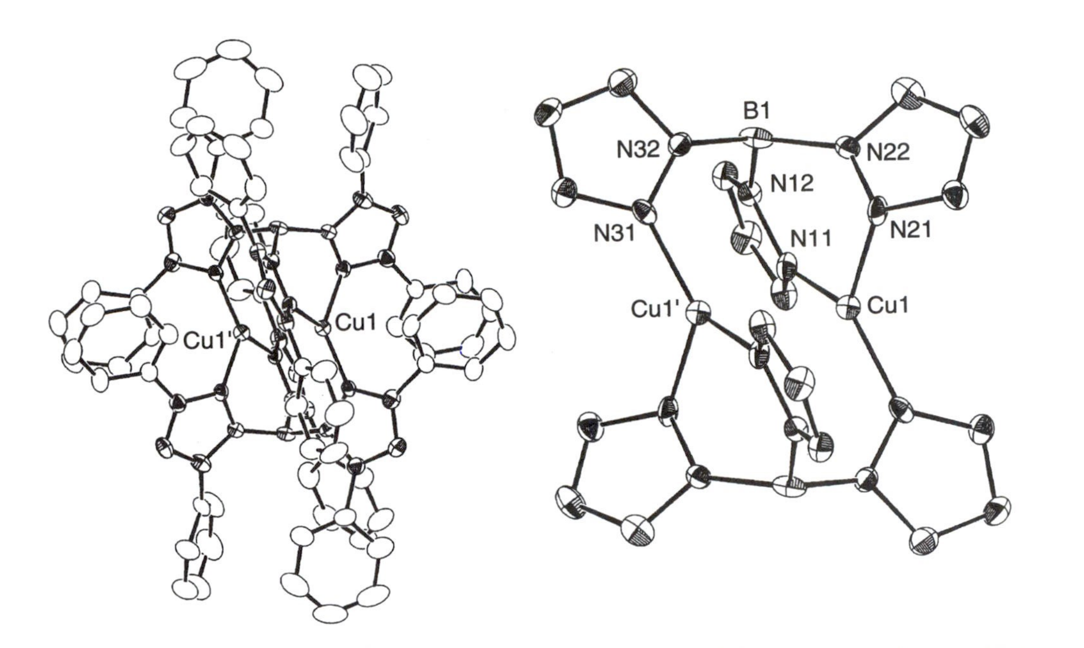

Figure 2. Left: ORTEP drawing of $[Tp^{Ph2}Cu]_2$, **5** *(hydrogen atoms omitted for clarity). Right: ORTEP drawing of core of* $[Tp^{Ph2}Cu]_2$, **5**, *with phenyl groups removed and atom labels shown for all noncarbon atoms on one of the two symmetry equivalent* Tp^{Ph2} *ligands (57). The structure is analogous to that of* $[Tp^{Me2}Cu]_2$, **4** *(58).*

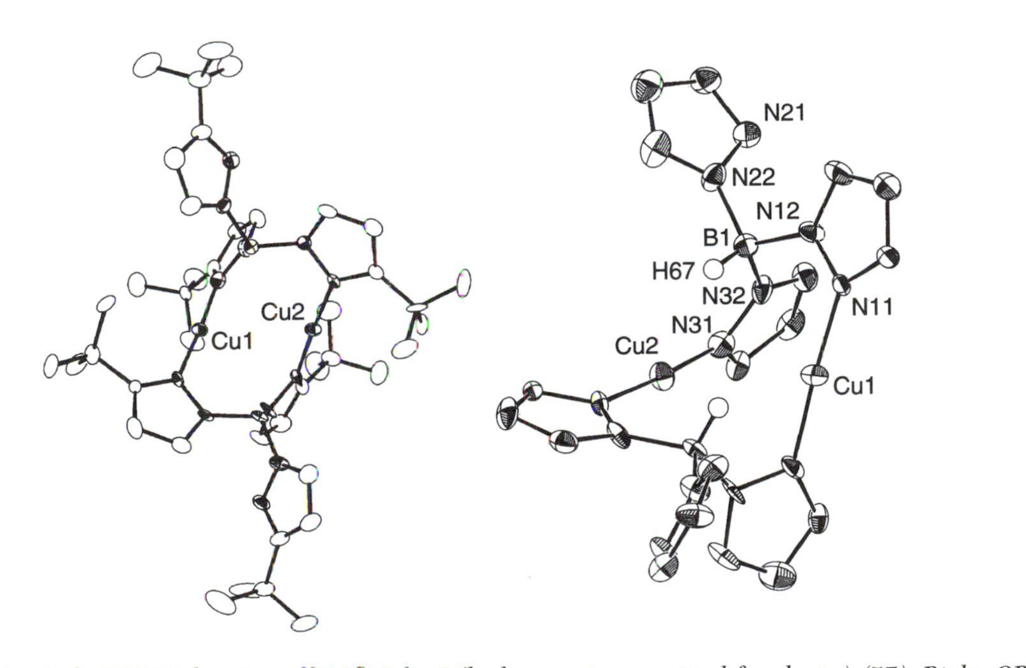

Figure 3. Left: ORTEP drawing of $[Tp^{t\text{-}Bu}Cu]_2$, **6** *(hydrogen atoms omitted for clarity) (57). Right: ORTEP drawing of core of* $[Tp^{t\text{-}Bu}Cu]_2$, **6**, *with tert-butyl groups removed and atom labels shown for the noncarbon atoms on one of the two inequivalent* $Tp^{t\text{-}Bu}$ *ligands.*

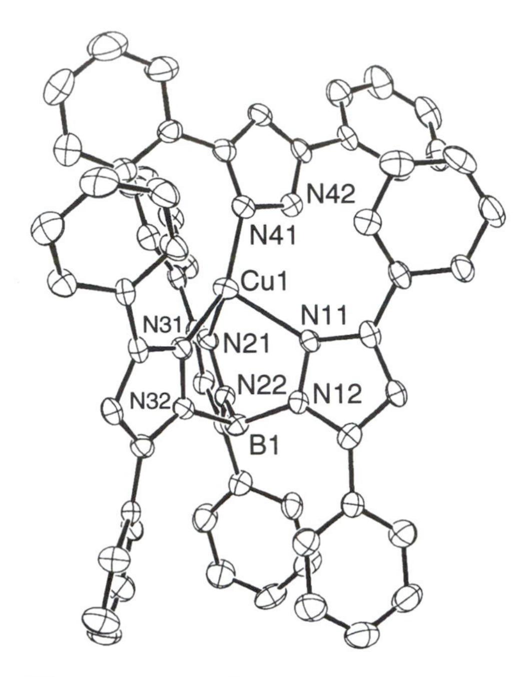

Figure 4. ORTEP drawing of $Tp^{Ph2}Cu(3,5\text{-}Ph_2pz)$, 7 (hydrogen atoms removed for clarity) (57).

solutions under a vacuum or purging with N_2 or Ar caused these absorption bands to bleach, and ^1H NMR analysis of the resulting colorless solutions indicated almost quantitative regeneration of the Cu(I) precursors. Thus, the binding of NO is reversible. Indeed, manometric measurements for the $Tp^{t\text{-}Bu}$ case indicated that only ~40% of the available copper ions are coordinated to NO in solution at 23 °C, with increasing amounts of NO bound as the temperature is lowered (53). These data suggest that the Cu(I) precursors and NO are in dynamic equilibrium with the respective nitrosyl adducts (Scheme 5), with an approximate $K_{eq} = 140\ M^{-1}$. Using the results of the manometry experiments, extinction coefficients for the electronic absorption features of the nitrosyl compounds were calculated to be ~1400 $M^{-1}cm^{-1}$, which suggests that the absorbances should be attributed to charge-transfer (CT) transitions. Moreover, a shift of the absorption to higher energy in the compound with the more electron-deficient copper ion (Tp^{Ph2}) favors a metal-to-ligand charge transfer (MLCT) rather than a ligand-to-metal charge transfer (LMCT) assignment, an observation of some importance in defining the electronic structure of the copper nitrosyl unit (*see* later).

By cooling an NO-saturated solution of **6** in aromatic solvent, X-ray-quality crystals of $Tp^{t\text{-}Bu}Cu(NO)$, **8**, were obtained and the resulting

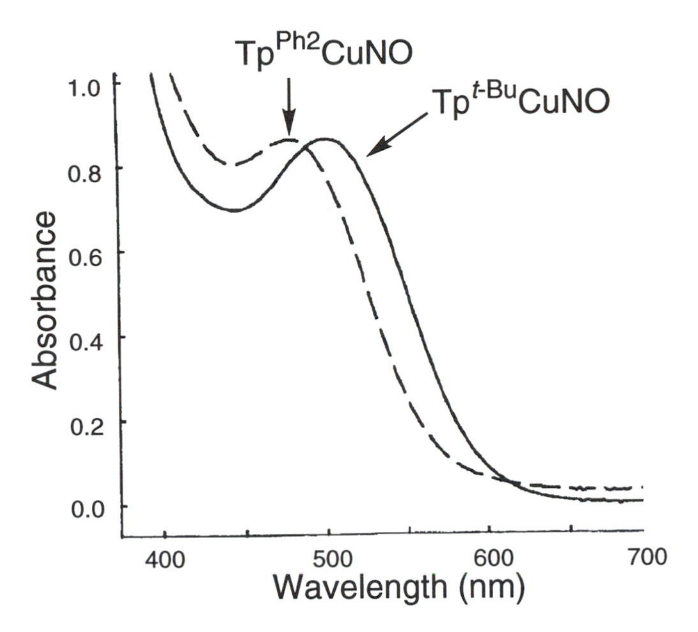

Figure 5. UV–vis spectra of CH_2Cl_2 solutions of the $Tp^{RR'}Cu(NO)$ complexes (52, 53).

structure (Figure 6) provided definitive proof that the Cu(I) dimer had been cleaved to form the first example of a mononuclear copper nitrosyl complex. Both the short Cu–NO and N–O bond lengths (1.76 and 1.11 Å, respectively) in **8** are typical for transition metal nitrosyl compounds (*60, 61*). The Cu–N–O bond angle is 163°, intermediate between the linear (180°) and bent (120°) metal nitrosyl extremes. In addition, the copper ion in **8** has an approximately tetrahedral geometry (close to local C_{3v} symmetry) similar to that found in other four-coordinate $Tp^{t\text{-Bu}}$ or $Tp^{i\text{-Pr2}}$ complexes of both Cu(I) and Cu(II) (*57, 62, 63*).

It is difficult to judge the extent to which electron transfer from Cu(I) to NO to form a Cu(II)–(NO^-) species had occurred upon adduct formation by considering just the IR [ν(NO) = 1720 cm^{-1}, in the region of overlap between linear and bent nitrosyl complexes (*60, 61*)] and X-ray crystal structure results (tetrahedral geometry and Cu–N–O angle = 163°). In order to understand the electronic structure of the CuNO unit in **8** and its Tp^{Ph2} congener, a series of EPR, magnetic circular dichroism (MCD), and near-IR absorbance spectroscopic studies were un-

$$\tfrac{1}{2}[Tp^{t\text{-Bu}}Cu]_2 + NO \rightleftarrows Tp^{t\text{-Bu}}Cu(NO)$$

Scheme 5

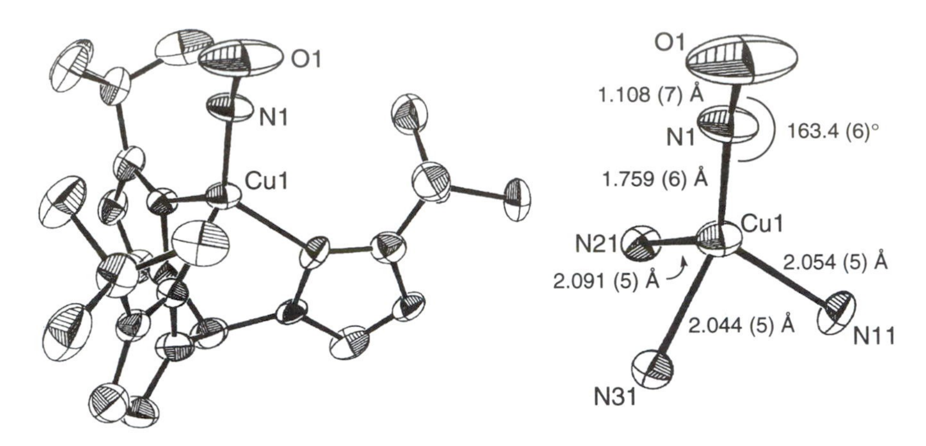

Figure 6. Left: ORTEP drawing of $Tp^{t\text{-}Bu}Cu(NO)$, 8 (hydrogen atoms removed for clarity) (52, 53). *Right: ORTEP drawing of core of $Tp^{t\text{-}Bu}Cu(NO)$, 8, with important metrical parameters noted.*

dertaken (*52, 53*). The combined results of these investigations are best rationalized by an electronic structural description in which the copper ion retains its full set of d electrons and the unpaired electron is substantially localized on the NO ligand. Using the valence bond formalism, which does not take into account the relatively high degree of covalency commonly attributed to transition metal nitrosyls (*60, 61, 64*), a Cu(I)–NO*(SD description is favored. A qualitative molecular orbital picture that takes into account the covalency intrinsic to the Cu–NO bonding can also be constructed and used to explain the spectroscopic data (*61, 64*) (Figure 7); this view has been supported by detailed ab initio calculations performed on the model $(NH_3)_3CuNO$ (*53*). According to the qualitative description the unpaired electron resides in a π^* orbital primarily localized on NO but having an antibonding interaction with the d_{zx},d_{yz} set on copper (4e in Figure 7, with degeneracy split by a Jahn–Teller distortion involving bending of the nitrosyl). The filled levels below have largely d orbital character, with contributions from either σNO or π^*NO as shown.

Each view of the bonding of NO to copper predicts that dd features in the visible and near-IR regions of the absorption and MCD spectra of the nitrosyl complexes will be absent, an hypothesis that was supported by experiment. The MLCT assignment for the ~500 nm optical absorption band is also consistent with the bonding pictures, because the filled, essentially d orbital set prohibits an alternative LMCT attribution. A metal d → π^*NO transition to yield an essentially Cu(II)–(NO^-) excited state seems reasonable—and is predicted by ab initio theory (*53*)—

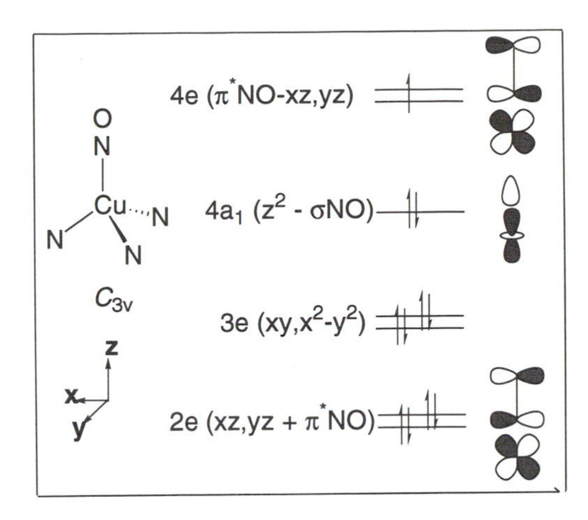

Figure 7. Qualitative molecular orbital diagram for the $Tp^{RR'}Cu(NO)$ complexes (based on analogous diagrams for $\{MNO\}^{10}$ systems described in references 61 and 64.) (Reproduced from reference 53. Copyright 1993 American Chemical Society.)

because of the presence of low-lying, only partially occupied, primarily π*NO orbitals. The generation of such an electronic excited state was supported by observation of an intense feature at ~500 nm in the MCD spectrum, spin-orbit coupling in the excited state Cu(II) ion resulting in a strong pseudo A-term (52, 53).

The substantial localization of the unpaired electron in the ground state of the copper nitrosyls in the NO-based orbitals also results in unusual EPR spectroscopic properties (Figure 8). Simulations of X- and S-band EPR data for ^{14}NO and ^{15}NO complexes yielded the common parameters $g_{\parallel} = 1.83$, $g_{\perp} = 2.00$, $A_{\parallel}^{Cu} = 116$ G, $A_{\perp}^{Cu} = 66$ G, and $A_{\perp}^{NO} = 30$ G (52, 53). Interestingly, the EPR signal is not observable at temperatures above 40 K, indicating the presence of a facile relaxation process which is currently under investigation (H. Koteiche, W. E. Antholine, C. E. Ruggiero, and W. B. Tolman, unpublished results). Observation of g values <2.0 and the lack of a signal above 30 K are both unusual for Cu(II) complexes, thus providing support for the Cu(I)–NO° formulation. In addition, the large $A_{\perp}^{NO}$ directly indicates extensive unpaired spin density on the NO ligand in the complexes, a conclusion also corroborated by ab initio calculations (53). One explanation for the low g value is a positive spin-orbit coupling constant (λ) arising from a less-than-half-filled, close-to-degenerate set of orbitals localized on NO (the 4e set in Fig-

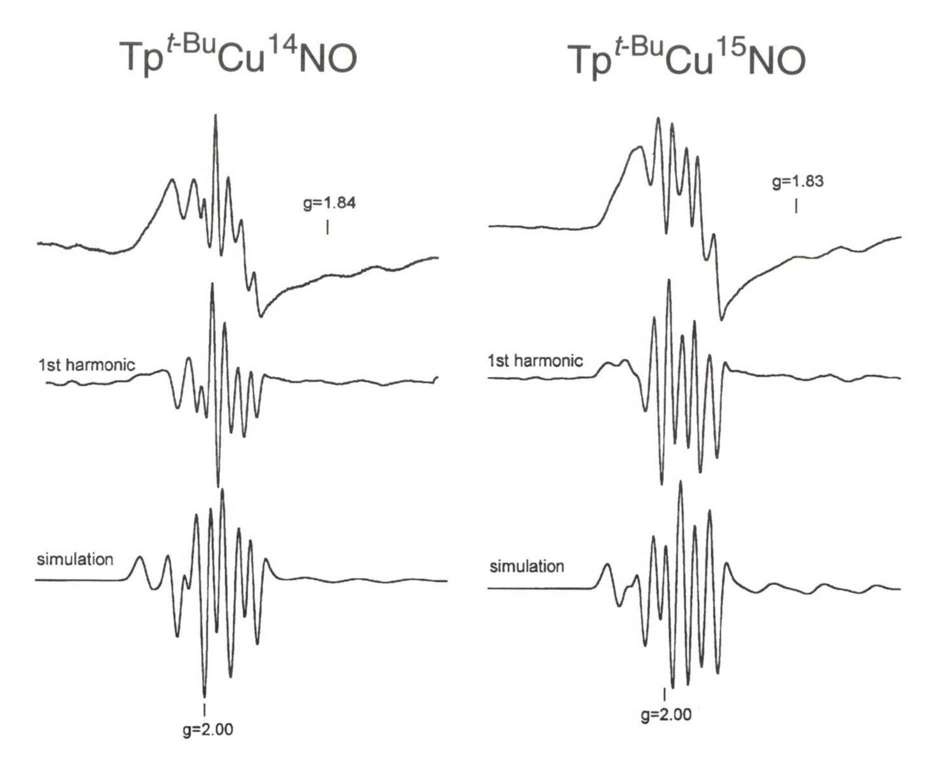

Figure 8. X-band EPR spectra (top), 2nd derivatives (middle), and simulations (bottom) of Tp$^{t\text{-Bu}}$Cu(^{14}NO) *(left) and* Tp$^{t\text{-Bu}}$Cu(^{15}NO) *(right) at* ~15 K.

ure 7) in the ground state of the complex. In contrast, positive g shifts normally result from the more-than-half-filled d orbital array in Cu(II) complexes, where for Cu(II) $\lambda = -830\ cm^{-1}$.

The combined spectroscopic evidence acquired to date thus supports binding of NO to the Tp$^{RR'}$Cu(I) complexes with little or no concomitant electron transfer, a result distinctly different from the situation seen for the dicopper polypyridyl compounds prepared by Karlin and co-workers (*47, 48*). Although the involvement of a second metal ion in the latter chemistry must certainly be invoked in any explanation of this divergent behavior, Cu(I)–Cu(II) redox potentials would be expected to correlate with the extent of electron transfer from Cu(I) to coordinated nitric oxide. In fact, cyclic voltammograms characterized by chemically reversible waves with large positive $E_{1/2}$ values for Tp$^{t\text{-Bu}}$Cu(CH_3CN) and **7** have been measured (+0.93 V and +0.63 V versus SCE, respectively, in CH_2Cl_2) (*57*). We speculate that the origin of these high potentials, and the resulting lack of electron transfer to coordinated NO in the

TpCu compounds, is the relative instability of the Cu(II) state due to the pseudotetrahedral metal ion geometry enforced by the sterically hindered Tp ligands. Analogous geometric effects on the redox potentials of Cu(I) complexes of sterically hindered multidentate N-donor ligands have been reported (*65*). In contrast, the polypyridyl ligands used by Karlin and co-workers (*47, 48*) can more readily support favorable five-coordinate Cu(II) geometries, resulting in decreased redox potentials and more facile electron transfer to NO.

Preliminary experiments have revealed that the reactivity of the $Tp^{RR'}Cu(NO)$ compounds is also unique. As noted previously, NO binding is weak and reversible in these complexes, and displacement of NO by CO or CH_3CN to yield the respective Cu(I) adducts is therefore readily accomplished. More importantly from the perspective of modeling NiR activity, we have recently found that by using $Tp^{Ph2}Cu(CH_3CN)$ or the less hindered $Tp^{Me2}Cu(I)$ starting material **4**, NO disproportionation to yield N_2O and $Tp^{R2}Cu(NO_2)$ (R = Ph or Me) can be induced (Scheme 6) (*66*). Specifically, treatment of CH_2Cl_2 solutions of the Cu(I) starting materials with excess NO resulted in rapid formation of red solutions of the respective copper nitrosyl complexes. However, in contrast to the case where the 3-pyrazolyl substituents are *t*-Bu, the red color changed to yellow-green over several hours and an equivalent of N_2O (quantified by GC) and the respective Cu(II)-nitrite complex (quantified by UV-vis and EPR spectroscopy and characterized by X-ray crystallography) were formed. The overall disproportionation reaction, although known for other metals (*60*), is unique in copper chemistry and provides precedent for NO reactions with copper proteins such as NiR (*23*) and hemocyanin (*8, 67*) that result in N_2O generation and concomitant copper oxidation.

Unlike in the polypyridyl system, where oxo-dicopper(II) complex formation accompanies N–N bond formation (*48*), a third NO accepts the oxygen atom in our case [the known (*68*) complex $Tp^{Me2}Cu$–(μ-O)–$CuTp^{Me2}$ was not observed for the case R = Me]. We hypothesize that in the initial stages of the disproportionation reaction the adduct $Tp^{Me2}Cu(I)$–$(NO)_2$ forms and either dimerizes to yield a dibridged dicopper complex analogous to intermediates postulated by Paul and Karlin (*48*) or adds a second NO to afford a mononuclear species with cis

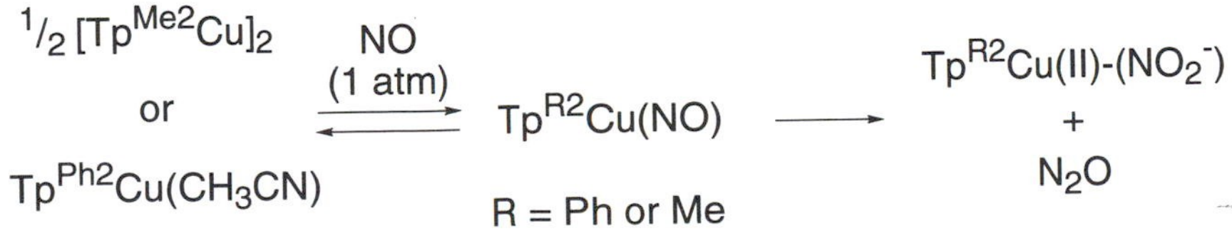

$$\begin{matrix} \tfrac{1}{2}[Tp^{Me2}Cu]_2 \\ \text{or} \\ Tp^{Ph2}Cu(CH_3CN) \end{matrix} \underset{}{\overset{NO\ (1\ atm)}{\rightleftharpoons}} \begin{matrix} Tp^{R2}Cu(NO) \\ R = Ph \text{ or } Me \end{matrix} \longrightarrow \begin{matrix} Tp^{R2}Cu(II)\text{-}(NO_2^-) \\ + \\ N_2O \end{matrix}$$

Scheme 6

nitrosyl ligands or a coordinated hyponitrite (Scheme 7, hyponitrite possibility not shown). The nitrosyls in either intermediate would then be well-disposed for facile N–N bond formation. Preliminary kinetic data indicate a first-order dependence of the reaction rate on Cu(I) ion concentration, consistent with the mononuclear pathway (*66*). How the subsequent steps, oxo-transfer from a third NO, release of N_2O, and electron transfer from Cu(I) to NO_2, occur remains a mystery.

Conclusions

The synthesis and characterization of the copper nitrosyl complexes **2** and **8** represent important first steps in efforts to provide an in-depth chemical understanding of CuNO–protein interactions. Profound differences exist between the model compounds' geometric and electronic structures and reactivity patterns, suggesting that divergent reaction pathways for copper-nitrosyl adducts in biological systems are probable. Most striking is the relatively strong binding of the nitrosyl ligand, with the accompanying electron transfer from copper to the NO group, in the dinuclear complex compared to the weak ligation, with little or no electron transfer, in the mononuclear $Tp^{RR'}Cu$ compounds. The $Tp^{RR'}Cu(NO)$ complexes represent the first characterized examples of NO coordinated to single copper sites and they model a possible intermediate in nitrite reduction by NiR, although they have one more electron than the previously postulated Cu^+–NO^+ denitrification in-

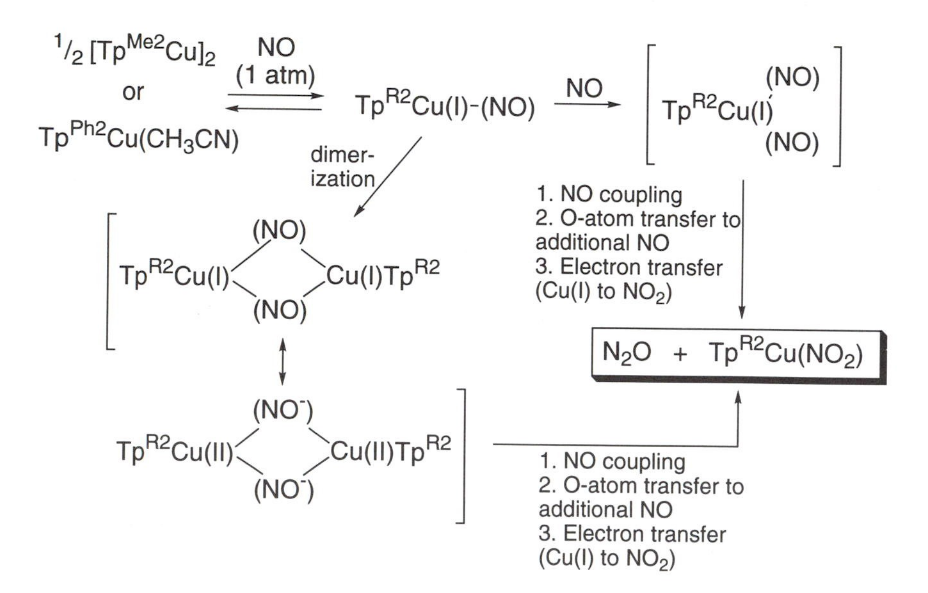

Scheme 7

termediate. Nevertheless, it is conceivable that the Cu(I)–NO° unit may form during nitrite reduction by binding of NO to the reduced Cu(II) site or by reduction of the Cu^+–NO^+ species. Both mechanisms are supported by the demonstrated (*23*) involvement of NO in N_2O generation by NiR.

Nitrous oxide generation has been induced from both **2** and **8**, but it is apparent from preliminary experiments that the mechanisms for the processes differ. In particular, oxygen atom transfer yields an oxo-bridged dicopper(II) species in the polypyridyl compounds but affords NO_2^- in the case of the NO reaction with **4** and $Tp^{Ph2}Cu(CH_3CN)$. Preliminary evidence suggests that N–N bond formation involves mononuclear intermediates in the latter reactions, but the subsequent steps remain unclear.

Finally, the unusual spectroscopic properties of the mononuclear copper nitrosyl complexes provide an important benchmark for comparison to data acquired for other copper proteins. Although examinations of the reactions of NO with reduced copper proteins are sparse (*8, 13, 67*), future studies may benefit from the modeling work (*see* Note Added in Proof). A Cu(I)–NO° adduct has been reported to form upon exposure of reduced laccase to NO on the basis of an EPR spectrum with $g = 2.0$ and $g = 1.8$ components with substantial A_{NO} (*13*). Similarities between the EPR spectra of our fully characterized nitrosyl complexes and that of the reduced laccase species corroborate the published Cu(I)–NO$^\bullet$ assignment, although the absence of Cu hyperfine in the laccase signal raises the possibility that NO interaction may occur elsewhere in the protein (*41, 42*). Interestingly, EPR signals with $g < 2.0$ with large A_{NO} have also been observed for NO adsorbed on Cu(I)-exchanged zeolites that have been identified as highly effective catalysts for environmentally important NO decomposition to N_2 and O_2, suggesting that Cu(I)–NO° moieties similar to those we have characterized may also exist in the heterogeneous systems (*69, 70*). Future investigations on model systems will continue to focus on mechanistic issues of relevance to the nitrogen oxide conversions catalyzed by copper in both biological and zeolite environments.

Abbreviations

The tris(pyrazolyl)hydroborate ($Tp^{RR'}$) ligand nomenclature used has been described by Trofimenko (*56*). The superscripts R and R′ refer to substituents attached to the 3- and 5-positions, respectively, of the pyrazolyl rings of the tris(pyrazolyl)hydroborate (Tp) ligand.

Note Added in Proof

Evidence from pulse radiolysis experiments for electron transfer from the type 1 to the type 2 copper site in *A. cycloclastes* NiR has been reported (*71*). Subsequent to the submission of this chapter, the preparation of a mononuclear copper(I)–η^1-*N*-nitrite complex that evolves NO upon protonation was accomplished (*72*). Electrochemical generation of N_2O and NO from a copper(II)–nitrite complex also has been reported (*73*). X-ray crystallographic and site-directed mutagenesis studies on NiR from *Alcaligenes faecalis* S-6 indicated copper active-site structures analogous to those identified in the enzyme from *A. cycloclastes* and further supported the electron transfer role of the type 1 site in NiR function (*74*). Finally, on the basis of comparisons of IR absorptions observed for NO-treated cytochrome *c* oxidase to those reported for **8**, the presence of a Cu–NO adduct in this protein was inferred (*75*).

Acknowledgments

I would like to thank my co-workers, C. E. Ruggiero, S. M. Carrier, and R. P. Houser and collaborators, W. E. Antholine (National EPR Center, Medical College of Wisconsin), J. W. Whittaker (Carnegie Mellon University), and C. J. Cramer (University of Minnesota) for their extensive contributions to the work cited herein. I also thank B. A. Averill and E. Adman for providing preprints prior to publication. Funding was provided by the Exxon Education Foundation, the National Institutes of Health, the Searle Scholars Program–Chicago Community Trust, the National Biomedical ESR Center in Milwaukee, and the University of Minnesota.

References

1. Kroneck, P. M. H.; Beuerle, J.; Schumacher, W. In *Degradation of Environmental Pollutants by Microorganisms and Their Metalloenzymes;* Sigel, H.; Sigel, A., Eds.; Metal Ions in Biological Systems; Marcel Dekker: New York, 1992; Vol. 28, pp 455–505.
2. Kroneck, P. M. H.; Zumft, W. G. In *Denitrification in Soil and Sediment;* Revsbech, N. P.; Sorensen, J., Eds.; Plenum: New York, 1990; pp 1–20.
3. Stouthamer, A. H. In *Biology of Anaerobic Microorganisms;* Zehnder, A. J. B., Ed.; Wiley: New York, 1988; pp 245–301.
4. *Denitrification, Nitrification, and Atmospheric Nitrous Oxide;* Delwiche, C. C., Ed.; John Wiley & Sons: New York, 1981.
5. Gorren, A. C. F.; Boer, E. D.; Wever, R. *Biochim. Biophys. Acta* **1987,** *916*, 38–47.
6. Uiterkamp, A. J. M. S. *FEBS Lett.* **1972,** *20*, 93–96.
7. Deen, H. v. d.; Hoving, H. *Biochemistry* **1977,** *16*, 3519–3525.
8. Verplaetse, J.; Tornout, P. V.; Defreyn, G.; Witters, R.; Lontie, R. *Eur. J. Biochem.* **1979,** *95*, 327–331.
9. Himmelwright, R. S.; Eickman, N. C.; Solomon, E. I. *Biochem. Biophys. Res. Commun.* **1978,** *81*, 237–242.
10. Brudvig, G. W.; Stevens, T. H.; Chan, S. I. *Biochemistry* **1980,** 5275–5285.

11. Boelens, R.; Rademaker, H.; Wever, R.; Gelder, B. F. V. *Biochim. Biophys. Acta* **1984,** *765*, 196–209.
12. Wever, R.; Boelens, R.; Boer, E. D.; Gelder, B. F. V.; Gorren, A. C. F.; Rademaker, H. J. *Inorg. Biochem.* **1985,** *23*, 227–232.
13. Martin, C. T.; Morse, R. H.; Kanne, R. M.; Gray, H. B.; Malmström, B. G.; Chan, S. I. *Biochemistry* **1981,** *20*, 5147–5155.
14. *The Biology of Nitric Oxide;* Moncada, S.; Marletta, M. A.; Hibbs, J. B.; Higgs, E. A., Eds.; Portland Press: London, 1992; Vol. 1 and 2.
15. Traylor, T. G.; Sharma, V. S. *Biochemistry* **1992,** *31*, 2847–2849.
16. Traylor, T. G.; Duprat, A. F.; Sharma, V. S. *J. Am. Chem. Soc.* **1993,** *115*, 810–811.
17. Godden, J. W.; Turley, S.; Teller, D. C.; Adman, E. T.; Liu, M. Y.; Payne, W. J.; LeGall, J. *Science (Washington, D.C.)* **1991,** *253*, 438–442.
18. Dooley, D. M.; Moog, R. S.; Liu, M.-Y.; Payne, W. J.; LeGall, J. *J. Biol. Chem.* **1988,** *263*, 14625–14628.
19. Iwasaki, H.; Noji, S.; Shidara, S. *J. Biochem.* **1975,** *78*, 355–361.
20. Liu, M.-Y.; Liu, M.-C.; Payne, W. J.; LeGall, J. *J. Bacteriol.* **1986,** *166*, 604–608.
21. Hulse, C. L.; Averill, B. A.; Tiedje, J. M. *J. Am. Chem. Soc.* **1989,** *111*, 2322–2323.
22. Ye, R. W.; Toro-Suarez, I.; Tiedje, J. M.; Averill, B. A. *J. Biol. Chem.* **1991**, *266*, 12848–12851.
23. Jackson, M. A.; Tiedje, J. M.; Averill, B. A. *FEBS Lett.* **1991,** *291*, 41–44.
24. Libby, E.; Averill, B. A. *Biochem. Biophys. Res. Commun.* **1992,** *187*, 1529–1535.
25. Suzuki, S.; Yoshimura, T.; Kohzuma, T.; Shidara, S.; Masuko, M.; Sakurai, T.; Iwasaki, H. *Biochem. Biophys. Res. Commun.* **1989,** *164*, 1366–1372.
26. Adman, E. T. *Adv. Protein Chem.* **1991,** *42*, 145–197.
27. Solomon, E. I.; Baldwin, M. J.; Lowery, M. D. *Chem. Rev.* **1992,** *92*, 521–542.
28. Solomon, E. I.; Lowery, M. D. *Science (Washington, D.C.)* **1993,** *259*, 1575–1581.
29. Messerschmidt, A.; Rossi, A.; Ladenstein, R.; Huber, R.; Bolognesi, M.; Gatti, G.; Marchesini, A.; Petruzzelli, R.; Finazzi-Agro, A. *J. Mol. Biol.* **1989,** *206*, 513–529.
30. Messerschmidt, A.; Huber, R. *Eur. J. Biochem.* **1990,** *187*, 341–352.
31. Masuko, M.; Iwasaki, H.; Sakurai, T.; Suzuki, S.; Nakahara, A. *J. Biochem.* **1984,** *96*, 447–454.
32. Zumft, W. G.; Gotzmann, D. J.; Kroneck, P. M. H. *Eur. J. Biochem.* **1987,** *168*, 301–307.
33. Abraham, Z. H. L.; Lowe, D. J.; Smith, B. E. *Biochem. J.* **1993,** *295*, 587–593.
34. Howes, B. D.; Abraham, Z. H. L.; Lowe, D. J.; Brüser, T.; Eady, R. R.; Smith, B. E. *Biochemistry* **1994,** *33*, 3171–3177.
35. Kohzuma, T.; Shidara, S.; Suzuki, S. *Bull. Chem. Soc. Jpn.* **1994,** *67*, 138–143.
36. Waleh, A.; Ho, N.; Chantranupong, L.; Loew, G. H. *J. Am. Chem. Soc.* **1989,** *111*, 2767–2772.
37. Ozawa, S.; Fujii, H.; Morishima, I. *J. Am. Chem. Soc.* **1992,** *114*, 1548–1554.
38. Aerssens, E.; Tiedje, J. M.; Averill, B. A. *J. Biol. Chem.* **1986,** *261*, 9652–9656.

39. Garber, E. A. E.; Hollocher, T. C. *J. Biol. Chem.* **1982,** *257,* 8091–8097.
40. Kim, C.-H.; Hollocher, T. C. *J. Biol. Chem.* **1984,** *259,* 2092–2099.
41. Whittaker, M. M.; Whittaker, J. W. *Biophys. J.* **1993,** *64,* 762–772.
42. Whittaker, J. W. In *Bioinorganic Chemistry of Copper;* Karlin, K. D.; Tyeklár, Z., Eds.; Chapman & Hall: New York, 1993; pp 447–458.
43. Fraser, R. T. M. *J. Inorg. Nucl. Chem.* **1961,** *17,* 265–272.
44. Mercer, M.; Fraser, R. T. M. *J. Inorg. Nucl. Chem.* **1963,** *25,* 525–534.
45. Doyle, M. P.; Siegfried, B.; Hammond, J. J. *J. Am. Chem. Soc.* **1976,** *98,* 1627–1629.
46. Yordanov, N. D.; Terziev, V.; Zhelyazkowa, B. G. *Inorg. Chim. Acta* **1982,** *58,* 213–216.
47. Paul, P. P.; Tyeklár, Z.; Farooq, A.; Karlin, K. D.; Liu, S.; Zubieta, J. *J. Am. Chem. Soc.* **1990,** *112,* 2430–2432.
48. Paul, P. P.; Karlin, K. D. *J. Am. Chem. Soc.* **1991,** *113,* 6331–6332.
49. Moser, W. R. In *The Catalytic Chemistry of Nitrogen Oxides;* Klimisch, R. L.; Larson, J. G., Eds.; Plenum: New York, 1975; pp 33–43.
50. Tolman, W. B. *Inorg. Chem.* **1991,** *30,* 4878–4880.
51. Carrier, S. M.; Ruggiero, C. E.; Tolman, W. B.; Jameson, G. B. *J. Am. Chem. Soc.* **1992,** *114,* 4407–4408.
52. Tolman, W. B.; Carrier, S. M.; Ruggiero, C. E.; Antholine, W. E.; Whittaker, J. W. In *Bioinorganic Chemistry of Copper;* Karlin, K. D.; Tyeklár, Z., Eds.; Chapman & Hall: New York, 1993; pp 406–418.
53. Ruggiero, C. E.; Carrier, S. M.; Antholine, W. E.; Whittaker, J. W.; Cramer, C. J.; Tolman, W. B. *J. Am. Chem. Soc.* **1993,** *115,* 11285–11298.
54. Kitajima, N. *Adv. Inorg. Chem.* **1992,** *39,* 1–77.
55. Looney, A.; Parkin, G.; Alsfasser, R.; Ruf, M.; Vahrenkamp, H. *Angew. Chem. Int. Ed. Engl.* **1992,** *31,* 92–93.
56. Trofimenko, S. *Chem. Rev.* **1993,** *93,* 943–980.
57. Carrier, S. M.; Ruggiero, C. E.; Houser, R. P.; Tolman, W. B. *Inorg. Chem.* **1993,** *32,* 4889–4899.
58. Mealli, C.; Arcus, C. S.; Wilkinson, J. L.; Marks, T. J.; Ibers, J. A. *J. Am. Chem. Soc.* **1976,** *98,* 711–718.
59. Kitajima, N.; Fujisawa, K.; Fujimoto, C.; Moro-oka, Y.; Hashimoto, S.; Kitagawa, T.; Toriumi, K.; Tatsumi, K.; Nakamura, A. *J. Am. Chem. Soc.* **1992,** *114,* 1277–1291.
60. Richter-Addo, G. B.; Legzdins, P. *Metal Nitrosyls*; Oxford University: New York, 1992.
61. Feltham, R. D.; Enemark, J. H. In *Topics in Stereochemistry;* Feltham, R. D.; Enemark, J. H., Eds.; Wiley: New York, 1981; Vol. 12; pp 155–215.
62. Kitajima, N.; Fujisawa, K.; Moro-oka, Y. *J. Am. Chem. Soc.* **1990,** *112,* 3210–3212.
63. Kitajima, N.; Fujisawa, K.; Fujimoto, C.; Moro-oka, Y. *Chem. Lett.* **1989,** 421–424.
64. Enemark, J. H.; Feltham, R. D. *Coord. Chem. Rev.* **1974,** *13,* 339–406.
65. Sorrell, T. N.; Jameson, D. L. *Inorg. Chem.* **1982,** *21,* 1014–1019.
66. Ruggiero, C. E.; Carrier, S. M.; Tolman, W. B. *Angew. Chem. Int. Ed. Engl.* **1994,** *33,* 895–897.
67. Himmelwright, R. S.; Eickman, N. C.; Solomon, E. I. *Biochem. Biophys. Res. Commun.* **1979,** *86,* 628–634.
68. Kitajima, N.; Koda, T.; Moro-oka, Y. *Chem. Lett.* **1988,** 347–350.
69. Chao, C. C.; Lunsford, J. H. *J. Phys. Chem.* **1972,** *76,* 1546–1548.

70. Giamello, E.; Murphy, D.; Magnacca, G.; Morterra, C.; Shioya, Y.; Nomura, T.; Anpo, M. *J. Catal.* **1992,** *136,* 510–520.
71. Suzuki, S.; Kohzuma, T.; Deligeer; Yamaguchi, K.; Nakamura, N.; Shidara, S.; Kobayashi, K.; Tagawa, S. *J. Am. Chem. Soc.* **1994,** *116,* 11145–11146.
72. Halfen, J. A.; Tolman, W. B. *J. Am. Chem. Soc.* **1994,** *116,* 5475–5476.
73. Komeda, N.; Nagao, H.; Adachi, G.; Suzuki, M.; Uehara, A.; Tanaka, K. *Chem. Lett.* **1993,** 1521–1524.
74. Kukimoto, M.; Nishiyama, M.; Murphy, M. E. P.; Turley, S.; Adman, E. T.; Horinouchi, S.; Beppu, T. *Biochemistry* **1994,** *33,* 5346.
75. Zhao, X.-J.; Sampath, V.; Caughey, W. S. *Biochem. Biophys. Res. Commun.* **1994,** *204,* 537–543.

RECEIVED for review June 10, 1993. ACCEPTED revised manuscript December 8, 1993.

8

Structural Characterization of Manganese Redox Enzymes

Results from X-ray Absorption Spectroscopy

Pamela J. Riggs-Gelasco[1], Rui Mei[2], and James E. Penner-Hahn[1,3]

[1] Willard H. Dow Laboratories, Department of Chemistry, University of Michigan, Ann Arbor, MI 48109–1055

[2] Department of Biology, University of Michigan, Ann Arbor, MI 48109–1055

X-ray absorption spectroscopy has proven to be a useful technique for characterizing both the local site structure and the oxidation state of the manganese in Mn redox enzymes. Quantitative analysis of the X-ray absorption near-edge structure has been used to show that Mn catalase cycles between Mn(II)–Mn(II) and Mn(III)–Mn(III) oxidation states and that a superoxidized Mn(III)–Mn(IV) derivative of Mn catalase is inactive. Similar studies of the oxygen-evolving complex (OEC) demonstrate that the resting enzyme has an average oxidation state of ca. 3.5 but that the Mn can be reduced by treatment in the dark with either hydroxylamine or hydroquinone. Extended X-ray absorption fine structure studies show that reduced Mn catalase has 1–2 imidazole ligands but no detectable Mn–Mn scattering. The lack of Mn scattering result implies that there is only weak, or perhaps no, bridging between the Mn in the dinuclear site. In contrast, the active OEC and the inactive superoxidized Mn catalase both have an oxo-bridged $Mn_2(\mu\text{-}O)_2$ core structure. For the OEC, this structure is disrupted by treatment with hydroquinone but not by treatment with hydroxylamine. Taken together, these findings provide support for the "dimer-of-dimers" model of the OEC.

DIVALENT Mn PLAYS AN IMPORTANT ROLE IN BIOLOGY, and there are numerous well-characterized examples of Mn(II)-activated proteins. However, in most cases, the Mn can be replaced by other divalent cat-

*Corresponding author

0065–2393/95/0246–0219/$09.98/0

ions, typically Mg(II), and the cation is not redox active. In the past decade, there has been increased interest in the role of Mn in biological redox enzymes (*1*). These enzymes include Mn superoxide dismutase, Mn catalase, and the Mn cluster in the oxygen-evolving complex (OEC) of Photosystem II (PSII). These enzymes mediate the one-, two-, and four-electron chemistry of dioxygen, respectively (reactions 1–3).

$$2HO_2 \rightarrow H_2O_2 + O_2 \quad (1)$$

$$2H_2O_2 \rightarrow O_2 + 2H_2O \quad (2)$$

$$2H_2O \rightarrow O_2 + 4H^+ + 4e^- \quad (3)$$

Reaction 3 occurs in all photosynthetic higher plants, algae, and cyanobacteria and is the source of the oxidizing atmosphere that sustains all other higher life forms. One result of an oxidizing atmosphere is the production of highly reactive reduced oxygen species such as hydrogen peroxide and superoxide. Both hydrogen peroxide and superoxide are potentially lethal, therefore reactions 1 and 2 play a vital role in protecting cells from oxidative damage. Depending on the organism, biological superoxide disproportionation (eq 1) is catalyzed by a copper–zinc enzyme, a mononuclear Fe enzyme, or a mononuclear Mn enzyme; whereas hydrogen peroxide disproportionation (eq 2) is catalyzed by a heme enzyme and or a dinuclear Mn enzyme. In contrast, biological oxygen evolution (eq 3) has a strict requirement for Mn.

The importance of reactions 1–3 in the biosphere is clear. However, relatively little is known about the catalytic mechanisms of these reactions, particularly reactions 2 and 3. In order to better understand the catalytic mechanisms of these enzymes, it is important to establish the correlation between metal site structure and enzymatic function. X-ray absorption spectroscopy is one of the premier tools for determining the local structural environment of metalloprotein metal sites. In the following, we summarize our results using X-ray absorption spectroscopy to characterize the structure of the Mn active site environments in manganese catalase and in the OEC and show how these structural results can be used to deduce details of the catalytic mechanism of these enzymes.

X-ray Absorption Spectroscopy

X-ray absorption spectroscopy (XAS) refers to the structured absorption occurring in the vicinity of a core-electron X-ray absorption threshold (an absorption "edge") (*2–5*). In the following, we restrict our attention to the Mn K-edge (1s initial state) absorption spectrum, which occurs at an energy of ca. 6500 eV (i.e., an X-ray wavelength of ca. 2 Å). An XAS spectrum is conventionally divided into two regions, the X-ray absorption near-edge structure (XANES) region and the extended X-ray absorption fine structure (EXAFS) region. XANES refers to the structured absorption within ca. 50 eV of the edge. For the Mn K-edge, this region extends from ca. 6525 to 6575 eV. The EXAFS region extends from ca. 50 eV above the edge to as much as 1000 eV above the edge. However, for biological Mn samples, the inevitable presence of Fe, which has an absorption edge at 7100 eV, limits the useful EXAFS region to 50–550 eV above the edge.

Although XANES and EXAFS features are often treated as distinct, both result from scattering of the X-ray-excited photoelectron. For energies above the X-ray edge, the incident X-ray has sufficient energy to eject a core electron. This photoelectron can be described as a wave with DeBroglie wavelength $\lambda = h/p$, where p is the photoelectron momentum. As it encounters neighboring atoms, this wave will be scattered back in the direction of the absorbing atom. Interference between the outgoing and the backscattered waves modulates the photoabsorption cross section, resulting in an oscillatory fine structure superimposed on the X-ray edge.

EXAFS Spectroscopy. The EXAFS, $\chi(E)$, is defined as the fractional modulation of the X-ray absorption coefficient:

$$\chi(E) = \frac{\mu(E) - \mu_0(E)}{\mu_0(E)} \approx \frac{\mu(E) - \mu_s(E)}{\mu_f(E)} \tag{4}$$

where $\mu(E)$ is the observed absorption cross section, and $\mu_0(E)$ is the cross section in the absence of any EXAFS effects. As μ_0 cannot be determined experimentally, it is approximated by μ_s, a calculated smooth background. The difference is then normalized to the theoretical falloff of the cross section with energy, $\mu_f(E)$.

If the independent variable, E, is replaced by the photoelectron wave-vector, $k = 2\pi/\lambda = \sqrt{2m_e(E - E_0)/\hbar^2}$, where m_e is the mass of the electron, and E_0 is the threshold energy for the excitation of a core electron, the EXAFS is given by:

$$\chi(k) = \sum \frac{N_s A_s(k)}{kR_{as}^{\ 2}} \exp(-2k^2\sigma_{as}^{\ 2}) \exp\left(\frac{-2R_{as}}{\lambda}\right) \cdot \sin\left[2kR_{as} + \phi_{as}(k)\right] \tag{5}$$

where R_{as} is the absorber-scatterer distance; N_S is the number of scattering atoms (the coordination number); σ_{as} is the mean-square deviation in R_{as}, the so-called Debye–Waller factor; λ is the mean free-path of the photoelectron; $\phi_{as}(k)$ is the phase shift that the photoelectron wave undergoes when passing through the potentials of the absorbing and scattering atoms; and $A_s(k)$ is the backscattering amplitude function. The expression is summed over all of the shells of scattering atoms, where a shell consists of one or more scatterers at a single (average) distance from the absorber.

From equation 5, it is apparent that each shell of scatterers will contribute a different frequency of oscillation to the overall EXAFS spectrum. A common method used to visualize these contributions is to calculate the Fourier transform (FT) of the EXAFS spectrum. The FT is a pseudoradial-distribution function of electron density around the absorber. Because of the phase shift $[\phi_{as}(k)]$, all of the peaks in the FT are shifted, typically by ca. −0.4 Å, from their true distances. The backscattering amplitude, Debye–Waller factor, and mean free-path terms make it impossible to correlate the FT amplitude directly with coordination number. Finally, the limited k range of the data gives rise to so-called truncation ripples, which are spurious peaks appearing on the wings of the true peaks. For these reasons, FTs are never used for quantitative analysis of EXAFS spectra. They are useful, however, for visualizing the major components of an EXAFS spectrum.

Quantitative analysis of EXAFS spectra requires that the $\phi_{as}(k)$, $A_s(k)$, and λ terms be determined for every absorber-scatterer pair. This determination can be done either empirically, using a model compound of known structure or ab initio. Several computer programs can be used to calculate accurate EXAFS parameters, and this method for obtaining the amplitude and phase shift functions is often preferred, particularly when model compounds may be difficult or impossible to prepare (6). When $\phi_{as}(k)$, $A_s(k)$, and λ have been determined, the bond-length, coordination number, and Debye–Waller factor for the unknown are refined in a least-squares sense. Bond lengths depend on the frequency of the EXAFS oscillations. Because several periods of oscillation are usually observable, the bond length can typically be determined with an accuracy of 1%; that is, ±0.02 Å. The accuracy for determining coordination number and chemical identity are generally worse. The amplitude of the backscattered wave is less well-defined than its phase and is a complex function of scatterer type, Debye–Waller factor, mean free-path, and coordination number. Consequently, the uncertainty in coordination number is typically ±25%. Both $\phi_{as}(k)$ and $A_s(k)$ depend on the chemical identity of the scattering atom. However, in practice this dependence is a slowly varying function of atomic number, and it is only possible to identify the scatterer to the nearest row of the periodic table. Despite these limitations, EXAFS remains an extremely powerful structural method. For noncrystalline samples, EXAFS is the only technique that can provide direct local structural information.

XANES Spectroscopy. In the XANES region the photoelectron has, by definition, low kinetic energy. Because low-energy electrons have a long mean free-path and are strongly scattered by the surrounding medium, the XANES region is very sensitive to the detailed arrangement of atoms around the absorbing atom. Multiple scattering, in which the photoelectron is scattered by two or more atoms before returning to the absorber, is believed to be important in the XANES region (although multiple scattering also occurs in the EXAFS region, it is generally less important here than in the XANES region). The advantage of multiple scattering is that it makes the XANES region sensitive to the geometry of the Mn site; the disadvantage is that detailed analysis becomes prohibitively difficult.

Representative XANES spectra for crystallographically characterized Mn(II), Mn(III), and Mn(IV) model compounds are shown in Figure 1. It is apparent that the Mn(II) XANES spectra are characterized by an intense, relatively narrow resonance at low energy. In contrast, the XANES spectra for Mn(III) are broader and shifted to higher energy, and those spectra for Mn(IV) are broader still and even higher in energy.

A variety of approaches have been used for interpreting Mn XANES spectra. An edge "energy" can be defined, for example, as the first inflection point on the rising edge or as the energy at the half-height of the edge jump. The edge energy shows a strong correlation to the coordination charge of the Mn, where the coordination charge is an attempt to correct the formal oxidation state for the electronegativity of the ligands (*7, 8*). One difficulty with this analysis is that it reduces the complex shape of the XANES spectrum to a single number (the edge energy).

We have adopted an alternative approach in which the observed XANES spectrum is fit using a linear combination of spectra drawn from a library of authentic Mn(II), Mn(III), and Mn(IV) model compounds (*9, 10*). Fits are constrained to use only positive coefficients, and the only variables are the fraction of a model spectrum that is used in the fit. This approach is reasonably sensitive to the presence of Mn(II) and, to a lesser extent, is able to distinguish between Mn(III) and Mn(IV). Given the similarity of the Mn(III) and Mn(IV) XANES spectra, it is not surprising that these oxidation states are more difficult to distinguish. In tests of crystallographically characterized mixed-valence complexes, it has proven possible to estimate the Mn(II) content with an uncertainty of ±10%. An analogous approach, based on fitting the first derivative of Mn XANES spectra, has recently been shown to give comparable results (*11*).

A weakness of this quantitation procedure is that it assigns all of the XANES changes to oxidation-state changes. It is well-known that XANES energies also depend somewhat on ligation type. In particular, there appears to be a reasonably good correlation between Mn structure (specifically Mn–ligand bond length) and XANES energy. In principle, this limits the utility of XANES for oxidation-state determination. However, in the models

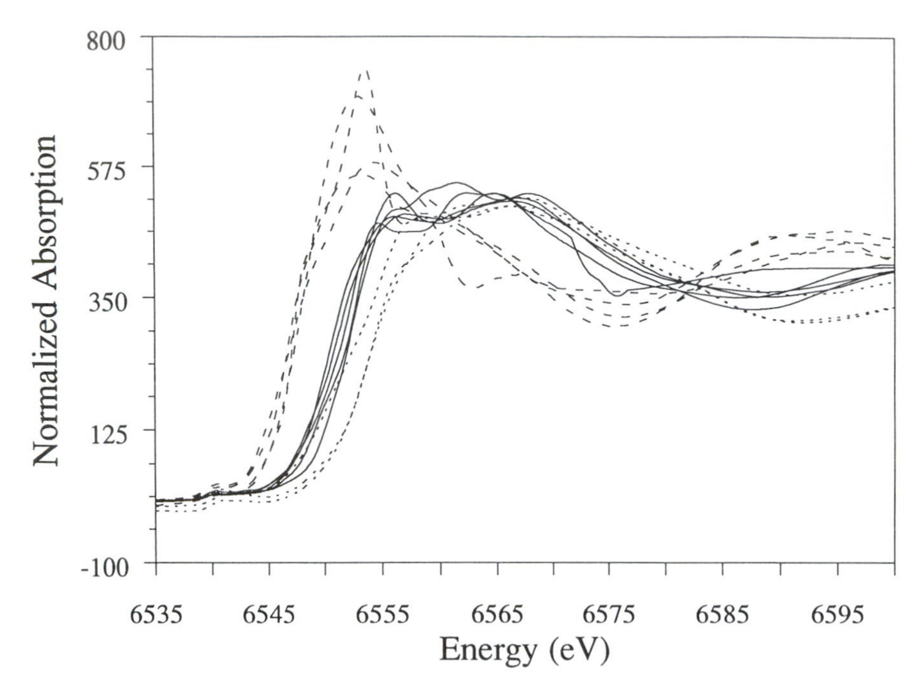

Figure 1. XANES spectra for crystallographically characterized Mn(II), Mn(III), and Mn(IV) model compounds. Long dashed lines indicate Mn(II) compounds; solid lines indicate Mn(III) compounds; short dashed lines indicate Mn(IV) complexes. The models chosen to illustrate this phenomenon are [Mn(II)HB(3,5-iPr$_2$pz)$_3$]$_2$(OH)$_2$ and (pyrazolyl)$_2$Mn(OC$_6$H$_5$)$_3$Mn(II)(HB 3,5-iPr$_2$pz)$_3$ [where HB(3,5-iPr$_2$pz)$_3$ is hydrotris(3,5-diisopropyl-1-pyrazolyl borate); Mn(II)(hexakisimidazole)Cl$_2$; Mn(III)(acac)$_3$; Mn(III)$_2$, [2-OH-(5-Cl-SAL)PN]$_2$(CH$_3$OH), [Mn(IV)(μ_2-O)(SALPN)]$_2$, and [Mn(IV)-(SALPN)]$_2$(μ_2-O)(μ_2-OH) [where SALPN is 1,3-bis(salicylideneimanato) propane]; [Mn(III)(SALAHP)(AcO)]$_2$ [where SALAHP is 3-(salicylideneiminato)-2-methyl-1,3-dihydroxypropane]; and Mn(IV)(SALADHP)$_2$ [where SALADHP is the dianion of 2-methyl-2-(salicylideneamino)-1,3-dihydroxypropane].

examined to date, we have found the XANES shape for Mn(II) to be a unique oxidation-state marker. This probably reflects the unique structure (i.e., long Mn–ligand bond lengths) of Mn(II) complexes. It should be noted, however, that a hypothetical Mn(III) complex having a Mn(II)-like structure (e.g., Mn–O distances at 2.2 Å) might have a XANES spectrum mimicking that of Mn(II). Likewise, a Mn(II) complex with unusually short bond lengths [e.g., a four-coordinate Mn(II)] might appear similar to Mn(III).

Manganese Catalase

Although most catalases contain the iron–protoporphyrin IX prosthetic group, it has been known for over 30 years that some bacteria are able to synthesize catalases that are not inhibited by millimolar concentrations of azide and cyanide (*12–14*). This ability suggests that some catalases are nonheme enzymes. Early work showed that the nonheme catalases contained either Fe or Mn, and the presence of Mn was confirmed in 1983 with the isolation and purification of the nonheme catalase from *Lactobacillus plantarum* (*15*). The nonheme catalases are frequently referred to as pseudocatalases. However, there is nothing pseudo about their catalase activity and indeed, if Mn catalase activity is destroyed (either genetically or through inactivation of the enzyme), cells show decreased viability, which suggests that Mn catalase plays a physiologically important role in H_2O_2 detoxification (*15, 16*).

Three Mn catalases have been purified and characterized, and all appear to have similar Mn structures (*17*). The Mn stoichiometry is ca. 2 Mn/subunit, suggesting a dinuclear Mn site. The optical spectrum of the as-isolated enzyme has a broad weak absorption band at ca. 450–550 nm in addition to the protein absorption at higher energies. This spectrum is similar to those observed for Mn(III) superoxide dismutase and for a variety of Mn(III) model complexes, thus implying that at least some of the Mn in Mn catalase is present as Mn(III). In particular, the absorption maximum at ca. 500 nm is similar in energy and intensity to the transitions seen for oxo–carboxylato-bridged Mn dimers, suggesting that a similar core structure may be seen for Mn catalase (*18*).

Definitive proof for a dinuclear Mn site comes from electron paramagnetic resonance (EPR) spectroscopy. The EPR spectrum of the as-isolated enzyme is a complex multicomponent signal (*19, 20*). At 50 K the EPR is dominated by a 16-line signal centered at ca. $g = 2.0$ (g is a spectroscopic splitting factor that is characteristic of the electronic structure of the paramagnetic center) that is ca. 1300 Gauss wide. This signal is characteristic of an $S = \frac{1}{2}$ (S is the total electron spin quantum number) mixed-valence Mn dimer and has hyperfine couplings typical of Mn(III)–Mn(IV) complexes. On treatment with NH_2OH, the 16-line signal disappears and a new, very broad signal, attributed to a Mn(II)–Mn(II) form of the enzyme, appears. This signal is quite sensitive to the buffer, pH, and anions that are present (i.e., SO_4^{2-}, N_3^-, F^-, CN^-). At temperatures below 50 K, a third EPR signal, consisting of ca. 18 lines centered at $g \approx 2$ is observed. This new signal is again characteristic of an $S = \frac{1}{2}$ mixed-valence dimer. On the basis of the hyperfine coupling and the temperature dependence of this signal, it has been attributed to a Mn(II)–Mn(III) derivative. All three signals contribute to the observed spectrum of the as-isolated enzyme but each can be enhanced by selective redox interconversions.

The crystal structure for the *T. thermophilus* Mn catalase (*21*) shows that this protein has a structure containing four parallel alpha-helices, similar to the structure found in dinuclear Fe proteins such as hemerythrin. Only low-resolution detail is available and no refined structural parameters have been reported. However, the structure does show two regions of enhanced electron density separated by ca. 3.6 Å, which may represent the two Mn ions. The analogy to hemerythrin is intriguing, because the dinuclear Fe site in hemerythrin has an oxo–carboxylato-bridged core similar to that suggested for Mn catalase (*18*).

XANES Spectroscopy of Mn Catalase. Although EPR spectroscopy has proven very useful for identifying the different oxidation states that are accessible to Mn catalase, it is limited by the fact that some forms of the enzyme do not have an EPR signal. In contrast, XANES spectroscopy is sensitive to all of the Mn in the sample, regardless of spin state, and has proven very useful for assigning the oxidation states of different Mn catalase derivatives. The XANES spectra for several different redox modifications of *L. plantarum* Mn catalase are shown in Figure 2 (22). On the basis of the pronounced shoulder at low energy, the as-isolated enzyme contains some Mn(II). This shoulder probably arises from the Mn(II)–Mn(II) dimer that was detected by EPR. As shown in Figure 2, there is a dramatic change in the XANES spectrum when Mn catalase is treated with NH_2OH. This change demonstrates that NH_2OH reduces a significant fraction of the Mn to Mn(II). Quantitative

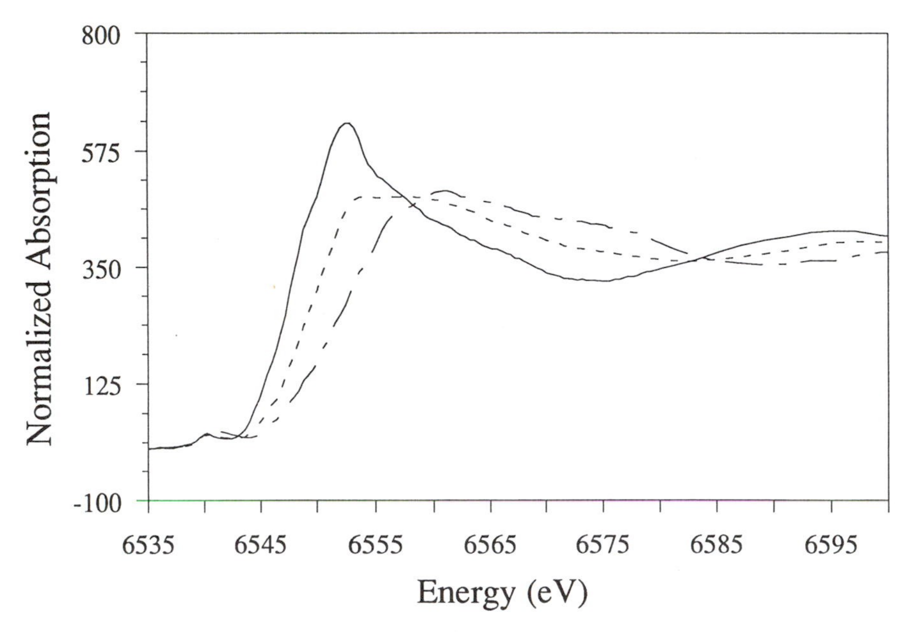

Figure 2. Normalized XANES spectra for Mn catalase. Dashed line indicates native (as-isolated); solid line indicates reduced (add 10 mM NH_2OH); and dot–dash line indicates superoxidized (add NH_2OH and H_2O_2).

analysis of these spectra suggests that ca. 60% of the Mn is present as Mn(II) in the as-isolated enzyme, whereas >90% is reduced to Mn(II) on treatment with NH_2OH. Because enzymatic activity is unaffected by treatment with NH_2OH, the Mn(II)–Mn(II) form of the enzyme must either be catalytically active or be readily convertible to a catalytically active form of the enzyme. If Mn catalase is reduced, dialyzed to remove NH_2OH, and then treated with H_2O_2, the XANES spectrum (not shown) is very similar to that observed for the as-isolated enzyme; that is, H_2O_2 treatment results in a net reoxidation of roughly half of the Mn (*23, 24*). These observations are consistent with a Mn(II)–Mn(II) ↔ Mn(III)–Mn(III) catalytic cycle as summarized by equations 6 and 7.

$$\text{Mn(II)–Mn(II)} + H_2O_2 + 2H^+ \rightarrow \text{Mn(III)–Mn(III)} + 2H_2O \quad (6)$$

$$\text{Mn(III)–Mn(III)} + H_2O_2 \rightarrow \text{Mn(II)–Mn(II)} + O_2 + 2H^+ \quad (7)$$

Although NH_2OH alone has no effect on Mn catalase activity, the enzyme is rapidly inactivated by treatment with a mixture of NH_2OH and H_2O_2 (*15, 25, 26*). The XANES features for the inactivated enzyme shift to higher energy (*see* Figure 2), demonstrating that the inactivated enzyme is a superoxidized derivative. In addition to the changes in the XANES, NH_2OH and H_2O_2 inactivation also leads to a ca. 10-fold enhancement in the amplitude of the 16-line catalase EPR signal (*22*). In fact, the enzymatic activity is negatively correlated with the intensity of this signal, demonstrating that the 16-line signal comes from inactive Mn catalase. Because the XANES spectra of the inactive derivative are consistent only with Mn(III) or Mn(IV), the 16-line signal must come from a Mn(III)–Mn(IV) dimer rather than a Mn(II)–Mn(III) dimer. EPR spectroscopy, in contrast, does not permit unique distinction between Mn(II)–Mn(III) and Mn(III)–Mn(IV) interpretations of this species.

Superoxidized Mn catalase can be completely reactivated by long term (>2 h) anaerobic incubation with 10 mM NH_2OH (*22*). Reactivation is accompanied by complete disappearance of the Mn(III)–Mn(IV) EPR signal. If the as-isolated protein is incubated with NH_2OH (as opposed to simply adding NH_2OH and freezing, as in the experiments described previously), there is a 10–20% increase in catalytic activity and the small 16-line EPR signal completely disappears. It is apparent that the as-isolated enzyme must contain a mixture of the Mn(II)–Mn(II), the Mn(III)–Mn(III), and the inactive Mn(III)–Mn(IV) derivatives.

The observation that Mn(III)–Mn(III) catalase is rapidly reduced by NH_2OH, whereas the Mn(III)–Mn(IV) catalase is only slowly reduced under equivalent conditions is intriguing. The Mn(III)–Mn(III) derivative is apparently a better oxidant than the Mn(III)–Mn(IV) enzyme. The slow oxidation of NH_2OH by the Mn(III)–Mn(IV) derivative may be either a thermodynamic effect (e.g., enhanced stability of higher oxidation states because of the presence of oxo bridges) or a kinetic effect (e.g., a major structural rearrangement between oxidized and reduced forms). In either case, the observation that the Mn(III)–Mn(IV) derivative is not an effective oxidant provides the likely explanation for the lack of catalase activity for this derivative.

As noted previously, reduced Mn catalase is rapidly oxidized by H_2O_2 to a species best described as a mixture of Mn(II) and Mn(III). Because the EPR signal characteristic of a Mn(II)–Mn(III) mixed-valence derivative is not observed, this finding is most consistent with production of a mixture of the reduced Mn(II)–Mn(II) enzyme and the oxidized Mn(III)–Mn(III) enzyme. A similar result is produced, albeit more slowly, if the enzyme is oxidized by long-term (12 h) exposure to O_2 (data not shown) (*23*, *24*). It is somewhat surprising that autooxidation does not produce 100% of the oxidized Mn(III)–Mn(III) enzyme. The most likely explanation is that H_2O_2, produced by equation 8, can react either with reduced enzyme, generating a second equivalent of Mn(III)–Mn(III), or with oxidized enzyme, regenerating the Mn(II)–Mn(II) form.

$$\text{Mn(II)–Mn(II)} + O_2 + 2H^+ \rightarrow \text{Mn(III)–Mn(III)} + H_2O_2 \qquad (8)$$

A variety of anions, including azide, chloride, and fluoride, are inhibitors of Mn catalase. It is difficult to define the effect of these inhibitors using EPR, because the Mn(III)–Mn(III) derivative is EPR-silent and the Mn(II)–Mn(II) derivative is only EPR-active in the presence of added anions. XANES is an ideal probe, however, because it is sensitive to all of the Mn in the system. As expected, treatment with halide alone has no effect on Mn oxidation state. However, treatment with fluoride or chloride in the presence of H_2O_2 gives complete reduction of the Mn to Mn(II) (data not shown) (*23*, *24*). The same result is obtained regardless of whether one starts with the reduced enzyme, the autooxidized enzyme (*see* eq 8) or the as-isolated enzyme. This result provides direct evidence that the halides inhibit the enzyme by trapping it in the reduced valence state.

EXAFS Spectra of Mn Catalase. The Fourier transforms for the reduced and the superoxidized derivatives of Mn catalase are shown in Figure 3 (*27*). The superoxidized enzyme has two principal peaks at $R + \alpha \approx 1.4$ and 2.3 Å, corresponding to Mn–(O,N) nearest neighbor and Mn–Mn scattering, respectively. The structure is altered dramatically on reduction, with longer Mn–(O,N) distances and complete disappearance of the 2.7 Å Mn–Mn interaction. The outer shell peaks for the reduced enzyme ($R + \alpha \approx 3.0$ and 3.8 Å) are typical of those observed for Mn–imidazole complexes (data not shown). However, the outer peaks for catalase are ca. 3-fold smaller than those for Mn(imidazole)_6, suggesting that catalase has at most 1–2 imidazole ligands per Mn.

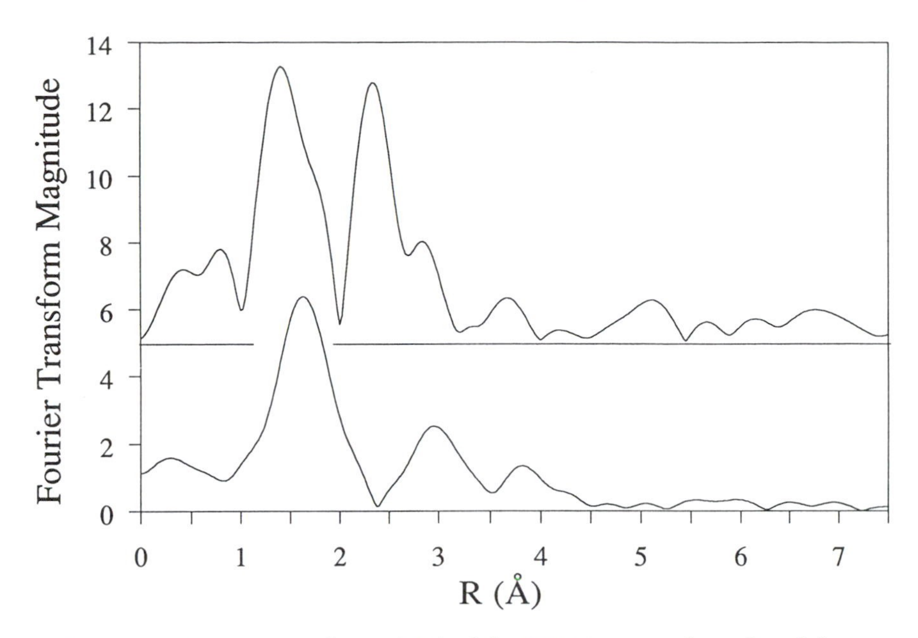

Figure 3. Fourier transforms (FTs) of the EXAFS spectra for reduced (bottom) and superoxidized (top) Mn catalase. Fourier transforms calculated over the range 3–11.5 Å^{-1} using k^3 weighted data. The FT for the superoxidized enzyme is offset vertically by 5 for clarity.

The quantitative curve-fitting results for Mn catalase are summarized in Table I (27). The data for the superoxidized enzyme cannot be fit without including two shells of nearest neighbor scatterers and a Mn–Mn interaction at 2.7 Å. It is not possible to distinguish between N and O scatterers using EXAFS. However, the 1.82 Å distance is assigned to an Mn–O interaction because this value is typical of Mn-bridging oxo distances. This assignment is further supported by the 2.7 Å Mn–Mn distance, which is only found in model compounds that have $Mn_2(\mu\text{-}O)_2$ cores. As noted previously, the Fourier transform suggests that there are additional low-Z scatterers at longer distances. This finding is supported by the curve-fitting analysis, and the observed distances are consistent with second and third shell N/C atoms in coordinated imidazoles. The apparent coordination numbers suggest an average of 1–2 imidazoles per Mn in the superoxidized catalase. However, this number is not well-defined because of the limited k range of the data and interference from the strong Mn–Mn scattering.

As expected from the Fourier transform, the EXAFS for reduced catalase is dominated by a single shell of low-Z scatterers at 2.19 Å. There is no evidence for a shell of scatterers at ca. 1.8 Å (i.e., a bridging oxo ligand). However, the fit quality is improved significantly if additional

Table I. EXAFS Curve Fitting Results for Mn Catalase

	Superoxidized			*Reduced*			$Mn^{II}(Im)_6$
Interaction	N	R(Å)	$\Delta\sigma^2 \cdot 10^3$	N	R(Å)	$\Delta\sigma^2 \cdot 10^3$	R(Å)
Mn–O	2	1.82	2.1	—	—	—	—
Mn–N/O	4	2.14	−1.9	6	2.19	0.0	2.27
Mn–Mn	1	2.67	1.4	—	—	—	—
Mn–C	4	3.00	0.0	8	3.16	6.9	3.27
Mn–N/C	4	4.33	−0.8	4	4.42	−1.9	4.43

shells of C and N/C scatterers are included at ca. 3.2 and 4.4 Å. The apparent coordination numbers suggest an average of 2–4 imidazoles per Mn, although again this number is not well defined. There is a small improvement in the fit if a shell of Mn is added at 3.55 Å. However, equivalent improvements are observed if instead the Mn is added at 3.99 Å, or if a shell of C is added at 3.64 Å. In no case is the improvement sufficient to support the conclusion that the reduced catalase contains an EXAFS-detectable Mn–Mn interaction. All of the Mn–scatterer distances in reduced catalase are slightly but significantly shorter than those in Mn(II)(imidazole)$_6$. This difference could be caused by one or both of the Mn having a coordination number less than six or by the substitution of imidazole for oxygen ligands (Mn–O distances are shorter then Mn–N distances) or by a combination of these effects. Overall, the Mn ligation appears similar to that of Fe in deoxyhemerythrin (*28*).

The short Mn–Mn distance in the superoxidized enzyme demonstrates that there are two oxo bridges in this derivative (*18, 29*). Although there are few examples, it appears that unsupported (μ-O)$_2$ bridges (*30, 31*) have Mn–Mn distances $\geq$2.70 Å, whereas additional bridging ligands lead to shorter Mn–Mn distances (*32–34*). The 2.67-Å Mn–Mn distance thus suggests that there may be an additional bridge, for example, (μ-O)$_2$ (μ-carboxylato). A carboxylate bridge conforms nicely with the suggestion that catalase and hemerythrin have analogous structures and is consistent with the proposal (*18, 35, 36*) that the oxidized Mn(III)–Mn(III) enzyme has a (μ-O)(μ-carboxylato)$_2$-bridged core. It is important to recognize, however, that EXAFS data alone cannot provide direct evidence for a bridging carboxylate ligand.

The EXAFS data for the reduced enzyme do not permit unambiguous definition of a Mn–Mn distance. The inability to detect distances that are known to be present is not an uncommon problem for EXAFS. Similar difficulties in defining metal-metal distances in binuclear iron proteins have been attributed to the loss of bridging ligands. Although single atom bridges (e.g., oxo ligands) provide a relatively rigid core structure and hence EXAFS-detectable metal-metal scattering, other bridges, such as carboxylate ligands, do not provide such constraints. Bridging structures consistent with the data for reduced catalase include (μ-carboxylato)$_n$ and (μ-OH)(μ-carboxylato)$_n$, where n is 1–3.

The structural results suggest an explanation for the inactivity of superoxidized Mn at the molecular level. If the oxidized enzyme has a [Mn(III)(μ-O)(μ-carboxylato)$_2$ Mn(III)] core, conversion to the superoxidized derivative involves addition of an oxo bridge. This addition is expected to stabilize Mn with respect to reduction thus converting the Mn(III)–Mn(III) derivative, which is a good oxidant, into a species that, although formally more oxidized, is a poor oxidant.

Mechanism. A possible mechanistic model, based on the known coordination chemistry of Mn dimers, is shown in Figure 4 (*17*). Details of the ligation (e.g., the protein ligands to the Mn) are not shown because they cannot be defined using the data available at present. In this model, formation of the oxo-bridge is envisioned as stabilizing the Mn(III)–Mn(III) derivative, thereby facilitating the peroxide reduction step. Similarly, the oxo-bridge can act as a Lewis base, facilitating the oxidation of peroxide to dioxygen. Also shown on this scheme is the inactive superoxidized derivative. This derivative is formed by an unknown mechanism in the presence of H_2O_2 and NH_2OH. The superoxidized derivative can be reactivated by anaerobic reduction with a variety of reductants, including NH_2OH and hydroquinone (*22*).

The Oxygen-Evolving Complex

The photosynthetic oxidation of water to dioxygen is catalyzed by the Photosystem II (PSII) reaction center. This center is a multipolypeptide complex of proteins embedded in the thylakoid membranes of chloroplasts. When photosynthetic membranes are illuminated with short (<5 μs), saturating flashes of light, oxygen is released on every fourth flash, starting with the third flash (*37*). This observation led to the conclusion that each PSII reaction center acts independently to sequentially acquire four oxidizing equivalents. The five kinetically resolvable intermediates are known as S states (S_i, i = 0–4) where the subscript indicates the number of stored oxidizing equivalents. The S-state scheme is illustrated in Figure 5. States S_0 through S_3 each donate one electron following photooxidation; S_4 decays rapidly to S_0 with the release of O_2. The S_1 state is the stable, dark-adapted form of the OEC (*38*) and accounts for the initial oxygen release on the third flash.

Four (*39–41*) specifically bound Mn ions, together with Ca^{2+} and Cl^-, are required for OEC activity (*42*). A multiline EPR signal consisting of >16 lines centered at $g \approx 2$ is specifically associated with the S_2 state. Because the hyperfine coupling to a single Mn ion (with nuclear spin quantum number $I = 5/2$) can give rise to only 6 lines in the EPR spectrum, the multiline signal must come from a magnetically coupled cluster containing at least two, and possibly all four, manganese ions. The multiline signal has the same period-four oscillation that is observed for oxygen production (*43, 44*), thus directly implicating Mn in the catalytic cycle. A second, broad EPR signal at $g \approx 4.1$ is also associated with the S_2 state (*45*). This signal has recently been shown to also have numerous Mn hyperfine lines (at least for ammonia-treated, oriented membranes) demonstrating that it too must come from a multinuclear Mn cluster (*46, 47*).

A third EPR signal has been reported for the OEC (*48*). This signal is attributed to the S_1 state and is only seen using parallel polarization.

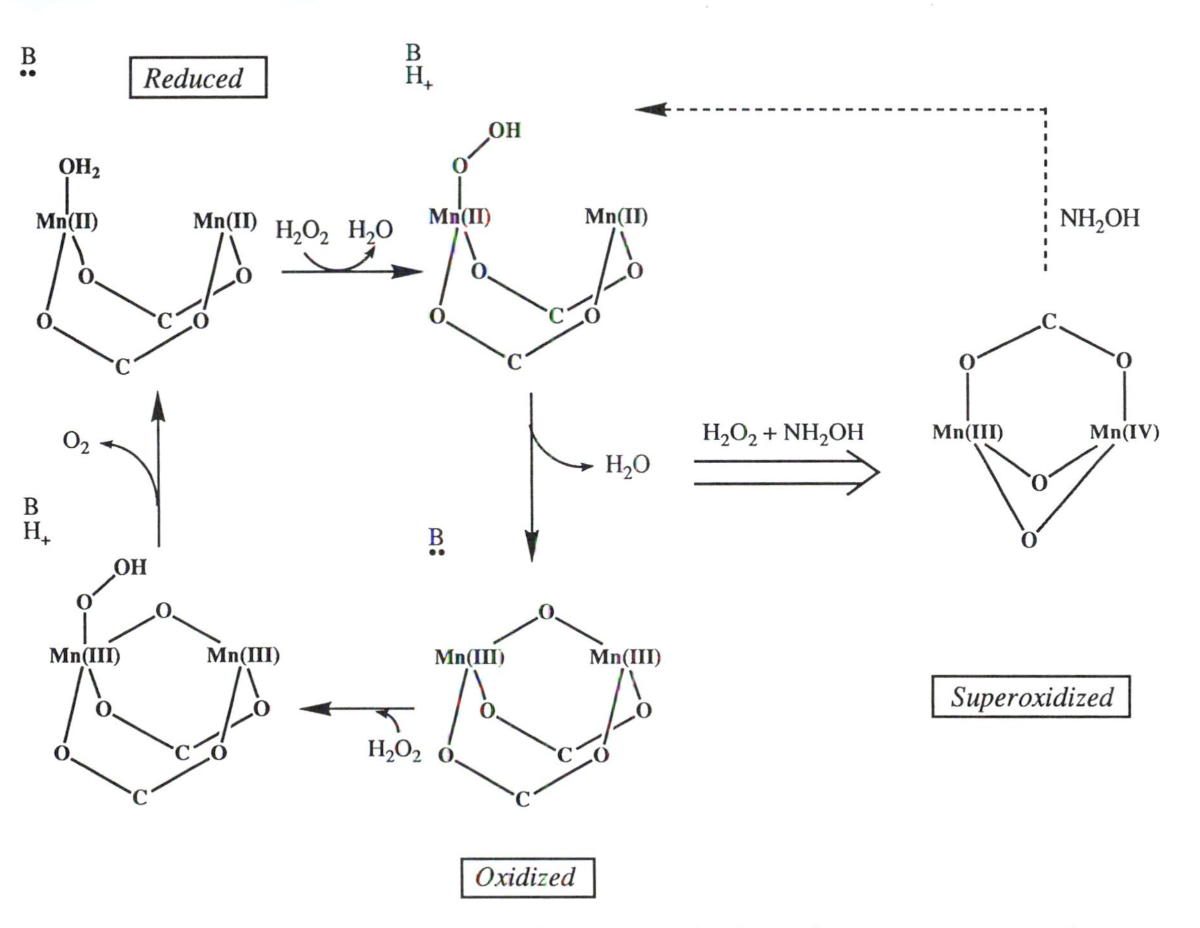

Figure 4. Schematic illustration of a possible mechanism for the catalase reaction in Mn catalase.

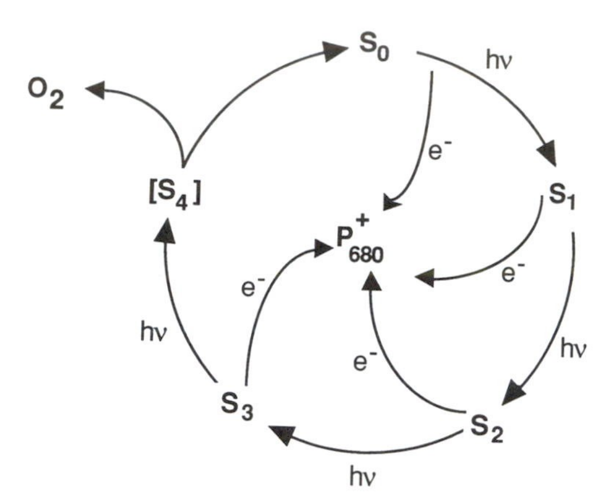

Figure 5. S-state scheme for oxygen evolution. S_1 is the stable species in the dark.

The signal has been attributed to an integer spin system, most likely an S = 1 state. The S_1 signal decays proportionally with the increase in the S_2 multiline signal, suggesting that it arises from a reduced paramagnetic precursor to the cluster that gives rise to the multiline signal. On the other hand, spin-lattice relaxation measurements have been interpreted to indicate that S_1 is diamagnetic in long-term dark-adapted samples (*49*). These apparently contradictory results can be reconciled if the OEC S_1 state can exist either in an "active" conformation, prepared by short-term dark adaptation or a "resting" conformation, prepared by long-term dark adaptation, where the "active" conformation is paramagnetic and the "resting" conformation is diamagnetic (*50*).

Chemically Reduced States of the OEC. Hydroxylamines, hydrazines, and hydrogen peroxide have been studied extensively as potential substrate analogues for the OEC. Treatment of thylakoid membranes with low concentrations of hydroxylamine results in a two-flash delay in oxygen evolution (*51, 52*). Similar effects are seen for hydrogen peroxide and hydrazine (*53, 54*). In addition to the delay in oxygen evolution, multiline signal formation is also delayed by two flashes upon treatment with low concentrations of NH_2OH (*39*). At higher concentrations, NH_2OH induces inhibitory loss of Mn^{2+} from the cluster (*41, 55*). Several interpretations of the two-flash delay have been suggested (*56, 57*). One involves the two-electron dark reduction of S_1 to a superreduced species, denoted S_{-1}; two quanta of light then oxidize the cluster back to the S_1 state. This model is supported by the observation that the two-flash delay persists even after a 10-h equilibration and gel filtration to remove NH_2OH (*57*). Alternatively, NH_2OH could bind to the S_1 cluster but not react with it. On oxidation to the S_2 state, NH_2OH would reduce the cluster to S_0.

XANES Studies. XANES spectroscopy has been used to investigate the oxidation state of the Mn in the OEC and to characterize the effect of S-state transitions and chemical treatments on Mn oxidation state. There is now general agreement that the XANES spectra for the resting OEC are most consistent with an oxidation state of ca. 3.5 (*10, 58, 59*). Given the similarity of Mn(III) and Mn(IV) XANES spectra (Figure 1), it is not realistic to expect that XANES can be used to define the precise ratio of Mn(III):Mn(IV). However, if the S_2-state EPR signal, which comes from an $S = 1/2$ cluster, does indeed represent four magnetically coupled Mn ions, then it can only come from a $Mn(III)_3Mn(IV)$ or a $Mn(III)Mn(IV)_3$ cluster. These imply a $Mn(III)_4$ or a $Mn(III)_2Mn(IV)_2$ cluster, respectively, in S_1. Of these, $Mn(III)_2Mn(IV)_2^-$ is most consistent with the S_1 XANES spectra.

The X-ray absorption edge shifts to higher energy when S_1 is converted to S_2 (*7*). This shift provides direct support for the conclusion (from EPR) that Mn is oxidized on going from S_1 to S_2. Although there were initial indications that Mn was not oxidized (as judged by XANES) on a cryogenically obtained $S_2 \rightarrow S_3$ transition (*7, 60*), more recent results suggest that the edge shifts, and hence that Mn is oxidized, for all three observable S-state transitions ($S_0 \rightarrow S_1$, $S_1 \rightarrow S_2$, and $S_2 \rightarrow S_3$) (*59*).

Conflicting results have been reported for the effect of NH_2OH on the XANES spectrum (*10, 61*). Guiles et al. report that there is no change in the XANES spectrum following treatment with low concentrations (40 μM) of NH_2OH in the dark. The edge did shift for higher concentrations ($\geq$100 μM); however, this shift was attributed to destruction of a small percentage of centers because a Mn(II) six-line EPR signal was observed at higher hydroxylamine concentrations (*61*). After exposure to NH_2OH, samples were illuminated at low temperature (low-temperature illumination limits the OEC to a single S-state advancement). After illumination the edge shifted to lower energy. These data were interpreted as support for the proposal that NH_2OH reacts with S_2 but not with S_1. The edge shift on illumination was attributed to formation of an S_0* species.

We have reexamined the NH_2OH reaction using an OEC "complex preparation" (*10*). In these samples, the 17 and 23 kDa extrinsic polypeptides and the light-harvesting complex have been removed. This removal increases the concentration of Mn, thus improving the quality of the XAS data, and in addition, increases the accessibility of the Mn site to chemical reagents (*62*). High (10 mM) concentrations of $CaCl_2$ were added to stabilize the OEC in the presence of reductants (*63*).

Under these conditions, the OEC is completely stable, as measured by oxygen evolution rate, following exposure to 100 μM NH_2OH for 3 min.

When the complex preparations are exposed to NH_2OH, the edge shifts to increasingly low energies as the exposure time or NH_2OH concentration increases. However, in contrast with the earlier results, we find that the edge is shifted to a lower energy even for samples that have been exposed to noninhibitory concentrations of NH_2OH (100 μM NH_2OH for 3 min) and that this shift occurs for samples that have been treated in the dark (Figure 6). If these samples are allowed to turn over under continuous illumination and are then dark-adapted again, the edge position returns to that of the oxidized control S_1 state (*10*). The complete reversibility and the minimal loss of activity (0–20%) demonstrate that the edge shift following dark treatment with NH_2OH cannot be due to the formation of inactive centers.

The differences between our results and those of Guiles et al. (*61*) may be due to the use of different samples or different experimental conditions. As noted previously, the complex preparations are more susceptible to chemical attack (*62*). To some extent, however, this susceptibility was corrected for by the use of different reaction times. A second difference is that excess NH_2OH was removed in our study (by 40-fold dilution followed by centrifugation), but not in the earlier work.

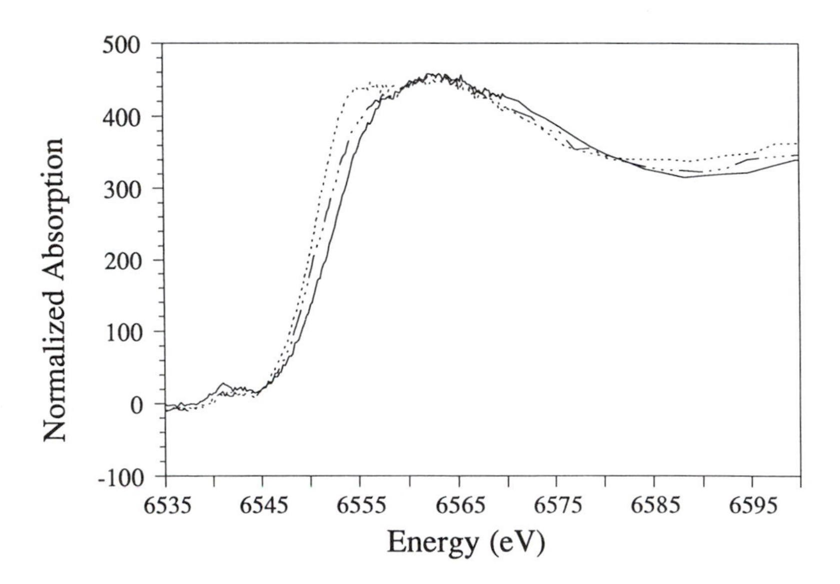

Figure 6. XANES spectra for the OEC. Solid line indicates dark-adapted, S_1 state; dotted–dashed line indicates 100 μM NH_2OH for a 3-min incubation quenched by dilution; and dashed line indicates 200 μM H_2Q for 30 min, quenched with ferricyanide.

This difference could not explain the differences in dark reactivity because the samples where NH_2OH was not removed were the samples that showed no reduction. However, it could affect the results following illumination if reactive reduced nitrogen species were produced by illuminating the samples in the presence of NH_2OH.

A third possibility is that the difference lies in interpretation rather than in the experimental results. Guiles et al. did report an edge shift for samples exposed in the dark to NH_2OH concentrations $\geq 100\ \mu M$. The magnitude of this shift is similar to that seen in our study of reaction center complexes. Guiles et al. attributed their shift to the formation of inactive centers, based on the appearance of an Mn(II) six-line EPR signal; however, no oxygen evolution activities were reported. A Mn(II) six-line signal does not necessarily indicate the presence of inactive centers (*see* discussion of H_2Q-treated sample). Moreover, inactive centers ordinarily lose Mn(II) to solution in the presence of Ca(II). This Mn should be in the supernatant after centrifugation and thus should not contribute to the XANES spectrum. It is possible that comparable dark reductions were observed in both experiments.

In an effort to understand the mechanism of NH_2OH reaction with PSII, we measured the XANES spectra of NH_2OH-treated samples under the following conditions: (1) Varying concentrations of NH_2OH (50 μM, 100 μM, 150 μM, 250 μM, and 400 μM) for a fixed incubation time of 3 min; and (2) varying incubation times (1′, 3′, 5′, 7′, 9′, and 12′) for a fixed NH_2OH concentration of 100 μM. These results are shown in Figures 7 and 8. An increase in either concentration or incubation time leads to an increase in the extent of reduction. As the edge shifts to lower energy, the activity of the sample drops dramatically. The amount of Mn remaining in the pellet (and thus presumably in the OEC) can be estimated from the magnitude of the Mn edge jump. As expected, harsher conditions result in increasing amounts of Mn(II) being released into the supernatant.

The correlation between step height (i.e., the amount of membrane-bound Mn) and activity is shown in Figure 9. This correlation between step height and activity illustrates several important points about the NH_2OH inactivation reaction. Most importantly, these data show that there is no significant loss of either Mn or activity for mild NH_2OH treatment conditions. This confirms our earlier finding that there is no significant inactivation. The plot of Mn content versus activity extrapolates to a y-intercept of ~ 0. This finding means that the amount of Mn remaining in the sample under harsh treatment conditions is the amount of Mn necessary to account for the observed activity. This confirms our assertion that inactive Mn is lost to the supernatant under these conditions. Under the most severe conditions, the edge becomes very reduced, indicating the presence of substantial amounts of Mn(II), despite

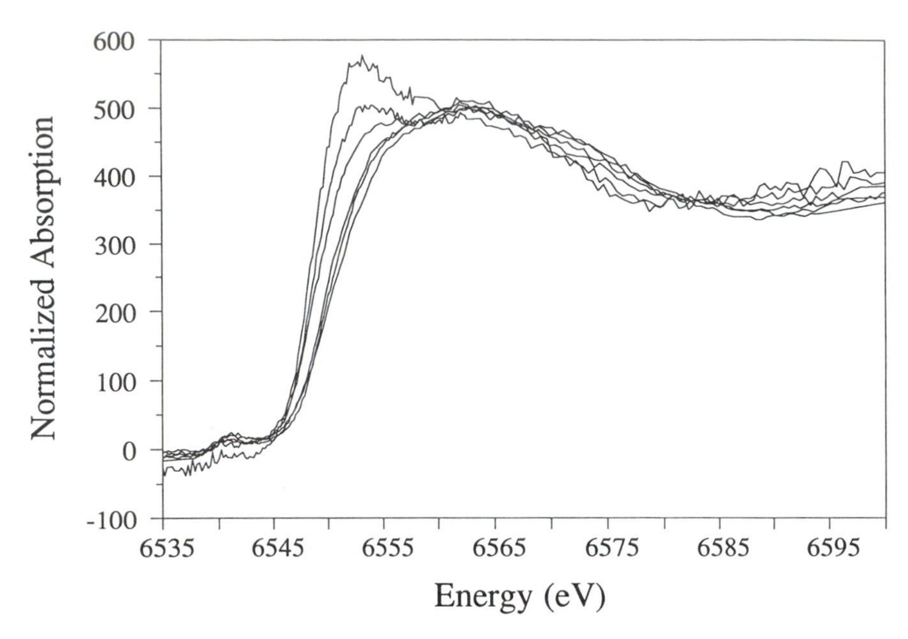

Figure 7. NH_2OH concentration dependence. From right to left: control S_1 state, 50 μM NH_2OH, 100 μM, 150 μM, 250 μM, and 400 μM. All reactions were incubated for 3 min and were quenched by dilution and centrifugation. Edges have been normalized.

the fact that the activity per Mn remains the same. Thus, it apparently is possible to prepare active, highly reduced derivatives of the OEC.

Mei and Yocum (*64*) have shown that a second reductant, hydroquinone (H_2Q), is also able to reduce the OEC. As shown in Figure 6, H_2Q gives a more dramatic edge shift than is observed for NH_2OH. The XANES spectrum for samples treated with 200 μM H_2Q can be fit with ca. 30% Mn(II) (Table II). These samples show the six-line EPR signal characteristic of Mn(II) and quantitation of this signal suggests ca. two Mn(II) per center (*64*). After continuous illumination for 2 min, the EPR six-line signal disappears and the XANES spectrum of the reduced-then-illuminated sample is indistinguishable from that of the control (*10*).

Figure 6 and Table II show that significantly less Mn(II) is produced by NH_2OH than by hydroquinone. In fact, it is possible to fit the XANES for the NH_2OH-reduced sample using a model that contains only Mn(III). The smaller extent of reduction is perhaps not surprising considering the smaller concentration and shorter incubation time used for NH_2OH. (These conditions were chosen to maximize the reduction reaction without compromising activity.) What is somewhat surprising, however, is that the Mn(II) produced by H_2Q treatment is not chelatable with

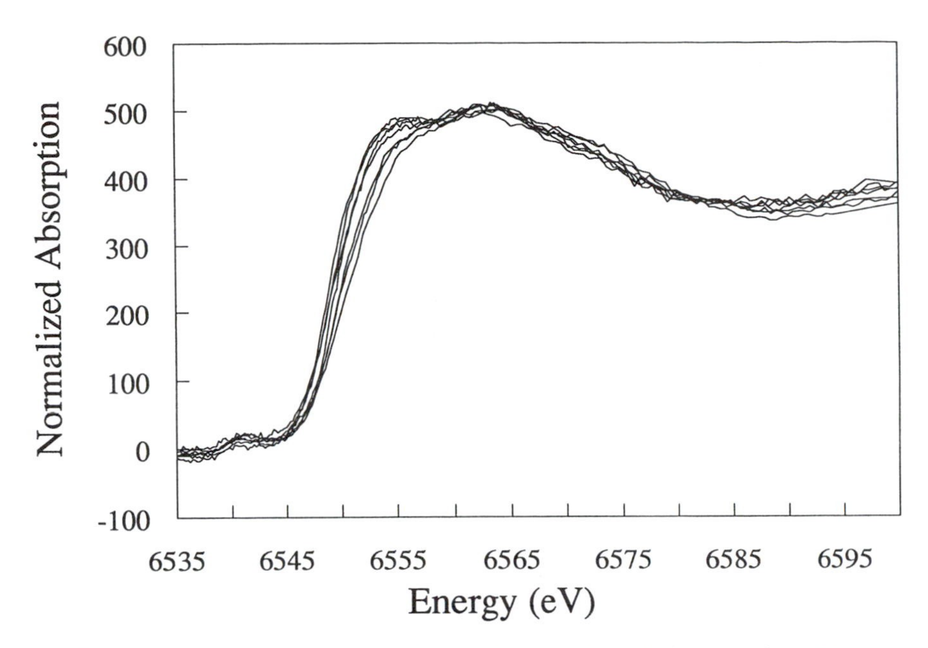

Figure 8. NH_2OH time dependence. From right to left: control S_1 state; NH_2OH incubation times of 1′, 3′, 5′, 7′, 9′, and 12′. Concentration of NH_2OH was 100 μM and reactions were quenched by dilution and centrifugation.

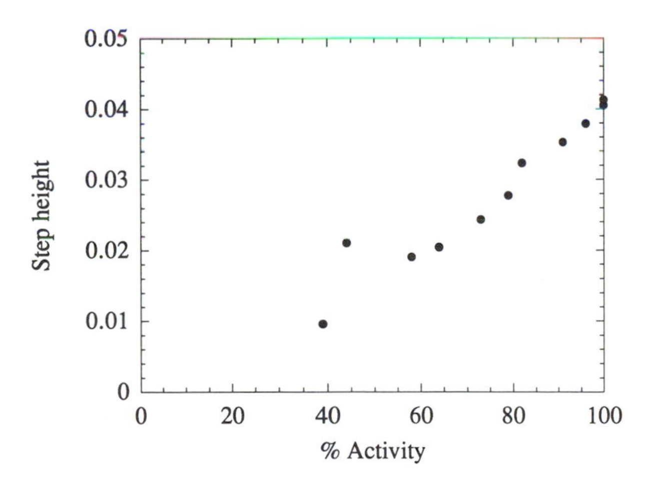

Figure 9. Step height versus activity. The data are for the samples shown in Figures 7 and 8. Step heights were calculated from unnormalized data ratioed to the incident X-ray intensity. Percent activity is relative to the control (ca. 1200 μM O_2/mg chlorophyll/h) using 2,6-dichlorobenzoquinone as an acceptor.

Table II. Mn Oxidation State Composition for OEC

Treatment	*Mn(II)*	*Mn(III)*	*Mn(IV)*
Control S_1	—	49 (20)	51 (20)
Hydroquinone	23 (7)	77 (7)	—
200 μM, 30′	41 (5)	—	59 (5)
NH_2OH	—	86 (12)	14 (12)
100 μM, 3′	4 (5)	96 (5)	—
	25 (7)	—	75 (7)

NOTE: Values are percent composition of different Mn oxidation states. Standard deviations (in parentheses) are for all combinations of models with the indicated oxidation states. For reduced samples, the Mn(II)–Mn(III) and Mn(II)–Mn(IV) models cannot be distinguished; the true percent of Mn(II) is likely between the two extremes.

EDTA nor does it enhance the NMR relaxation of the bulk solvent (*64*). These observations suggest that the Mn(II) produced by hydroquinone reduction remains in a sequestered domain in PSII near the binding sites from which the metal was released.

In contrast to hydroquinone, which gives an EDTA-insensitive product, NH_2OH treatment (100 μM, 3 min) produces a derivative that is quickly inactivated by EDTA. Moreover, the hydroquinone-reduced sample is quickly inactivated by the addition of concentrations of NH_2OH (20 μM) that are not by themselves inhibitory (*64*). This synergism, coupled with differences in EDTA sensitivity and in the XANES spectra, led Mei and Yocum to suggest that NH_2OH and hydroquinone attack different sites within the Mn cluster (*64*). In this model, NH_2OH, perhaps because of its chemical similarity to substrate, can penetrate to a Mn site that is not sensitive to dihydroquinone. The larger size of dihydroquinone relative to NH_2OH may be the factor that limits access of dihydroquinone to this site. Samples reduced by NH_2OH for short times are stable. However, further reduction either by long-term treatment with NH_2OH or by addition of dihydroquinone, causes reduction of a second site leading to cluster decomposition and release of Mn(II).

A necessary condition for this two-site model is that electron transfer not occur between the two sites under the reducing conditions. If electron transfer is rapid, the two different reduction pathways would lead to a single, stable product. Several circumstances could block intersite electron transfer. One possibility is that the two sites are physically separated by a protein domain that prevents rapid redox equilibration. An alternative possibility is that a structural rearrangement takes place when the lower potential site is reduced. This rearrangement then prevents the electron transfer to the higher potential site. Our EXAFS results, show that just such a dramatic rearrangement occurs following dihydroquinone reduction.

EXAFS Spectroscopy. The Fourier transforms of the EXAFS spectra for the OEC (65) are shown in Figure 10. The Fourier transform for the S_1 state is strikingly similar to that observed for superoxidized Mn catalase (Figure 3). The OEC FT has three principal features: A nearest neighbor peak that can be fit as a shell of O/N at ca. 1.9 Å; a next-nearest peak corresponding to Mn–Mn scattering at 2.7 Å; and a third peak corresponding to a Mn–scatterer distance of ca. 3.3 Å. The third peak can be modeled by either Mn–Ca or Mn–Mn scattering.

The Fourier transforms of the EXAFS data for the hydroquinone-reduced OEC derivative is also shown in Figure 10 (65). Hydroquinone reduction leads to dramatic changes in the EXAFS. The nearest neighbor (1.9 Å Mn–O/N) peak decreases in amplitude, and a new shell of low-Z ligands appears at ca. 2.2 Å. The peak at 2.2 Å most likely represents the longer Mn–O bonds associated with the Mn(II). The peak due to 2.7-Å Mn–Mn distances decreases to $\frac{1}{2}$ to $\frac{2}{3}$ of its original amplitude, and the 3.3-Å feature is no longer discernible above the noise level.

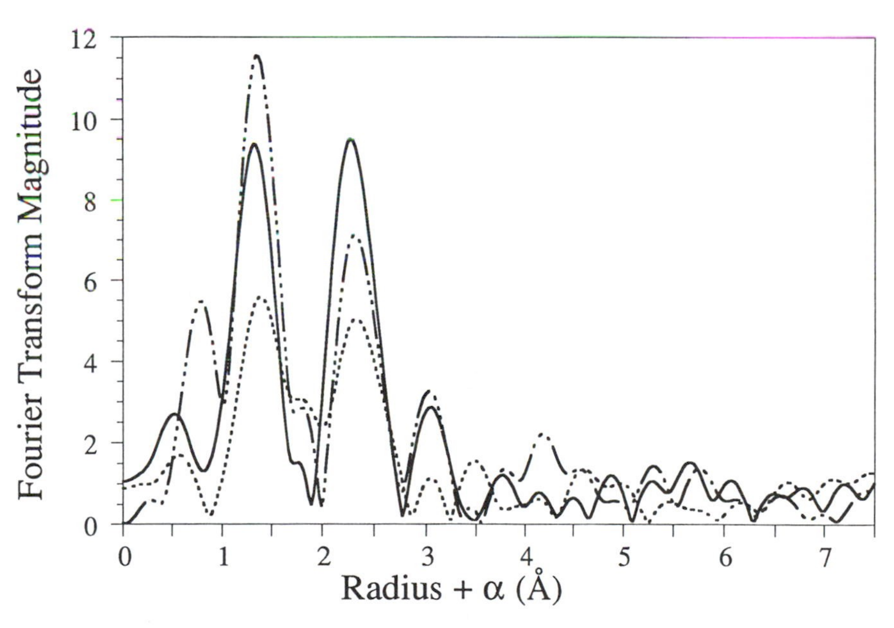

Figure 10. Fourier transforms of the EXAFS spectra for the OEC. Solid line indicates control, dark-adapted S_1 spectrum; dotted–dashed indicates 100 μM NH_2OH for 3′, quenched by dilution; and dashed line indicates 200 μM H_2Q, 30′ quenched with ferricyanide. Fourier transforms calculated over the range k = 3–11.5 $Å^{-1}$ using k^3 weighted data.

These structural changes are completely reversed by continuous illumination (data not shown).

In contrast to H_2Q, the NH_2OH-reduced sample shows only subtle changes in its EXAFS spectrum. The primary difference in the NH_2OH sample relative to the control is that there is an increase in the disorder of the 2.7-Å peak. The minor structural changes that are observed for the NH_2OH-reduced sample relative to those for the hydroquinone-reduced sample are consistent with the smaller edge shift observed after NH_2OH treatment. As with the H_2Q-treated sample, the changes in the EXAFS spectra following NH_2OH treatment are completely reversed by continuous illumination for 3 min and subsequent dark adaptation (*65*).

Relevance to Photoactivation. A striking property of these reduced samples is the ease with which they are fully oxidized back to the native structure. Samples were diluted with a 10 mM $CaCl_2$/50 mM MES (pH 6) buffer in the presence of an acceptor (0.31 mM dichlorobenzoquinone) and illuminated for 3 min with room light or sunlight at 0 °C. This protocol gives complete recovery of the native structure as judged by EXAFS. Indeed, because oxygen evolution measurements indicate 80–100% retention of activity, the conditions used in a standard assay with a Clark-type oxygen electrode are also sufficient to recover the native structure.

Much research has been devoted to photoactivation, the process by which a completely Mn-depleted sample recovers activity by photoligating and photooxidizing exogenous Mn(II). Samples are inactivated with NH_2OH or other treatments, incubated in light (either continuous illumination or with actinic flashes) in the presence of Ca(II), Cl^-, and Mn(II), and eventually recover activity, albeit with a very low quantum yield (*66, 67*). It is generally agreed that photoactivation requires at least two photoevents to ligate and oxidize two Mn(II) ions to Mn(III) (*66–70*). The process is thought to involve several unstable intermediates that decay rapidly back to the inactive apoenzyme unless a light-independent rearrangement occurs, followed by a second photoevent. A fully oxidized cluster than must bind Ca(II) to confer water oxidation activity (*66, 67*).

It is possible that the hydroquinone-reduced species with its trapped Mn(II) corresponds to an intermediate in the photoactivation process. There appears to be one oxidized Mn dimer in the hydroquinone-reduced sample, as indicated by the 2.7-Å feature in the EXAFS. This feature could correspond to the Mn that is photooxidized in the photoactivation process. Oxidation and photoligation of the first two Mn are the rate-limiting steps in photoactivation. Thus, as long as hydroquinone reduction does not affect these two critical Mn, the effects of hydroquinone should be easily and rapidly reversed, as is observed. In this model, further reduction of the hydroquinone-treated sample would reduce the critical dimer, presumably leading to loss of Mn and activity. This model is consistent with the synergism reported by Mei and Yocum (*64*) for NH_2OH and H_2Q inactivation.

Conclusions

It is instructive to compare the Mn site in the Mn catalase with that found in the OEC, particularly because the OEC has been suggested to have Mn-dependent catalase activity (*71*, *72*). The Mn ions in the OEC appear to be present primarily as Mn(III) and Mn(IV). Although it is not clear which oxidation states are responsible for the catalase activity, it is clear that the OEC functions using higher oxidation states of Mn. The OEC is reduced by low concentrations of NH_2OH and is inactivated by high concentrations of NH_2OH. In contrast, Mn catalase appears to function in the lower oxidation states of Mn and is stable to NH_2OH treatment. When oxidized to the oxidation states believed to be present in the OEC, the Mn catalase is inactive. These differences in oxidation state and reactivity most likely reflect differences in ligation between the enzymes.

The similarity between the EXAFS for superoxidized catalase and for the OEC suggests that they contain similar, probably di-μ-oxo-bridged, Mn structures. In the catalase, the $Mn(\mu\text{-}O)_2Mn$ core is a relatively ineffective oxidant, perhaps because of the stabilizing effect of the di-μ-oxo bridge. It may be that the presence of $Mn(\mu\text{-}O)_2Mn$ units in the OEC is important in stabilizing the OEC against premature oxidation of water. That is, as a relatively ineffective oxidant, the Mn_2O_2 core may help to prevent the OEC from oxidizing water prior to formation of the S_4 state.

Mechanistic Implications

The available data do not permit an unambiguous definition of either the OEC structure or the mechanism of water oxidation. It is possible, however, to construct a model consistent with both the known chemistry of Mn and the available data and to use this model to make testable predictions concerning OEC structure. The present structural results are consistent with, if not proof of, the "dimer-of-dimers" model of the OEC (*11*). Within this model, we attribute the 2.7-Å feature to the intradimer Mn–Mn distances and the 3.3-Å feature to an interdimer Mn–Mn distance. The EXAFS finding that there are two to three (per 4 Mn) 2.7-Å Mn–Mn distances (*65*) is consistent with the presence of two $Mn(\mu\text{-}O)_2Mn$ dimers. The 3.3-Å feature, which disappears on hydroquinone treatment (*see* Figure 10), is attributed to an interdimer Mn–Mn distance. This distance is typical of oxo- or oxocarboxylatobridged Mn–Mn separations. Hydroquinone reduction appears to disrupt approximately one-half of the Mn(μ-O)$_2$Mn units with production of Mn(II). This disruption would be consistent with the two-electron reduction of a Mn(III)–Mn(III) dimer. Hydroxylamine, in contrast, appears to cause only minor structural changes and gives little, if any, Mn(II). This finding would be consistent with reduction of a Mn(IV)(μ-O)$_2$Mn(IV) dimer to a Mn(III)(μ-O)$_2$Mn(III) dimer. The increased disorder in the 2.7-Å Mn–Mn feature following NH_2OH reduction may indicate that one or more of the oxo bridges is protonated in the reduced dimer, because Mn(μ-O)(μ-OH)Mn cores have longer Mn–Mn distances than Mn(μ-O)$_2$Mn cores (*73*). Protonation of a μ-oxo bridge on reduction would be consistent with the strong dependence of oxo-bridge acidity of Mn oxidation state (*74, 75*). Hydroxylamine is able to further reduce the OEC, as indicated by the data in Figures 7 and 8 and by the ultimate loss of activity.

A schematic structural model consistent with these observations is shown for S_1 in Figure 11. The remainder of Figure 11 is a hypothetical scheme that could account for the catalase and water oxidation reactions of the OEC. Only the Mn core structure is shown because no information is available to define the remaining ligands (average Mn coordination number is ca. 5–6). In this model, water oxidation is envisioned to take place through protonation of a Mn(IV)(μ-O)$_2$Mn(IV) dimer, giving Mn(III) and H_2O_2. The H_2O_2 then reacts with a second Mn dimer, giving O_2 and the S_0 state of the OEC. The postulated S_{-1}^- structure is based on the EXAFS results for NH_2OH-treated samples. In contrast, the reduced species formed on hydroquinone treatment (not shown) appears to contain Mn(II) and only retains a single Mn(μ-O)$_2$Mn unit.

If the catalase chemistry involves an S_1–S_{-1} cycle, this cycle could use either the same Mn(II)–Mn(II) $\leftrightarrow$ Mn(III)–Mn(III) reaction found for Mn catalase or, as shown in Figure 11, a Mn(III)–Mn(III) $\leftrightarrow$ Mn(IV)–Mn(IV) scheme similar to that demonstrated for Mn model compounds (*76*). The water oxidation portion of Figure 11 is based on the suggestion (*77*) that this reaction could involve two sequential two-electron oxidations. This suggestion is reasonable given the observations that there appear to be two different sites of reduction within the OEC. If Mn is oxidized on both the $S_1 \rightarrow S_2$ and the $S_2 \rightarrow S_3$ transitions (*59*), the dimer-of-dimers model would suggest that S_3 contains two Mn(IV)(μ-O)$_2$Mn(IV) dimers. This suggestion leads to the prediction that S_2 and S_3 should have EXAFS spectra that are very similar to that for S_1. In Figure 11, the trigger for water oxidation reaction is proposed to be the protonation of one of the Mn dimers. The protonation could result from oxidation of an organic radical (e.g., histidine) (*78*) on the $S_3 \rightarrow S_4$ transition. Both the formation of H_2O_2 by protonation of a Mn(IV)(μ-O)$_2$Mn(IV) core

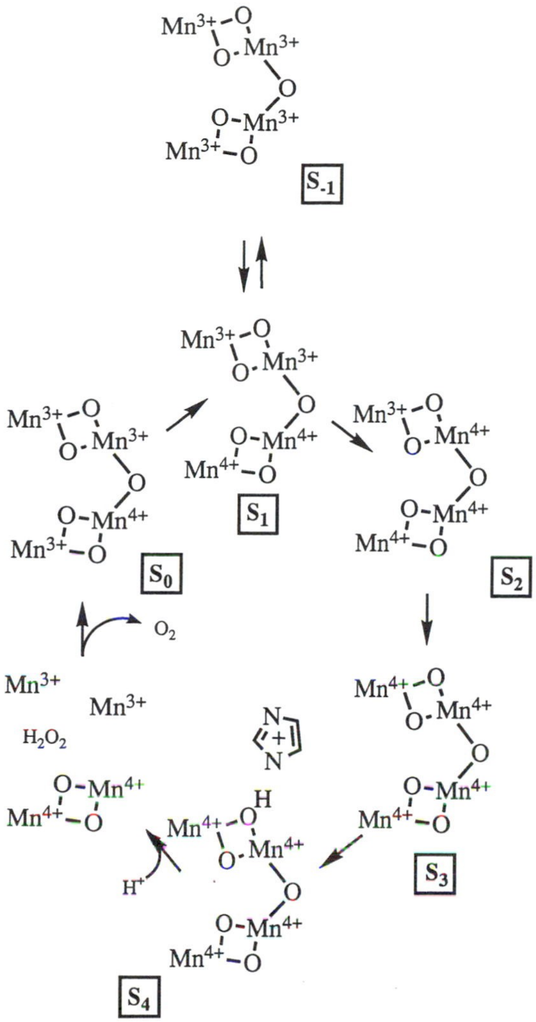

Figure 11. A working hypothesis for the OEC reaction mechanism.

and the formation oxygen from H_2O_2 and Mn(IV)(μ-O)$_2$Mn(IV) are known reactions for synthetic Mn models.

Tests (and no doubt refinement) of this model will require better definition of the structure of the oxidized and reduced derivatives of the OEC. Efforts along these lines are in progress in several laboratories and the prospects for ultimately understanding the molecular details of this reaction seem quite good.

Acknowledgments

The authors thank Charles F. Yocum and Vincent L. Pecoraro for their helpful discussions. Model compounds were prepared in the laboratories of Vincent L. Pecoraro and Nobumassa Kitajima. This work was supported by the National Institutes of Health through grants GM-38047 and GM-45205 to J. E. Penner-Hahn, through National Institutes of Health training grant support for P. J. Riggs-Gelasco, and through the United States Department of Agriculture National Research Initiative Competitive Grants Program (to C. F. Yocum). The XAS spectra described in this paper were measured at the Stanford Synchrotron Radiation Laboratory (SSRL) and the National Synchrotron Light Source (NSLS). SSRL is operated by the Department of Energy, Office of Basic Energy Sciences, Division of Chemical Sciences; with additional support from the National Institutes of Health, Biomedical Resource Technology Program, Division of Research Resources; and the Department of Energy, Office of Health and Environmental Research. NSLS is supported by the U.S. Department of Energy. Data at NSLS were measured at beamlines X9 and X19A. X9 is supported by the National Institutes of Health, Division of Research Resources. Beamline X19A was supported in part by the Office of the Vice President for Research, University of Michigan.

References

1. *Manganese Redox Enzymes;* Pecoraro, V. L., Ed.; VCH Publishers: New York, 1992.
2. Cramer, S. P.; Hodgson, K. O. *Prog. Inorg. Chem.* **1979,** *25*, 1–39.
3. Scott, R. A. *Methods Enzymol.* **1985,** *117*, 414–459.
4. Teo, B. K. EXAFS: *Basic Principles and Data Analysis;* Springer-Verlag: New York, 1986; Vol. 9.
5. Bart, J. C. *J. Adv. Catal.* **1986,** *34*, 203–296.
6. Rehr, J. J.; de Leon, J. M.; Zabinsky, S. I.; Albers, R. C. *J. Am. Chem. Soc.* **1991,** *113*, 5135–5140.
7. Goodin, D. B.; Yachandra, V. K.; Britt, R. D.; Sauer, K.; Klein, M. *Biochim. Biophys. Acta* **1984,** *767*, 209–216.
8. Kirby, J. A.; Goodin, D. B.; Wydrzynski, A. S.; Robertson, A. S.; Klein, M. P. *J. Am. Chem. Soc.* **1981,** *103*, 5537–5542.
9. Penner-Hahn, J. E.; Fronko, R. H.; Pecoraro, V. L.; Yocum, C. F.; Betts, S. D.; Bowlby, N. R. *J. Am. Chem. Soc.* **1990,** *112*, 2549–2557.
10. Riggs, P. J.; Mei, R.; Yocum, C. F.; Penner-Hahn, J. E. *J. Am. Chem. Soc.* **1992,** *114*, 10650–10651.
11. Sauer, K.; Yachandra, V. K.; Britt, R. D.; Klein, M. P. In *Manganese Redox Enzymes;* Pecoraro, V. L., Ed.; VCH Publishers: New York, 1992; pp 141–176.
12. Delwiche, E. A. *J. Bacteriol.* **1961,** *81*, 416–418.
13. Johnston, M. A.; Delwiche, E. A. *J. Bacteriol.* **1965,** *90*, 352–356.
14. Jones, D.; Diebel, D. H.; Niven, C. F., Jr. *J. Bacteriol.* **1965,** *88*, 602–610.
15. Kono, Y.; Fridovich, I. *J. Biol. Chem.* **1983,** *258*, 6015–6019.

16. Beyer, W. F. J.; Fridovich, I. In *Manganese in Metabolism and Enzyme Function;* Schramm, V. L.; Wedler, F. C., Eds.; Academic: Orlando, FL, 1986; pp 193–219.
17. Penner-Hahn, J. E. In *Manganese Redox Enzymes;* Pecoraro, V. L., Ed.; VCH Publishers: New York, 1992; pp 29–46.
18. Wieghardt, K. *Angew. Chem.* **1989,** *101,* 1179–1198.
19. Khangulov, S. U.; Barynin, V. V.; Antonyak-Barynina, S. U. *Biochim. Biophys. Acta* **1990,** *1020,* 25–33.
20. Khangulov, S. U.; Goldfeld, M. G.; Gerasimenko, V. V.; Andreeva, N. E.; Barynin, V. V.; Grebenko, A. I. *J. Inorg. Biochem.* **1990,** *40,* 279–292.
21. Vainshtein, B. K.; Melik-Adamyan, W. R.; Barynin, V. V.; Vagin, A. A.; Grebenko, A. I. *Proc. Int. Symp. Biomol. Struct. Interactions, Suppl. J. Biosci.* **1985,** *8,* 471–479.
22. Waldo, G. S.; Fronko, R. M.; Penner-Hahn, J. E. *Biochemistry* **1991,** *30,* 10486–10490.
23. Waldo, G. S. Ph.D. thesis, University of Michigan; 1991.
24. Waldo, G. S.; Penner-Hahn, J. E. *Biochemistry* **1995,** *34,* 1507–1512.
25. Kono, Y. In *Superoxide and Superoxide Dismutase in Chemistry, Biology, and Medicine;* Potilio, G., Ed.; Elsevier: 1986; pp 231–233.
26. Kono, Y. *Biochem. Biophys. Res. Commun.* **1984,** *124,* 75–79.
27. Waldo, G. S.; Yu, S.; Penner, H. J. E. *J. Am. Chem. Soc.* **1992,** *114,* 5869–5870.
28. Sanders-Loehr, J. In *Iron Carriers and Iron Proteins;* VCH: New York, 1989; pp 373–466.
29. Larson, E.; Lah, M. S.; Li, X.; Bonadies, J. A.; Pecoraro, V. L. *Inorg. Chem.* **1992,** *31,* 373–378.
30. Stebler, M.; Ludi, A.; Bürgi, H.-B. *Inorg. Chem.* **1986,** *25,* 4743–4750.
31. Plaksin, P. M.; Stoufer, R. C.; Mathew, M.; Palemik, G. J. *J. Am. Chem. Soc.* **1972,** *94,* 2121.
32. Wieghardt, K.; Bossek, U.; Zsolnai, L.; Hattner, G.; Blandin, G.; Girerd, J.-J.; Babonneau, F. *J. Chem. Soc. Chem. Commun.* **1987,** 651–653.
33. Bashkin, J. S.; Schake, A. R.; Vincent, J. B.; Chang, H. R.; Li, Q.; Huffmann, J. C.; Christou, G.; Hendrickson, D. N. *J. Chem. Soc. Chem. Commun.* **1988,** 700–702.
34. Hagen, K. S.; Armstrong, W. H.; Hope, H. *Inorg. Chem.* **1988,** *27,* 967–969.
35. Vincent, J. B.; Christou, G. *Adv. Inorg. Chem.* **1989,** *33,* 197–257.
36. Sheats, J. E.; Czernuszewicz, R. S.; Dismukes, G. C.; Rheingold, A. L.; Petrouleas, V.; Stubbe, J.; Armstrong, W. H.; Beer, R. H.; Lippard, S. J. *J. Am. Chem. Soc.* **1987,** *109,* 1435–1444.
37. Kok, B.; Forbush, B.; McGloin, M. *Photochem. Photobiol.* **1970,** *11,* 457.
38. Styring, S.; Rutherford, A. W. *Biochemistry* **1987,** *26,* 2401–2405.
39. Sivaraja, M.; Dismukes, G. C. *Biochemistry* **1988,** *27,* 3467–3475.
40. Cheniae, G.; Martin, I. *Biochim. Biophys. Acta* **1970,** *197,* 219–239.
41. Yocum, C. F.; Yerkes, C. T.; Blankenship, R. E.; Sharp, R. R.; Babcock, G. T. *Proc. Natl. Acad. Sci. U.S.A.* **1981,** *78,* 7507–7511.
42. Pauly, S.; Witt, H. T. *Biochim. Biophys. Acta* **1992,** *1099,* 211–218.
43. Zimmerman, J. L.; Rutherford, A. W. *Biochemistry* **1986,** *25,* 4609–4615.
44. Brudvig, G. W.; Casey, J. L.; Sauer, K. *Biochim. Biophys. Acta* **1982,** *723,* 366–371.
45. Casey, J.; Sauer, K. *Biochim. Biophys. Acta* **1984,** *767,* 21–28.

46. Kim, D. H.; Britt, R. D.; Klein, M. P.; Sauer, K. *J. Am. Chem. Soc.* **1990,** *112,* 9389–9391.
47. Kim, D. H.; Britt, R. D.; Klein, M. P.; Sauer, K. *Biochemistry* **1992,** *31,* 541–547.
48. Dexheimer, S. L.; Klein, M. P. *J. Am. Chem. Soc.* **1992,** *114,* 2821–2826.
49. Koulougliotis, D.; Hirsh, D. J.; Brudvig, G. W. *J. Am. Chem. Soc.* **1992,** *114,* 8322–8323.
50. Beck, W. F.; dePaula, J. C.; Brudvig, G. W. *Biochemistry* **1985,** *24,* 3035–3043.
51. Kok, B.; Velthuys, B. In *Research in Photobiology;* Castellaini, A., Ed.; Plenum: New York, 1977; p 119.
52. Bouges, B. *Biochim. Biophys. Acta* **1971,** *234,* 103–112.
53. Messinger, J.; Wacker, U.; Renger, G. *Biochemistry* **1991,** *30,* 7852–7862.
54. Velthuys, B.; Kok, B. *Biochim. Biophys. Acta* **1978,** *502,* 211–221.
55. Cheniae, G. M.; Martin, I. F. *Plant Physiol.* **1971,** *47,* 568–575.
56. Beck, W. F.; Brudvig, G. W. *J. Am. Chem. Soc.* **1988,** *110,* 1517–1523.
57. Kretschmann, H.; Pauly, S.; Witt, H. T. *Biochim. Biophys. Acta* **1991,** *1059,* 208–214.
58. Yachandra, V. K.; DeRose, V. J.; Latimer, M. J.; Mukerji, I.; Sauer, K.; Klein, M. P. *Science (Washington, D.C.)* **1993,** *260,* 675–679.
59. Ono, T.-A.; Noguchi, T.; Inoue, Y.; Kusunoki, M.; Matsushita, T.; Oyanagi, H. *Science (Washington, D.C.)* **1992,** *258,* 1335–1337.
60. Guiles, R. D.; Zimmerman, J.-L.; McDermott, A. E.; Yachandra, V. K.; Cole, J. L.; Dexheimer, S. L.; Britt, R. D.; Wieghardt, K.; Bossek, U.; Sauer, K.; Klein, M. P. *Biochemistry* **1990,** *29,* 471–485.
61. Guiles, R. D.; Yachandra, V. K.; McDermott, A. E.; Cole, J. L.; Dexheimer, S. L.; Britt, R. D.; Sauer, K.; Klein, M. P. *Biochemistry* **1990,** *29,* 486–496.
62. Ghanotakis, D. F.; Topper, J. N.; Yocum, C. F. *Biochim. Biophys. Acta* **1984,** *767,* 524–531.
63. Mei, R.; Yocum, C. F. *Biochemistry* **1991,** *30,* 7836–7842.
64. Mei, R.; Yocum, C. F. *Biochemistry* **1992,** *31,* 8449–8454.
65. Riggs-Gelasco, P. J.; Mei, R.; Yocum, C. F.; Penner-Hahn, J. E. unpublished results.
66. Tamura, N.; Cheniae, G. *Biochim. Biophys. Acta* **1987,** *890,* 179–194.
67. Miller, A. F.; Brudvig, G. W. *Biochemistry* **1989,** *28,* 8181–8190.
68. Miller, A.-F.; Brudvig, G. W. *Biochemistry* **1990,** *29,* 1385–1392.
69. Miyao, M.; Inoue, Y. *Biochim. Biophys. Acta* **1991,** *1056,* 47–56.
70. Miyao-Tokutomi, M.; Inoue, Y. *Biochemistry* **1992,** *31,* 526–532.
71. Mano, J.; Takahashi, M.; Asada, K. *Biochemistry* **1987,** *26,* 2495–2501.
72. Frasch, W. D.; Mei, R. *Biochemistry* **1987,** *26,* 7321–7325.
73. Larson, E. J.; Riggs, P. J.; Penner-Hahn, J. E.; Pecoraro, V. L. *J. Chem. Soc. Chem. Commun.* **1992,** 102–103.
74. Baldwin, M. J.; Gelasco, A.; Pecoraro, V. L. *Photosynth. Res.* **1993,** *38,* 303–308.
75. Thorp, H. H.; Sarneski, J. E.; Brudvig, G. W. *J. Am. Chem. Soc.* **1989,** *111,* 9249–9250.
76. Larson, E. J.; Pecoraro, V. L. *J. Am. Chem. Soc.* **1991,** *113,* 7809–7810.
77. Pecoraro, V. L. In *Manganese Redox Enzymes;* Pecoraro, V. L., Ed.; VCH Publishers: New York, 1992; pp 197–232.
78. Boussac, A.; Rutherford, A. W. *Biochemistry* **1992,** *31,* 7441–7445.

RECEIVED for review July 19, 1993. ACCEPTED revised manuscript May 5, 1994.

9

Structure and Function of Manganese in Photosystem II

Gary W. Brudvig

Department of Chemistry, Yale University, New Haven, CT 06511

The tetranuclear Mn complex in photosystem II functions to accumulate oxidizing equivalents and to catalyze the four-electron oxidation of water to molecular oxygen. In the water-oxidation cycle, the Mn complex is advanced through five intermediate oxidation states called S_i *states (*i = 0–4*), where* i *denotes the number of stored oxidizing equivalents. Mechanistic studies of the photosynthetic water oxidation process have been aimed at characterizing the structure and properties of the Mn complex in each of the different S states. Electron paramagnetic resonance and X-ray absorption spectroscopies have been especially powerful methods to probe the Mn complex. However, to use these spectroscopic methods, it is necessary to prepare highly concentrated samples in a specific S state. The photochemical methods used to prepare the different S states are described and the results of studies of the* S_1 *and* S_2 *states are summarized. The structure of the Mn complex is considered in light of recent studies.*

ONE OF THE MOST IMPORTANT ROLES OF MANGANESE in biology is in the photosynthetic oxidation of water to molecular oxygen. This reaction is catalyzed by a membrane-bound protein complex called photosystem II (PSII). PSII uses light to produce a charge-separation reaction that results in the reduction of plastoquinone, the oxidation of water, and the generation of a proton gradient across the chloroplast thylakoid membrane. The overall reaction, which requires four photochemical charge-separation reactions, is given in equation 1:

$$2H_2O + 2PQ + 4H_{out}^+ \rightarrow O_2 + 2PQH_2 + 4H_{in}^+ \qquad (1)$$

0065–2393/95/0246–0249/$08.00/0

where PQ denotes plastoquinone, PQH_2 denotes plastoquinol, and in and out refer to the inside and outside of the thylakoid vesicle, respectively (for reviews, *see* references 1–5).

The mechanism of O_2 evolution in photosynthesis has been the focus of much research. In a classic experiment, Joliot and co-workers (6) showed that O_2 is evolved in a periodic pattern when a dark-adapted sample is given a series of short, saturating light flashes (Figure 1). This pattern of O_2 evolution was explained by Kok and co-workers (7, 8) by the S-state model shown in equation 2:

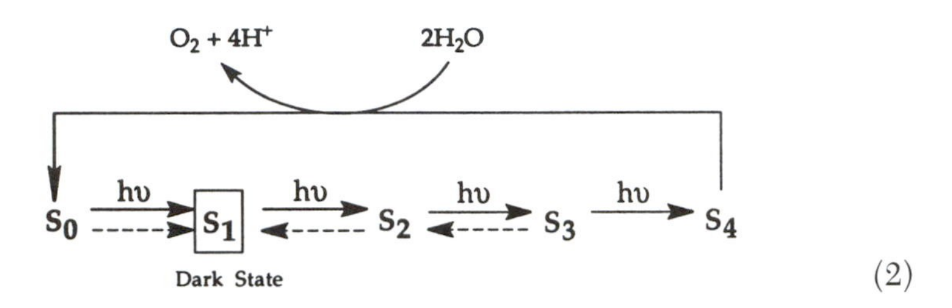

(2)

where the solid lines denote light-driven reactions and the dashed lines denote dark reactions. Each photochemical charge separation of PSII produces one oxidizing equivalent, which is stored in the O_2-evolving center (OEC) of PSII. The intermediate oxidation states of the OEC are referred to as S ("store") states with a subscript to denote the number of oxidizing equivalents that have been stored. To account for the maximal yield of O_2 on the third flash (Figure 1), it is required that the S_1 state is the dark-stable state. In subsequent flashes, the yield of O_2 is maximal on every fourth flash due to the requirement of four oxidizing equivalents to produce O_2 from water. Because of "misses" (in which the OEC does not turn over during a flash) or "double hits" (in which a single flash causes a two-step advancement of the OEC), the yield of

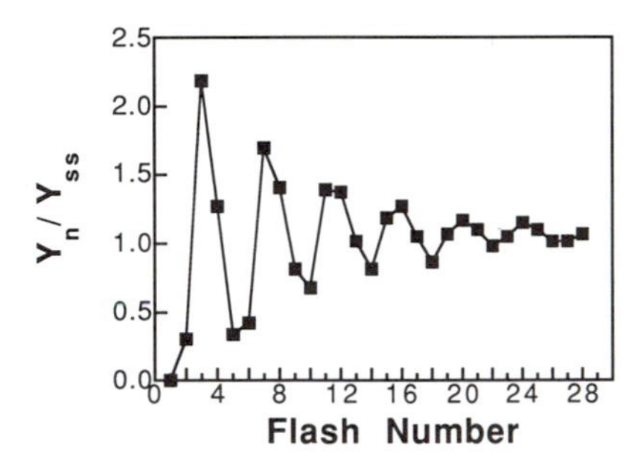

Figure 1. Normalized O_2 yields from thylakoid membranes given a train of saturating flashes. Y_n denotes the yield of O_2 on flash n and Y_{ss} denotes the steady-state yield of O_2. Conditions are as described in reference 44.

O_2 reaches a steady-state value after about 30 flashes. In the steady state, the S_0–S_3 states are assumed to be present in approximately equal amounts. The S_4 state does not accumulate in the steady state because it rapidly decays to the S_0 state by releasing O_2; the S_4 state is not, in fact, even observed as a kinetic intermediate because formation of the S_4 state is the rate-limiting step in the S_3 to S_0 conversion. The S_2 and S_3 states are unstable and decay in seconds to the S_1 state in the dark. The S_0 state also decays in the dark to the S_1 state on a minutes time scale. Hence, a dark-adapted sample contains predominantly the S_1 state.

The Kok model for O_2 evolution is now generally accepted. However, this model does not address the molecular basis of the intermediates. Thus, the Kok model only provides a starting point from which more detailed mechanistic questions can be posed.

A considerable body of evidence indicates that the active site for water oxidation contains a polynuclear Mn complex. Ca^{2+} and Cl^- are also required for water oxidation and may be associated with the Mn complex. Four Mn ions are required for maximal O_2-evolution activity, but the organization of the Mn ions in the OEC has been debated. It is clear, however, that oxidation of Mn occurs during the S-state transitions. X-ray absorption edge studies have been reported on samples in which the S states were advanced by flashes of light (9). The Mn K-edge energy increases from the S_0 to the S_3 state and shows a period-four oscillation. These results support the view that the S_0 to S_3 states are increasingly higher oxidation states of the Mn complex.

To understand the mechanism of water oxidation, it is necessary to characterize each of the S states. A variety of spectroscopic methods have been brought to bear on this problem. Electron paramagnetic resonance (EPR) and X-ray spectroscopies have been especially useful because these techniques allow the Mn complex to be probed directly. EPR spectroscopy has the restriction that the Mn complex must be paramagnetic to be studied. The S_2 state is an odd-electron state, and EPR spectroscopy has been used extensively to study the Mn complex in the S_2 state. X-ray spectroscopy has the advantage that any state of the Mn complex is observable. However, the successful application of EPR and X-ray spectroscopies requires that a specific S state be prepared in high yield in highly concentrated samples.

Methods for Preparation of Specific Oxidation States of the Mn Complex

In most redox enzymes, the different oxidation states of a metal center can be produced by poising the ambient redox potential of the sample. However, the intermediate oxidation states of the Mn complex in PSII have very high reduction potentials (in the range of 0.8–1.2 V vs. the normal hydrogen electrode). To date, it has not been possible to carry out a redox titration of any of the species on the oxidizing side of PSII. Therefore, one must use photochemical methods that allow the controlled turnover of PSII to prepare a specific oxidation state of the Mn complex. A key to the success of photochemical methods is the preparation of an initially homogeneous S state.

Dark adaptation will produce a sample that contains primarily the S_1 state. The S_2 and S_3 states are unstable and decay back to the S_1 state in the dark (equation 2). The S_0 state also decays to the S_1 state in the dark via a redox reaction with the oxidized tyrosine residue, $Y_D^{\bullet}$ (equation 3):

$$S_0 + Y_D^{\bullet} \xrightarrow{\text{dark}} S_1 + Y_D \tag{3}$$

However, the reaction of the S_0 state with $Y_D^{\bullet}$ is fairly slow (tens of minutes (*10*)) and, depending on the prior history of illumination of the sample, $Y_D^{\bullet}$ may not be present in all of the PSII centers. Therefore, the initial yield of the S_1 state can be maximized by giving a previously dark-adapted sample one flash (or continuous illumination at 200 K; *see* subsequent paragraphs) before further dark adaptation. These illuminations will cause the oxidation of S_0 to S_1 or Y_D to $Y_D^{\bullet}$; any S_2 state produced by the illumination will decay back to the S_1 state in the dark.

Once a homogeneous initial S-state population is produced, it is possible to advance the S states synchronously by using short, saturating flashes of light (as shown in Figure 1). For samples that are sufficiently dilute or have a very short path length, flashes work well for preparing each of the S states. Provided that the sample is dark adapted to ensure that the initial S_1 state is homogeneous, the S_2, S_3, or S_0 state can be generated by using one, two, or three flashes, respectively.

Unfortunately, turnover control of PSII is more complicated than the above description would indicate. Because turnover of the S states is achieved via a photochemical reaction, the yield of the reaction depends on both the electron donors and the electron acceptors. The overall picture of electron transfer in PSII is shown in Figure 2 (*11*). Light induces a series of electron-transfer reactions that lead to the formation of progressively more stable charge-separated states. The dominant reaction under physiological conditions leads to a one-step advancement of the S state and reduction of the secondary quinone electron acceptor (Q_B). In purified PSII preparations, however, the quinones are depleted and the Q_B site will mostly be unoccupied unless exogenous quinones are added.

In flashing light, there are limitations on the minimum and maximum times between flashes. The rate-limiting step in turnover of PSII is exchange of quinone for quinol, which occurs at the Q_B site. Consequently, the electron acceptors will produce a bottleneck if the time between

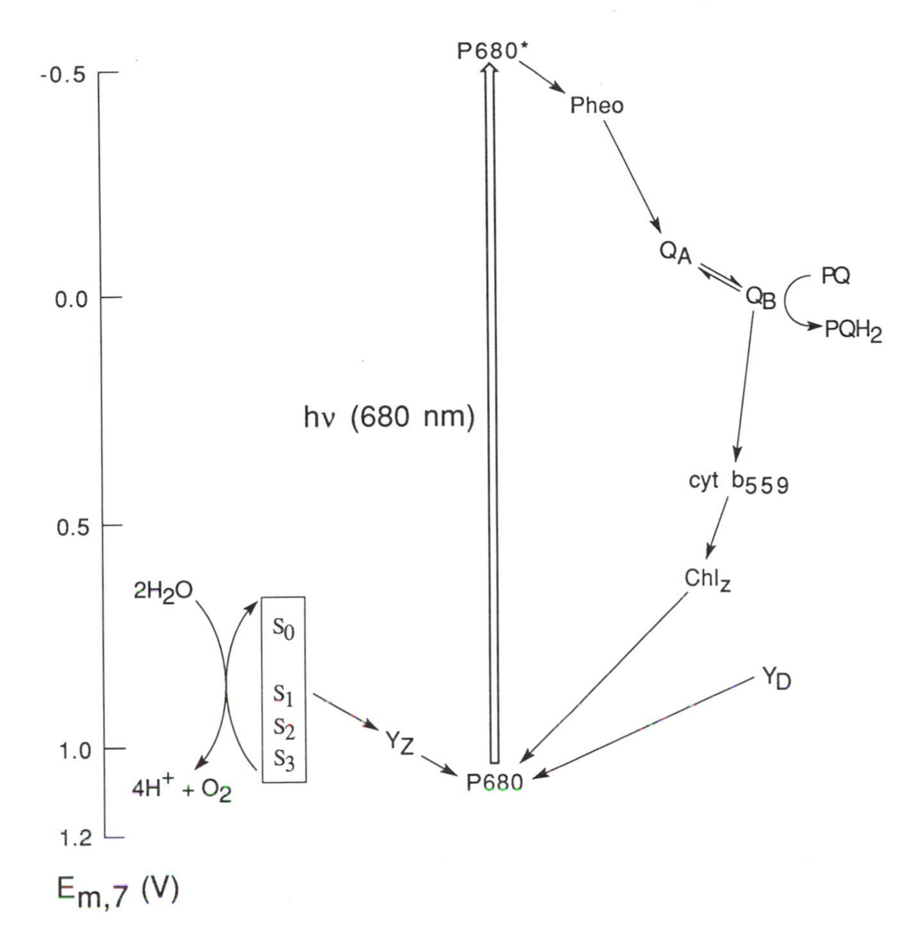

Figure 2. Paths of electron transfer in PSII: P680, reaction-center chlorophyll that functions as the primary electron donor; P680, first excited singlet state of P680; Pheo, pheophytin; Q_A, primary quinone electron acceptor; Q_B, secondary quinone electron acceptor; cyt b_{559}, cytochrome b_{559}; Chl_Z, redox-active chlorophyll that mediates electron transfer between cytochrome b_{559} and P680; Y_D, redox-active tyrosine that gives rise to the dark-stable tyrosine radical; Y_Z, redox-active tyrosine that mediates electron transfer from the Mn complex to P680.*

flashes is shorter than about 10 ms. On the other hand, the time between flashes cannot be too long because the S_2 and S_3 states are unstable and decay on a time scale of seconds, either by charge recombination or by oxidizing cytochrome b_{559} or Y_D (*11*). The time scales of these reactions leave open a window of time between flashes for which the turnover of the S states is optimal, although "misses" on the order of 5–10% per flash are unavoidable.

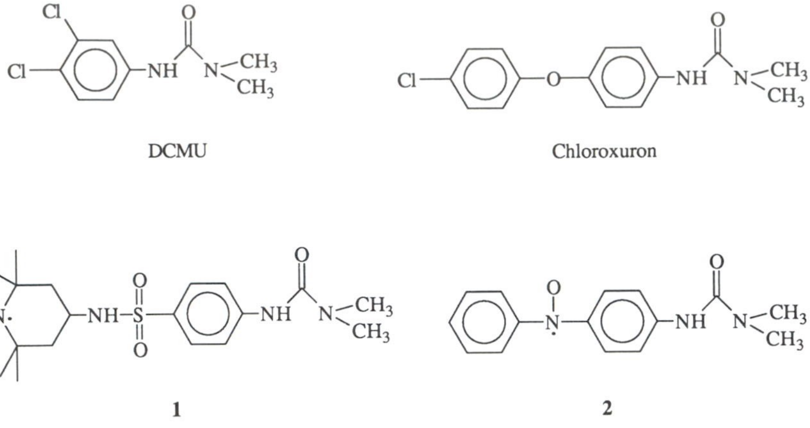

DCMU

Chloroxuron

1

2

Figure 3. Structures of herbicides that bind to the Q_B site in PSII.

One significant drawback of using flashes to advance the S states is that the sample must be very dilute or the path length must be very short to allow saturating light to penetrate throughout the sample. (A typical PSII membrane preparation has about 200 chlorophylls per PSII (*12*), which means that a sample containing 50 μM PSII has an optical density of 50–1000 throughout the visible spectrum.) EPR and extended X-ray absorption fine structure (EXAFS) measurements require a relatively large volume of concentrated sample. In such samples, flashes are not effective to produce a high yield of the S_2, S_3, or S_0 states.

Fortunately, the electron-acceptor side of PSII can be exploited to allow turnover control of the S states in highly concentrated samples. A number of herbicides are known that bind tightly to the Q_B site and block electron transfer past the primary quinone electron acceptor (Q_A) (*13*). Some examples are shown in Figure 3. Equations 4 and 5 show the reactions of PSII in the presence of 3-(3,4-dichlorophenyl)-1,1-dimethylurea (DCMU, Figure 3).

$$S_1Q_AQ_B + \text{DCMU} \rightleftarrows S_1Q_A\text{DCMU} + \text{PQ} \tag{4}$$

$$S_1Q_A\text{DCMU} \xrightarrow[\text{}]{\text{continuous illumination}} S_2Q_A^-\text{DCMU} \tag{5}$$

Q_A can only accept a single electron and, at low temperature, the $S_2Q_A^-$ charge separation is stable. Because PSII is restricted to only one turnover by DCMU, continuous light can be used to produce a single charge separation (*14*). The sample can be illuminated for as long as necessary to achieve a charge separation in every PSII complex. Electron transfer from Q_A^- to Q_B is also blocked at temperatures at or below 200 K. Therefore, illumination of a dark-adapted sample at 200 K will produce the S_2 state, even without adding DCMU (*15*). These methods have been extensively used to produce the S_2 state for EPR or EXAFS studies.

This protocol has been extended to produce the S_3 state by using the redox-active herbicides **1** and **2** (Figure 3) (*16*). In this case, two sequential charge-separation reactions occur as shown in equation 6:

$$S_1Q_ARR'NO + H^+ \xrightarrow[\text{illumination}]{\text{continuous}} S_2Q_ARR'NOH \xrightarrow[\text{illumination}]{\text{continuous}} S_3Q_A^-RR'NOH \quad (6)$$

where R and R′ are the groups bound to the nitroxyl group of molecule **1** or **2** (Figure 3). The first charge separation results in a one-electron reduction of the nitroxyl group of the herbicide and the second results in the one-electron reduction of Q_A. By using the redox-active herbicide, **2**, and continuous illumination at 250 K, a yield of the S_3 state of about 80% was achieved which is similar to what is obtained by using flashes and more dilute samples.

EPR Spectroscopic Studies of the S_1 and S_2 States

The two S states that can be produced most readily in high concentration are the S_1 and S_2 states. Consequently, these two states have been most accessible to spectroscopic study. In general, one expects to observe an EPR signal from a transition metal complex whenever the complex possesses an odd number of unpaired electrons. If the complex possesses an even number of unpaired electrons, EPR signals are often undetectable with conventional instrumentation due to large zero-field splittings. Because each S-state transition involves the removal of one electron from the O_2-evolving center, alternate S states will have an odd number of electrons. These odd-electron S states should, in principle, be detectable by EPR spectroscopy. The S_2 state is an odd-electron state and exhibits a number of EPR signals from the Mn complex (Figure 4; for reviews, *see* references 4 and 17). Although the S_1 state is an even-electron state, recent results indicate that it can exist in a paramagnetic form that can be studied by EPR spectroscopy (*18*, *19*).

EPR Studies of the S_2 State. Two types of EPR signals have been observed from the S_2 state (Figure 4). The first is a multiline EPR signal centered at about $g = 2$ (Figure 4a). This EPR signal arises from an $S = 1/2$ state of a multinuclear Mn complex; the hyperfine lines arise from the coupling of the unpaired electron to the nuclear spins of several ^{55}Mn ions (each ^{55}Mn ion has $I = 5/2$). The second EPR signal from the S_2 state exhibits a turning point at $g = 4.1$ (Figure 4b). The $g = 4.1$ EPR signal can be induced by illumination at 130 K in untreated samples,

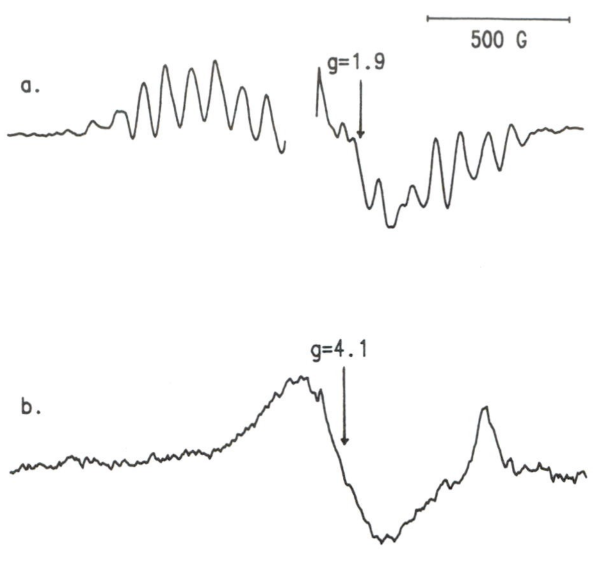

Figure 4. *S_2 state EPR spectra. (a) Multiline EPR signal produced by illumination at 200 K. (b) g = 4.1 EPR signal produced by illumination at 130 K. (Reproduced from reference 45. Copyright 1988 American Chemical Society.)*

but is unstable and is converted into the multiline EPR signal upon warming to 200 K. A number of treatments such as the addition of sucrose, ammonia, or fluoride; depletion of chloride; or replacement of Ca^{2+} by Sr^{2+} cause the $g = 4.1$ EPR signal to be stabilized at 200 K (reviewed in reference 4). In untreated samples, the $g = 4.1$ EPR signal does not exhibit resolved ^{55}Mn hyperfine structure. However, in ammonia-treated samples that are oriented by drying onto Mylar sheets, the $g = 4.1$ EPR signal exhibits resolved ^{55}Mn hyperfine structure (*20*), indicating that this EPR signal also arises from a multinuclear Mn complex. A variety of modified forms of the multiline and $g = 4.1$ EPR signals have been observed after the addition of various exogenous molecules including amines and fluoride or by depletion of Ca^{2+} and replacement by Sr^{2+}.

The S_2-state multiline and $g = 4.1$ EPR signals provide important information on the structure and organization of the Mn ions in the OEC. However, the assignment of these signals is still not fully resolved, although recent results have significantly limited the range of possibilities. Originally, two interpretations of the identity of the species that give rise to the multiline and $g = 4.1$ EPR signals were proposed.

Hansson et al. (*21*) proposed that the two EPR signals arise from distinct Mn centers: an antiferromagnetically exchange-coupled mixed-valence binuclear Mn center giving the multiline EPR signal from its ground $S = 1/2$ state and a mononuclear high-spin Mn^{IV} center giving the $g = 4.1$ EPR signal from the $S = 3/2$ state with axial symmetry. This proposal followed from the similarity of the multiline and $g = 4.1$ EPR signals to those of binuclear Mn^{III}–Mn^{IV} and mononuclear Mn^{IV} model complexes, respectively. The interconversion of the S_2-state multiline and $g = 4.1$ EPR signals was explained by a redox equilibrium between the binuclear and mononuclear centers.

de Paula et al. (*22, 23*) proposed that both the multiline and $g = 4.1$ EPR signals arise from the same tetranuclear Mn complex. The multiline and $g = 4.1$ EPR signals were attributed to two conformations of the complex having different exchange couplings between the Mn ions. This proposal was made by analogy to the $S = 1/2$ and $S = 3/2$ forms that have been observed for Fe_4S_4 cubanelike clusters. Based on this analogy and on studies of the temperature dependence of the two EPR signals, the multiline and $g = 4.1$ EPR signals were proposed to arise from an excited $S = 1/2$ state and a ground $S = 3/2$ state of an exchange-coupled Mn tetramer, respectively.

More recent studies have clarified the assignments of the S_2-state multiline and $g = 4.1$ EPR signals. Studies of the temperature dependence of the multiline EPR signal have shown that the signal follows Curie-law temperature dependence from 1.4 to 20 K (*24, 25*). These results, which we reproduced in our lab, are best explained if the multiline EPR signal arises from an $S = 1/2$ ground state. The earlier results of de Paula et al. (*23, 26*), indicating that the multiline EPR signal came from an excited state, were artifactual because of microwave power saturation. An $S = 1/2$ ground state could arise from an exchange-coupled mixed-valence Mn dimer, trimer, or tetramer. These possibilities have been compared by EPR spectral simulations (*27*). The breadth of the multiline EPR signal is only accounted for by a Mn tetramer, if one assumes that the S_2 state contains only Mn^{III} and Mn^{IV} ions without large magnetic anisotropy.

The S_2-state $g = 4.1$ EPR signal was originally assumed (*21–23*) to arise from the perpendicular component of an axially symmetric $S = 3/2$ state based on the position of the EPR resonance. However, measurements of the $g = 4.1$ EPR signal at several microwave frequencies (*28*) supports an assignment of the $g = 4.1$ signal to the middle Kramers doublet of an $S = 5/2$ state with near-rhombic symmetry. The orientation dependence of the $g = 4.1$ EPR signal has also been measured in PSII membrane samples that have been oriented by drying onto mylar sheets (*20, 29*). The observation of a least 16 hyperfine lines on the $g = 4.1$ EPR signal from oriented, ammonia-treated PSII membranes requires, with reasonable assumptions, that the $g = 4.1$ signal originates from a cluster of at least three Mn ions. This result, together with the requirement that the S_2-state multiline EPR signal must also arise from a cluster of at least two Mn ions, provides strong support for the proposal that both the g = 4.1 and multiline EPR signals arise from the same Mn cluster.

The remaining question is whether the redox-active center observed by EPR in the S_2 state is a trinuclear or tetranuclear Mn cluster. If the g = 4.1 and multiline EPR signals both arise from a trinuclear cluster, then the remaining Mn ion would have to be Mn^{III} in both the S_1 and S_2 states to escape EPR detection. However, recent EPR spin relaxation studies have shown that the Mn cluster is diamagnetic in the S_1 resting state [(*19*), *see* Structure of the Manganese Complex.]. Because an isolated Mn^{III} ion would be paramagnetic, this result rules out the possibility that the four Mn ions in PSII are arranged as a trimer plus a monomer. One is left with the conclusion from the EPR studies that the Mn ions in PSII are organized into an exchange-coupled tetranuclear complex.

EPR Studies of the S_1 State. The S_1 state is an even-electron state. In such an integer-spin system, it is often the case that large zero-field splittings prevent the observation of conventional EPR signals. Although no conventional EPR signals have been observed from the S_1 state, two approaches have been used to study the S_1 state by EPR spectroscopy. A number of papers have been published on the spin relaxation enhancement of the stable tyrosine radical, $Y_D^•$, by the Mn cluster (*19, 30–33*). These studies provide information on the magnetic properties of the Mn cluster. Another approach is to use parallel-mode EPR spectroscopy, which selects for so-called "$\Delta m_s = 2$" and "$\Delta m_s = 4$" transitions of integer-spin systems (*34*). An integer-spin EPR signal at g = 4.8 has been reported and assigned to Mn in the S_1 state (*18*).

The Mn complex has been found to enhance markedly the spin-lattice relaxation rate of the stable tyrosine radical, $Y_D^•$. Consequently, the spin-lattice relaxation rate of $Y_D^•$ can serve as a probe of the magnetic properties of the Mn complex in each of the S states. A detailed study of the S_1 state was reported (*19*). It was found that the S_1 state can exist in two forms with very different magnetic properties: a resting state or an active state, depending on whether the sample was subjected to long (4 h) or short (6 min) periods of dark adaptation at 0 °C, respectively. The S_1 resting state has a diamagnetic ground state. In addition, from the temperature dependence of the spin-lattice relaxation rate enhancement, it was found that a paramagnetic excited state is populated at higher temperatures. However, the splitting between the ground and excited state is very large (at least 100 cm^{-1}), indicating that the Mn ions are strongly antiferromagnetically exchange coupled in the S_1 resting state. The S_1 active state has a paramagnetic ground or low-lying excited state because its spin-lattice relaxation rate enhancement of $Y_D^•$ was significant even at 4 K. The different magnetic properties of the S_1

active and resting states can be explained by two conformations of the Mn complex that have different exchange couplings between the Mn ions, as has also been invoked to explain the different forms of the S_2 state.

Dexheimer and Klein (*18*) have reported an integer-spin EPR signal at $g = 4.8$ from the S_1 state in long dark-adapted PSII membranes. These dark-adapted samples would be expected to be in the S_1 resting state, which was found to be diamagnetic (*19*) and should not exhibit any EPR signals. However, the parallel-mode EPR signal was very weak; this weakness could be explained if the signal arises from a small amount of the S_1 active state in the long dark-adapted samples. To address the question of the origin of the parallel-mode EPR signal reported by Dexheimer and Klein (*18*), we attempted to measure this signal in both S_1 active-state and S_1 resting-state PSII membrane samples. Measurements were made by using conventional perpendicular-mode EPR (*19*) and also by using parallel-mode EPR (D. Koulougliotis and G. Brudvig, Yale University; M. Hendrick, University of Minnesota, unpublished results). We did not observe any EPR signal at $g = 4.8$, or other signals in the scans from 0 to 5000 G that could be attributed to the S_1 state.

Structure of the Manganese Complex

On the basis of the EPR studies described in the previous paragraphs, the Mn ions in the OEC are determined to be organized into an exchange-coupled tetranuclear cluster. The exchange interactions between the Mn ions must be variable to account for the different types of EPR signals that have been observed from the S_2 state and the different magnetic properties of the S_1 active and resting states. Although the magnitudes of the exchange interactions between the Mn ions are not well-defined by the EPR studies, these couplings must be large enough to produce a splitting between the ground and first excited spin states of over 100 cm^{-1} in the S_1 resting state (*19*) and approximately 30 cm^{-1} in the multiline form of the S_2 state (*21*). The structures for the Mn cluster that are considered must account for such large exchange interactions between the Mn ions.

The magnetic properties of the S_1 resting state place a significant restriction on the possible structures of the Mn complex. A mononuclear Mn ion is ruled out by the observation that the S_1 resting state has a diamagnetic ground state; a mononuclear Mn ion is expected to be paramagnetic in any of its possible oxidation states. This leaves only two possible models for the organization of the Mn ions in the S_1 resting state. Either the Mn ions are arranged as a dimer-of-dimers in which the Mn ions in each dimer have the same oxidation state and are strongly antiferromagnetically exchange-coupled, or the Mn ions are arranged as a tetramer in which strong exchange interactions between the Mn ions produce a net spin of zero in the ground state.

Further information is available from X-ray absorption near-edge structure (XANES) and extended X-ray absorption fine structure (EXAFS) studies. From analyses of the Mn XANES, it is possible to obtain information on the oxidation states of the Mn ions. The Mn XANES of the S_1 state have been analyzed by several groups (*35–39*). The results indicate that the S_1 state does not contain Mn^{II} and most probably contains two Mn^{III} and two Mn^{IV}. An alternate approach for assigning oxidation states involves an analysis of metal-ligand bond lengths. Average metal-ligand bond lengths for the Mn ions in PSII have been obtained from EXAFS studies. These results have been used to calculate that the S_1 state has two Mn^{III} and two Mn^{IV} by using the bond valence sum method (*40*). Thus, it seems probable that the oxidation states of the Mn ions in the S_1 state are $Mn^{III}{}_2Mn^{IV}{}_2$. However, the analysis of the XANES data and the bond valence sum method both require a comparison to model complexes. Because the ligation of the Mn ions in PSII is not known, it is also possible that the S_1 state has four Mn^{III}.

Mn EXAFS data provide more detailed structural information. Several groups have collected and analyzed Mn EXAFS data from the S_1 state (*35, 36, 38, 39*). Although some disagreements remain, a consensus seems to be emerging on several points. Perhaps most significant is the observation of a Mn backscatterer at about 2.7 Å. Simulations of the EXAFS data indicate that each Mn atom has 1–1.5 neighboring Mn atoms at this distance. This distance is significant because, to date, it has only been observed in model complexes having di-μ_2-oxo or di-μ_3-oxo bridges between two Mn ions (*41*). It seems probable that two or three di-μ-oxo bridged Mn_2 units exist in PSII. Another heavy-atom backscatterer has been identified at 3.3 Å. It remains unclear, however, whether this backscatterer is a Mn ion, Ca^{2+}, or both. Based on these EXAFS data, a dimer-of-dimers model for the Mn cluster in the S_1 state has been proposed (*39*) in which two di-μ-oxo bridged Mn dimers (each having a 2.7 Å Mn–Mn separation) are connected via a single μ-oxo bridge (giving a 3.3 Å separation between two of the Mn ions). Alternatively, the EXAFS data could be explained by a distorted cubane-like tetranuclear Mn complex. To compare these two possibilities, one can consider the results of studies of Mn model complexes.

One important consideration is the type of structure that could produce the strong antiferromagnetic exchange interactions needed to account for the magnetic properties of the S_1 and S_2 states. There is a substantial body of data on the magnetic properties of binuclear Mn complexes. It has been observed that di-μ-oxo bridged Mn dimers have antiferromagnetic exchange couplings that decrease exponentially with the distance between the two Mn ions; this could be explained if direct exchange makes an important contribution for these structures (*41*). All examples of di-μ-oxo bridged Mn^{IV} dimers have large antiferromagnetic exchange interactions. However, there are very few examples of di-μ-oxo bridged Mn^{III} dimers because the ability of Mn^{III} to support a di-μ-oxo ligation requires that the ancillary ligands be poor donors. Magnetic data are available only for the $[Mn^{III}_2(O)_2(6\text{-}Me_2bispicen)_2]^{2+}$ complex (*42*), which shows fairly strong antiferromagnetic exchange coupling ($J = 172.8\ cm^{-1}$, $H = JS_1 \cdot S_2$). However, other di-μ-oxo bridged Mn^{III} dimers are also found as units in higher nuclearity Mn complexes. In such systems for which magnetic studies have been done, the exchange coupling between the Mn^{III} ions is weak (reviewed in reference 41). The exchange interactions are also invariably weak between Mn ions that are bridged by groups other than the di-μ-oxo ligands.

These results place some constraints on the dimer-of-dimers model. To produce the magnetic properties of the S_1 resting state with a $Mn^{III}_2Mn^{IV}_2$ system, one dimer should be Mn^{III}_2 and the other Mn^{IV}_2. If both dimers have a di-μ-oxo bridged structure, as indicated from the EXAFS data, then the required magnetic properties should be present. However, the remaining ligands to the Mn^{III} pair would have to be poor donors to account for the stabilization of the di-μ-oxo bridged structure at the Mn^{III} oxidation level. A combination of poor donor ligands to one pair of Mn and stronger donor ligands to the other pair could explain the necessary asymmetric arrangement of the oxidation states among two Mn dimers. It is unclear, however, whether a di-μ-oxo bridged Mn^{III} dimer could be bridged to a Mn^{IV} dimer via an oxo group as proposed by Yachandra et al. (*39*). This would place three oxo ligands on one Mn^{III}, which would be unfavorable unless this Mn^{III} were less than six-coordinate or were coordinated to especially poor donor ligands. It is also possible to explain the different magnetic properties of the S_1 resting and active states with a dimer-of-dimers model. A small conformational change of the OEC could cause the ligands to the Mn^{III} pair to become better donor ligands. This could lead to protonation of one of the oxo bridges between the Mn^{III} ions, which would greatly weaken their exchange coupling and could give rise to a paramagnetic state.

Much less information is available on the magnetic properties of other tetranuclear structures, such as a distorted cubane-like structure, that could account for the spectroscopic data of the OEC. A series of distorted cubane-like high-valent Mn complexes having a $Mn^{IV}Mn^{III}_3O_3Cl$ core have been characterized (reviewed in reference 41). The magnetic properties of the $Mn_4O_3Cl_4(O_2CCH_3)(py)_3$ complex, which has C_3 symmetry, have been analyzed in detail (*43*). Although such a complex with C_3 symmetry can, in principle, have a ground state with any spin from $S = 1/2$ to $S = 15/2$, depending on the values of the exchange interactions, the $Mn_4O_3Cl_4(O_2CCH_3)(py)_3$ complex has a ground state with $S = 9/2$. This complex has an oxidation state that is equal to the probable oxidation state of the S_0 state. Unfortunately, the magnetic properties of the S_0 state are not well-characterized. Therefore, it is not possible to make a very definitive comparison between the model complex data and possible tetranuclear structures of the Mn complex in PSII. Further work to prepare and characterize new model complexes and to better characterize the magnetic properties of the S_0 state is needed.

Acknowledgments

I thank Dionysios Koulougliotis and Michael Hendrick for making the parallel-mode EPR measurements on PSII and Chao Lin for help with preparation of the figures. This work was supported by the National Institutes of Health (GM32715 and GM36442).

References

1. Babcock, G. T. In *New Comprehensive Biochemistry: Photosynthesis;* Amesz, J., Ed.; Elsevier: Amsterdam, 1987; pp 125–158.
2. Andréasson, L.-E.; Vänngård, T. *Annu. Rev. Plant Physiol. Plant Mol. Biol.* **1988,** *39,* 379–411.
3. Brudvig, G. W.; Beck, W. F.; de Paula, J. C. *Annu. Rev. Biophys. Biophys. Chem.* **1989,** *18,* 25–46.
4. Debus, R. J. *Biochim. Biophys. Acta* **1992,** *1102,* 269–352.
5. Rutherford, A. W.; Zimmermann, J.-L.; Boussac, A. In *The Photosystems: Structure, Function and Molecular Biology;* Barber, J., Ed.; Elsevier Science Publishers B. V.: Cedex, France, 1992; pp 179–229.
6. Joliot, P.; Barbieri, G.; Chabaud, R. *Photochem. Photobiol.* **1969,** *10,* 309–329.
7. Kok, B.; Forbush, B.; McGloin, M. *Photochem. Photobiol.* **1970,** *11,* 457–475.
8. Forbush, B.; Kok, B.; McGloin, M. P. *Photochem. Photobiol.* **1971,** *14,* 307–321.
9. Ono, T.-a.; Noguchi, T.; Inoue, Y.; Kusunoki, M.; Matsushita, T.; Oyanagi, H. *Science (Washington, D.C.)* **1992,** *258,* 1335–1337.
10. Styring, S.; Rutherford, A. W. *Biochemistry* **1987,** *26,* 2401–2405.
11. Buser, C. A.; Diner, B. A.; Brudvig, G. W. *Biochemistry* **1992,** *31,* 11449–11459.
12. Berthold, D. A.; Babcock, G. T.; Yocum, C. F. *FEBS Lett.* **1981,** *134,* 231–234.
13. Draber, W.; Tietjen, K.; Kluth, J. F.; Trebst, A. *Angew. Chem., Int. Ed. Engl.* **1991,** *30,* 1621–1633.
14. Beck, W. F.; de Paula, J. C.; Brudvig, G. W. *Biochemistry* **1985,** *24,* 3035–3043.
15. Brudvig, G. W.; Casey, J. L.; Sauer, K. *Biochim. Biophys. Acta* **1983,** *723,* 366–371.
16. Bocarsly, J. R.; Brudvig, G. W. *J. Am. Chem. Soc.* **1992,** *114,* 9762–9767.
17. Miller, A.-F.; Brudvig, G. W. *Biochim. Biophys. Acta* **1991,** *1056,* 1–18.
18. Dexheimer, S. L.; Klein, M. P. *J. Am. Chem. Soc.* **1992,** *114,* 2821–2826.

19. Koulougliotis, D.; Hirsh, D. J.; Brudvig, G. W. *J. Am. Chem. Soc.* **1992,** *114,* 8322–8323.
20. Kim, D. H.; Britt, R. D.; Klein, M. P.; Sauer, K. *J. Am. Chem. Soc.* **1990,** *112,* 9389–9391.
21. Hansson, Ö.; Aasa, R.; Vänngård, T. *Biophys. J.* **1987,** *51,* 825–832.
22. de Paula, J. C.; Innes, J. B.; Brudvig, G. W. *Biochemistry* **1985,** *24,* 8114–8120.
23. de Paula, J. C.; Beck, W. F.; Brudvig, G. W. *J. Am. Chem. Soc.* **1986,** *108,* 4002–4009.
24. Aasa, R.; Andréasson, L.-E.; Lagenfelt, G.; Vänngård, T. *FEBS Lett.* **1987,** *221,* 245–248.
25. Britt, R. D.; Lorigan, G. A.; Sauer, K.; Klein, M. P.; Zimmermann, J.-L. *Biochim. Biophys. Acta* **1992,** *1040,* 95–101.
26. de Paula, J. C.; Brudvig, G. W. *J. Am. Chem. Soc.* **1985,** *107,* 2643–2648.
27. Bonvoisin, J.; Blondin, G.; Girerd, J.-J.; Zimmermann, J.-L. *Biophys. J.* **1992,** *61,* 1076–1086.
28. Vänngård, T.; Hansson, Ö.; Haddy, A. In *Manganese Redox Enzymes;* Pecoraro, V. L., Ed.; VCH Publishers: New York, 1992; pp 105–118.
29. Rutherford, A. W. *Biochim. Biophys. Acta* **1985,** *807,* 189–201.
30. Styring, S.; Rutherford, A. W. *Biochemistry* **1988,** *27,* 4915–4923.
31. Innes, J. B.; Brudvig, G. W. *Biochemistry* **1989,** *28,* 1116–1125.
32. Evelo, R. G.; Styring, S.; Rutherford, A. W.; Hoff, A. J. *Biochim. Biophys. Acta* **1989,** *973,* 428–442.
33. Beck, W. F.; Innes, J. B.; Brudvig, G. W. In *Current Research in Photosynthesis;* Baltscheffsky, M., Ed.; Kluwer Academic Publishers: Dordrecht, 1990; pp 817–820.
34. Hendrick, M. P.; Debrunner, P. G. *J. Magn. Reson.* **1988,** *78,* 133–141.
35. George, G. N.; Prince, R. C.; Cramer, S. P. *Science (Washington, D.C.)* **1989,** *243,* 789–791.
36. Penner-Hahn, J. E.; Fronko, R. M.; Pecoraro, V. L.; Yocum, C. F.; Betts, S. D.; Bowlby, N. R. *J. Am. Chem. Soc.* **1990,** *112,* 2549–2557.
37. Kusunoki, M.; Ono, T.; Matsushita, T.; Oyanagi, H.; Inoue, Y. *J. Biochem.* **1990,** *108,* 560–567.
38. MacLachlan, D. J.; Hallahan, B. J.; Ruffle, S. V.; Nugent, J. H. A.; Evans, M. C. W.; Strange, R. W.; Hasnain, S. S. *J. Biochem.* **1992,** *285,* 569–576.
39. Yachandra, V. K.; DeRose, V. J.; Latimer, M. J.; Mukerji, I.; Sauer, K.; Klein, M. P. *Science (Washington, D.C.)* **1993,** *260,* 675–679.
40. Thorp, H. H. *Inorg. Chem.* **1992,** *31,* 1585–1588.
41. Thorp, H. H.; Brudvig, G. W. *New J. Chem.* **1991,** *15,* 479–490.
42. Goodson, P. A.; Oki, A. R.; Glerup, J.; Hodgson, D. J. *J. Am. Chem. Soc.* **1990,** *112,* 6248–6254.
43. Schmitt, E. A.; Noodleman, L.; Baerends, E. J.; Hendrickson, D. N. *J. Am. Chem. Soc.* **1992,** *114,* 6109–6119.
44. Lin, C.; Brudvig, G. W. *Photosynth. Res.* **1993,** *38,* 441–448.
45. Brudvig, G. W. In *Metal Clusters in Proteins*; Que, L., Jr., Ed.; ACS Symposium Series 372; American Chemical Society: Washington, DC, 1988; pp 221–237.

RECEIVED for review August 11, 1993. ACCEPTED revised manuscript December 27, 1993.

10

Reactivity and Mechanism of Manganese Enzymes

A Modeling Approach

Vincent L. Pecoraro, Andrew Gelasco, and Michael J. Baldwin

The Willard H. Dow Laboratories, Department of Chemistry,
The University of Michigan, Ann Arbor, MI 48109–1055

Manganese satisfies a wide variety of biological activities that range from Lewis acid–base reactions to redox catalysis. This chapter summarizes the proposed chemical mechanisms for a wide variety of manganese-based enzymes that fulfill both nonredox and redox roles. We follow this minireview of manganese enzymes with specific examples drawn from our laboratory showing how small molecule coordination complexes can be prepared that mimic structural, spectroscopic, and reactivity properties of two redox-based manganese enzymes: the manganese catalase and the oxygen evolving complex. We describe two relatively simple systems composed of manganese dimers that catalyze H_2O_2 disproportionation using either low-valent ($Mn(II)_2 \rightarrow Mn(III)_2$) or high-valent ($Mn(III)_2 \rightarrow Mn(IV)_2$) cycles. From this work we illustrate how understanding the mechanism of well-defined model compounds may provide new mechanistic insight to more complex and ill-defined biological catalytic centers.

OVER THE PAST DECADE inorganic chemists, biochemists, and biophysicists have gained an enhanced appreciation for the role of manganese in the biosphere. Although once a curiosity, manganese is now recognized to be an essential component in photosynthesis and is found to be an alternative to iron in a wide variety of redox enzymes. In addition, the metal is an excellent Lewis acid catalyst and often confers stability to protein structure. In this chapter we summarize the known chemistry of manganese in biological systems and in compounds that have been synthesized in our laboratory to mimic the reactivity of these fascinating

0065–2393/95/0246–0265/$11.24/0

proteins as an example of the modeling approach for understanding their mechanisms.

Biological Chemistry of Manganese

Manganese plays an essential and versatile role in the biochemistry of many microorganisms, plants, and animals. Manganese is considered an essential trace element in humans with total body metal being ~20 mg (*1, 2*). In plants and oxygen-evolving photosynthetic bacteria, manganese is an essential component of the oxidative end of photosynthesis. Numerous bacterial sources use manganese in place of iron in reactions involving oxygen detoxification. Manganese is an extremely versatile element in biology because this element is relatively abundant, has multiple accessible redox states, and, in the +2 oxidation level, is similar in size and properties to Mg(II). Therefore, manganese may serve as a Lewis acid catalyst or be directly involved in multielectron redox conversions. In the following section we describe the varied functions of manganoenzymes and cofactors placing particular emphasis on the mechanistic aspects of this chemistry. Enzymes that use manganese in a purely structural role will not be discussed.

The simplest use of manganese is as part of a metal cofactor such as manganese(II) adenosine triphosphate, [Mn(II)ATP^{2-}]. The metal complex is usually found as the bidentate [Mn(II)β,γ-ATP^{2-}] (*3, 4*) as illustrated in Figure 1. Enzyme specificity for Mg(II), Mn(II), or Ca(II) ATP complexes is dependent on a variety of factors; however, once selected, each metal apparently functions in an analogous manner. The metal ion serves two main purposes. First, based on the coordination geometry

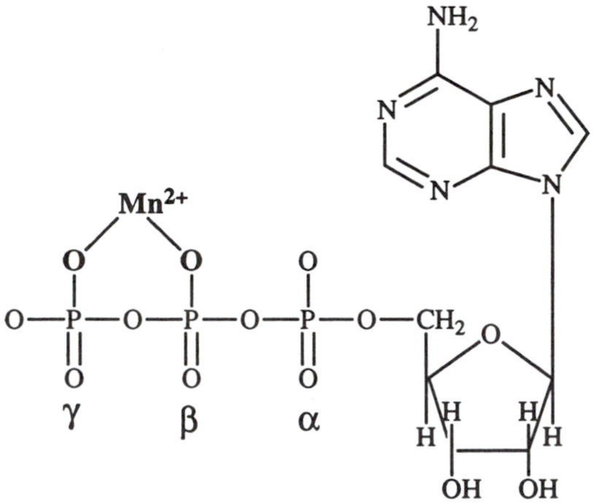

Figure 1. β,γ-bidentate Mn(II) adenosine triphosphate dianion. The Mn(II) coordination sphere includes four additional water molecules.

of the complex, the phosphate oxygens and the bound water molecules can form hydrogen bonds that stabilize a specific conformation of the $M(II)ATP^{2-}$ complex. This is important in leading to productive binding of the $M(II)ATP^{2-}$ to allow phosphoryl group transfer. Second, the metal polarizes the O–P bond so as to move electron density away from the phosphorus atom. The generation of this δ^+ makes the phosphorus a better site for attack by an entering nucleophile such as an alcohol, amine, or phenol. This S_N2 mechanism leads to a trigonal bipyramidal phosphorus intermediate that collapses to $[Mn(II)(H_2O)_4(\beta\text{-}ADP)(O_3POR)]^{3-}$ and subsequently forms in solution the products O_3POR^{2-} and $[Mn(II)(H_2O)_4(\alpha,\beta\text{-}ADP)]^-$. The reversibility of a kinase (phosphoryl transfer enzyme) is correlated with the metal ATP complex that is recognized as a substrate (*4*). Irreversible enzymes such as hexokinase (reaction: glucose + $[M(II)ATP]^{2-}$ → glucose-6-phosphate + $[M(II)ADP)]^-$) use bidentate $[M(II)\beta,\gamma\text{-}ATP]^{2-}$ as substrate in the forward reaction and monodentate $[M(II)\beta\text{-}ADP]^-$ for the reverse reaction. This substrate preference strongly drives the reaction toward products because the equilibrium concentration of $[M(II)\beta,\gamma\text{-}ATP]^{2-}$ is large, whereas the concentration of $[M(II)\beta\text{-}ADP)]^-$ is extremely small. In contrast, reversible enzymes such as creatine kinase (reaction: creatine + $[M(II)ATP]^{2-}$ → phosphocreatine + $[M(II)ADP)]^-$) use tridentate $[M(II)\alpha,\beta,\gamma\text{-}ATP]^{2-}$ in the forward reaction and $[M(II)(\alpha,\beta\text{-}ADP)]^-$ in the reverse reaction. In this case the concentration of $[M(II)(\alpha,\beta\text{-}ADP)]^-$ exceeds that of $[\alpha,\beta,\gamma$-tridentate $[M(II)ATP]^{2-}$. Another example of α,β,γ-tridentate $[M(II)ATP]^{2-}$ is the β-subunit of the chloroplast ATP synthase, which apparently recognizes tridentate Mn(II)ATP as substrate.

The interaction of manganese or magnesium with nucleic acids provides additional interesting examples of charge neutralization or polarization effects of the type described in the preceding paragraph. Cech (*5*) has shown that metal cofactors are needed for the proper functioning of *ribozymes*. In addition to a possible structural role, it is now established that either Mn(II) or Mg(II) will function as essential cofactors for the cleavage of exogenous RNA or DNA by the *Tetrahymena* ribozyme. Apparently, the M(II) cation binds directly to the ribozyme in the active site, serving to stabilize the negative charge on the oxygen atom of the 3′O–P bond in the transition state. Thus, it was concluded that the ribozymes are metalloenzymes that have mechanistic characteristics that are similar to protein enzymes. Although Mn(II) may form direct inner sphere complexes with nucleotides, DNA, or RNA, there are cases in which charge, not Lewis acidity, is the dominant factor for catalysis. Cowan (*6*) suggested that in some cases, the only apparent function of the metal is charge neutralization. In this case, the fact that the topoisomerase I hydrolysis of DNA could be affected by $Co(NH_3)_6{}^{3+}$ in place of Mg(II) suggests that an inner sphere complex is not necessary for catalysis. Rather, the appropriate functional groups can be oriented in close proximity once the phosphate backbone is partially neutralized.

Although the majority of proteins containing multinuclear centers composed partially or completely with manganese are involved in redox reactions, there are a significant number of substrate binding or hydrolytic activities that rely on metal clusters that include manganese. Rat liver *arginase* catalyzes the hydrolysis of L-arginine to give L-ornithine and urea. Manganese(II) is the essential cofactor for this enzyme (*7*, *8*). Recent electron paramagnetic resonance (EPR) studies conclusively demonstrate that this enzyme has a dinuclear active site that binds two exchange-coupled Mn(II) ions (*9*). A more detailed understanding of the structure and mechanism of these hydrolytic enzymes has been gathered for *glutamine synthetase*. An X-ray structure of this enzyme reveals that the two Mn(II) ions are in close proximity (separation 6 Å) with asymmetric ligation (*10*). The first manganese ion has three coordinated carboxylate moieties, whereas the second has mixed ligation using two carboxylates and imidazole. There is no obvious endogenous or exogenous ligand that bridges the two ions such as one observes in the X-ray structure of *conconavalin A* (*11*). Glutamine synthetase exhibits a ping-pong kinetic mechanism in which glutamate is first phosphorylated by ATP, forming a γ-glutamyl phosphate intermediate and ADP (*12*). This activated phosphate group is then displaced by ammonia, forming glutamine and inorganic phosphate. The precise role of manganese in this process has not been fully explained but is likely to be as a Lewis acid to make the substrate more nucleophilic.

Manganese Redox Enzymes

There are numerous enzymes that exploit the redox capabilities of manganese. Included in this group are the Mn-superoxide dismutases, catalases, manganese peroxidases, manganese ribonucleotide reductase, and perhaps the most complex and important manganoenzyme, the oxygen-evolving complex (OEC). Many of these enzymes act directly, or peripherally, with dioxygen or one of its reduced forms such as peroxide or superoxide. In each of these cases, the stoichiometry of electron equivalents equals the number of manganese found in the enzyme. A brief description that emphasizes the present understanding of the catalytic mechanism of these systems is now presented, alphabetically, by enzyme.

Mn Catalase. The majority of known catalases contain a heme prosthetic group to catalyze the disproportionation of hydrogen peroxide. These enzymes are long established (e.g., the beef liver catalase was crystallized in 1937 (*13*)) and are believed to protect respiring cells from unwanted oxidative damage that inevitably arises from the Fenton chemistry that is initiated by hydrogen peroxide. For the iron enzymes, the presently accepted mechanism (*14*) involves the oxidation of an Fe(III) porphyrin by hydrogen peroxide to a ferryl π-cation radical (often abbreviated $(Fe^{IV}{=}O)P^{+\bullet}$) called "compound 1." This highly oxidized enzyme form subsequently reacts with another molecule of hydrogen peroxide to generate oxygen and regenerate the Fe(III) species. Neutral hydrogen peroxide is the substrate that is bound (as opposed to HO_2^- or O_2^{2-}); however, an active-site acid–base catalyst serves to deprotonate the hydrogen peroxide in the first step of the reaction to form an $Fe(III)P(O_2H)$ complex that collapses to compound 1 and releases water. Heme-based catalases are particularly sensitive to inhibition by micromolar concentrations of azide or cyanide that efficiently bind to the iron porphyrin.

It was known for many years that some bacteria contain catalases that are insensitive to these inhibitors. In 1983, a nonheme catalase containing manganese was isolated and purified from *Lactobacillus plantarum* (*15, 16*). Since this discovery two other similar manganese catalases have been isolated from *Thermoleophilum album* (*17*) and *Thermus thermophilus* (*18*). The *T. thermophilus* enzyme has since been crystallized and a low-resolution structure demonstrates that two manganese ions are in very close proximity being separated by $\approx$3.6 Å (*18, 19*). The *T. thermophilus* and the *L. plantarum* enzymes have been studied using EPR spectroscopy, and in both cases a weakly exchange-coupled dinuclear site has been confirmed. EPR spectroscopic signatures for Mn(II,II), Mn(II,III), and Mn(III,IV) forms have been observed (*20*). The X-ray structure of the *T. thermophilus* enzyme is most likely in the Mn(II,II) form, although Mn(II,III) or Mn(III,III) formulations cannot be excluded on the basis of published data. X-ray absorption spectra of the Mn(III,IV) form conclusively demonstrate that the manganese ions are in much closer proximity ($\approx$2.7 Å). This close proximity suggests that the metals have a di-μ_2-oxo structure (*21*). The low valent forms of the *L. plantarum* and *T. thermophilus* enzymes exhibit optical absorption spectra similar to that seen for Mn(III) model complexes with a broad, weak band at 450–500 nm. On the basis of these data, a bis-μ-carboxylato, μ-oxo-bridged structure has been postulated for the active site of the Mn(III,III) enzyme form.

Considering all of the manganese catalases together, there have been four cluster oxidation levels that are established: Mn(II,II), Mn(II,III), Mn(III,III), and Mn(III,IV). The as-isolated enzyme contains a mixture of these states. The Mn(II,II) enzyme can be prepared by the addition of hydroxylamine to the isolated enzyme. If hydrogen peroxide is added to this sample, without removing the hydroxylamine, the enzyme is converted to the Mn(III,IV) form; however, if the hydroxylamine is first removed, the enzyme turns over normally and large amounts of the Mn(III,IV) form are not detected. A minimal four-step mechanism has been proposed by Penner-Hahn (22) for the catalytic cycle as shown in Figure 2. The Mn(II,II) enzyme first releases water as it binds hydrogen peroxide. The substrate is coordinated directly to the manganese as the hydroperoxide anion with the released proton being absorbed by an active-site acid–base catalyst. The cluster is next oxidized by two elec-

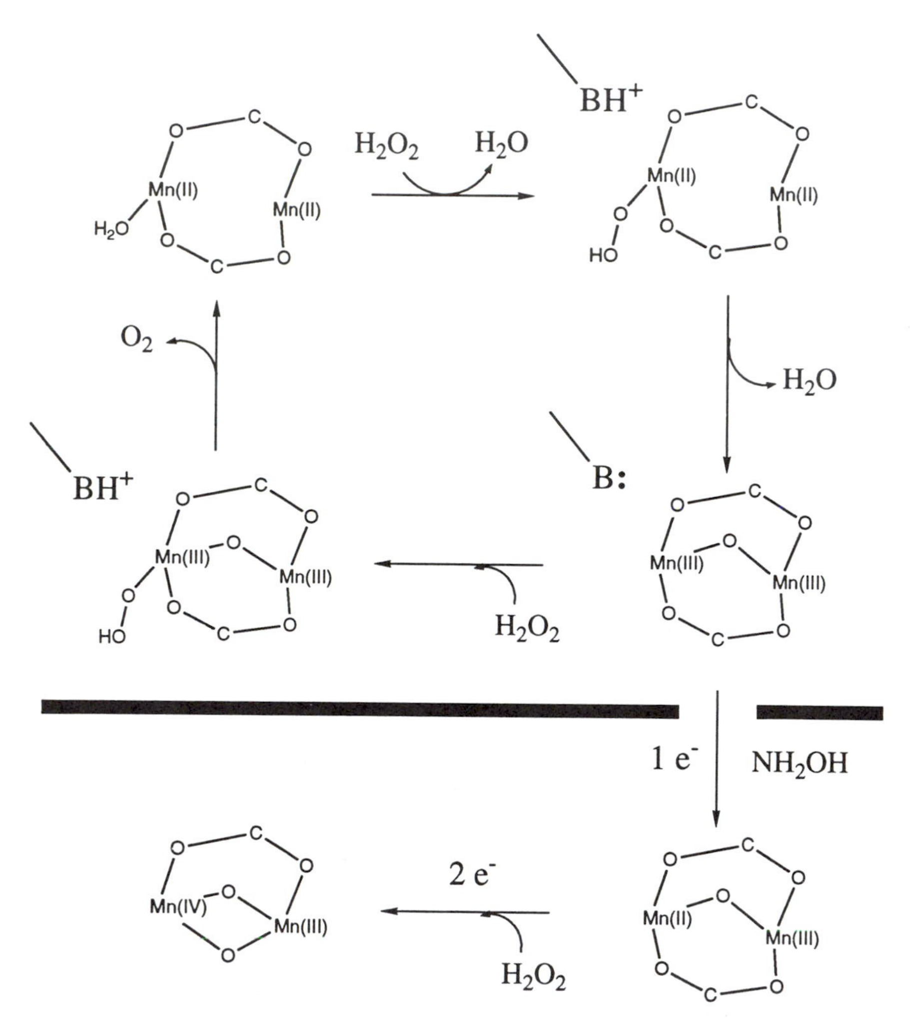

Figure 2. The four-step mechanism for hydrogen peroxide disproportionation by the Mn catalse proposed by Penner-Hahn (22) is shown above the solid line. Our proposed mechanism for the inactivation of the Mn catalase by hydroxylamine in the presence of hydrogen peroxide is shown below the solid line.

trons to generate a bis-μ-carboxylato, μ-oxo-bridged Mn(III,III) structure. The enzyme concurrently releases the first equivalent of water. A second equivalent of hydrogen peroxide is then bound (again as the hydroperoxide anion) and subsequently oxidized to dioxygen. The second equivalent of water is also released at this point.

The Mn(III,IV) enzyme, generated during turnover conditions when hydroxylamine is present, is not catalytically active. One possible path for the formation of this "superoxidized" form is through a Mn(II,III) intermediate as illustrated in the lower part of Figure 2. Hydroxylamine may act as both an oxidant and a reductant. Additionally, it may serve as either a one- or two-electron transfer agent. If, during the course of the reaction, hydroxylamine intercepts the functional two-electron cycle between Mn(II,II) and Mn(III,III) to generate through a one-electron reaction a Mn(II,III) form, then subsequent two-electron oxidation of this species by hydrogen peroxide will result in the inactive Mn(III,IV) enzyme (*see* subsequent paragraphs).

Although manganese catalases have often been referred to as "azide insensitive," these enzymes actually are inhibited by azide and related molecules albeit at higher concentrations than are necessary for the heme enzymes. Penner-Hahn and co-workers (*22*) have shown that HN_3 is the likely protonation state of the inhibitor and have calculated an apparent K_i of 80 mM. Slope replots of the pH dependence of azide inhibition are linear with a slope of 1. These data can be used to calculate a true K_i of $\approx$300 mM. Because azide is a competitive inhibitor with respect to peroxide, it is likely that azide is bound directly to the manganese center. Recent EPR and ^{1}H paramagnetic relaxation enhancement studies support this viewpoint. Other inhibitors include fluoride and thiocyanide. All of the reported inhibition studies are consistent with the catalase cycle and hydroxylamine inhibition of the catalase cycle.

Oxygen-Evolving Complex. The enzyme responsible for the light-driven oxidation of water to molecular oxygen in photosynthesis can be found in the thylakoid membrane of chloroplasts in algae, green plants, and some cyanobacteria. The OEC forms a part of photosystem II (PSII), which consists of a complex set of polypeptides containing light-harvesting pigments, a water oxidation center, and electron-transfer components. The site of water oxidation contains four manganese ions, probably associated in a single center, and may also include calcium and chloride ions. Although manganese is almost certainly necessary for the oxidizing equivalents to convert water to dioxygen, the role of calcium and chloride ions for proper functioning of the OEC is unclear (*23*).

The kinetic mechanism of the OEC was deduced by Bessel Kok (*24*), who used an oxygen electrode to monitor yields of dioxygen associated with single flashes of light. The OEC exhibits a periodicity in oxygen evolution corresponding to one molecule of molecular oxygen released every four quanta of light absorbed by the light-harvesting complex. Five distinct states (S states) of the complex can be identified and are designated as S_0–S_4, going from the most reduced (S_0) to oxidized (S_4) forms of the enzyme. It is now believed that each S-state transition is associated with a change in redox state of the manganese. Proton release within the thylakoids has been shown to be oscillatory, apparently following the Kok S-state cycle. Certain complications, such as the possibility of proton release under conditions when no oxygen is evolved, make determination of the stoichiometry of proton release difficult. The initial experiment using the amphiphilic dye Neutral red indicated that a 1:0:1:2 stoichiometry of proton release occurs through the $S_0 \rightarrow S_1$, $S_1 \rightarrow S_2$, $S_2 \rightarrow S_3$, $S_3 \rightarrow S_4$, S_0 cycle in rigorously dark-adapted sites (*25*, *26*). However a reinterpretation of the proton release experiments with Neutral red has implicated a strongly pH-dependent noninteger stoichiometry (*27*). In yet another experiment Jahns et al. (*28*) found a pattern of 1:1:1:1.

The dark-stable resting state is the S_1 oxidation level. A single flash of light leads to the S_2 state, which has been intensively studied because of its ease of access and EPR spectral features. The EPR spectrum obtained at liquid helium temperatures of the S_2 state shows a 19–26 line signal centered at $g \approx 2$. The breadth of the signal, the magnitude of the nuclear hyperfine coupling, and the energy of the transition conclusively identifies this signal as arising from a mixed-valence manganese cluster (*29*). The observation of this signal was the first demonstration that at least some of the manganese form a cluster in S_2. The multiline EPR signal exhibits the same four-flash periodicity as observed for the oxygen yield, but is displaced by two oxidizing equivalents within the cycle (i.e., O_2 production follows $S_0 \rightarrow S_4$, whereas multiline intensity cycles at S_2). This observation directly links the manganese ions to the oxidative chemistry of this enzyme. Recent X-ray absorption data (*30*, *31*) also implicates manganese in the redox cycle. The presently accepted oxidation states for the manganese ions in each oxidation level, as well as likely candidates for metal ligands are provided in Table I.

Another EPR signal at $g \approx 4.1$ can be observed under certain conditions (e.g., adding ammonia to the sample) (*32–36*). This signal can be made to show a multiline feature, which suggests that it is also part of a cluster of manganese (*34*). Extended X-ray absorption fine structure (EXAFS) studies have been performed by various groups on numerous S states of the enzyme. This method yields detailed information about the first coordination sphere of manganese. It is generally agreed that at least two Mn–Mn vectors at 2.7 Å are present as well as a 3.3-Å vector corresponding to a Mn-M (Mn or Ca) interaction in S_1. Also, the Mn atoms are bound to either oxygen or nitrogen donors with distances

Table I. Summary of Conclusions for Mn in OEC

Spin	*Oxidation State*	*Manganese Separation*	*Ligands*
S_0	$Mn^{III}{}_3Mn^{IV}{}_1$	—	—
S_1	$Mn^{III}{}_2Mn^{IV}{}_2$	2.7 Å Mn–Mn 3.3 Å Mn–Mn or Mn–Ca	O, N: carboxylates, imidazole
S_2	$Mn^{III}{}_1Mn^{IV}{}_3$	2.7 Å Mn–Mn 3.3 Å Mn–Mn or Mn–Ca	O, N: carboxylates, imidazole, chloride?
S_3	$Mn^{IV}{}_4$	?	O, N: carboxylates, imidazole, chloride?, water
S_4	$Mn^{IV}{}_3Mn^{V}{}_1$	?	?

between 1.8 and 1.9 Å. The 2.7-Å distance further implicates a cluster of manganese because such close Mn–Mn separations are only observed in model compounds that have structural units such as $[Mn(\mu_2\text{-}O)]_2$ (*36*). Nonprotein-based ligands to the manganese, in at least some of the S states, almost certainly include water and possibly chloride. To determine whether nonexchangeable water-derived ligands are bound to the Mn site prior to the oxygen evolution step, experiments were performed in which thylakoid membranes are prepared in $H_2{}^{18}O$ (in the S_1 state) and then washed with $H_2{}^{16}O$, after subsequent flashes of light. The evolved oxygen was then checked for isotopic content by mass spectrometry. In cases in which the $H_2{}^{16}O$ wash was done after either one or two preflashes, no $^{18}O_2$ was found in the evolved oxygen (*37*). This implies that partially oxidized intermediates are not formed prior to the $S_4 \rightarrow S_0 + O_2$ conversion. The reader should be aware that this experiment does not rule out water ligands binding to manganese at earlier stages of the cluster oxidation (in rapid exchange with the bulk solvent) but does suggest that no oxidized forms of water (peroxide or superoxide) are observed before or at the S_3 step.

Although manganese is almost certainly the site of redox advancement in the OEC, water oxidation will not occur unless both Ca^{2+} and Cl^- ions are present (*38*). Other nonphysiological ions (e.g., Sr^{2+} and Br^-) can substitute for these ions, but lead to far lower activities. In the absence of Ca^{2+} or Cl^- the manganese ions may still be oxidized, leading to S-state advancement; however, O_2 is not formed. Proposals for both calcium and chloride being directly associated with the manganese cluster have been advanced (*39*). Unfortunately, there is no compelling data to verify the propositions. Little is known of the function of calcium and chloride. Suggestions for the role of calcium have included maintaining structural integrity, modifying the redox potential of the manganese cluster, and binding and then activating water for catalysis. The role of chloride is equally obscure and may include stabilizing the manganese cluster, acting as an inner sphere electron-transfer mediator and protecting the cluster from unwanted oxidative side reactions (*see* subsequent paragraphs).

Inhibition of the "S" cycle has been shown with substituted amines by causing a two-flash delay of the cycle. The reaction of NH_2OH with dark-adapted PSII samples in the S_1 state leads to a two-flash delayed S_1 state (*40*), with reduced Mn centers (*41, 42*). Higher concentrations of NH_2OH lead to the irreversible reduction and loss of three of four Mn ions. Two of these ions are released cooperatively, probably from the same reaction site. Another binding site, presumably the ferro-semiquinone acceptor site, can be hit by very high concentrations (≥ 6 NH_2OH/PSII) and cause irreversible structural change (*42*). Hydroquinone can also reduce the manganese in a reversible manner; however, recent extended X-ray absorption fine structure spectra show that the site of reduction may be different for hydroxylamine and hydroquinone (*41, 43*). These two reductants act synergistically at low concentrations to deactivate the enzyme completely. The use of NH_2NH_2, a two-electron reductant, has also been shown to reduce the system by a two-flash delay, although this mechanism is less well-understood. Unlike hydroxylamine, hydroquinone, or hydrazine, ammonia has been shown to inhibit the enzyme in a nonredox fashion. Ammonia will bind reversibly to Mn cluster in the S_2 state of PSII leading to a new EPR signal at $g \approx 4.1$ (*44*). When oriented samples of this protein derivative are examined, hyperfine structure can be observed. These hyperfine structures suggest that this low-field EPR signal is associated with a cluster of manganese ions (*34*). That ammonia binding is competitive with Cl^- and inhibits oxygen evolution might suggest that chloride also binds directly to the manganese cluster; however, ammonia apparently inhibits the enzyme in a noncompetitive manner versus chloride, making interpretations more difficult.

Frasch (*45, 46*) has shown that the OEC can catalyze an azide-insensitive catalase reaction in the dark. The activity can be directly associated with the OEC because (1) competitive inhibitors of water oxidation are also competitive inhibitors of the catalase activity and (2) the K_i for water oxidation and catalase activity are essentially identical. The enzyme apparently cycles in this case between S_0 and S_2. Mano and co-workers (*47*) showed that the S_1/S_{-1} states are also competent to carry out catalase reactions; however, this reaction is highly pH-dependent. For example, at pH 8.8, the S_1 state can oxidize H_2O_2 to O_2, but S_{-1} is incapable of completing the reaction cycle; however, if the pH is lowered to pH 6 steady-state measurements of oxygen evolution can be gathered. Just as is the case with water oxidation, these catalase reactions are accelerated by Ca^{2+}. Steady-state kinetic analyses of both S_{-1}/S_1 and S_0/S_2 cycles support an ordered mechanism with Ca^{2+} binding prior to hydrogen peroxide. The Michaelis constant for H_2O_2 (177 mM) is comparable to the K_m for the Mn catalase.

Numerous workers have suggested that H_2O_2 may be generated as an intermediate during the water oxidation reactions of photosynthesis. The $H_2{}^{18}O$ experiments of Radmer (*37*) speak against this proposition, at least for the lower S states. However, there is an accumulating body of evidence that indicates that the OEC can, under certain circumstances, generate H_2O_2 by light-driven reactions. Depletion of chloride at pH 7.2 leads to optimal hydrogen peroxide production and virtually no O_2 evolution. In contrast, at pH 6 dioxygen is the exclusive product. The rate of peroxide formation at pH 7.2 is inversely related to the chloride concentration. On the basis of these studies, Frasch (*46, 48*) has proposed that one role of chloride may be to protect the OEC from discharging its oxidizing equivalents at too low an S state to form peroxide rather than be further oxidized to gather sufficient oxidizing equivalents to convert water to dioxygen. Addition of certain lipid analogues (e.g., lauroylcholine chloride) to PSII membranes has also been shown to inhibit oxygen formation and to induce transient peroxide formation (*49*). It is unclear from this experiment how the lauroylcholine chloride affects the manganese cluster and if it causes an alternative reaction pathway to occur for peroxide formation.

Mn Peroxidase. The manganese peroxidase (MnP) is one of the two known enzymes capable of the oxidative degradation of lignin, an amorphous, random, aromatic polymer synthesized from *p*-hydroxycinnamyl alcohol, 4-hydroxy-3-methoxycinnamyl alcohol, and 3,5,-dimethoxy-4-hydroxycinnamyl alcohol precursors by woody plants. Both enzymes contain the protoporphyrin IX heme prosthetic group, similar to the heme peroxidases with an L5-histidine and both use hydrogen peroxide as a substrate. However, the manganese peroxidase has an absolute requirement for Mn(II) to complete its catalytic cycle (*50*). The X-ray structure of this protein has recently appeared (*51*).

The presently accepted mechanism (*52*) involves the oxidation of an Fe(III) porphyrin by hydrogen peroxide to form an $(Fe^{IV}{=}O)P^{+\bullet}$ analogous to the previously mentioned "compound 1" of the heme catalase. This highly oxidized enzyme form subsequently reacts with an equivalent of Mn(II) to give "compound 2," $(Fe^{IV}{=}O)P$, and Mn(III), which can diffuse off of the enzyme and into the medium. There is little restriction for the type of Mn(II) required in the first reductive step; however, the subsequent reduction of compound 2 to resting enzyme requires an Mn(II) dicarboxylate or α-hydroxyacid complex. Studies suggest that the enzyme prefers the 1:1 Mn(II) oxalate complex as substrate. The oxidized Mn(III) oxalate is once again released into the medium, presumably forming [bis(oxalato)Mn(III)]$^-$, to wreak oxidative damage on its polymeric substrate through the initiation of phenolic radicals in the lignin superstructure.

Although [bis(oxalato)Mn(III)]$^-$ can only be formed transiently because of its oxidative instability, model chemistry has shown that Mn(III) and Mn(IV) α-hydroxy acids (e.g., lactic acid) can be prepared as stable solids that can then be used to evaluate lignin degradative reactions (*53*). The [Mn(III)(lactate)$_2$]$^-$ and related compounds such as [Mn(III)$_2$(lactate)$_3$]$^+$ will oxidatively degrade vanillin acetone through a radical mechanism to form pyruvaldehyde and vanillin. The reaction requires two oxidizing equivalents provided by at least two Mn(III) complexes. Although [bis(oxalato)Mn(III)]$^-$ is thought to be the natural substrate, [bis(malonato)Mn(III)]$^-$ does not appear to catalyze the vanillin acetone degradation. Presumably this is due to the greater stability of the [bis(malonato)Mn(III)]$^-$ complex. Surprisingly, the more highly oxidized manganese complex [Mn(IV)(α-hydroxybutyric acid)$_3$]$^{2-}$ also does not oxidize vanillin acetone. This may in part be due to the extreme water sensitivity of this complex (*54*).

Mn Ribonucleotide Reductase. Iron-containing ribonucleotide reductases play an important role in the transfer of genetic material in most organisms. However, certain Coryneform bacteria such as *Brevibacterium ammoniagenes* and *Micrococcus luteus* require manganese for growth and specifically for DNA synthesis. A manganese-containing ribonucleotide reductase was isolated and purified from *B. ammoniagenes* (*55, 56*). The active enzyme appears to consist of a dimer (B2) of MW = 100,000 daltons acting as the catalytic subunit along with a nucleotide binding monomer (B1) of MW = 80,000 daltons. The enzyme is believed to contain a dinuclear active site. The presence of Mn(III) is suggested by the similarity of the optical absorption to the same dimeric Mn(III) model complexes that have been used in relation to the manganese catalases (*see* Mn Catalases). Furthermore, the active enzyme is EPR silent, which is consistent with either Mn(III) or a strongly coupled Mn(IV) dimer. On the basis of these limited observations, it has been suggested that the manganese and iron ribonucleotide reductases may function in analogous ways.

The site in the active Fe ribonucleotide reductase contains two Fe(III) ions 3.3 Å apart, bridged by one carboxylate from a glutamate residue and a water-derived oxo bridge (*57*). The function of this iron center appears to be the formation and stabilization of a free radical on a tyrosine about 5 Å away. This radical is formed by reaction of the reduced, diferrous center with O_2, probably through peroxide and ferryl intermediates. This unusually stable tyrosyl radical is thought to participate in the reduction of ribonucleotides by electron transfer of the free radical, possibly through the Fe center, to the substrate interaction surface in which the ribose ring is oxidized by one electron to activate it for reaction with redox-active thiols in the protein (*58*).

Mn Superoxide Dismutase. The manganese superoxide dismutase (SOD) contains a single Mn ion that shuttles between the +2 and +3 oxidation level during the catalytic cycle. The Mn SOD is the best understood of all the manganese redox enzymes. This is partially due to the high-resolution X-ray structures of the Mn(II) and Mn(III) enzymes from *Thermus thermophilus* (*59, 60*), the detailed kinetic analysis of the enzyme that has been undertaken (*61, 62*), and (once again) the similarity of this manganoenzyme to an iron enzyme of like function (*63*). The X-ray crystal structures of the Mn and Fe SODs indicate that these two proteins are very similar, both in terms of the active-site geometry and a high degree of sequence homology (*63*). Both the major global fold and the metal-active center are essentially unaltered when the enzyme is reduced to the Mn(II) form. The metal ion in both the Mn and Fe SODs have a trigonal bipyramidal ligand array, as shown in Figure 3, with two histidines and a carboxylate in the equatorial positions and an additional histidine and a solvent-derived hydroxide or water in the axial positions. A crystal structure of the azide-bound derivative of Fe(SOD) shows that small molecules may bind to the active-site metal making it six-coordinate (*64*). This suggests direct binding of superoxide to the metal site during the catalytic disproportion cycle.

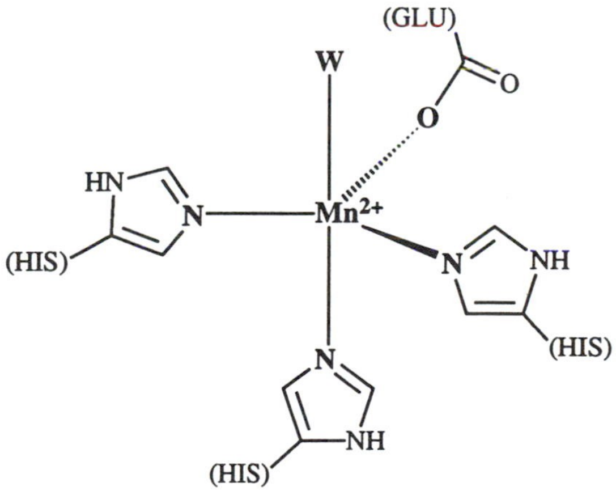

Figure 3. The active site of Mn superoxide dismutase on the basis of the X-ray structure of Ludwig et al. (59). *The "W" represents a water-derived ligand, either* OH^- *or* H_2O.

Kinetic studies of the Fe(SOD) indicate a fairly simple oxidation–reduction cycle in which O_2^- is bound to the Fe(III) form of the protein and oxidized to O_2, followed by binding of a second O_2^- to the resulting Fe(II) form and reduction to H_2O_2. The kinetics of Mn(SOD) are more complicated and the turnover number (1300 s^{-1} at 25 °C) is much lower than for Fe(SOD) (26,000 s^{-1}); however, a similar catalytic cycle is believed to occur. The manganese enzyme kinetics are complicated by a side reaction to form a "dead end complex," possibly a Mn(III) peroxide complex (*61, 62*).

Biomimetic Reactions Using Synthetic Model Compounds

The research groups interested in the reaction chemistry of manganese have focused on understanding either the assembly reactions of manganese clusters or the reactivity of manganese compounds with H_2O, HO_2, H_2O_2, O_2, or iodosylbenzene. In the majority of the cluster assembly reactions one is examining acid–base substitution reactions that interconvert stable or meta-stable aggregations of manganese oxo cores. We will not dwell on these reactions in this section but rather direct the interested reader to the literature in this area (*65–67*). More pertinent to the point of enzyme mimetic chemistry is the reactivity of small molecules with Mn(II) through Mn(V). The following section describes the work that we have done in our laboratory as an example of how the study of inorganic model complexes can lead to an understanding of possible mechanisms for the reactivity found in biological systems. For more details on work done by other groups in this field, the reader is directed to a recent review on this chemistry (*68*).

Models for the Reactivity of [Mn(IV)(μ_2-O)]$_2$ and [Mn(III)(μ_2-OR)]$_2$ Cores in Photosynthesis. As discussed in the previous paragraphs, the oxygen-evolving complex has an assembly of four manganese ions that form the heart of the water oxidation reaction. The manganese ions appear to be arranged in groups of [Mn(μ_2-O)]$_2$ units (Mn–Mn separation $\approx$ 2.7 Å) with the oxidation state of the manganese varying between Mn(III) and Mn(IV) depending on the enzyme oxidation level (S_n states). In S_2, the enzyme is believed to contain 1 Mn(III) and 3 Mn(IV) ions (*23*). This state can turnover catalytically with S_0, [Mn(III)$_3$Mn(IV)$_1$], in a catalase reaction (*45, 48*). On the basis of these observations, we believed that it was important to understand how [Mn(IV)(μ_2-O)]$_2$ could be prepared and how they behave with respect to catalase reactivity.

Our first task was to understand precisely how Mn(III) dimers could be converted to [Mn(IV)(μ_2-O)]$_2$ cores. We prepared [Mn(III)(3,5-diCl-salpn)(μ_2-OCH_3)]$_2$ (salpn is 1,3-bis(salicylideneimanato)propane), the first example of a dialkoxide - bridged Mn(III) dimer having Jahn–Teller distortions along two of the Mn–OR bonds (*36*). The tetradentate salpn ligand adopted the *cis-β* configuration in this isomer. This molecule is air-sensitive, converting after several hours to [Mn(IV)(salpn)(μ_2-O)]$_2$. This [Mn(IV)(μ_2-O)]$_2$ dimer has a Mn–Mn separation of 2.72 Å, whereas in the [Mn(III)(μ_2-OR)]$_2$ dimer the separation is 3.19 Å (*36, 69*). These molecules, which form the basis of this study, are shown in Figure 4.

Formation of [Mn(IV)(salpn)(μ_2-O)]$_2$ Using Hydrogen Peroxide. The complex [Mn(III)(salpn)(CH_3OH)$_2$]ClO_4 is unreactive with H_2O_2 when dissolved in methanol, DMF, or acetonitrile; however, addition of a base such as NaOH or NaOMe leads to rapid formation of [Mn(IV)(salpn)(μ_2-O)]$_2$ (*70*). We showed that [Mn(III)(salpn)(CH_3OH)$_2$]ClO_4 converts to [Mn(III)(salpn)(μ_2-OCH_3)]$_2$ if NaOMe is added to a degassed acetonitrile solution. Reaction of this Mn(III) dimer with peroxide in acetonitrile is rapid and quantitative in the production of [Mn(IV)(salpn)(μ_2-O)]$_2$. Dioxygen will also react with [Mn(III)(salpn)(μ_2-OCH_3)]$_2$ to give [Mn(IV)(salpn)O]$_2$; however, the reaction is ~1000 times slower and gives ≈70% product. Maslen and Waters (*71*) characterized the product of dioxygen reaction with Mn(II)(salpn) by X-ray crystallography. They interpreted it as [Mn(III)(salpn)(μ_2-OH)]$_2$ because Mn(IV) complexes were considered unlikely at the time. Boucher and Coe (*72*) suggested a [Mn(IV)(μ_2-O)]$_2$ formulation for that structure and that the product was the same as for the oxidation of [Mn(III)(salpn)]$^+$. They were shown by our studies (*36, 70*) and Armstrong's (*69*) to be correct.

An important question in the [Mn(IV)(salpn)(μ_2-O)]$_2$ formation reaction is "which molecule(s) is(are) the source of the bridging oxides in the Mn(IV) dimer?" We addressed this question by following the incorporation of ^{18}O from labeled O_2, H_2O_2, and water into [Mn(IV)-(salpn)(μ_2-O)]$_2$ (*69*). The molecular ion in the negative ion FAB mass spectrum of [Mn(IV)(salpn)(μ_2-O)]$_2$ allowed for detection and quantitation of labeled product [Mn(IV)(salpn)(μ_2-^{16}O)]$_2$, 702; [Mn(IV)$_2$-(salpn)$_2$(μ_2-^{16}O, μ_2-^{18}O)], 704; [Mn(IV)(salpn)(μ_2-^{18}O)]$_2$, 706 amu. Dropwise addition of ~0.1 M $H_2{}^{18}O_2$ in $H_2{}^{16}O$ to an acetonitrile solution of [Mn(III)(salpn)(μ_2-OCH_3)]$_2$ resulted exclusively in the doubly labeled [Mn(IV)(salpn)(μ_2-^{18}O)]$_2$. Similar results were obtained under aerobic ($^{18}O_2$) and anaerobic conditions. No rise is observed above the statistical distribution of the mass peak at 704 in spite of an 550-fold molar excess of $H_2{}^{16}O$. This clearly demonstrates that the primary reaction mechanism excludes oxygen contained in water from entering the bridges in [Mn(IV)(salpn)(μ_2-O)]$_2$. To test whether both oxide atoms are generated from the same molecule of hydrogen peroxide, we examined mixtures of 0.1 M aqueous ($H_2{}^{16}O$) $H_2{}^{16}O_2$, and $H_2{}^{18}O_2$ with

A

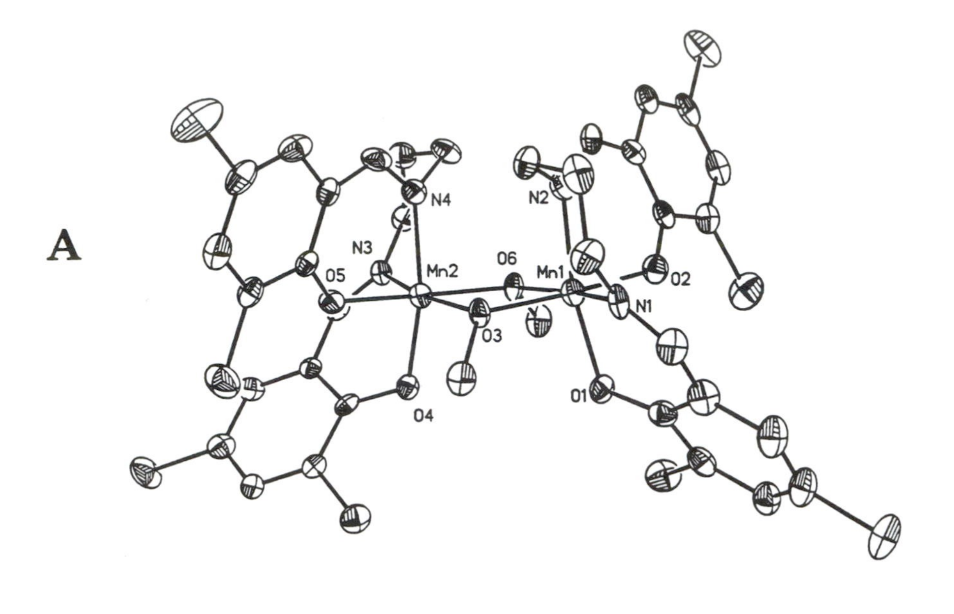

B

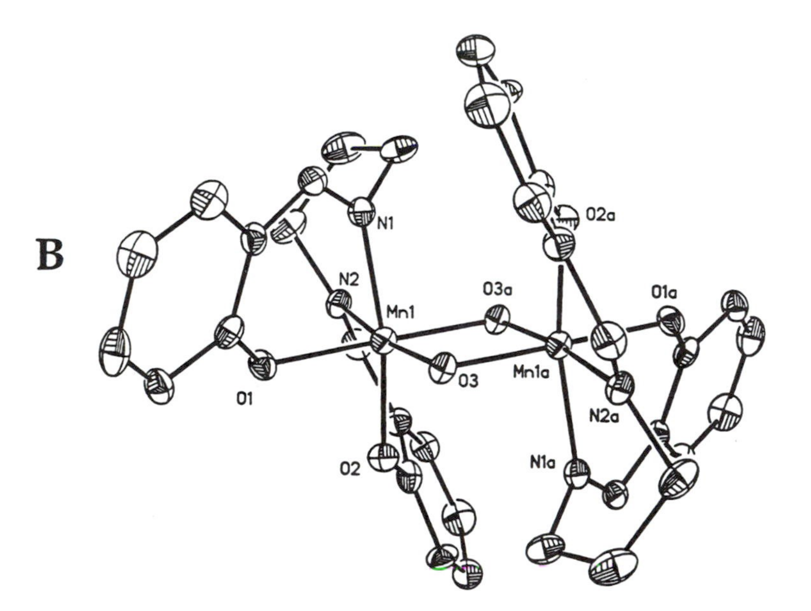

Figure 4. (A) ORTEP diagram of $[Mn(IV)(salpn)(\mu_2\text{-}O)]_2$. *(B) ORTEP diagram of* $[Mn(III)(salpn)(\mu_2\text{-}OCH_3)]_2$.

$[Mn(III)(salpn)(CH_3O)]_2$. The primary mass peaks are observed at 702 and 706 amu, respectively, with no observed increase in the mixed dimer mass peak at 704. From these results we conclude that hydrogen peroxide is the source of the oxide oxygens in $[Mn(IV)(salpn)(\mu_2\text{-}O)]_2$ and that both oxide-bridging atoms originate from the same peroxide molecule. This isotopic distribution demonstrates that monomeric intermediates such as LMn(IV)O or LMn(V)O are not part of the reaction mechanism in this case.

Although high-valent, monomeric complexes that might form after peroxide bond breaking have been eliminated as intermediates in the formation of $[Mn(IV)(salpn)(\mu_2\text{-}O)]_2$, the existence of a monomeric intermediate prior to hydrogen peroxide cleavage must also be addressed. This point was tested by reacting a 50:50 mixture of $[Mn(III)(salpn)(\mu_2\text{-}OCH_3)]_2$ and $[Mn(III)(3,5\text{-diCl-salpn})(\mu_2\text{-}OCH_3)]_2$ with H_2O_2 in acetonitrile. In this case, a mixture of $[Mn(IV)(salpn)(\mu_2\text{-}O)]_2$, $[Mn(IV)_2(3,5\text{-diCl-salpn})(salpn)(\mu_2\text{-}O)]$, and $[Mn(IV)(O)(3,5\text{-diCl-salpn})(\mu_2\text{-}O)]_2$ was obtained. When mixed, neither $[Mn(IV)(salpn)(\mu_2\text{-}O)]_2$ and $[Mn(IV)(3,5\text{-diCl-salpn})(\mu_2\text{-}O)]_2$ nor $[Mn(III)(salpn)(\mu_2\text{-}OCH_3)]_2$ and $[Mn(III)(3,5\text{-diCl-salpn})(\mu_2\text{-}OCH_3)]_2$ will scramble on the time scale of the experiment to give $[Mn(III)_2(3,5\text{-diCl-salpn})(salpn)(\mu_2\text{-}OCH_3)_2]$ or $[Mn(IV)_2(3,5\text{-diCl-salpn})(salpn)(\mu_2\text{-}O)_2]$, respectively. Thus, insertion of hydrogen peroxide requires the dissociation of the Mn(III) dimer prior to cleavage of the peroxide bond.

Another important consideration for the evaluation of reactivity involving formal oxidation state changes is the electrochemical potential of the reactants. The monomeric $[Mn(III)(salpn)(CH_3OH)]ClO_4$ shows an irreversible one-electron reduction ($Mn(III) \rightarrow Mn(II)$), confirmed by rotating platinum electrode voltammetry, in acetonitrile at −106 mV versus SCE. We have observed no oxidative electrochemistry for this compound out to potentials of +1 V. In contrast, $[Mn(III)(salpn)(\mu_2\text{-}OCH_3)]_2$ show a quasireversible, one-electron oxidation around +550 mV and an irreversible one-electron reduction around −650 mV. The dramatic stabilization of the Mn(IV) oxidation level is likely a result of the additional basic oxyanion per manganese, which provides another negative charge to the Mn ion.

A Proposed Mechanism for the Formation of $[Mn(IV)(salpn)(\mu_2\text{-}O)]_2$. If hydrogen peroxide is added to $[Mn(III)(salpn)(\mu_2\text{-}OCH_3)]_2$ in degassed acetonitrile, an instantaneous reaction occurs to give $[Mn(IV)(salpn)(\mu_2\text{-}O)]_2$ in >97% yield. A mechanistic proposal for the production of $[Mn(IV)(salpn)(\mu_2\text{-}O)]_2$ by reaction of H_2O_2 with $[Mn(III)(salpn)(\mu_2\text{-}OCH_3)]_2$ is shown in Figure 5. This scheme is consistent with both the reactivity and isotopic-labeling studies presented. It features H_2O_2 reacting via two successive deprotonation steps that use

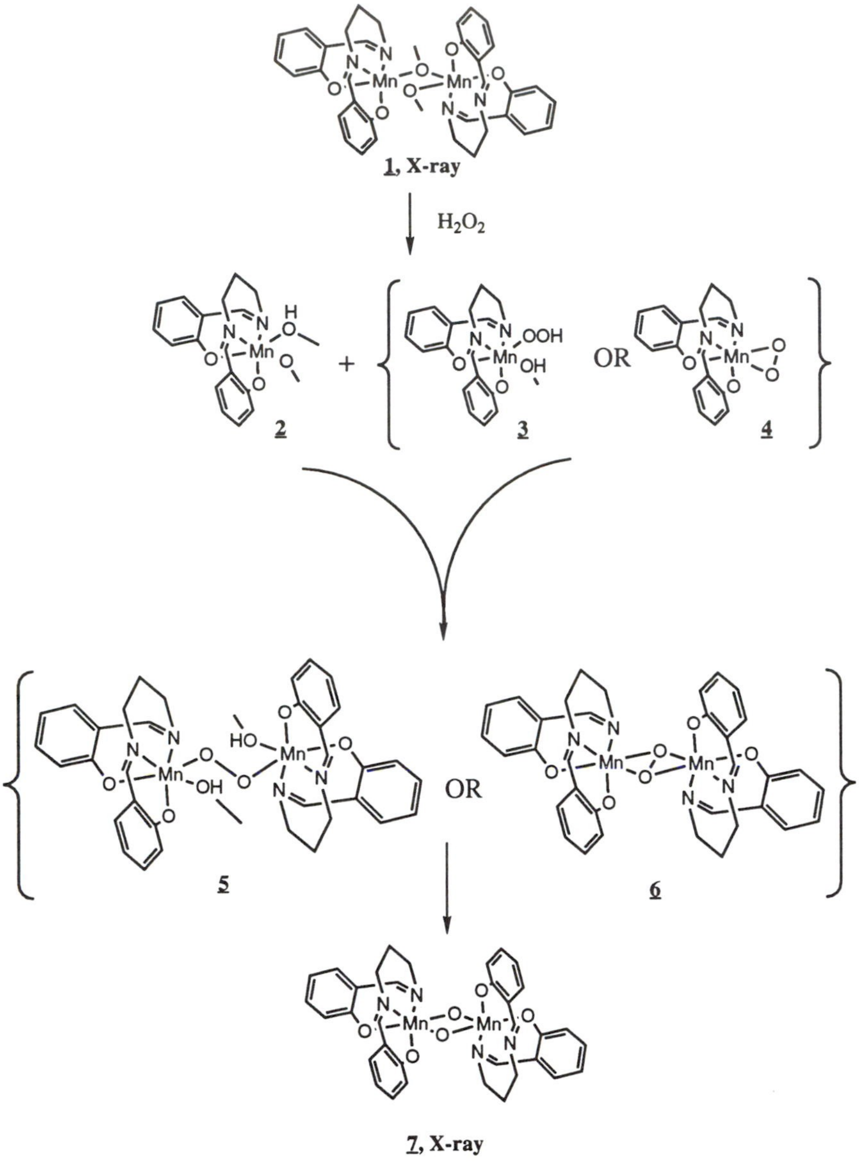

Figure 5. The proposed mechanism for the formation of [Mn(IV)(salpn)-(μ_2-O)]$_2$ from [Mn(III)(salpn)(μ_2-OCH$_3$)]$_2$ and hydrogen peroxide.

basic oxyanions associated with the Mn(III) precursors. The presence of some form of base is an absolute requirement as illustrated by the differential reactivity of [Mn(III)(salpn)(CH_3OH)]ClO_4 versus [Mn(III)(salpn)(μ_2-OCH_3)]$_2$ (69). Also, the ^1H NMR and electrochemical studies using triethylamine indicate that the abilities of oxyanions to stabilize the Mn(IV) oxidation level or to promote formation of cis-β "pretemplated" complexes are not the necessary characteristics that impart reactivity. We conclude that it is the proton-accepting feature of these anions that allows oxidation at the metal center and that deprotonation must, therefore, occur before cleavage of the peroxide bond. Although H_2O_2 may transiently coordinate to [Mn(III)(salpn)(CH_3OH)]ClO_4, metal-centered oxidation probably does not occur prior to deprotonation. It is notable that association of the hydroperoxide anion to Mn(III) creates a ligand environment similar to that seen in [Mn(III)(salpn)(μ_2-OCH_3)]$_2$. Thus, deprotonation leads to stabilization of the Mn(IV) oxidation level at the metal center and replacement of LM^{3+} for H^+ on H_2O_2 (giving [LM^{3+}(O_2H)]) may make the coordinated peroxide molecule more easily reduced. Both of these factors should lead to an increase in the driving force for internal oxidation of the intermediate. Our observations argue for dissociation that is instigated by H_2O_2 rather than a predissociation step. We prefer the formulation of **5** or **6** converting to **7** as the intermediates in this process; however, on the basis of these data, we can not exclude a Stomberg-type peroxo monomer (e.g., [Mn(III)(salpn)(O_2)]$^-$ as seen for V(V) complexes) that could then react with a monomeric Mn(salpn)$^+$ to form [Mn(IV)(salpn)(μ_2-O)]$_2$. A pathway as shown in Figure 5 is consistent with these observations.

The formation of [Mn(IV)(salpn)(μ_2-O)]$_2$ from [Mn(III)(salpn)(μ_2-OCH_3)]$_2$ using *t*-butyl hydroperoxide as the oxidant appears to follow a pathway very different from that of reaction with H_2O_2 (69). The reaction is not nearly as fast, requiring a number of hours (at room temperature) and the presence of water. *t*-Butyl hydroperoxide lacks a second dissociable proton. To form [Mn(IV)(salpn)(μ_2-O)]$_2$, it could lose *t*-butyl cation, leading to conservation of both unlabeled peroxide oxygens in [Mn(IV)(salpn)(μ_2-O)]$_2$. However, the operative mechanism allows extensive mixing of labeled water. The isotopic pattern is consistent with the initial formation of Mn(V)(salpn)O, which then reacts with an equivalent of Mn(III)salpn(OH) to give [Mn(IV)(salpn)(μ_2-O)]$_2$, although numerous pathways could be envisioned that might fit these data. Finally, the reaction of [Mn(III)(salpn)(μ_2-OCH_3)]$_2$ with dioxygen apparently follows a different (or at least additional) pathway because direct oxidation by $^{16}O_2$ in the presence of a small amount of $^{18}OH^-$ leads to scrambling of the label in the oxo bridges. This is in contrast to the hydrogen peroxide oxidations for which labeling of the bridging oxides was not dependent on the presence of either $H_2{}^{18}O$ or $^{18}OH^-$.

Catalase Reactions Using $[Mn(IV)(salpn)(\mu_2\text{-}O)]_2$. The reactions described previously investigated the mechanism of formation of $[Mn(IV)(\mu_2\text{-}O)]_2$ units that had structures appropriate to model a single dimer in the "dimer of dimers" structural proposal for the OEC. The observation that the OEC could undergo a catalase reaction by using such Mn(IV) dimers inspired us to examine the reactivity of $[Mn(IV)(salpn)(\mu_2\text{-}O)]_2$ with hydrogen peroxide. The observed catalase reactions were completed by the addition of a small aliquot of $\approx$5 mM H_2O_2 in acetonitrile to a dichloromethane solution containing $[Mn(IV)(salpn)(\mu_2\text{-}O)]_2$. Turnover of hydrogen peroxide to yield dioxygen was monitored by manometry and shown to be quantitative in less than 1 min using as much as 1000-fold molar excess H_2O_2. Furthermore, greater than 98% of the starting catalyst was recovered. These observations clearly demonstrated that $[Mn(IV)(salpn)(\mu_2\text{-}O)]_2$ was an effective "catalase" (*73*).

At this point, we applied our isotope-labeling methodology to an examination of this catalase reaction. The reaction of $[Mn(IV)(salpn)(\mu_2\text{-}{}^{16}O)]_2$ with $H_2{}^{18}O_2$ yielded exclusively $^{18}O_2$, whereas the reverse combination $[Mn(IV)(salpn)(\mu_2\text{-}{}^{18}O)]_2$ and $H_2{}^{16}O_2$ gave the expected $^{16}O_2$. This demonstrates that dioxygen is derived exclusively from hydrogen peroxide and not from the μ_2-oxo linkages of the dimer. The reaction of $[Mn(IV)(salpn)(\mu_2\text{-}{}^{16}O)]_2$ with a 1:1 mixture of $H_2{}^{16}O_2$ and $H_2{}^{18}O_2$ gave a mixture of $^{16}O_2$ and $^{18}O_2$ with only the predicted statistical distribution of $^{16,18}O_2$ based on residual $H_2{}^{16,18}O_2$. Thus, both oxygen atoms of dioxygen must come from the same hydrogen peroxide molecule. This isotope-labeling pattern is identical to that reported for the *L. plantarum* Mn catalase and the oxygen-evolving complex. The isotopic composition of $[Mn(IV)(salpn)(\mu_2\text{-}O)]_2$ recovered from the catalase experiments showed substantial enrichment of label into the $\mu_2\text{-}O^{2-}$. Single turnover reactions completed under high dilution conditions using a 2:1 ratio of $H_2{}^{18}O_2$ and $[Mn(IV)(salpn)(\mu_2\text{-}{}^{16}O)]_2$ gave predominantly (>95%) $[Mn(IV)(salpn)(\mu_2\text{-}{}^{18}O)]_2$, demonstrating that the $\mu_2\text{-}O_2$ ligands are stoichiometrically exchanged during the course of the reaction. The reaction of the 1:1 mixture of $H_2{}^{16}O_2$:$H_2{}^{18}O_2$ in a twofold excess over $[Mn(IV)(salpn)(\mu_2\text{-}{}^{18}O)]_2$ gives $[Mn(IV)(salpn)(\mu_2\text{-}{}^{16}O)]_2$ and $[Mn(IV)(salpn)(\mu_2\text{-}{}^{18}O)]_2$ in equal ratios, but no increase in $[Mn(IV)(salpn)(\mu_2\text{-}{}^{16,18}O)]_2$. These experiments strongly suggest that the catalase reaction results in both bridging oxo groups originating from the same peroxide molecule. The reaction of a 1:1 mixture of $[Mn(IV)(salpn)(\mu_2\text{-}O)]_2$ and $[Mn(IV)(3,5\text{-diCl-salpn})(\mu_2\text{-}O)]_2$ with hydrogen peroxide gave a mixture of $[Mn(IV)(salpn)(\mu_2\text{-}O)]_2$, $[Mn(IV)_2(3,5\text{-diCl-salpn})(salpn)(\mu_2\text{-}O)]_2$

and [Mn(IV)(3,5-diCl-salpn)(μ_2-O)]$_2$. A 1:1 mixture of the parent [Mn(IV)(salpn)(μ_2-O)]$_2$ and [Mn(IV)(3,5-diCl-salpn)(μ_2-O)]$_2$ is stable to ligand exchange under these conditions in the absence of hydrogen peroxide. Therefore, scrambling of the MnL units requires that a monomeric intermediate must be formed during the reaction.

These observations suggested that the first redox step of the process was initial oxidation of hydrogen peroxide to give a Mn(III) species. The electrochemistry of [Mn(IV)(salpn)(μ_2-O)]$_2$ shows reversible, one-electron reductive electrochemistry at −480 mV versus SCE to form [Mn(III/IV)(salpn)(μ_2-O)]$_2$. A second reduction wave was not observed. Protonation of [Mn(IV)(salpn)(μ_2-O)]$_2$ at the oxo bridge to form [Mn(IV)$_2$(salpn)$_2$(μ_2-O)(μ_2-OH)]$^+$ can also be achieved. The electrochemistry of this species is less reversible than for [Mn(IV)(salpn)(μ_2-O)]$_2$ and occurs nearly 800 mV to more positive potential, consistent with a decrease in donation of the bridging ligand. Analysis of EXAFS spectra of [Mn(IV)$_2$(salpn)$_2$(μ_2-O)(μ_2-OH)]$^+$ reveal that the Mn–Mn separation increases by 0.1 Å to 2.81 Å, again consistent with a poorer donating bridging ligand (72). In addition, [Mn(IV)$_2$(salpn)$_2$(μ_2-O)(μ_2-OH)]$^+$ is not competent in the catalase reaction with H_2O_2 but will react with $NaHO_2$ (72). These observations suggest that protons may be very important in these types of catalytic reactions. In fact, Boucher and Coe (72) claim to have detected hydrogen peroxide produced in solutions of the Schiff base complex, [Mn(IV)(Bu-salpn)(μ_2-O)]$_2$, upon addition of $HClO_4$. They suggest that this hydrogen peroxide production occurs by protonation of both oxo bridges, followed by decomposition to H_2O_2 and Mn(III) complexes. A pyrogallol solution was used to test for oxygen evolution upon addition of acid; however, none was detected.

It is worth considering why [Mn(IV)(salpn)(μ_2-O)]$_2$ can be protonated and is an efficient catalase, whereas previous [Mn(IV)(μ_2-O)]$_2$ complexes do not show similar chemistry. In most cases, the [Mn(IV)(μ_2-O)]$_2$ core is found in systems that contain neutral nitrogen donor ligands. Recent electrochemical studies of such compounds have shown pKa values of $\approx$11 for [Mn(III)(μ_2-O)]$_2$ and $\approx$2 for Mn(III)Mn(IV)(μ_2-O)$_2$ (75, 76). The anionic charge and stronger donor ability of phenolates compared to pyridine and related ligands leads to more electron density on the metal centers. Therefore, the oxo bridges are not required to donate as much electron density to the metal center and can be involved in proton acceptor chemistry. Norton (77) showed that protonation rates of molecules such as [Mn(IV)(salpn)(μ_2-O)]$_2$ are relatively slow. This appears to be due to π donation from the oxygen lone pair. Thus, relative thermodynamic and kinetic considerations appear to be important in determination of the reactivity of oxo manganese clusters. This would suggest that enzymes that incorporate a highly oxygen-rich ligand environment require a high valent catalytic cycle, whereas those with a nitrogen-rich coordination sphere proceed through lower valent metal centers. Such appears to be the case with the OEC and manganese catalase.

Reactivity of $[Mn(IV)(salpn)(\mu_2\text{-}O)]_2$ with Hydrochloric Acid. Although the addition of a weak acid with a noncoordinating anion such as pyridinium perchlorate to $[Mn(IV)(salpn)(\mu_2\text{-}O)]_2$ leads to the singly protonated $[Mn(IV)(salpn)_2(\mu_2\text{-}O),(\mu_2\text{-}OH)]ClO_4$ (*74*), strong acids with coordinating anions give drastically different products. The monomeric $Mn(IV)(salpn)Cl_2$ can be prepared by the addition of four equivalents of HCl to a dichloromethane solution of $[Mn(IV)(salpn)(\mu_2\text{-}O)]_2$. This deep green complex, shown in Figure 6, has *trans* chloride ligation and exhibits a highly axial EPR spectrum. The details of this conversion have not been thoroughly elaborated; however, the monoprotonated $[Mn(IV)(salpn)_2(\mu_2\text{-}O)(\mu_2\text{-}OH)]^+$ appears to be an intermediate on the basis of low-temperature studies. If HCl is added to a dichloromethane solution of $[Mn(IV)(salpn)(\mu_2\text{-}O)]_2$ at −50 °C, the purple monoprotonated derivative is formed and is stable for several hours. This solution converts directly and rapidly to $Mn(IV)(salpn)Cl_2$ when the solution is warmed to 0 °C. This dichloro complex is stable in CH_2Cl_2 for days at room temperature if precautions are taken to exclude contact with water.

The $Mn(IV)(salpn)Cl_2$ is a good oxidant, showing a reduction potential at +890 mV versus SCE. Other highly oxidized Mn complexes that contain chlorine have been shown to chlorinate alkanes, alkenes, and

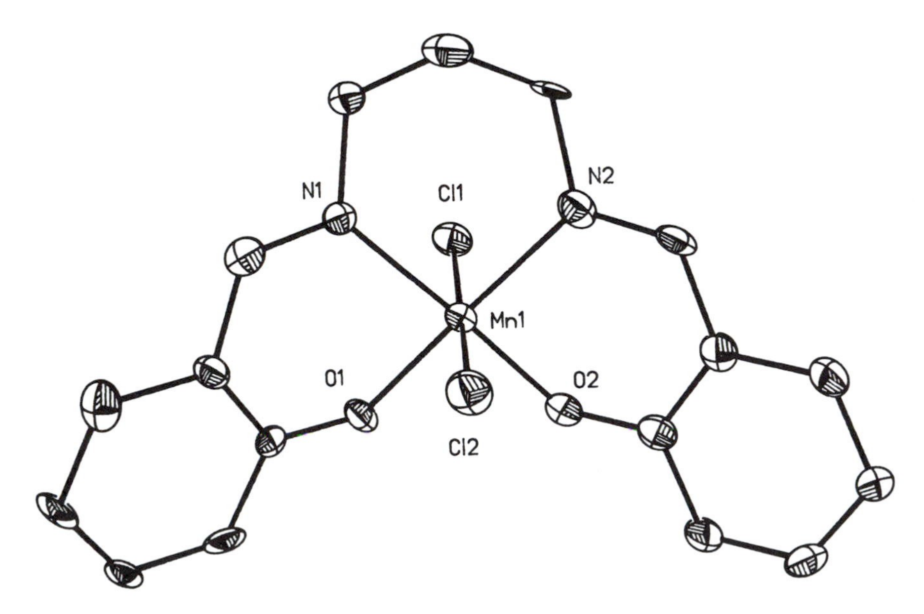

Figure 6. ORTEP diagram of Mn(IV)(salpn)Cl₂.

aromatics. For this reason, we decided to explore the reactivity of Mn(IV)(salpn)Cl_2 with a variety of substrates including cyclohexene, 1-methylcyclohexene, cyclohexane, norbornene, heptyne, and toluene. Cyclohexene can be converted to *trans*-dichlorocyclohexane in 58% yield after 36 h by Mn(IV)(salpn)Cl_2 in CH_2Cl_2. The Mn compound recovered from this reaction is Mn(III)(salpn)Cl. Norbornene is converted in CH_2Cl_2 to dichloronorbornane in 63% overall yield after 26 h. The product *cis*:*trans* ratio is 1:4.7. Most important, skeletal rearrangement to give nortricyclenyl chloride is not observed. Because these carbon migrations are observed when a carbocation intermediate is generated, this eliminates an ionic chlorination pathway for the Mn(salpn)-based reaction. Neither toluene, cyclohexane, or heptyne are halogenated by Mn(IV)(salpn)Cl_2. Therefore, chlorine radical ($Cl^\bullet$) is excluded as an intermediate because this agent can abstract hydrogen and will chlorinate these substrates. Additional evidence speaking against a solvent-derived halide radical is that chlorination of olefins in CH_2Br_2 proceeds in high yield without concomitant bromination of substrate. We favor a radical process in which the metal directs the final stereochemistry of the product. A proposed reaction cycle is provided as Figure 7. It is particularly noteworthy that each of the molecules specified in this cycle were isolated and characterized using diffraction techniques.

Figure 7 also shows an important competing side reaction that lowers the chlorination yield. This is the reaction of water with Mn(IV)(salpn)Cl_2 giving a Mn(III) product. Although this decomposition of the complex is an annoyance in the chlorination reactions, it may be of considerable interest as a process to generate dioxygen from water.

Matsushita and Shono (*78, 79*) demonstrated that Mn(IV)L_2Cl_2 complexes (where L is *N*-alkyl-3-nitrosalicylideneimines) can produce oxygen from water. The reaction of water with these high-valent Schiff–base complexes produced oxygen with a maximum yield of 0.27 mol O_2/Mn(IV) complex. Using various *N*-substituted derivatives, they correlated the reduction potentials of the Mn(IV) complexes and the hydrophobicity of the Mn(IV) center to the yield of oxygen production (*79*). Matsushita and Shono also demonstrated that a maximum yield is obtained at neutral pH and that by using ^{18}O-labeled H_2O, the oxygen produced does come from the water reacting with the Mn(IV)L_2Cl_2 complexes. They characterized the products of all of their reactions and propose that the reaction with water proceeds according to the following equation:

$$2Mn(IV)L_2Cl_2 + 2H_2O \rightarrow 2Mn(II)L_2 + O_2 + 4H^+ + 4Cl^-$$

Details of the water oxidation mechanism are unavailable; however, it is likely that one could generate an intermediate HOCl that could go on, under acid conditions in a second step, to form O_2. This is reminiscent

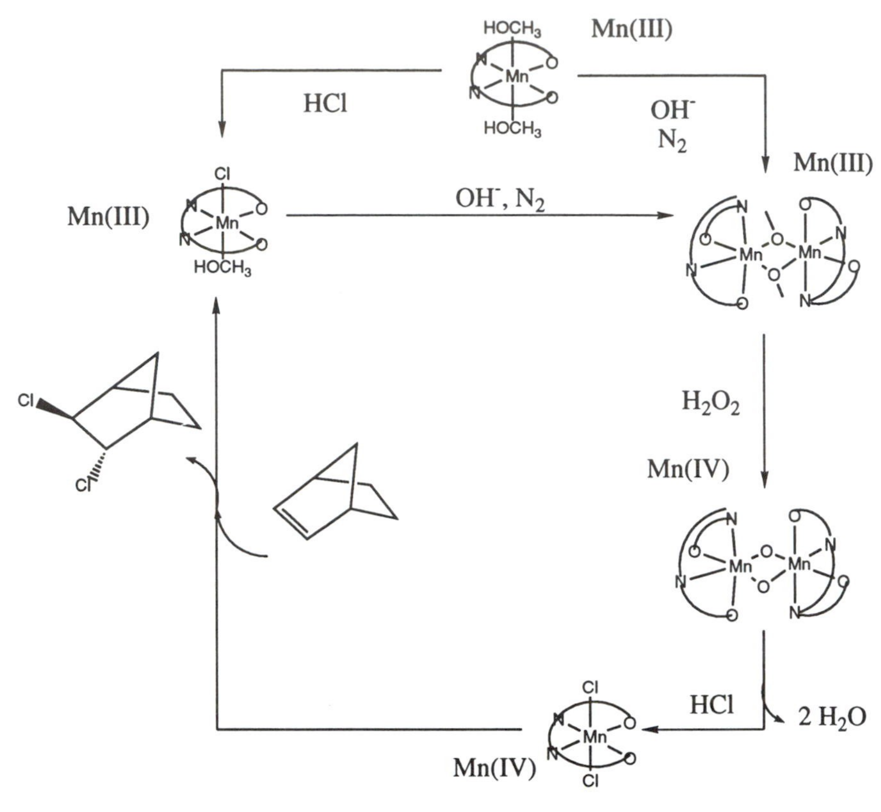

Figure 7. The proposed chlorination cycle for alkenes by Mn(IV)(salpn)Cl_2.

of the Ru-based water oxidation system of Meyer, in which both O_2 and Cl_2 were detected (*80, 81*).

Recognizing the relevance of both the Boucher and Coe (*72*) ($[Mn(IV)(Bu\text{-}salpn)(\mu_2\text{-}O)]_2 + HClO_4 \rightarrow H_2O_2$) and Matshushita et al. (*78, 79*) ($Mn(IV)(N\text{-}Pr\text{-}3\text{-}NO_2\text{-}sal)_2Cl_2 + 2H_2O \rightarrow O_2 + 4H^+ + 4Cl^-$) reactions, one can deduce two distinct mechanisms for producing dioxygen from a $[Mn(IV)(\mu_2\text{-}O)]_2$ core. In addition, the catalase activity of $[Mn(IV)(salpn)(\mu_2\text{-}O)]_2$ described by Larson and Pecoraro (*73*) provides a tidy explanation for the alternate catalase activity observed for the OEC. These three processes are shown in Figure 8.

The OEC catalase activity is summarized in Figure 8A, assuming a dimer of dimers formulation for the manganese cluster. Because the cycle is between S_2 and S_0, the relevant enzyme oxidation levels are $Mn(III)Mn(IV)_3$ and $Mn(III)_3Mn(IV)$, respectively. The catalase chemistry is illustrated at the $Mn(IV)_2$ dimer. The first equivalent of hydrogen peroxide is oxidized giving dioxygen and the reduced cluster. A subsequent molecule of H_2O_2 reoxidizes the center to regenerate S_2. In this

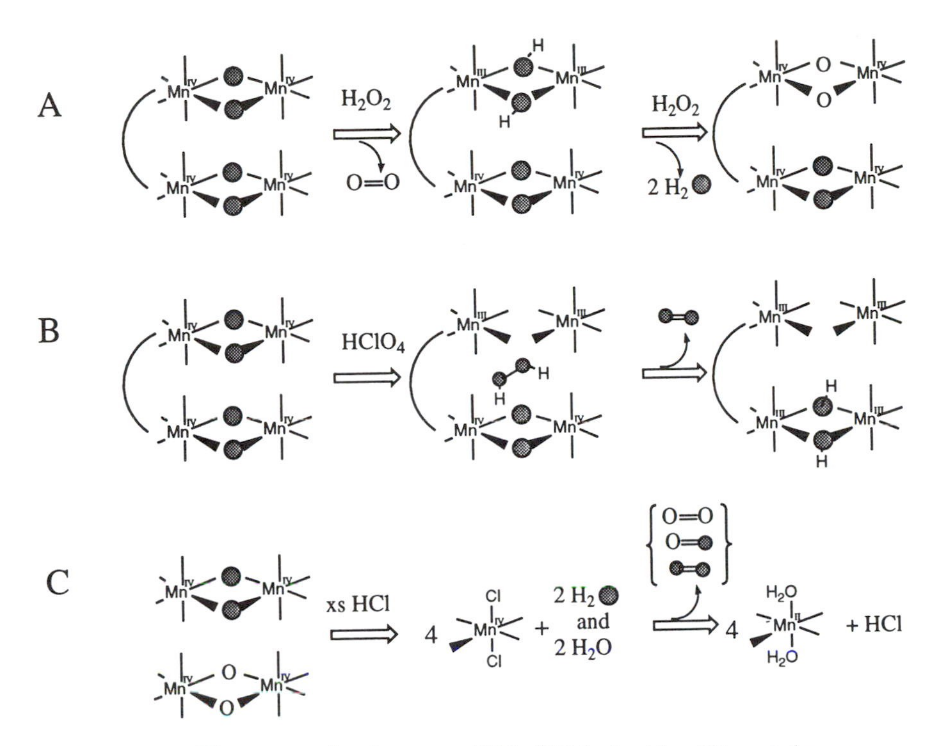

Figure 8. The potential relevance of $[Mn(IV)(salpn)(\mu_2\text{-}O)]_2$ *catalase activity to the reactions of the oxygen-evolving complex. (A) An explanation for the alternate catalase activity* ($S_0 \rightarrow S_2$ *cycle of the oxygen-evolving complex). (B) A mechanism for water oxidation to dioxygen using a combination of the Boucher and Coe (72) hydrogen peroxide producing reaction and the oxidation of peroxide by* $[Mn(IV)(salpn)(\mu_2\text{-}O)]_2$. *(C) A mechanism for water oxidation to dioxygen using the* $Mn(IV)(salpn)Cl_2$ *formation reaction and the oxidative chemistry of Matsushita et al. (78). The isotope-labeling pattern for the resulting dioxygen differs significantly between mechanisms B and C. The filled oxygens represent* ^{18}O*, and the open oxygens represent* ^{16}O.

cycle dioxygen retains the label found in peroxide, as has been shown both for the $[Mn(IV)(salpn)(\mu_2\text{-}O)]_2$ and OEC reactions (*47, 72*).

The second process, illustrated in Figure 8B, combines the Boucher and Coe acidification chemistry (*72*) with the $[Mn(IV)(salpn)(\mu_2\text{-}O)]_2$ catalase chemistry (*73*) to generate dioxygen. It relies on a two-step reaction sequence in which a $[Mn(IV)(\mu_2\text{-}O)]_2$ reacts with acid to generate 2Mn(III) and H_2O_2. Although Boucher and Coe did not follow the reaction using labeled $[Mn(IV)(Bu\text{-}salpn)(\mu_2\text{-}O)]_2$, it is likely that this is simply the back reaction of the $[Mn(IV)(salpn)(\mu_2\text{-}O)]_2$ formation with hydrogen peroxide (i.e., $2Mn(III) + H_2O_2 \rightarrow Mn(IV)(\mu_2\text{-}O) + 2H^+$). In this case, both oxygens in the resultant hydrogen peroxide would originate from the same $[Mn(IV)(\mu_2\text{-}O)]_2$ molecule. Subsequent oxidative chemistry of hydrogen peroxide with a second $[Mn(IV)(\mu_2\text{-}O)]_2$ would give O_2 + 2Mn(III) along the lines reported by Larson and Pecoraro (*73*). Functionally, this corresponds to conversion of O^{2-} (H_2O) to dioxygen at the expense of four manganese-oxidizing equivalents. Furthermore, if the initial core was labeled as $[Mn(IV)(\mu_2\text{-}^{18}O)]_2$, the liberated dioxygen should be exclusively $^{18}O_2$.

The dioxygen-producing reaction that proceeds through a monomeric $Mn(IV)Cl_2$ is mechanistically dissimilar to the peroxide pathway described in the preceding paragraph and should lead to mixed-labeled O_2. In the first step, acid and halide attack the $[Mn(IV)(\mu_2\text{-}O)]_2$ core as we described in preceding paragraphs; however, rather than liberating H_2O_2 in a redox reaction, the $Mn(IV)Cl_2$ intermediate is formed and water is released. This H_2O can then back react with the $Mn(IV)Cl_2$ to give O_2 according to the chemistry of Matsushita et al. (*78*). An oxygen label would scramble in this reaction because all of the oxygen atoms are released prior to reoxidation. Another important distinction between the Boucher–Coe and Matsushita intermediates is that in the first case two sequential two-electron oxidations at manganese dimers lead to product, whereas in the second, acid–base chemistry precedes oxidation that occurs exclusively at mononuclear centers.

Dinuclear Manganese Complexes as Models for the Manganese Catalase. As discussed previously, the manganese catalase has a dinuclear active site that is thought to function by cycling redox states between $Mn(II)_2$ and $Mn(III)_2$. Although the $[Mn(IV)(salpn)(\mu_2\text{-}O)]_2$ chemistry nicely explains the alternate catalase reactions of the OEC, this system is an inappropriate model for the Mn catalase because the redox cycle in that enzyme is lower and the core structure is believed to be dramatically different. In fact, a $[Mn(III/IV)(\mu_2\text{-}O)]_2$ superoxidized state of the Mn catalase has been identified and shown to be inactive.

We observed that a slight modification of the salpn ligand led to dramatically different coordination chemistry that has provided significant insight into the manganese catalase reactions. Substitution of a hydroxide group at the 2 position of the propane backbone leads to the pentadentate ligand 2-OH-salpn. Three distinct dinuclear structure types, illustrated in Figure 9, have been defined for manganese with this ligand. The first contains a di-μ-oxo bridged core that is analogous to $[Mn(III/IV)(salpn)(\mu_2\text{-}O)]_2^-$. The second and third categories have the 2-hydroxyl group bound as an alkoxide. Group two corresponds to symmetric structures, so designated because the molecules contain two, equivalent bridging alkoxide donors that are part of the ligand. The third category contains asymmetric complexes that have only one of the

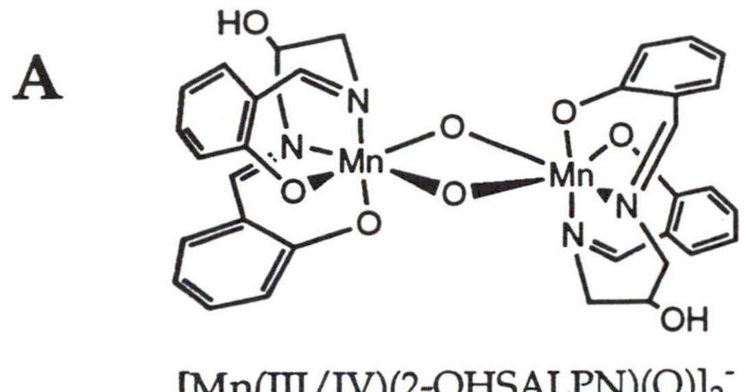

$[Mn(III/IV)(2\text{-}OHSALPN)(O)]_2^-$

B

$[Mn(III)(2\text{-}OHSALPN)]_2$

C

$[Mn(III)(2\text{-}OHSALPN)(CH_3OH]_2$

Figure 9. Three distinct structural types are available for Mn complexes of 2-OH-salpn. (A) A di-μ-oxo bridged core structure represented by $[Mn(III,IV)(2\text{-}OH\text{-}salpn)(\mu_2\text{-}O)]_2^-$. (B) Symmetric dialkoxide compounds represented by the $[Mn(III)(2\text{-}OH\text{-}salpn)]_2$. (C) Asymmetric dialkoxide compounds represented by $[Mn(III)_2(2\text{-}OH\text{-}salpn)_2(CH_3OH)]$.

alkoxide oxygen atoms participating in a bridge between the manganese ions (*82*, *83*). Although category 1 and 2 complexes have all manganese coordination sites filled by the ligand or μ_2-O bridges, the sixth coordination site of one of the manganese ions in the category 3 structures is filled by solvent. These three structure types show a substantial range of Mn–Mn separations. The $[Mn(III/IV)(2\text{-}OH\text{-}salpn)(\mu_2\text{-}O)]_2^-$ structure is expected to have a Mn–Mn separation on the order of 2.7 Å. In contrast, the Mn(III/IV) distance in the asymmetric complex $[Mn(III/IV)_2(2\text{-}OH\text{-}salpn)_2THF]^+$ is 3.61 Å (*84*). Reducing this complex by one electron leads to the asymmetric $[Mn(III)_2(2\text{-}OH\text{-}salpn)_2CH_3OH]$ (Mn–Mn 3.83 Å). This is over 0.5 Å longer than the symmetric dimer $[Mn(III)(2\text{-}OH\text{-}salpn)]_2$ (Mn–Mn 3.23 Å) (*85*). Stepwise reduction of the symmetric Mn(III) dimer results in the lengthening of the Mn–Mn distance, first to $[Mn(II,III)(2\text{-}OH\text{-}salpn)]_2^-$ (Mn–Mn 3.37 Å) and then a small decrease to $[Mn(II)(2\text{-}OH\text{-}salpn)]_2^{2-}$(Mn–Mn 3.33 Å).

The symmetric $[Mn(II)(2\text{-}OH\text{-}salpn)]_2^{2-}$ is air-sensitive and will convert to $[Mn(III)(2\text{-}OH\text{-}salpn)]_2$ in acetonitrile and $[Mn(III)_2(2\text{-}OH\text{-}salpn)_2CH_3OH]$ in methanol. A wide variety of phenyl ring derivatives can be prepared that allow for the fine tuning of the redox potentials of these manganese complexes. If the $[Mn(II)(2\text{-}OH\text{-}(3,5\text{-}Cl\text{-}sal)pn)]_2^{2-}$ is air-oxidized at −40 °C in ethanol, the intermediate $[Mn(II,III)(2\text{-}OH\text{-}(3,5\text{-}Cl\text{-}sal)pn)]_2^-$ can be isolated in high yield. Alternatively, bleeding a small amount of O_2 into a reaction vessel at room temperature will provide $[Mn(II,III)(2\text{-}OH\text{-}salpn)]_2^-$, but in low yield. The $[Mn(III)(2\text{-}OH\text{-}salpn)]_2$ can be converted into the asymmetric complex, $[Mn(III)_2(2\text{-}OH\text{-}salpn)_2CH_3OH]$, by dissolving in methanol. The desolvation of $[Mn(III)_2(2\text{-}OH\text{-}salpn)_2CH_3OH]$ to give the symmetric dimer can be achieved by redissolving in acetonitrile.

The reaction of hydrogen peroxide with $[Mn(II)(2\text{-}OH\text{-}salpn)]_2^{2-}$ or $[Mn(III)(2\text{-}OH\text{-}salpn)]_2$ in acetonitrile causes the evolution of dioxygen, apparently cycling between the two complexes (*85*). Oxygen evolution experiments indicate that these complexes can complete the catalase reaction for at least 1000 turnovers without significant decomposition of the catalyst. Isolation of dioxygen after the addition of $H_2{}^{18}O_2$ yields exclusively $^{18}O_2$. If $H_2{}^{18}O_2$ and $H_2{}^{16}O_2$ are added to either manganese complex, $^{18}O_2$ and $^{16}O_2$ are recovered but no $^{16,18}O_2$ is detected. This isotope-labeling follows the isotopic O_2 composition shown for the *L. plantarum* catalase. The reaction of hydrogen peroxide with a 50:50 mixture of $[Mn(II)(2\text{-}OH\text{-}salpn)]_2^{2-}$ and $[Mn(II)(2\text{-}OH\text{-}(5\text{-}NO_2\text{-}sal)pn)]_2^{2-}$ gives mass peaks from the recovered material only for $[Mn(III)(2\text{-}OH\text{-}salpn)]_2$ and $[Mn(III)(2\text{-}OH\text{-}(5\text{-}NO_2\text{-}sal)pn)]_2$, and no $[Mn(III)_2(2\text{-}OH\text{-}(salpn)(2\text{-}OH\text{-}(5\text{-}NO_2\text{-}sal)pn)]$. This provides strong evidence that the dimers are not dissociating into monomers during the catalytic process. The UV-visible (UV-vis) spectra of the catalase reaction upon addition of 1 eq H_2O_2 to [Mn(III)(2-OH-salpn)]$_2$ are shown as Figure 10B. Figures 10A and 10C show the spectra for [Mn(II)(2-OH-salpn)]$_2^{2-}$ and [Mn(III)(2-OH-salpn)]$_2$, respectively. The isosbestic point demonstrates a clean conversion from the [Mn(III)(2-OH-salpn)]$_2$ to the [Mn(II)(2-OH-salpn)]$_2^{2-}$. This conversion shows no evidence for an intermediate.

The initial rate of reaction of these complexes with H_2O_2 was calculated using a modified Clark-type O_2 electrode. Table II compares the initial rate of different derivatives of [Mn(II)(2-OH-salpn)]$_2^{2-}$ and [Mn(III)(2-OH-salpn)]$_2$. Clearly, the observed rates for all of the derivatives were very similar. We decided to undertake a detailed kinetic study of the [Mn(III)(2-OH-(5-Cl-sal)pn)]$_2$ because this molecule has the greatest solubility in acetonitrile. Figure 11A shows that this complex exhibits saturation kinetics with respect to hydrogen peroxide, whereas Figure 12 demonstrates that at saturating hydrogen peroxide concentrations the reaction is first order in [Mn(III)(2-OH-(5-Cl-sal)pn)]$_2$. The saturation-type kinetics suggest that there is a reversibly formed "peroxy[Mn(2-OH-(5-Cl-sal)pn)]$_2$" intermediate in the turnover limiting step;

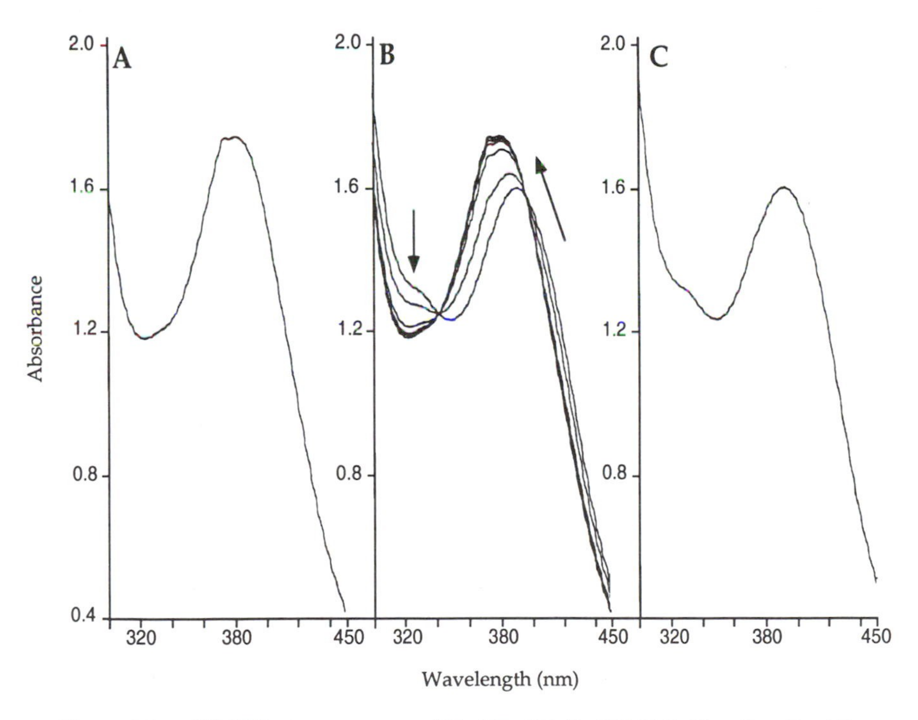

Figure 10. (A) UV–vis spectrum of Na_2[Mn(II)(2-OH-(3,5-Cl-sal)pn)]$_2$. (B) Conversion of [Mn(III)(2-OH-(3,5-Cl-sal)pn)]$_2$ to [Mn(II)(2-OH-(3,5-Cl-sal)pn)]$_2^-$ by addition of 1 equiv of H_2O_2. (C) UV–vis spectrum of [Mn(III)(2-OH-(3,5-Cl-sal)pn)]$_2$.

Table II. Initial Rate Data for Mn Catalase and Mn Models

Catalyst	*Rate (mol H_2O_2/mol catalyst s^{-1})*
L. plantarum Mn catalase	2×10^5
$Na_2[Mn(II)(2\text{-}OH\text{-}(3,5\text{-}Cl\text{-}sal)pn)]_2$	11.0 ± 0.9
$[Mn(III)(2\text{-}OH\text{-}(3,5\text{-}Cl\text{-}sal)pn)]_2$	10.0 ± 0.8
$[Mn(III)(2\text{-}OH\text{-}(5\text{-}Cl\text{-}sal)pn)_2]$	12.4 ± 0.5
$Na_2[Mn(II)(2\text{-}OH\text{-}(5\text{-}Cl\text{-}sal)pn)]_2$	13.1 ± 0.4
$[Mn(III)(2\text{-}OH\text{-}salpn)]_2$	13.0 ± 1.1
$[Mn(III)(2\text{-}OH\text{-}(5\text{-}NO_2\text{-}sal)pn)]_2$	11.2 ± 0.5
$Mn(II)(ClO_4)_2 \cdot 6H_2O$	$6.25 \pm 1.5 \times 10^{-3}$

however, the UV–vis spectra show no evidence of this intermediate. A double reciprocal plot of the data in Figure 11A, shown as Figure 11B, demonstrates a linear relationship from which k_{cat} and K_m may be calculated. The $K_m(H_2O_2)$ = 37 ± 10 mM and k_{cat} = 13 ± 3 mol H_2O_2 consumed/s. In comparison, the Mn catalase from *L. plantarum* has a $K_m(H_2O_2)$ = 200 mM and $k_{cat} = 2 \times 10^5$ mol H_2O_2 consumed/s, whereas $Mn(II)(ClO_4)_2$ has $k_{cat} = 6.3 \times 10^{-3}$ mol H_2O_2 consumed/s. Relative values for k_{cat}/K_m for the Mn catalase and $[Mn(III)(2\text{-}OH\text{-}(5\text{-}Cl\text{-}sal)pn)]_2$ are 1×10^6 and 3.5×10^2, respectively, indicating that the enzyme is ~3000 times more efficient than our synthetic catalyst.

The *L. plantarum* catalase was described initially as azide insensitive because the reaction is not inhibited at azide concentrations that would easily destroy heme catalase activity. However, Penner-Hahn (22) has shown that this enzyme can be inhibited by azide if the concentrations are sufficiently high. The K_i = 80 mM (pH 7) and is pH dependent. The $[Mn(III)(2\text{-}OH\text{-}(5\text{-}Cl\text{-}sal)pn)]_2$ is not inhibited in acetonitrile to saturating concentrations of azide (≈50 mM).

Hydroxylamine is another peroxide analogue that has been used as a probe of enzyme activity. If the Mn catalase is exposed to NH_2OH in the absence of hydrogen peroxide, the enzyme is reduced to the $Mn(II)_2$ form. However, if hydrogen peroxide is present, the enzyme is trapped in a catalytically inactive, superoxidized Mn(III,IV) form. Treatment with NH_2OH in the absence of hydrogen peroxide will regenerate the reduced form and activity. The $[Mn(II)(2\text{-}OH\text{-}salpn)]_2^{2-}$ system can also be driven to a catalytically inactive Mn(III,IV) form that has an EPR spectrum that is strikingly similar to the spectra of both the superoxidized Mn catalase and $[Mn(III,IV)(salpn)(\mu_2\text{-}O)]_2^-$. It is important to note that the EPR spectrum of this complex is different from that of the asymmetric complex $[Mn(III,IV)_2(2\text{-}OH\text{-}salpn)_2THF]^+$. The four EPR spectra are compared in Figure 13. As is the case with the enzyme, the "$[Mn(III,IV)(2\text{-}OH\text{-}(salpn)]_2$" (this molecule is either [Mn(III,IV)(2-OH-

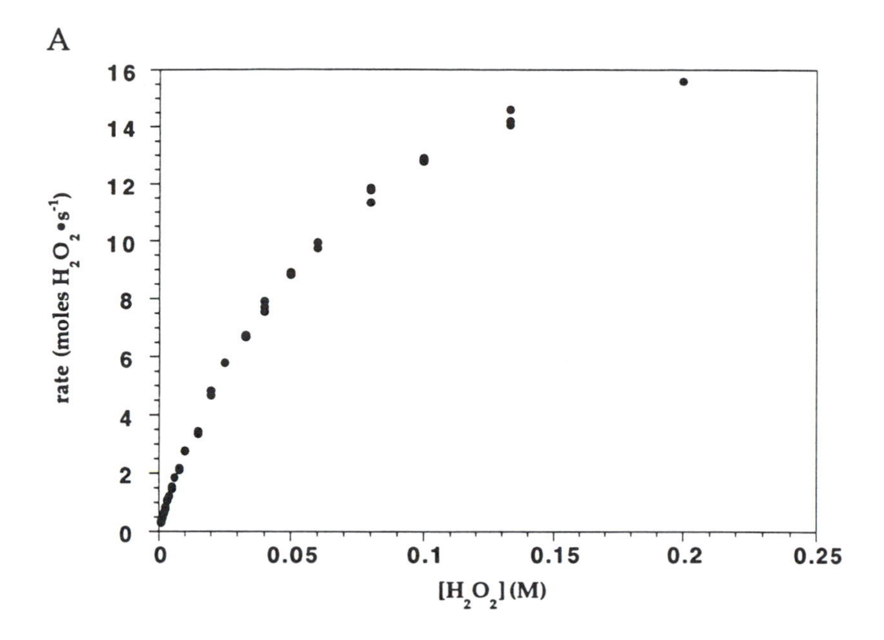

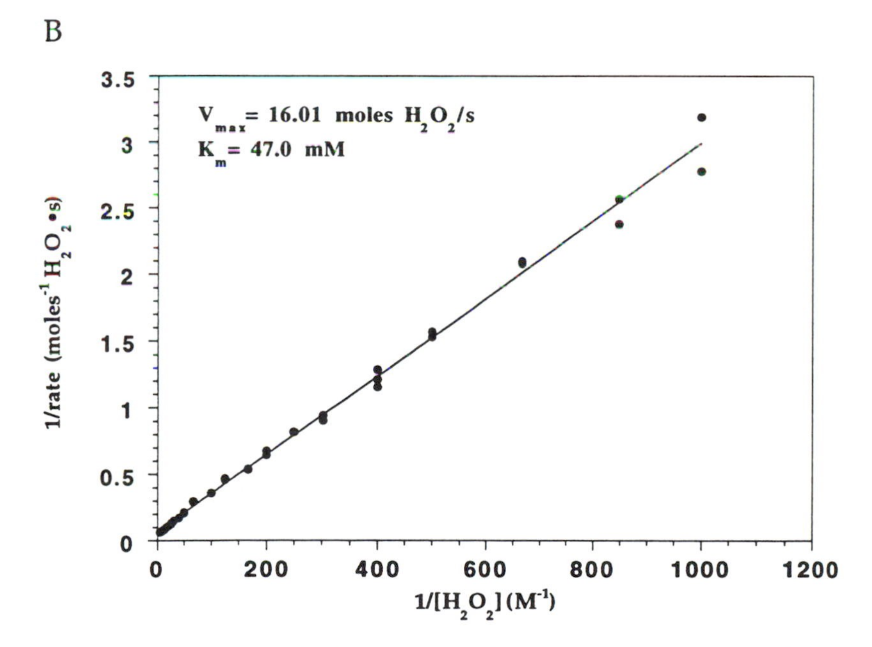

Figure 11. (A) Rate of hydrogen peroxide disproportionation vs. hydrogen peroxide concentration in acetonitrile. The plot shows saturation kinetics for $Na_2[Mn_2(2\text{-}OH\text{-}(5\text{-}Cl\text{-}sal)pn)]_2$. (B) Lineweaver–Burke plot (double reciprocal plot) of the data in A.

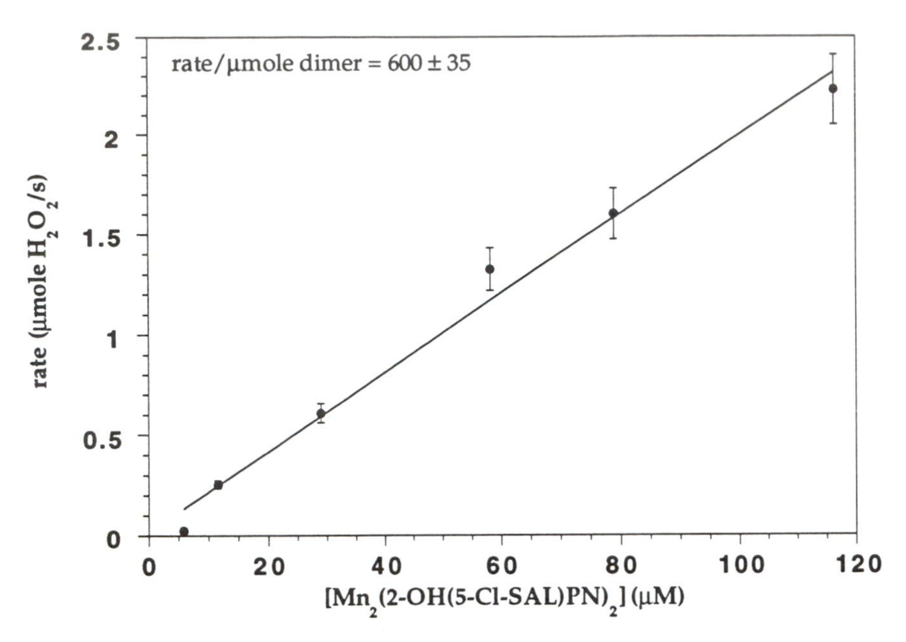

Figure 12. Plot of the hydrogen peroxide disproportionation rate vs. the concentration of $Na_2[Mn_2(2\text{-}OH\text{-}(5\text{-}Cl\text{-}sal)pn)]_2$ in acetonitrile.

salpn)(μ_2^--O)$]_2$ or [Mn(III,IV)$_2$(2-OH-salpn)$]_2^+$) is reduced by NH_2OH to a catalytically active form.

The [Mn(III)(2-OH-salpn)$]_2$ system is very exciting because it is the first crystallographically characterized dinuclear Mn model that begins to approach some of the structural, spectroscopic, and functional properties of the Mn catalases. The catalyst cycles between the Mn(II)$_2$ and Mn(III)$_2$ oxidation levels and maintains its dinuclear integrity throughout the process, shows respectable turnover rates, has saturation kinetics with hydrogen peroxide, is reasonably stable showing greater than 1000 turnovers, can be reduced by hydroxylamine, is azide insensitive under conditions that the enzyme is not inhibited by azide, forms a catalytically inactive Mn(III,IV) form (with an almost identical EPR to the enzyme) that can be reduced to a catalytically active state by hydroxylamine, and has the same $^{18}O_2$ labeling as the enzyme.

In addition to mimicking the function of the enzyme reasonably well, this system may provide some insight into the inactivation of the Mn catalase by the mixture of hydroxylamine and hydrogen peroxide. It is known that hydroxylamine is a hydrogen peroxide mimic as a reductant of the Mn catalase. However, unlike hydrogen peroxide, hydroxylamine reduces the enzyme by only one electron. If this is done in the Mn(III,III) oxidation level, an intermediate Mn(II,III) enzyme is

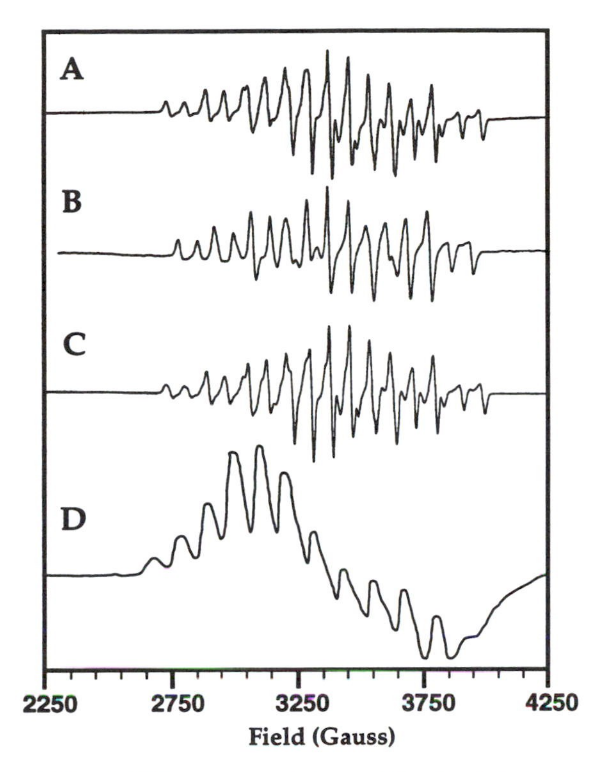

Figure 13. X-band EPR spectra of Mn(III/IV) dimers. (A) [Mn(III,IV)-(2-OH-(5-NO_2-sal)pn)]$_2$. (B) The inactive form of the Mn catalase. (C) [Mn(III,IV)(salpn)(μ_2-O)]$_2^-$. (D) [Mn(III,IV)$_2$(2-OH-(3,5-Cl-sal)pn)$_2$(THF)]$^+$.

formed. If the next molecule to bind is hydroxylamine, then the enzyme will be reduced to Mn(II,II) and the cycle can be continued; however, if hydrogen peroxide binds, the enzyme will be oxidized by two electrons to the Mn(III,IV) form. Because reduction of the Mn(III,IV) enzyme is slow, this superoxidized level will accumulate and eventually be the primary species. Similar chemistry is observed with [Mn(II,III)(2-OH-salpn)]$_2^-$ inasmuch as this molecule can be oxidized to a Mn(III,IV) by hydrogen peroxide and can be reduced to [Mn(II,II)(2-OH-salpn)]$_2^{2-}$ by hydroxylamine.

Summary

In this chapter we attempted to convey both the knowledge and the excitement that surrounds the study of manganese in biological systems. For many years, the field was focused on designing the most accurate structural models for these centers. As studies in these areas have matured, a functional and mechanistic understanding of these processes has become a more appreciated goal. The model studies presented herein have helped to explain the alternate catalase activity of the oxygen-evolving complex and the mechanism of inhibition of the Mn catalase by a mixture of hydroxylamine and hydrogen peroxide. We also commented on reasonable pathways for the production of dioxygen in water-splitting reactions using Mn dimers as possible catalysts. As this work continues to develop, we hope that an even greater level of understanding will allow us functionally to mimic oxygen evolution and catalase reactions with rates and efficiencies approaching the biological systems.

Acknowledgments

We thank Erlund Larson, Salman Saadeh, Joseph Bonadies, and Tim Machonkin for their scientific contributions to this work. We also thank Pamela Riggs-Gelasco and James Penner-Hahn for supplying unpublished X-ray absorption data and Tim Stemmler for providing the EPR spectrum of the Mn catalase in Figure 13. In addition, Charles Yocum, Gerald Babcock, Jack Norton, William Hatfield, and Martino Kirk have provided useful insight into the chemistry presented here. Funding for this work was provided by NIH grant GM39406, an Alfred P. Sloan Foundation fellowship (for Vincent L. Pecoraro), and an NIH postdoctoral fellowship (for Michael J. Baldwin; GM15102).

References

1. *Biochemistry of Essential Trace and Ultratrace Elements*; Frieden, E., Ed.; Plenum: New York, 1984.
2. Frieden, E. *J. Chem. Educ.* **1985,** *62*, 917.
3. Leyh, T. S.; Goodhart, P. J.; Nguyen, A. C.; Kenyon, G. L.; Reed, G. H. *Biochemistry* **1985,** *24*, 308–316.
4. Moore, J. M.; Reed, G. H. *Biochemistry* **1985,** *24*, 5328–5333.
5. Piccirilli, J. A.; Vyle, J. S.; Caruthers, M. H.; Cech, T. R. *Nature (London)* **1993,** *361*, 85–88.
6. Kim, S.; Cowan, J. A. *Inorg. Chem.* **1992,** *31*, 3495–3496.
7. Green, S. M.; Ginsburg, A.; Lewis, M. S.; Hensley, P. *J. Biol. Chem.* **1991,** *266*, 21474–21481.
8. Green, S. M.; Eisenstein, E.; McPhie, P.; Hensley, P. *J. Biol. Chem.* **1990,** *265*, 1601–1607.
9. Reczkowski, R. S.; Ash, D. E. *J. Am. Chem. Soc.* **1992,** *114*, 10992–10994.
10. Yamashita, M. M.; Almassy, R. J.; Janson, C. A.; Cascio, D.; Eisenberg, D. *J. Biol. Chem.* **1989,** *264*, 17681–17690.
11. Hardman, K. D.; Agarwal, R. C.; Freiser, M. J. *J. Mol. Biol.* **1982,** *157*, 69–86.
12. Abell, L. M.; Villafranca, J. J. *Biochemistry* **1991,** *30*, 6135–6141.
13. Sumner, J. B.; Dounce, A. L. *Science (Washington, D.C.)* **1937,** *85*, 366–367.
14. Chance, B.; Greenstein, D. S.; Roughton, F. J. W. *Arch. Biochem. Biophys.* **1952,** *37*, 301–321.

15. Kono, Y.; Fridovich, I. *J. Biol. Chem.* **1983,** *258,* 6015–6019.
16. Beyer, W. F., Jr.; Fridovich, I. *Biochemistry* **1985,** *24,* 6460–6467.
17. Algood, G. S.; Perry, J. J. *J. Bacteriol.* **1986,** *168,* 563–567.
18. Varynin, V. V.; Grebenko, A. I. *Dokl. Akad. Nauk SSSR* **1986,** *286,* 461–464.
19. Varynin, V. V.; Vagin, A. A.; Melik-Adamyan, V. R.; Grebenko, A. I.; Khangulov, S. V.; Popov, A. N.; Andrianova, M. E.; Vainshtein, B. K. *Dokl. Akad. Nauk SSSR* **1986,** *288,* 877–880.
20. Khangulov, S. V.; Voyevodskaya, N. V.; Varynin, V. V.; Grebenko, A. I.; Melik-Adamyan, V. R. *Biofizika* **1987,** *32,* 960–966.
21. Waldo, G. S.; Fronko, R. M.; Penner-Hahn, J. E. *Biochemistry* **1991,** *30,* 10486–10490.
22. Penner-Hahn, J. E. In *Manganese Redox Enzymes;* Pecoraro, V. L., Ed.; VCH Publishers: New York, 1992; pp 29–45.
23. Debus, R. J. *Biochim. Biophys. Acta* **1992,** *1102,* 269–352.
24. Kok, B.; Forbush, B.; McGloin, M. *Photochem. Photobiol.* **1970,** *11,* 457–475.
25. Wille, B.; Lavergne, J. *Photobiochem. Photobiophys.* **1982,** *4,* 131–144.
26. Förster, V.; Hong, Y.-Q.; Junge, W. *Biochim. Biophys. Acta* **1981,** *638,* 141–152.
27. Jahns, P.; Lavergne, J.; Rappaport, F.; Junge, W. *Biochim. Biophys. Acta* **1991,** *1057,* 313–319.
28. Jahns, P.; Haumann, M.; Bögershausen, O.; Junge, W. In *Research in Photosynthesis, Vol II;* Murata, N., Ed.; Kluwer Academic Publishers: Netherlands, 1992; Vol. II; pp 333–336.
29. Dismukes, G. C.; Siderer, Y. *Proc. Natl. Acad. Sci. U.S.A.* **1981,** *78,* 274–278.
30. Ono, T.-a.; Noguchi, T.; Inoue, Y.; Kusunoki, M.; Matsushita, T.; Oyanagi, H. *Science (Washington, D.C.)* **1992,** *258,* 1335–1337.
31. Riggs-Gelasco, P. J.; Mei, R.; Penner-Hahn, J. E. In *Mechanistic Bioinorganic Chemistry;* Advances in Chemistry Series 246; American Chemical Society: Washington, D.C., 1995; Chapter 8.
32. Casey, J. L.; Sauer, K. *Biochim. Biophys. Acta* **1984,** *767,* 21–28.
33. Zimmerman, J.-L.; Rutherford, A. W. *Biochim. Biophys. Acta* **1984,** *767,* 160–167.
34. Kim, D. H.; Britt, R. D.; Klein, M. P.; Sauer, K. *J. Am. Chem. Soc.* **1990,** *112,* 9389–9391.
35. Haddy, A.; Dunham, W. R.; Sands, R. H.; Aasa, R. *Biochim. Biophys. Acta* **1992,** *1099,* 25–34.
36. Larson, E.; Lah, M. S.; Li, X.; Bonadies, J. A.; Pecoraro, V. L. *Inorg. Chem.* **1992,** *31,* 373–378.
37. Radmer, R.; Ollinger, O. *FEBS Lett.* **1986,** *195,* 285–289.
38. Yocum, C. F. In *Manganese Redox Enzymes;* Pecoraro, V. L., Ed.; VCH: New York, 1992; pp 71–83.
39. Ghanotakis, D. F.; Topper, J. N.; Babcock, G. T.; Yocum, C. F. *FEBS Lett.* **1984,** *170,* 169–173.
40. Bouges, B. *Biochim. Biophys. Acta* **1971,** *234,* 103–112.
41. Riggs, P. J.; Mei, R.; Yocum, C. F.; Penner-Hahn, J. E. *J. Am. Chem. Soc.* **1992,** *114,* 10650–10651.
42. Sivaraja, M.; Dismukes, G. C. *Biochemistry* **1988,** *27,* 3467–3475.
43. Mei, R.; Yocum, C. F. *Biochemistry* **1991,** *30,* 7836–7842.

44. Britt, R. D.; Zimmerman, J.-L.; Sauer, K.; Klein, M. P. *J. Am. Chem. Soc.* **1989,** *111*, 3522–3532.
45. Frasch, W. D.; Mei, R. *Biochim. Biophys. Acta* **1987,** *891*, 8–14.
46. Fine, P. L.; Frasch, W. D. *Biochemistry* **1992,** *31*, 12204–12210.
47. Mano, J.; Takahashi, M.; Asada, K. *Biochemistry* **1987,** *26*, 2495–2499.
48. Frasch, W. D. In *Manganese Redox Enzymes*; Pecoraro, V. L., Ed.; VCH: New York, 1992; pp 47–70.
49. Ananyev, G.; Wydrzynski, T.; Renger, G.; Klimov, V. *Biochim. Biophys. Acta* **1992,** *1100*, 303–311.
50. Glenn, J. K.; Gold, M. H. *Arch. Biochem. Biophys.* **1985,** *242*, 329–341.
51. Sundaramoorthy, M.; Kishi, K.; Gold, M. H.; Poulos, T. L. *J. Biol. Chem.* **1994,** *263*, 32759–32767.
52. Wariishi, H.; Valli, K.; Gold, M. H. *Biochemistry* **1989,** *28*, 6017–6023.
53. Saadeh, S. M.; Lah, M. S.; Pecoraro, V. L. *Inorg. Chem.* **1991,** *30*, 8–15.
54. Saadeh, S. M., Ph.D. Thesis, University of Michigan, 1992.
55. Willing, A.; Follmann, H.; Auling, G. *Eur. J. Biochem.* **1988,** *170*, 603–611.
56. Willing, A.; Follmann, H.; Auling, G. *Eur. J. Biochem.* **1988,** *175*, 167–173.
57. Nordlund, P.; Sjöberg, B.-M.; Ekund, H. *Nature (London)* **1990,** *345*, 593–598.
58. Stubbe, J. *J. Biol. Chem.* **1990,** *265*, 5329–5332.
59. Ludwig, M. L.; Metzger, A. L.; Pattridge, K. A.; Stallings, W. C. *J. Mol. Biol.* **1991,** *219*, 335–358.
60. Borgstahl, G. E. O.; Parge, H. E.; Hickey, M. J.; Beyer, W. F.; Hallewell, R. A.; Tainer, J. A. *Cell* **1992,** *71*, 107–118.
61. Bull, C.; Fee, J. A. *J. Am. Chem. Soc.* **1985,** *107*, 3295–3304.
62. Bull, C.; Niederhoffer, E. C.; Yoshita, T.; Fee, J. A. *J. Am. Chem. Soc.* **1991,** *113*, 4069–4076.
63. Stallings, W. C.; Pattridge, K. A.; Strong, R. K.; Ludwig, M. L. *J. Biol. Chem.* **1984,** *259*, 10695–10699.
64. Stoddard, B. L.; Ringe, D.; Petsko, G. A. *Protein Eng.* **1991,** *4*, 113–119.
65. Armstrong, W. H. In *Manganese Redox Enzymes;* Pecoraro, V. L., Ed.; VCH Publishers: New York, 1992; pp 261–286.
66. Christou, G. *Acc. Chem. Res.* **1989,** *22*, 328–335.
67. Wieghardt, K. *Angew. Chem. Int. Ed. Engl.* **1989,** *28*, 1153–1172.
68. Pecoraro, V. L.; Baldwin, M. J.; Gelasco, A. *Chem. Rev.* **1994,** *94*, 807–826.
69. Gohdes, J. W.; Armstrong, W. H. *Inorg. Chem.* **1992,** *31*, 368–373.
70. Larson, E. J.; Pecoraro, V. L. *J. Am. Chem. Soc.* **1991,** *113*, 3810–3818.
71. Maslen, H. S.; Waters, T. N. *J. Chem. Soc., Chem. Commun.* **1973,** 760–761.
72. Boucher, L. J.; Coe, C. G. *Inorg. Chem.* **1975,** *14*, 1289–1295.
73. Larson, E. J.; Pecoraro, V. L. *J. Am. Chem. Soc.* **1991,** *113*, 7809–7810.
74. Larson, E. J.; Riggs, P. J.; Penner-Hahn, J. E.; Pecoraro, V. L. *J. Chem. Soc., Chem. Commun.* **1992,** 102–103.
75. Cooper, S. R.; Calvin, M. *J. Am. Chem. Soc.* **1977,** *99*, 6623.
76. Thorp, H. H.; Sarneski, J. E.; Brudvig, G. W.; Crabtree, R. H. *J. Am. Chem. Soc.* **1989,** *111*, 9249–9250.
77. Carroll, J. M.; Norton, J. R. *J. Am. Chem. Soc.* **1992,** *114*, 8744–8745.
78. Matsushita, T.; Fujiwara, M.; Shono, T. *Chem. Lett.* **1981,** 631–634.
79. Fujiwara, M.; Matsushita, T.; Shono, T. *Polyhedron* **1985,** *4*, 1895–1900.
80. Gilbert, J. A.; Eggleston, D. S.; Murphy, W. R., Jr.; Gelselowitz, D. A.; Gersten, S. W.; Hodgson, D. J.; Meyer, T. J. *J. Am. Chem. Soc.* **1985,** *107*, 3855–3864.

81. Geselowitz, D.; Meyer, T. J. *Inorg. Chem.* **1990,** *29*, 3894–3896.
82. Bonadies, J. A.; Kirk, M. L.; Lah, M. S.; Kessissoglou, D. P.; Hatfield, W. E.; Pecoraro, V. L. *Inorg. Chem.* **1989,** *28*, 2037–2044.
83. Bonadies, J. A.; Maroney, M. J.; Pecoraro, V. L. *Inorg. Chem.* **1989,** *28*, 2044–2051.
84. Larson, E.; Haddy, A.; Kirk, M. L.; Sands, R. H.; Hatfield, W. E.; Pecoraro, V. L. *J. Am. Chem. Soc.* **1992,** *114*, 6263–6265.
85. Gelasco, A.; Pecoraro, V. L. *J. Am. Chem. Soc.* **1993,** *115*, 7928–7929.

RECEIVED for review June 10, 1993. ACCEPTED revised manuscript May 10, 1994.

Application of NMR Spectroscopy to Studies of Aqueous Coordination Chemistry of Vanadium(V) Complexes

Debbie C. Crans, Paul K. Shin, and Kathleen B. Armstrong

Department of Chemistry, Colorado State University, Fort Collins, CO 80523–1872

The structure and dynamic processes of a series of oxovanadium(V) complexes with multidentate ethanolamine- or EDTA (ethylenediaminetetraacetic acid)-derived ligands are characterized using multinuclear NMR spectroscopy. Coordination-induced shifts (CIS) are used to determine the connectivity in the complex. ^{17}*O NMR spectroscopy is used to identify geometries in aqueous solution.* ^{51}*V NMR spectroscopy is used to determine the* H^+*-dependent formation constants of vanadate oxoanions. Dynamic processes are conveniently studied using 1D and 2D magnetization transfer techniques. Examples of applications of qualitative and quantitative 2D exchange spectroscopy (EXSY) are given. The major source of error in the rate constants originates in the volume integrations of the EXSY spectra, which are very sensitive to baseline correction. For exchange between strong off-diagonal resonances, the error in the site-to-site rate constants can approximate 10%, but for exchange between weak signals the error can be 100% or more. Consequently, 2D EXSY spectroscopy is a promising tool with which mechanistic problems in both chemistry and biology may be investigated.*

NUCLEAR MAGNETIC RESONANCE (NMR) spectroscopy is increasingly being applied to structural, thermodynamic, mechanistic, and dynamic studies of bioinorganic systems. The large number of NMR-active nuclei provide several alternative handles with which to study metal complexes by NMR spectroscopy (*1*). Proton and carbon-13 are particularly useful nuclei because the interpretation of their NMR spectra allows determination of the connectivity between metal and ligand in diamagnetic metal complexes. Recent advances in the NMR spectroscopy of para-

0065–2393/95/0246–0303$09.08/0

magnetic species show that such metal complexes can also be studied (*2–4*). In this chapter we describe our studies of the solution structures of a number of vanadium(V) complexes through the use of NMR techniques, including coordination-induced shifts (CIS) of NMR resonances, heteronuclear correlation experiments, and isotopic labeling. We also have used NMR spectroscopy in the determination of formation constants and as a tool to monitor a reaction's progress. Finally, we have used 2D NMR exchange spectroscopy (EXSY) to examine qualitatively and quantitatively intra- and intermolecular processes in aqueous vanadium(V) complexes.

NMR spectroscopy is a valuable tool in bioinorganic chemistry because such a large number of elements have NMR-active nuclei (*1*). Table I lists spin ½ nuclei and those nuclei with spin larger than ½ up to atomic number 209 and their natural abundances (*1*). The practical aspects of studying a particular nucleus by NMR depend on the properties of the nucleus and on the experimental feasibility of the NMR experiment. One important property of the nucleus is its natural abundance (*1*). Both ^{1}H and ^{51}V are present to greater than 99%, making them particularly amenable to study by NMR spectroscopy. Low natural abundance nuclei such as ^{13}C and ^{17}O (1.1 and 0.03%, respectively), require longer acquisition times for detection by NMR, and in the case of ^{17}O, samples enriched in ^{17}O must generally be prepared. The spin quantum number of the particular nucleus is an equally important property. The quadrupole moments of nuclei with $I > ½$ result in rapid relaxation of their NMR resonances with a concomitant loss in resolution and sensitivity. Additional properties affecting the quality of the NMR spectrum of a given nucleus include its magnetogyric ratio, its relative receptivity, and its quadrupole moment (which affects the signal width) (*1*). These nuclear parameters dictate the sensitivity of a nucleus and the information that may be derived from its NMR spectrum. Of practical importance is the NMR frequency at which the nucleus resonates. Spectrometers can be equipped with broadband probes, which enable the detection of most nuclei in addition to ^{1}H. Nuclei of weak magnetic strength such as ^{89}Y, ^{103}Rh, and ^{57}Fe, however, do require special instrumentation. Also of practical importance to an NMR experiment is the presence of paramagnetic species in the sample that can adversely affect the spectrum.

Research Problems Suitable for NMR Spectroscopic Study

There are a number of general problems of interest to a bioinorganic chemist that can be addressed by NMR spectroscopy. NMR spectroscopic studies can be instrumental in elucidating the structure of proteins and nucleic acids. The interaction of substrates with cofactors or other cel-

Table I. The NMR-Active Nuclei of Spin $^1/_2$ and Those with Spin $>^1/_2$

NMR nuclei spin $^1/_2$ nuclei[a]

^{1}H(99.985)	^{3}H(–)	^{3}He(1.3×10^{-4})	^{13}C(1.108)	^{15}N(0.37)	^{19}F(100)	^{29}Si(4.70)	^{31}P(100)	^{57}Fe(2.19)
^{77}Se(7.58)	^{89}Y(100)	^{103}Rh(100)	^{109}Ag(48.18)	^{113}Cd(12.26)	^{119}Sn(8.58)	^{125}Te(6.99)	^{129}Xe(26.44)	^{169}Tm(100)
^{171}Yb(14.31)	^{183}W(14.40)	^{187}Os(1.64)	^{195}Pt(33.8)	^{199}Hg(16.84)	^{205}Tl(70.50)	^{207}Pb(22.6)		

Spin $>^1/_2$ spin given in parentheses after natural abundance[b,c]

^{2}H(1.5×10^{-2}, 1)	^{6}Li(7.42, 1)	^{7}Li(92.58, $^3/_2$)	^{9}Be(100, $^3/_2$)	^{10}B(19.58, 3)	^{11}B(80.42, $^3/_2$)
^{14}N(99.63, 1)	^{17}O(3.7×10^{-2}, $^5/_2$)	^{21}Ne(0.257, $^3/_2$)	^{23}Na(100, $^3/_2$)	^{25}Mg(10.13, $^5/_2$)	^{27}Al(100, $^5/_2$)
^{33}S(0.76, $^3/_2$)	^{35}Cl(75.53, $^3/_2$)	^{37}Cl(24.47, $^3/_2$)	^{39}K(93.10, $^3/_2$)	[^{41}K(6.88, $^3/_2$)]	^{43}Ca(0.145, $^7/_2$)
^{45}Sc(100, $^7/_2$)	^{47}Ti(7.28, $^5/_2$)	^{49}Ti(5.51, $^7/_2$)	^{51}V(99.76, $^7/_2$)	^{53}Cr(9.55, $^3/_2$)	^{55}Mn(100, $^5/_2$)
^{59}Co(100, $^7/_2$)	^{63}Cu(69.09, $^3/_2$)	^{65}Cu(30.91, $^3/_2$)	^{67}Zn(4.11, $^5/_2$)	[^{69}Ga(60.4, $^3/_2$)]	^{71}Ga(39.6, $^3/_2$)
^{73}Ge(7.73, $^9/_2$)	^{75}As(100, $^3/_2$)	[^{79}Br(50.54, $^3/_2$)]	^{81}Br(49.46, $^3/_2$)	^{83}Kr(11.55, $^9/_2$)	[^{85}Rb(72.15, $^5/_2$)]
^{87}Rb(27.85, $^3/_2$)	^{87}Sr(7.02, $^9/_2$)	^{93}Nb(100, $^9/_2$)	^{95}Mo(15.72, $^5/_2$)	[^{97}Mo(9.46, $^5/_2$)]	[^{113}In(4.28, $^9/_2$)]
^{115}In(95.72, $^9/_2$)	^{121}Sb(57.25, $^5/_2$)	[^{123}Sb(42.75, $^7/_2$)]	^{127}I(100, $^5/_2$)	^{131}Xe(21.18, $^3/_2$)	^{133}Cs(100, $^7/_2$)
[^{135}Ba(6.59, $^3/_2$)]	^{137}Ba(11.32, $^3/_2$)	^{139}La(99.911, $^7/_2$)	^{181}Ta(99.988, $^7/_2$)	[^{185}Re(37.07, $^5/_2$)]	^{187}Re(62.93, $^5/_2$)
^{189}Os(16.1, $^3/_2$)	^{201}Hg(13.22, $^3/_2$)	^{209}Bi(100, $^3/_2$)			

NOTE: The boxes indicate nuclei used in NMR studies described in this chapter.
[a] If spin is $^1/_2$, no further information is provided. The natural abundance is indicated in parentheses.
[b] The spin ($>^1/_2$) are indicated in parentheses after the natural abundance.
[c] If the entire nucleus is in brackets, this nucleus is not the most favorable nucleus for a particular element.

lular components in a biological system can be defined (*5–7*). In addition, functional and structural models can be characterized (*7–9*). Although NMR spectroscopy is a very powerful tool, corroborative studies with other experimental methods are essential to many applications. In this chapter, we focus on multinuclear NMR studies related to the biochemistry of vanadium(V).

The discovery of the vanadium-containing haloperoxidases and nitrogenases, in conjunction with the known insulin mimetic activity of vanadium, has increased the interest in vanadium coordination chemistry (*10–12*). Because many proteins exist and function in an aqueous environment, the elucidation of the chemistry of vanadium in aqueous solution is essential for mimicking and hence understanding the functions of these proteins. Complexes containing vanadium(V) are d^0, making them diamagnetic. As such, they are ideal candidates for study by NMR spectroscopy (*10, 12*). The properties of vanadium(V) are in contrast to those of the paramagnetic vanadium(IV) complexes, which can be studied effectively by electron paramagnetic resonance (EPR) spectroscopy (*13, 14*). Solid-state characterization of vanadium(V) complexes by X-ray crystallography reveals that their structures in aqueous solution do not always correspond to their solid-state structures (*15*). These findings demonstrate the need for structure elucidation studies in solution. Vanadium(V) and vanadium(IV) derivatives show both similarities and differences in their coordination by multidentate ligands (*16–23*). These differences further illustrate the need for solution-state investigation of vanadium(V) derivatives. We describe here some applications of ^{1}H, ^{13}C, ^{17}O, and ^{51}V NMR spectroscopy to the characterization of the structural and dynamic properties of a series of aqueous vanadium(V) complexes.

Experimental Parameters

All samples for study by NMR spectroscopy contained D_2O and their preparation have been described in detail elsewhere (*24–26*). Proton NMR spectra were acquired on a Bruker ACP-300 NMR spectrometer (7.0 T) using standard parameters. Carbon-13 NMR spectra were acquired with a 200-ppm spectral window, a 90° pulsewidth and a relaxation delay of 700 ms. Exponential line-broadening (2 Hz) was applied to the free induction decay (FID) prior to Fourier transformation. Phase-sensitive correlation spectroscopy (COSY) and heteronuclear correlation (HETCOR) spectra were acquired on a Bruker ACP-300 NMR spectrometer (7.0 T) using standard parameters. The mixing time was 300 ms for the 2D ^{13}C EXSY experiment and 8 ms for the 2D ^{51}V EXSY experiment. Both experiments were acquired using parameters described previously (*25, 26*). The results from the 2D EXSY experiment, using these parameters, were examined by comparing the exchange rate constants from the 2D EXSY experiment with exchange rate constants from a 1D magnetization transfer experiment (*27*). The variable temperature experiments were conducted after calibrating the probe using a methanol standard (*28*). The ^{17}O NMR spectrum of a sample enriched in ^{17}O by the addition of 15% $H_2{}^{17}O$ was recorded at 41 MHz (7.0 T) using a 90° pulsewidth and no relaxation delay (*28*). The ^{51}V NMR spectra were recorded at 52.6 MHz (4.7 T) or 79.0 MHz (7.0 T) using 90° pulse angles and no relaxation delay.

Structural Studies

One-dimensional NMR studies provide information about the nuclear environment surrounding the nucleus. The theory of NMR and further experimental considerations have been described in detail elsewhere (*29*). The chemical shift, δ, of an NMR signal, provides structural information, as does its integration (relative to other signals) and its coupling pattern. The NMR spectra of most molecules exhibit complex coupling patterns. Although coupling information can be valuable, analysis of a spectrum may be simplified, and its signal-to-noise ratio (S:N) increased through decoupling. As a result, ^{13}C NMR spectra are frequently recorded decoupled from their attached protons. The combination of a ^{1}H spectrum and a decoupled ^{13}C spectrum is very powerful in exploring the solution-state structure of metal complexes with organic ligands. The following examples illustrate the use of NMR experiments in probing the structures of aqueous vanadium(V) complexes.

Application of ^{1}H and ^{13}C NMR Spectroscopy for Structural Studies of Vanadium(V) Complexes. Vanadate can potentially form a series of complexes with multidentate ligands containing various functionalities. For example, vanadate can interact with ethanolamine and ethylenediaminetetraacetic acid (EDTA)-derived ligands to form complexes with up to six moieties chelated to the metal. Chelation to the metal will shift the chemical shifts of the ^{1}H and ^{13}C nuclei in the ligand, resulting in distinct and characteristic CISs, defined as $\delta_{complex} - \delta_{ligand}$. As a result, ^{1}H and ^{13}C NMR spectroscopy are well-suited tools with which to study the solution structures of metal complexes. The ^{1}H NMR spectrum and the ^{13}C NMR spectrum of EDTA and its vanadate complex (V-EDTA) are shown in Figure 1. Also shown in Figure 1 are the ^{1}H and ^{13}C spectra of EDTA alone. Integration of the ^{1}H NMR resonances suggests that about 75% of the EDTA ligand is tied up as V-EDTA. This percentage is confirmed by ^{51}V NMR spectroscopy. The ^{13}C NMR spectrum of the free ligand contains three signals, whereas that of the ligand–complex mixture contains eight signals. The five remaining signals are consistent with a complex that contains two types of acetate arms, as shown in the structure in Figure 1. The carboxylate group coordinated to the vanadium (C_{1b}) has a resonance with a positive CIS value (5.1 ppm), and the carboxylate in the pendent arm (C_{1f}) has a negative CIS value (−1.4 ppm). The methylene groups have CIS values of 5.3 and

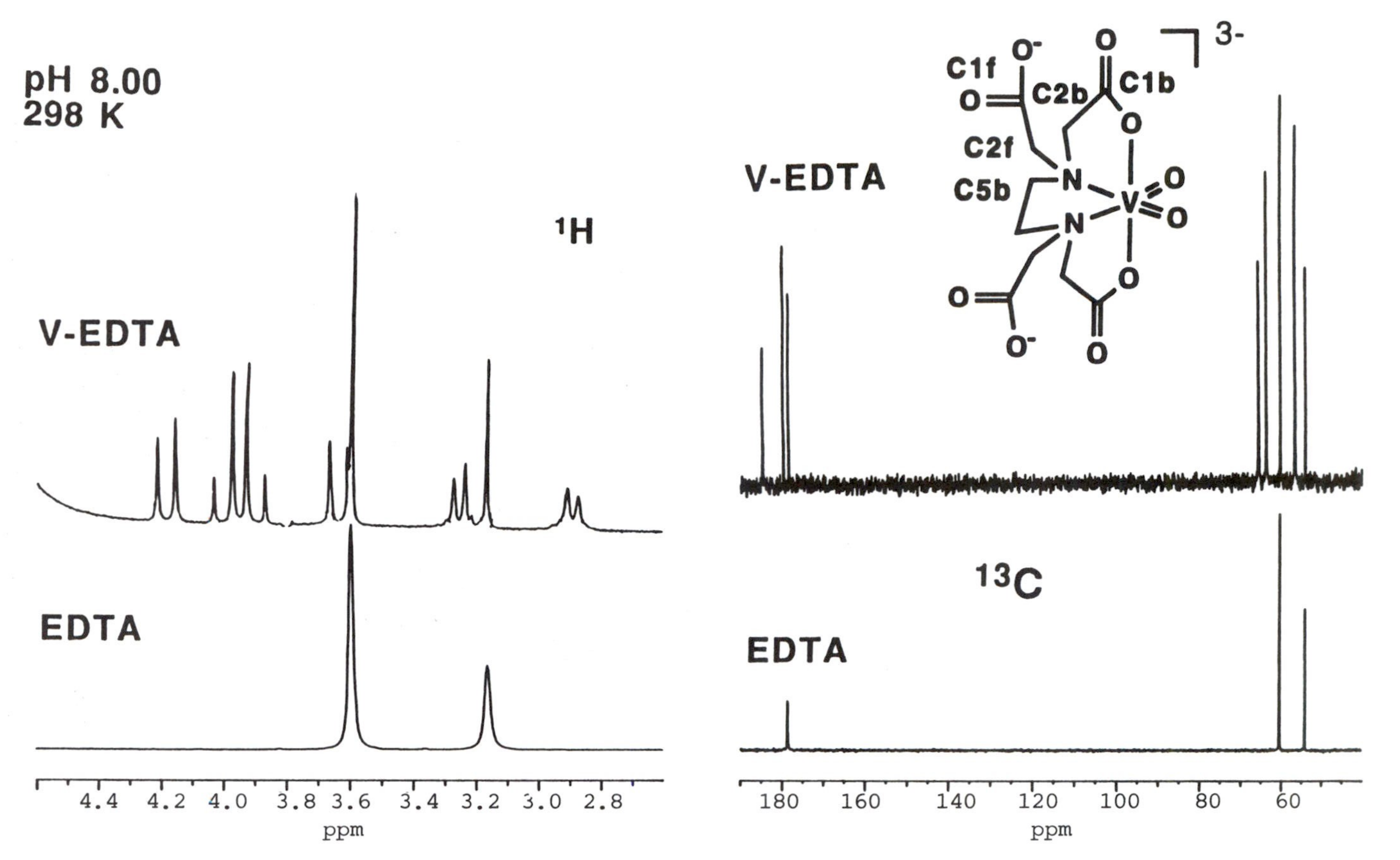

Figure 1. The ^{1}H and ^{13}C NMR spectra of EDTA (bottom) and a mixture of vanadate (125 mM) and EDTA (167 mM) at pH 8.00 and 298 K (top).

3.4 ppm, for the bound arm (C_{2b}) and the free arm (C_{2f}), respectively, whereas the positive CIS value of 2.6 ppm for the diamine portion of the complex also provides evidence of chelation of the vanadium by the ethylenediamine backbone. The chemical shifts and coupling patterns observed in the ^{1}H NMR spectra provide additional information to support this interpretation in agreement with the X-ray structure of this compound (*16, 17*). Detailed discussion of the assignments of the ^{1}H NMR signals, along with their coupling pattern, has been described previously (*30*).

Throughout this chapter we show partial ^{13}C NMR spectra that illustrate the shifts in the resonance frequencies that occur upon complexation. The carboxylate ^{13}C signals appear from 170 to 180 ppm, so that their observation requires that the NMR spectrum be recorded with a much greater window than shown in the figures. Given the low sensitivity of the ^{13}C nucleus and the longer relaxation time of the carboxylate group, ^{13}C NMR spectra can be recorded in much less time if the spectral width is reduced to exclude the carboxylate resonances. The 2D ^{13}C NMR spectra discussed in subsequent paragraphs were therefore recorded with reduced spectral widths. Foldover of the carboxylate resonances was not apparent in these spectra. The chemical shifts for the carboxylates of the complexes described here have been detailed elsewhere, but in all cases, they support the analysis described (*26*).

An NMR spectrum can be complicated by dynamic processes taking place on the timescale of the NMR experiment. The ^{13}C NMR spectrum of the *N*-(2-hydroxyethyl)iminodiacetic acid (HIDA)–vanadium complex at pH 8.53 (V-HIDA 1) provides an example of this. Both the C_4 and C_2 resonances exhibit a temperature dependence between 276 and 298 K. As the temperature increases, the C_4 signal broadens, and the C_2 signal sharpens. The spectra are shown at various temperatures in Figure 2, along with the assignments of carbon resonances C_2, C_3, and C_4 in the V-HIDA 1 complex. The corresponding resonances from the free ligand are labeled L_2, L_3, and L_4. The carboxylate groups, C_1 and L_1, are not shown. Because only the C_4 and C_2 signals show temperature dependence in these spectra and intermolecular exchange processes would have induced similar changes in all carbon resonances, we conclude that the observed process in this temperature range is an intramolecular process.

The large CIS value for C_4 (11.2 ppm) shows that this arm is coordinated to the vanadium atom. Only one chemical shift is observable for C_2, and its magnitude (5.6 ppm) suggests that it is also coordinated. Figure 3 shows a schematic representation of five possible structures for the V-HIDA 1 complex. Given the fact that six-coordinate complexes are most likely, the discussion here is limited to such structures (*26*). Although the observed equivalency of the carboxylate arm resonances

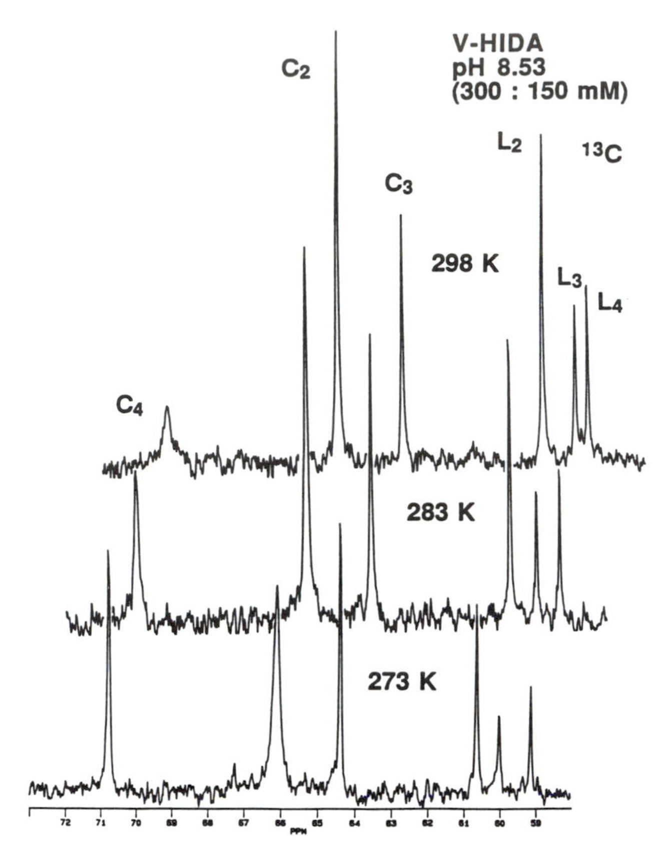

Figure 2. Partial ^{13}C NMR spectra at 298, 283, and 273 K of 300 mM vanadate and 150 mM HIDA at pH 8.53. The V-HIDA 1 complex concentrations were 72 mM at 298 K, 93 mM at 283 K, and 115 mM at 273 K. The carbons assigned to the complex are labeled C_x and for the ligand L_x.

(C_1 and C_2) is consistent with structure B, any of the other four structures might also reveal only four resonances in their ^{13}C spectra if rapid intramolecular exchange is taking place between the inequivalent carboxylates. The trans effect commonly exhibited by these complexes favors structure A if the complex contains a tetradentate ligand and structure D if the complex contains a tridentate ligand. The pendent arm in structures C, D, and E must be exchanging rapidly with the complexed arm for the geometry to be consistent with the observed NMR data. Low-temperature ^{1}H NMR spectra reveal an apparent decoalescence, consistent with the observation by ^{13}C NMR spectroscopy.

The ^{13}C spectrum for the V-HIDA complex at high pH (V-HIDA 1 in Figure 2) is very different from its spectrum at low pH (V-HIDA 2

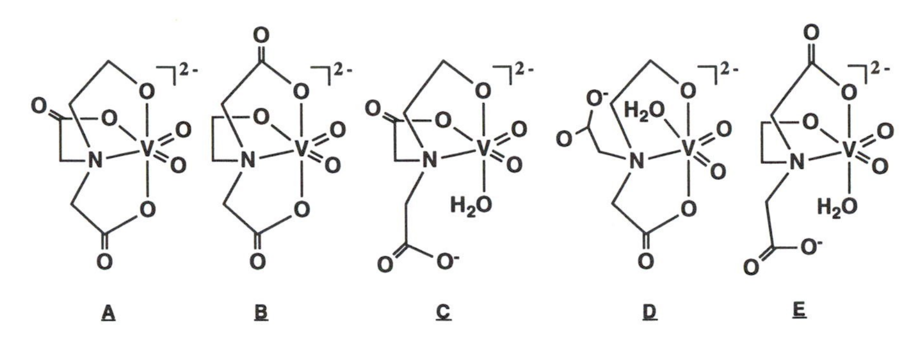

Figure 3. Five structural representations illustrating the connectivity in a dianionic 1:1 complex of V-HIDA 1. Further studies are necessary to determine the stoichiometry and the charge on the complex in solution because we have isolated and characterized (from solution) a 2:2 V-HIDA mixed-valence complex by X-ray crystallography (Mahroof-Tahir and Crans, Colorado State University, unpublished results).

in Figure 4). One complex is strongly favored at high pH, the other at low pH. The presence of the two complexes is confirmed by ^{51}V NMR spectroscopy which shows two different signals at neutral pH. At pH 5.13, the C_4 CIS value is 4 ppm smaller than its value at pH 8.53. Because stronger coordination shifts ^{13}C resonances downfield, the small, positive C_4 CIS value (2.4 ppm) at pH 5.13 suggests weak interaction between the hydroxyl group and the vanadium center. The CIS for the C_2 signal is very similar in both complexes. This similarity suggests that the amine nitrogens are similarly coordinated in the two complexes.

Figure 4 is a 2D ^{13}C-^{1}H HETCOR spectrum of the V-HIDA 2 sample at pH 5.13. The HETCOR experiment is one example of the many NMR experiments now in routine use (others include insensitive nuclei enhanced by polarization transfer (INEPT), distortionless enhanced polarization transfer (DEPT), and nuclear Overhauser effect (NOE)) that allow one to identify NMR-active nuclei that are either coupled to one another or in close proximity (*1, 5, 6, 29*). The broad resonance assigned to C_4 in Figure 4 correlates with a ^{1}H resonance assigned to a methylene proton of a pendent (or less tightly held) arm of the complex. This cross-peak therefore enables the ^{13}C signal to provide corroborative evidence for the ^{1}H assignment and vice versa. Variable-temperature ^{1}H NMR spectra were recorded, and they support the interpretation that the V-HIDA 2 complex contains a pendent or weakly coordinated arm. Considering the modest positive CIS of C_4, it may be a hydroxyethyl arm in this complex that is still protonated and pendent or weakly coordinated.

Application of ^{17}O NMR Spectroscopy for Structural Studies of Vanadium(V) Complexes. Isotopic-labeling experiments with

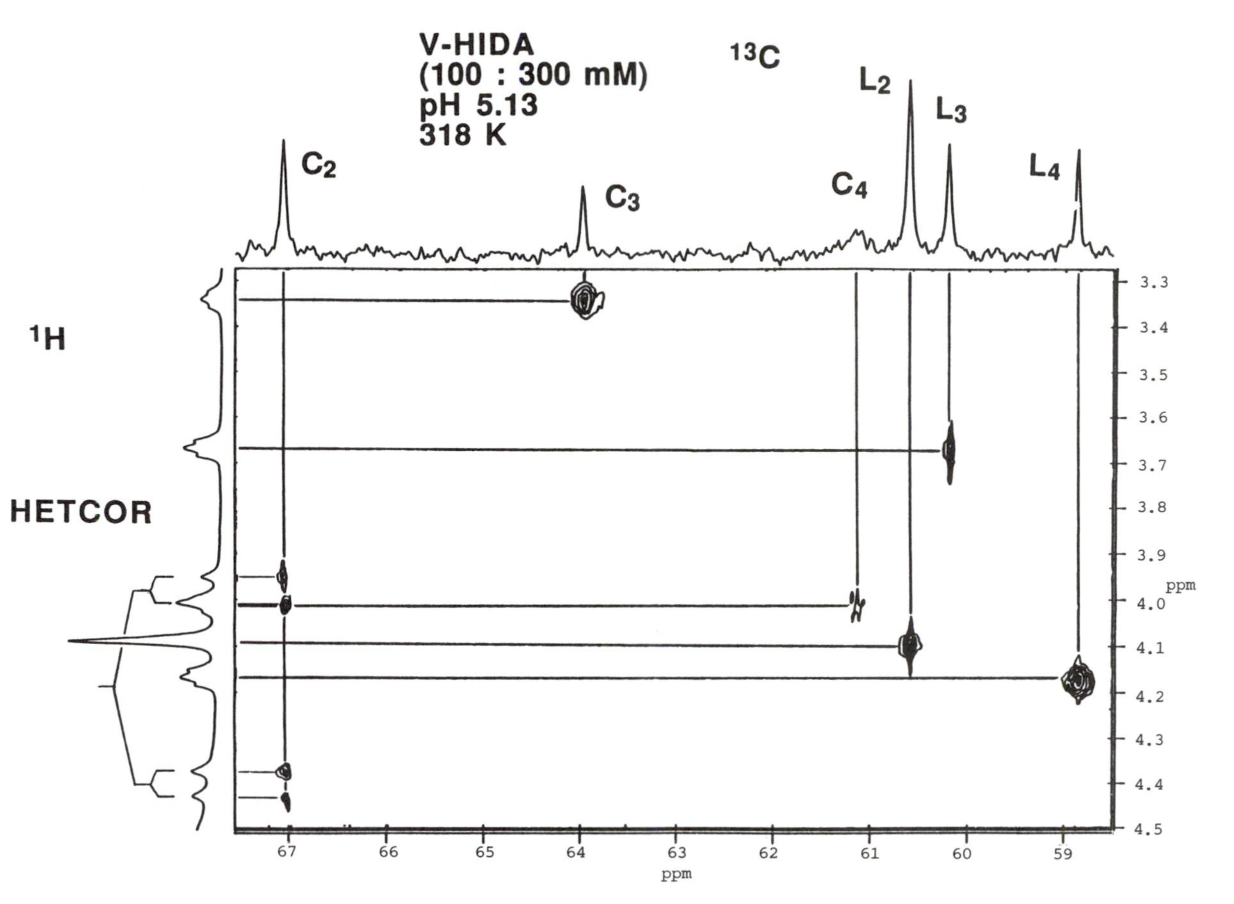

Figure 4. A HETCOR spectrum of the V-HIDA 2 complex 100 mM vanadate and 300 mM HIDA at pH 5.13 and 318 K. The carbons assigned to the complex are labeled C_x and those assigned to the ligand L_x.

NMR-active nuclei can often assist the chemist in distinguishing between two or more likely structures for a molecule. One such example is the use of ^{17}O NMR spectroscopy to distinguish between four possible structures for the complex of vanadium and tri-2-propanolamine (TPA) in aqueous solution (*15*). Analogous analysis was carried out with the complex of vanadium and triethanolamine (TEA), and the solution structure was compared to the structure determined by X-ray crystallography (*15, 31*). Analysis of the CIS ^{13}C values for this system shows that the V-TPA complex contains two 2-propanol arms along with the central amine chelating the vanadium center with one-arm pendent (*15*). X-ray crystallography shows that the solid-state V-TPA complex isolated from organic solvents contains tetradentate TPA with five-coordinate vanadium, in which the nitrogen is bound in an axial position. The structure for the aqueous V-TPA complex could, however, contain either five- or six-coordinate vanadium with either an axial or an equatorial bond to the nitrogen (see Figure 5). Addition of ^{17}O-labeled water to a solution containing TPA and vanadate would exchange the ^{17}O into the oxo portion of the complex. An ^{17}O NMR spectrum of the solution would allow the four structural possibilities to be distinguished from one another. The two complexes with five-coordinate vanadium would have two ^{17}O signals in a 1:1 intensity ratio if the nitrogen is axial and only one signal if the nitrogen is equatorial (assuming the cis or trans hy-

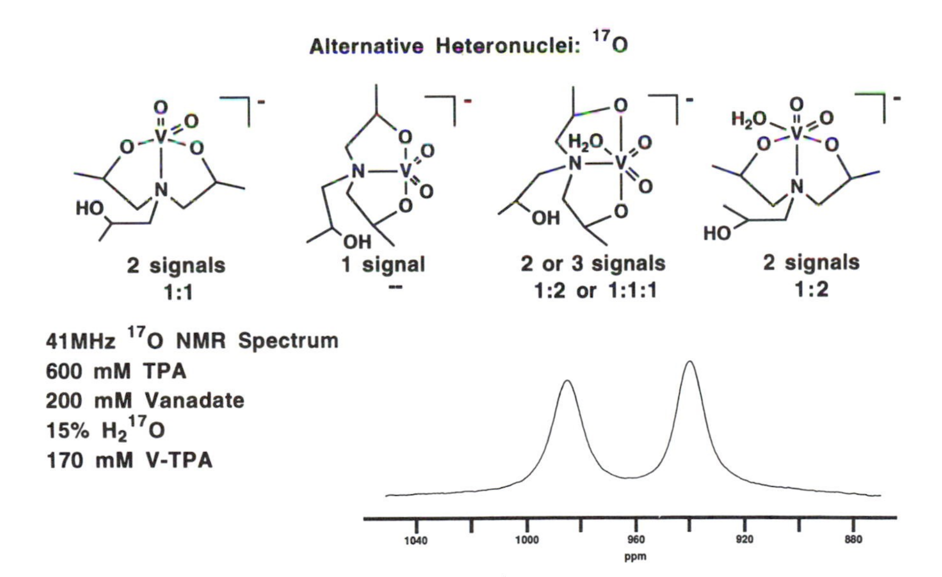

Figure 5. Four structural possibilities for the V-TPA complex. The number of ^{17}O NMR signals predicted for each structure and the experimentally observed ^{17}O NMR spectrum are shown below the structures.

droxypropyl group does not affect the chemical shift of the oxygen atom enough to overcome the broad ^{17}O signals). The exchange between labeled oxygens in the V-TPA complex is presumed to be fairly slow on the ^{17}O NMR timescale as shown for related complexes (*25*). If the complex was six-coordinate, two or three resonances would be expected in the ^{17}O NMR spectrum. If two resonances were observed, their intensity ratio should be 1:2, and if three resonances were observed, the intensity ratio should be 1:1:1. Because the spectrum shown in Figure 5 shows two signals in a 1:1 ratio with a chemical shift difference of more than 40 ppm, we conclude that the aqueous structure of the V-TPA complex is five-coordinate with the vanadium–nitrogen bond axial.

Thermodynamic Studies

NMR spectroscopy can also be used to determine the formation constants of various complexes. We illustrate this by quantificating the distribution of vanadate oligomers in aqueous solution by ^{51}V NMR spectroscopy. Moreover, the chemical shifts are very sensitive to protonation of the oxovanadium species, and ^{51}V spectra can provide information on the specific protonation state of the anions in solution.

Aqueous solutions of vanadate contain a mixture of monomer (V_1), dimer (V_2), tetramer (V_4), and pentamer (V_5) (*32*). Only V_4 has been characterized by X-ray crystallography (*33*, *34*). The speciation in aqueous solution is well-characterized by ^{51}V NMR spectroscopy because each anionic species has a distinct signal (in contrast to other methods used for quantification of vanadate oxoanions) (*12*). Because vanadium is a quadrupole and each of its oligomers relax with similar relaxation times (*32*), integration of ^{51}V NMR spectra gives accurate relative percentages of the various vanadate species present in solution. Assuming that all of the vanadium is present in the form of vanadium(V), we can calculate the concentration of each species. Figure 6 shows the ^{51}V NMR spectra recorded at 0.5, 1.0, and 5 mM total vanadate concentrations in the presence of 150 mM imidazole and at pH 8.0. Graphical representation of the pH-dependent equilibria between V_1 and V_2 and between V_1 and V_4 are also included in Figure 6. Plotting the V_2 and V_4 concentrations determined from the integrated ^{51}V NMR spectra as a function of $[V_1]^2$ and $[V_1]^4$, respectively, gives straight lines, with the H^+-dependent formation constant as the slope (*35*).

Quantitative solution-state NMR studies can be very important for bioinorganic studies because the active species cannot often be characterized by other means in solution (*15*, *32*, *35*, *36*).

Kinetics and Dynamic Processes

Metal–ligand complexes can undergo many types of dynamic processes (including complex formation). Dynamic processes can vary from ligand–

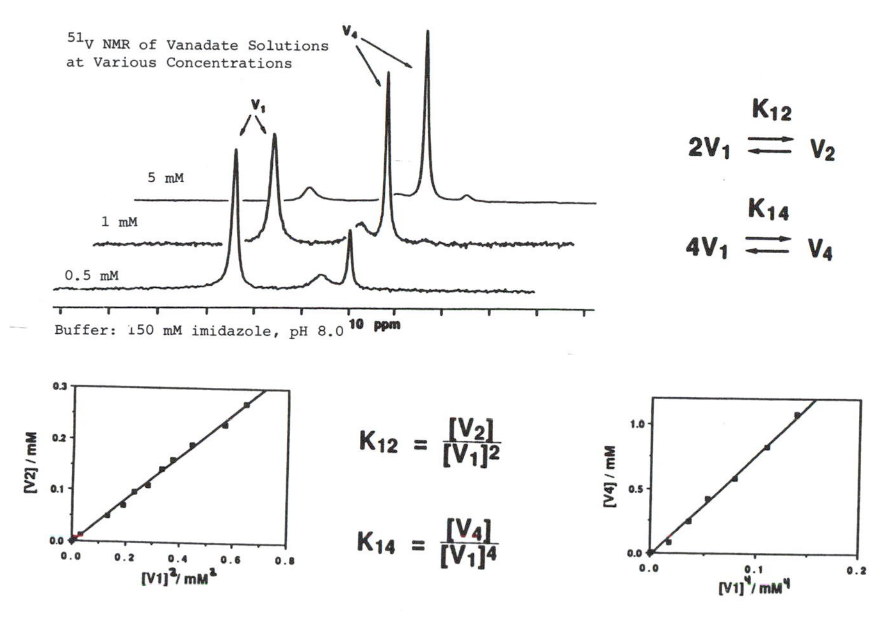

Figure 6. ^{51}V *NMR spectra of 0.5, 1.0, and 5.0 mM total vanadate in 150 mM imidazole at pH 8.0. The V_1 (−555 ppm) and V_4 (−579 ppm) resonances are indicated by arrows. The V_2 resonance (−573 ppm) is upfield of the V_1 and downfield of the V_4 resonance, and the V_5 resonance (−585 ppm) is furthest upfield. The H^+-dependent equilibrium constants K_{12} and K_{14} are defined, and plots of $[V_2]$ as a function of $[V_1]^2$ and $[V_4]$ as a function of $[V_1]^4$ are shown for a study carried out at constant ionic strength. The data shown in these plots were reported previously as assay conditions for studies of 6-phosphogluconate dehydrogenase from sheep liver (Data are from reference 35).*

complex exchange to conformational interconversions within the complex. NMR techniques exist with which these processes may be investigated, including variable temperature NMR spectroscopy, as discussed in **Thermodynamic Studies.** These NMR techniques include monitoring the time-course of a reaction by one-pulse NMR experiments, coalescence point determination, lineshape analysis, and 1D and 2D magnetization transfer techniques (*37*). This section focuses mainly on the application of qualitative and quantitative 2D EXSY NMR methods to study intra- and intermolecular processes (*38*).

Monitoring Complex Formation by NMR Spectroscopy. The rates of slow reactions are measured by either monitoring the formation of product or the disappearance of starting materials. When complex formation (or appropriate dynamic process) is slow on the time scale of the NMR experiment, its kinetics may be studied by NMR. In this case, slow complex formation may be followed by the observation of distinct resonances for the complex (product), intermediates, or starting materials. If the reaction progresses over the course of hours, days, or weeks, the reaction progress toward completion can be monitored by recording the 1D NMR spectra as a function of time (*39*). In such cases, the NMR analysis simply represents an analytic technique that allows quantification of the various species involved in the reaction. These reactions are all studied in their early stages before they reach completion.

Chemical Exchange Processes at Equilibrium. Chemical exchange processes are often fast, and when the system is at equilibrium, their kinetics can also be studied by NMR. These types of processes can be studied by classic line-shape analysis (rate constants approximating $10^4\ s^{-1}$) or magnetization transfer methods (rate constants approximating $10^{-2}\ s^{-1}$) (*37*). In general, methods that require the fewest assumptions about relaxation times, relaxation mechanisms, chemical shift differences, and temperature dependence are most reliable. Coalescence-point and band-shape simulation methods often assume temperature invariance of chemical shifts. Band-shape analysis requires knowledge of linewidths in the absence of exchange. Magnetization transfer techniques and 2D exchange techniques do not require knowledge of temperature-dependent chemical shifts or linewidths. Although one of these methods may present advantages over the others for a given system, it is advisable to determine any given rate constant by more than one method or by using more than one nucleus.

The 1D magnetization transfer method is well-known and is most effective for determining exchange rate constants for two- and three-site exchange systems (*40*). The 1D magnetization transfer experiment may be conducted by using either selective or nonselective pulse methods. Because selective pulses (*41*) often spill over to nearby resonances, less convenient nonselective methods are preferable in cases in which there is overlap in the spectral region of interest and hardware prohibits selective irradiation (*42, 43*). We used a 1D magnetization transfer experiment to measure the chemical exchange between vanadate monomer and dimer. These results are detailed elsewhere (*27*).

For exchange in multisite systems, both line-shape analysis and 1D magnetization transfer methods become tedious and time-consuming. Because 2D magnetization transfer methods (2D EXSY) provide simultaneous quantitative information about every exchange process taking place in a system, they present an attractive alternative to studying the dynamics of complex systems.

Qualitative 2D EXSY Spectroscopy of Vanadium(V) Complexes. The 2D EXSY experiment is similar to the 2D nuclear Overhauser effect (NOESY) spectroscopy experiment) used to estimate internuclear distances in a molecule. Both NOESY and EXSY experiments use the same pulse sequence ($90°-t_1-90°-t_m-90°-t_2$) with NOEs and chemical exchange taking place, respectively, during the second delay time (the mixing time t_m). The 2D EXSY spectrum thereby provides off-diagonal cross-peaks between shifts for each pair of exchanging sites in a multisite system. The signal intensities of the cross-peaks, being proportional to the rate of exchange, will lead to simultaneous quantitative information about each exchange process in a potentially complex multisite exchange system (*44*).

We first illustrate the use of 2D ^{13}C EXSY spectra to examine the exchange between ligand and complex in an aqueous solution of HIDA and V-HIDA 2 complex at pH 5.13. Figure 7 shows the partial 2D ^{13}C EXSY spectrum recorded at 318 K. The temperature most appropriate for recording the 2D EXSY spectrum was determined by a preliminary series of variable temperature ^{13}C NMR spectra. The temperature at which significant exchange is observed during the timescale of the experiment without excessive line-broadening is the preferred temperature to perform the 2D EXSY experiment. These preliminary studies are critical, because the dynamic processes are very sensitive to temperature, and the temperature range for favorable measurements is limited. Furthermore, experimental parameters such as sweep width, relaxation delay, and pulse width must be defined in preliminary experiments. The 2D experiment yields a 2D map with ^{13}C spectra along each dimension. If exchange occurs on the timescale of the experiment, an off-diagonal resonance between the exchanging sites will be observed. In Figure 6 off-diagonal resonances are observed between C_2 and L_2, between C_3 and L_3, and between C_4 and L_4. The integrated intensities of these off-diagonal resonances are directly proportional to the exchange rate between the ligand and the complex. The spectrum in Figure 6 clearly illustrates the exchange between the carbon atoms in the ligand and those in the complex.

In many complexes, intramolecular exchange is also taking place. A 2D EXSY spectrum that reveals the presence of both inter- and intramolecular exchange processes is shown in Figure 8. The structure of the aqueous V-TEA complex is analogous to the V-TPA complex described in previous paragraphs, with two bound (C_{4b} and C_{3b}) arms, and one free (C_{4f} and C_{3f}) arm (*15*). The ^{13}C EXSY spectrum shown in Figure 8 was acquired at 278 K and reveals several off-diagonal peaks. The solid lines connecting C_{4b} and L_4, C_{4f} and L_4, C_{3b} and L_3, and C_{3f} and L_3 indicate exchange between the complex and the free ligand. Off-diagonal resonances are also observed between C_{4b} and C_{4f} and between C_{3b} and C_{3f}. Quantification of this exchange process and the intermolecular exchange process show that the $C_{4b} \rightarrow C_{4f}$ and $C_{3b} \rightarrow C_{3f}$ off-diagonal signals are the result of intramolecular exchange. The V-TEA

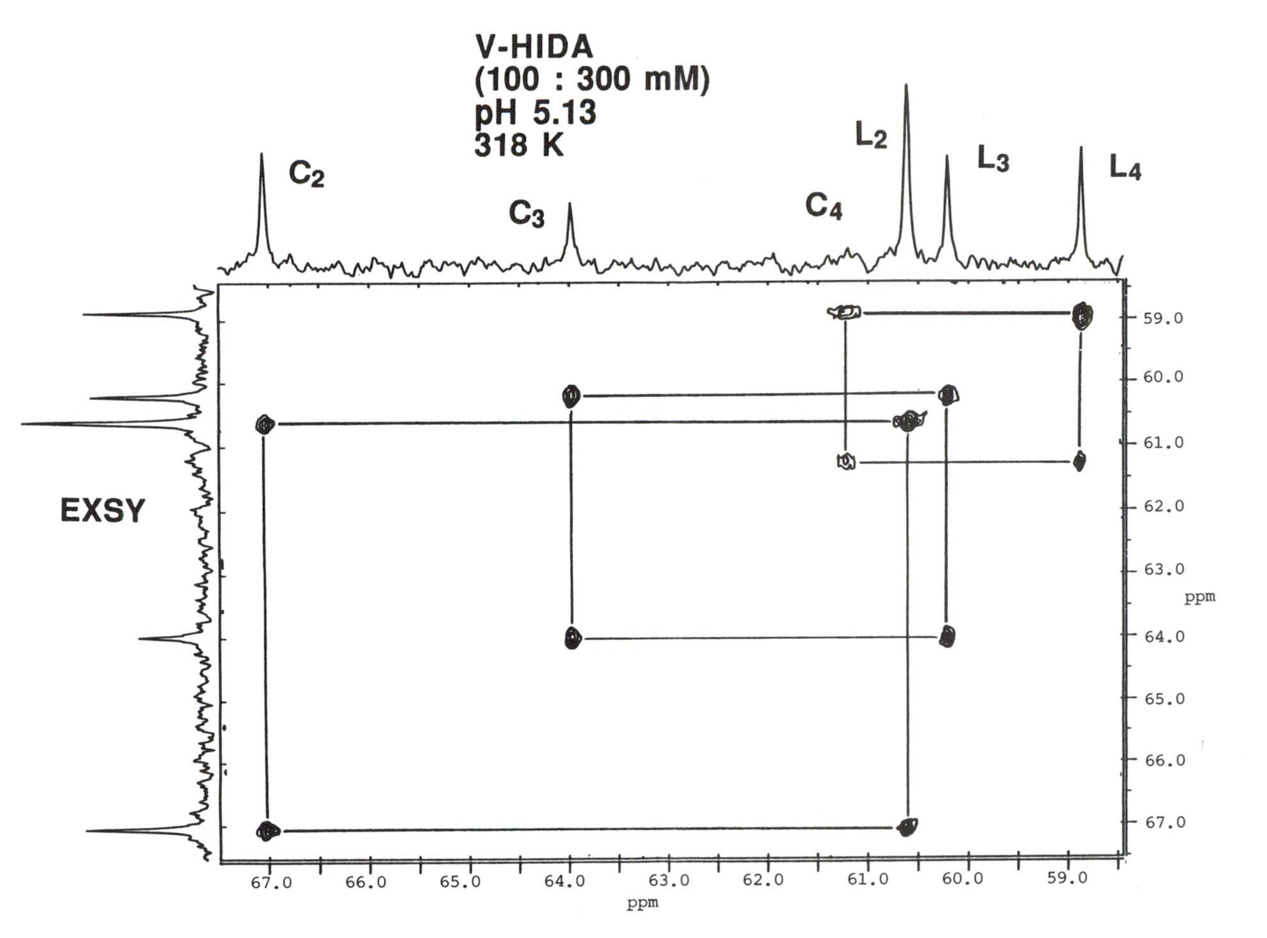

Figure 7. The ^{13}C *EXSY spectrum of the V-HIDA 2 complex formed in 100 mM vanadate and 300 mM HIDA at pH 5.13 and 318 K. The carbons assigned to the complex are labeled* C_x *and those of the ligand as*

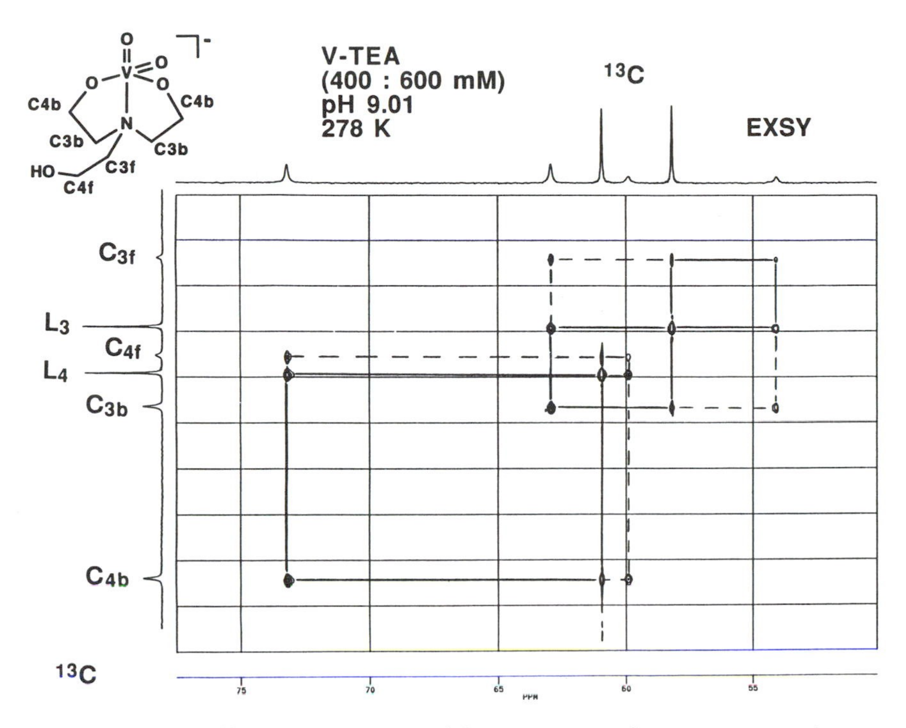

Figure 8. The ^{13}C EXSY spectrum of the V-TEA complex 400 mM vanadate and 600 mM HIDA at pH 9.01 and 278 K. The carbons assigned to the complex are labeled C_x and those of the ligand as L_x.

complex is an example of a complex in which intra- and intermolecular processes are occurring on the same timescale. Observation of both processes often requires that EXSY spectra be recorded at more than one temperature. The scientist has some flexibility in choosing the temperature at which to study a specific system. Decreasing the temperature could decrease the intermolecular exchange beyond detection and could possibly allow observation of the faster intramolecular process without interference from the slower intermolecular reaction also. In the case of the V-TEA system, temperatures much lower than 278 K could not be accessed, given the freezing point of water. Spectra recorded at higher temperatures show greater contribution from the intermolecular process relative to the intramolecular process. Even without quantification, this is evidence that all of the observed exchange is not due to one process. Other complexes such as the V-EDTA complex (*45*) and the V-Tricine [*N*-[tris(2-hydroxymethyl)methyl]glycine]–vanadium complex (*25*) allow observation of intra- and intermolecular processes without the interference of the other process.

Finally, we will describe a complex multisite intermolecular exchange system between the vanadate oligomers. This system is studied by 2D ^{51}V EXSY NMR spectroscopy. A 2D EXSY spectrum is shown in Figure 9. Vanadium-51 relaxes much more rapidly than carbon-13 and as a result, the 2D ^{51}V EXSY experiment is performed with mixing times of approximately 10 ms (compared to the mixing time of 300 ms for the 2D ^{13}C EXSY experiment). At pH 8.6, off-diagonal resonances are observed between all of the resonances. We therefore conclude that the V_1, V_2, V_4, and V_5 all exchange with one another (32). Because the vanadate oxoanions are all different species, this solution represents an example of a multisite intermolecular exchange system.

Quantification of EXSY data produces a "map" of individual site-to-site rate constants. Because all of the data are derived from a single experiment, this method is preferable to analogous 1D methods whenever the number of exchanging sites in the system exceeds two.

Quantitation and Interpretation of 2D EXSY Spectra. Quantification of the EXSY experiment is accomplished through a series of

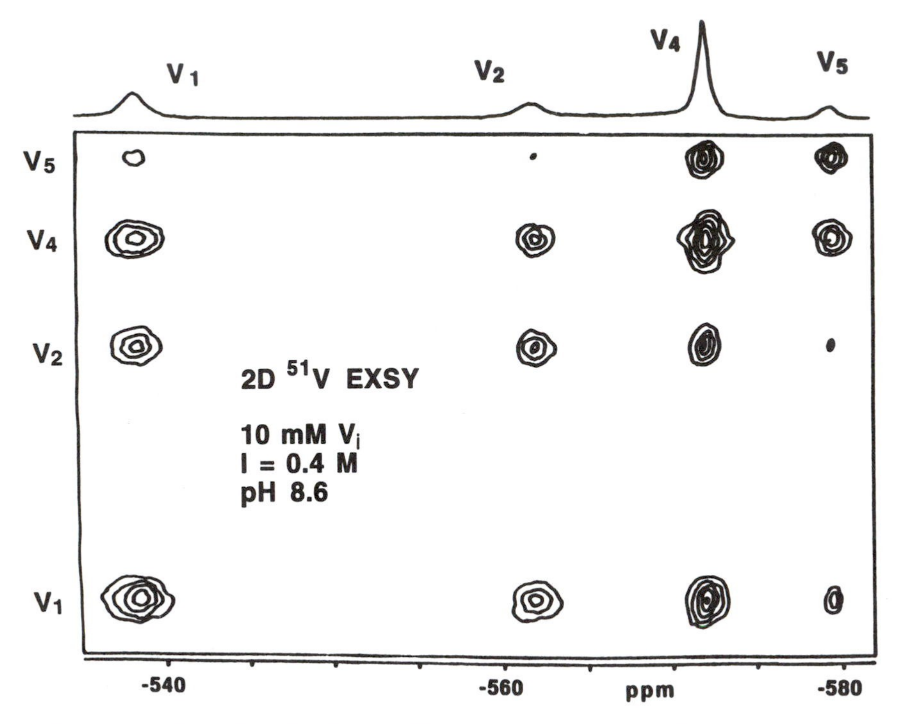

Figure 9. The 2D ^{51}V EXSY NMR spectrum of a 10 mM total vanadate solution containing KCl to obtain an ionic strength of 0.4 at pH 8.6 (32). The vanadate oligomers V_1 (−542 ppm), V_2 (−562 ppm), V_4 (−574 ppm), and V_5 (−579 ppm) are indicated.

matrix manipulations carried out by computer. Each of the magnetizations m_i of an EXSY spectrum may be expressed by equation 1, where m_i is the magnetization's deviation from its equilibrium value ($M_i - M_{i,0}$) and **m** is the corresponding matrix (*38*). This expression includes relaxation terms in addition to exchange terms for each of the exchange pathways from site *i*. The *n* magnetizations of an *n*-site exchange system may be expressed as rows in the $n \times n$ magnetization matrix **M** shown in equation 2. Calculation of the exchange matrix **R** (equation 3) is generally accomplished by a matrix diagonalization method, which linearizes the sum of exponentials in the integrated form of equation 2. In equation 3, **R** is the exchange matrix, t_m is the mixing time, **M** is the magnetization matrix, $\mathbf{M}_0$ is the matrix of equilibrium magnetizations, **X** is the matrix that diagonalizes **M**, $\mathbf{X}^{-1}$ its inverse, and Λ is the diagonal matrix of exchange and relaxation rate constants.

$$\frac{\mathrm{d}m_i}{\mathrm{d}t} = k_{1i}m_1 + k_{2i}m_2 + \cdots - \left(T_1^{\,i-1} + \sum_{j=1}^{N} k_{ij}\right)m_i \tag{1}$$

$$\frac{\mathrm{d}\mathbf{m}}{\mathrm{d}t} = -\mathbf{Rm} \tag{2}$$

$$\mathbf{R} = -t_m^{-1}[\ln(\mathbf{MM}_0^{-1})] = -t_m^{-1}[\mathbf{X}(\ln \Lambda)\mathbf{X}^{-1}] \tag{3}$$

Matrix representation allows the components of the resulting exchange matrix **R** to be readily interpreted. Coupling terms (off-diagonal elements) of the exchange matrix provide site-to-site rate constants. Each site-to-site rate constant k_{ij} (which may represent a series of elemental steps) is provided directly from the off-diagonal elements of the exchange matrix $\mathbf{R}_{ji}$. Relaxation information is contained only in the diagonal elements ($\mathbf{R}_{ii} = T_1^{\,i-1} + \sum_{j=1}^{N} k_{ij}$). Thus unlike in 1D dynamic methods (magnetization-transfer and line-width analysis), quantification of a single EXSY experiment separates each of the exchange rate constants from each other and from the relaxation rate constants.

Note, however, that the exchange rate constants in the rate matrix are not necessarily rate constants for elemental chemical reactions. The observed pseudo-first-order rate constants in **R** are dependent on the fractional populations at various sites and are often made up of several elemental rate constants. The rate constants for the elemental steps of a chemical reaction must therefore be derived from the observed rate constants with a given mechanism in mind. Considerations for interpreting the measured magnetization transfer rates have been discussed (*38*) for both intra- (*46*) and intermolecular systems (*32*). In the following section we show a few examples.

In the case of the V-TEA complex, the two equivalent coordinated arms exchange with the pendent arm in an intramolecular process (Figure 8). The population of the carbons in the coordinated arms are twice that of the carbons in the free arm and because mass balance must be maintained, the forward and reverse rate constants are related as shown in equation 4:

$$2k(C_{4b} \rightarrow C_{4f}) = k(C_{4f} \rightarrow C_{4b}) \quad (4)$$

Neither $k(C_{4b} \rightarrow C_{4f})$ nor $k(C_{4f} \rightarrow C_{4b})$ represents the "true rate constant for exchange" (k_{chem}) because such a rate constant should not depend on the population of each site. Furthermore, only those pathways that result in magnetization transfer will be observed and so the rate of magnetization transfer is not the same as the rate of the underlying process. In this case the measured rate constant $k(C_{4b} \rightarrow C_{4f})$ expresses the rate by which C_{4b} is converted to C_{4f}. In degenerate chemical exchange processes, the intermediate species (whether this be an intermediate or a transition state) can decay back to the ground-state structure or proceed to product. The exchange taking place in the V-TEA complex is such a process but is only observed if C_{4b} is converted to C_{4f}. The "true" rate constant characterizes a chemical event, which should reflect every passage to the transition state whether or not the passage continues on to observable product. The chemical event has been defined as the formation of a species on the multidimensional reaction manifold from which the final connectivity of those nuclei undergoing exchange are determined statistically (*46*). In the case of the V-TEA complex, likely intramolecular exchange mechanism would require that all three arms (one pendent and two chelated arms) have an equal likelihood of exchanging during the chemical event. Accordingly k_{chem} will be related to $k(C_{4b} \rightarrow C_{4f})$ and $k(C_{4f} \rightarrow C_{4b})$ as shown in equation 5:

$$k_{chem} = 6k(C_{4b} \rightarrow C_{4f}) = 3k(C_{4f} \rightarrow C_{4b}) \quad (5)$$

A second example is illustrated by the V-Tricine system (*25*). Its EXSY spectrum, revealing intramolecular exchange, is shown in Figure 10. The V-Tricine complex differs from the V-TEA complex in that the two-pendent hydroxymethyl arms are inequivalent (an ABC system) (*46*). The rate constants determined for the coordinated arm exchanging with each pendent arm [$k(C_{4b} \rightarrow C_{5f})$ and $k(C_{4b} \rightarrow C_{6f})$] are very similar (*25*). Assuming the chemical event is the formation of a single species from which the two pendent arms and the chelated arm is derived, each rate represents only one-third of the exchange rate as shown in equation 6:

$$k_{chem} = 3k(C_{4b} \rightarrow C_{5f}) = 3k(C_{4f} \rightarrow C_{6b}) = 3k(C_{5f} \rightarrow C_{4b}) = 3k(C_{6f} \rightarrow C_{4b}) \quad (6)$$

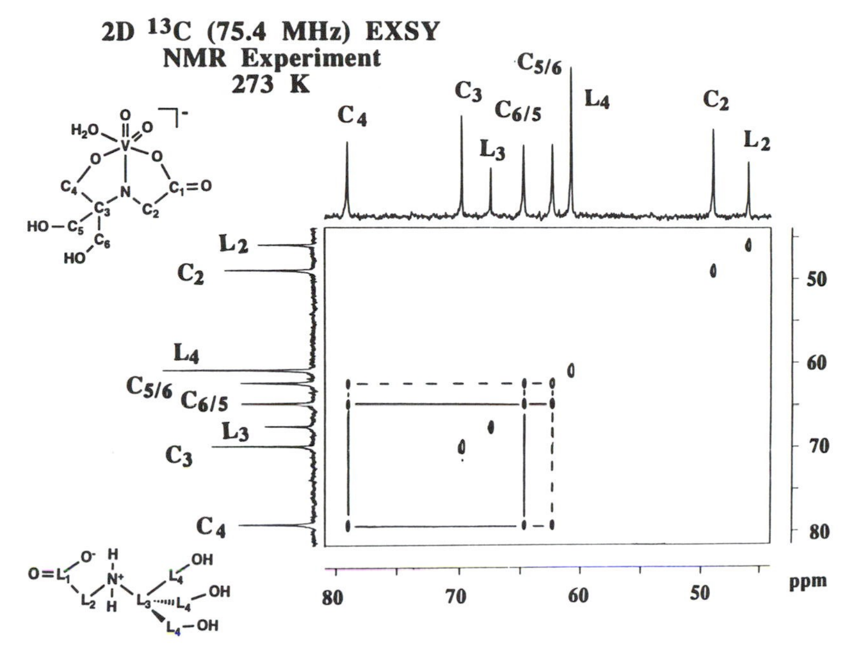

Figure 10. 2D ^{13}C EXSY NMR spectrum recorded at 75.4 MHz (7.05 T) at 0 °C (273 K) of a solution containing 412 mM total vanadate and 500 mM total Tricine at pH 7.1. This solution contains 406 mM V-Tricine, 0.38 mM V_1, and 94 mM free Tricine. The microscopic exchange rates were determined from the integrations. (Adapted from reference 25.)

However, as seen in Figure 10, exchange between the two pendent arms is also supported by another off-diagonal resonance ($k(C_{5f} \rightarrow C_{6f})$). This cross-peak would not be observed if there were only a single intramolecular mechanism for exchange as assumed above. This cross-peak is therefore likely to result from yet another dynamic process taking place for which several mechanisms can be proposed (25). Quantification of the rate constants is important when evaluating the merit of specific competing exchange pathways.

As described for the intramolecular systems, analyses of intermolecular exchange processes are also dependent on the mechanism of the reaction. A simple intermolecular exchange process is that between vanadate dimer (V_2) and vanadate monomer (V_1). Two likely mechanisms for formation of dimer are shown in equations 7 and 8. The first involves the combination of two monomers (equation 7). The second involves the combination of monomer and dimer (equation 8). Recording 2D ^{51}V EXSY spectra at a number of vanadate concentrations allows distinction between the two. A plot of the forward rate ($k_{(V1 \rightarrow V2)}[V_1]$) as a function of $[V_1]^2$ would be linear for equation 7 but not for equation 8. A linear relationship was in fact observed (*27*). This illustrates the use of quantitative 2D EXSY for mechanistic analyses. Such analyses require quantitative estimates of the uncertainties in the rate constant calculation. Only a few quantitative attempts have been made to estimate the errors in the rate constants obtained from 2D EXSY experiments (*47, 48*). Application of 2D EXSY to mechanistic studies requires that progress be made in this area, and we will describe briefly our efforts estimating errors from ^{51}V EXSY spectra.

$$V_1 + V_1 \rightleftarrows V_2 \tag{7}$$

$$V_1 + V_2 \rightleftarrows V_2 + V_1 \tag{8}$$

Error Estimation on 2D EXSY Experiments and Data Analysis. Estimating errors in a 2D EXSY measurement is complicated by the fact that the experiment is designed to provide all of the exchange rate constants from a single experiment. Statistical errors are frequently not reported nor determined, because the time and expense of the EXSY experiment often prohibits generation of a statistically significant number of repetitions (*48*). As a result, error estimation has focused on modeling random sources of error (*47, 48*). We examined various sources of both random and systematic error to determine the major contributors to error and to evaluate their magnitudes in the rate constants for exchange in a number of vanadate systems. We describe here our results on both two- and four-site systems.

Errors are propagated to the rate constants from the volume integrals. There are a number sources of error in the measurement of the volume integrals from which the site-to-site rate constants for exchange are derived. These have been discussed in the literature and include errors due to noise, baseline imperfection, incorrect phasing, experimental artifacts, inadequate digitization, and truncation (*49*). Our error estimates were obtained from an empirical estimation of the integration error and from the error in the integrals due to noise in the spectrum. We found that the precision of volume integration was limited primarily by the need for baseline correction. This is especially true in the case of ^{51}V NMR spectra, because the spectra of this nucleus are susceptible to rolling baselines (*50*).

Figure 11 shows the contour plots of the 2D ^{51}V EXSY spectrum of a two-site system (40 mM total vanadate at pH 10.9) in which V_1 exchanges with V_2. Quantification of the EXSY spectrum and calculation of the error propagated to the rate constants from the integration precision gives a 25% error on the rate constant. The results (both the rate constants and the errors) correspond nicely to the results obtained from a 1D magnetization transfer experiment on the same sample (*27*). The EXSY spectrum of a sample containing 12.5 mM total vanadate at 1.0

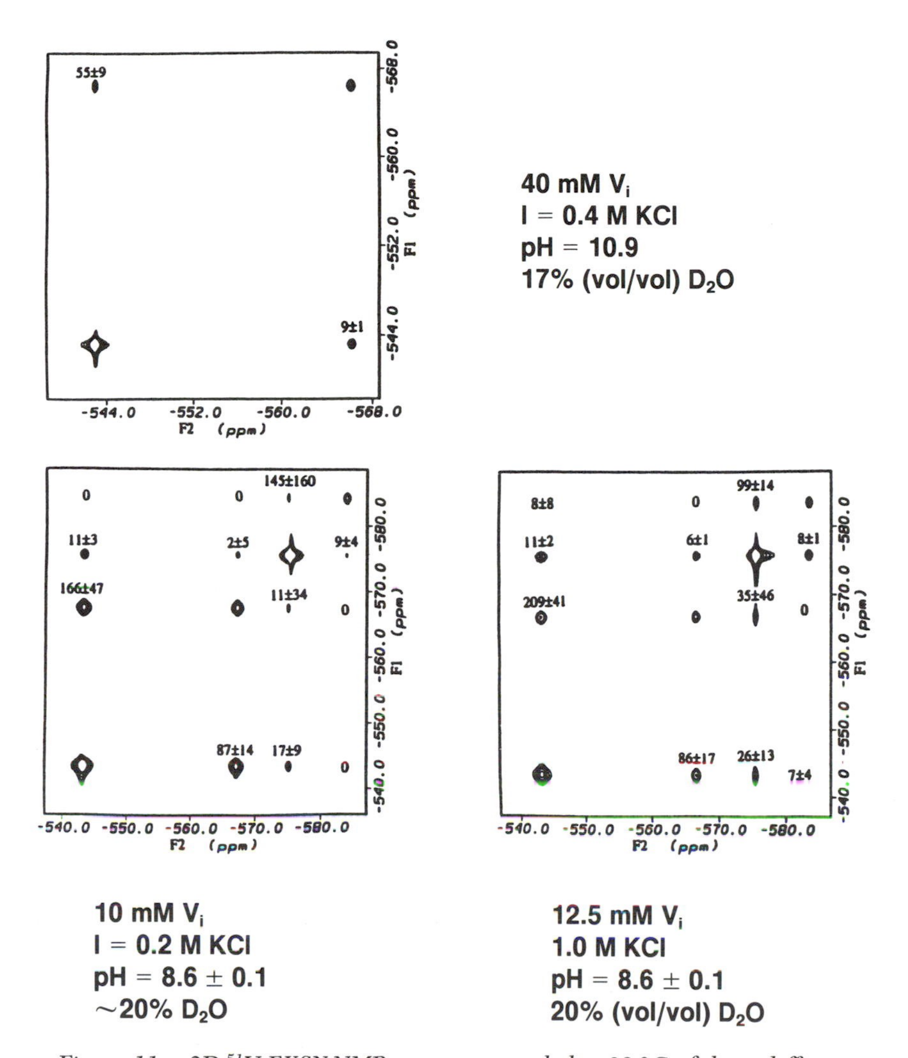

Figure 11. 2D ^{51}V EXSY NMR spectra recorded at 23 °C of three different vanadate solutions. A two-site system was observed in a solution containing 40 mM total vanadate at ionic strength of 0.4 M and pH 10.9. Two four-site systems were recorded at 10 and 12.5 mM total vanadate. The calculated rate constants are shown next to the off-diagonal resonances in all three maps. The errors were propagated to the rate constants from the integration precision and are also shown next to the off-diagonal resonance.

M ionic strength and pH 8.6 is shown in Figure 11 along with the calculated rate constants and their errors. The exchange between vanadate oligomers was anticipated to be very similar to that in the sample shown in Figure 9 (32). Nevertheless, less exchange is observed with the minor resonances (V_2 and V_5) in this sample, and several off-resonance cross-peaks are zero. The errors, determined by repeated volume integrations, on the rate matrix ranged from 20% to >100%. As anticipated, the errors associated with large off-resonance signals were small ($V_1 \rightarrow V_2$, $V_2 \rightarrow V_1$, $V_1 \rightarrow V_4$, $V_4 \rightarrow V_1$), whereas weak off-resonance signals result in large errors from 50% to 100% ($V_1 \rightarrow V_5$, $V_5 \rightarrow V_1$, $V_2 \rightarrow V_4$).

A sample containing 10.0 mM total vanadate in 0.2 M KCl at pH 8.6 was also analyzed. The spectrum and its associated rate constant matrix is shown in Figure 11. Although the 1D spectrum might lead one to conclude that the quality of this spectrum is the same as that of the previous sample, close inspection of the rate constants and errors reveal that higher errors were obtained on this measurement. Although the range of errors were similar, a greater fraction of small cross-resonances led to overall greater uncertainties. In addition to weaker off-diagonal signals, greater baseline distortion contributed significantly to the higher relative errors.

These examples illustrate the source and size of the experimental errors on a two- and a four-site system. They also highlight the importance of estimating the errors, because mechanistic interpretations often require fairly precise estimates of the rate parameters involved. If care is taken to minimize the noise in the spectrum (through signal averaging) and if the electronics of the system are set up to minimize baseline distortion, relative errors ≤25% should easily be obtained in even complex multisite systems.

Summary

Applications of NMR spectroscopy to structural, thermodynamic, and dynamic processes have been described. A brief discussion of the types of problems appropriate for study by this technique has been included. ^{1}H and ^{13}C NMR spectroscopy has been applied to define the ligand coordination in complexes. These experiments, combined with ^{17}O-labeling experiments, allowed deduction of the coordination number of the vanadium atom. Integration of NMR spectra allowed measurement of the formation constants and equilibrium constants. 2D ^{13}C and ^{51}V EXSY experiments were used in a qualitative and quantitative manner to examine intra- and intermolecular dynamic processes, of which several examples are discussed. The interpretation of the rate matrix and its relationship to the chemical processes under examination were also described. 2D EXSY spectroscopy has great potential as a tool with which to probe mechanisms in complex reactions; however, such uses often requires estimation of errors. The major source of error in 2D ^{51}V EXSY NMR studies on a two- and four-site vanadate system were found to be baseline distortion and the errors were estimated. Our results suggest that quantification of 2D EXSY experiments can be sufficiently accurate to make it a powerful mechanistic tool in studies of dynamic processes.

Acknowledgment

We thank the National Institutes of Health and the American Heart Association for funding most of this work. We also thank Christopher D. Rithner and Christopher R. Roberts for stimulating discussions and Christopher D. Rithner for technical assistance.

References

1. Harris, R. K.; Mann, B. E. *NMR and the Periodic Table;* Harris, R. K.; Mann, B. E., Eds.; Academic: New York, 1978.
2. Holz, R. C.; Que, L., Jr.; Ming, L.-J. *J. Am. Chem. Soc.* **1992,** *114,* 4434–4436.
3. Beer, R. H.; Lippard, S. J. *Inorg. Chem.* **1993,** *32,* 1030–1032.
4. Dec, S. F.; Davis, M. F.; Maciel, G. E.; Bronnimann, C. E.; Fitzgerald, J. J.; Han, S.-S. *Inorg. Chem.* **1993,** *32,* 955–960.
5. Wüthrich, K. *NMR in Biological Research: Peptides and Proteins;* Wüthrich, K., Ed.; Elsevier: New York, 1976.
6. Oppenheimer, N. J.; James, T. L. In *Spectral Techniques and Dynamics. Methods of Enzymology;* Oppenheimer, N. J.; James, T. L., Eds.; Academic: San Diego, CA, 1989; Vol. 176, Part A.
7. Gorenstein, D. G. *Phopshorus-31 NMR. Principles and Applications;* Gorenstein, D. G., Ed.; Academic: Orlando, FL, 1984.
8. Berkowitz, B. A.; Balaban, R. S. *Methods Enzymol.* **1989,** *176,* 330–494.
9. Boyd, J.; Brindle, K. M.; Campbell, I. D.; Radda, G. K. *J. Magn. Reson.* **1984,** *60,* 149–155.
10. Chasteen, N. D. *Vanadium in Biological Systems: Physiology and Biochemistry;* Chasteen, N. D., Ed.; Kluwer Academic: Boston, MA, 1990.
11. Butler, A.; Carrano, C. J. *Coord. Chem. Rev.* **1991,** *109,* 62–105.
12. Rehder, D. *Angew. Chem. Int. Ed. Engl.* **1991,** *30,* 148–167.
13. Chasteen, N. D. In *Biological Magnetic Resonance;* Reuben, J., Ed.; Plenum Press: New York, 1981; pp 53–119.
14. Eaton, S. S.; Eaton, G. R. In *Vanadium in Biological Systems: Physiology and Biochemistry;* Chasteen, N. D., Ed.; Kluwer Academic Publishers: Boston, 1990; p. 199–222.
15. Crans, D. C.; Chen, H.; Anderson, O. P.; Miller, M. M. *J. Am. Chem. Soc.* **1993,** *115,* 6769–6776.
16. Scheidt, W. R.; Countryman, R.; Hoard, J. L. *J. Am. Chem. Soc.* **1971,** *93,* 3878–3882.
17. Scheidt, W. R.; Collins, D. M.; Hoard, J. L. *J. Am. Chem. Soc.* **1971,** *93,* 3873–3877.
18. Kangjing, Z.; Xiaoping, L. *J. Struct. Chem.* **1987,** *6,* 14–16.
19. Kojima, A.; Okazaki, K.; Ooi, S.; Saito, K. *Inorg. Chem.* **1983,** *22,* 1168–1174.
20. Nishizawa, M.; Hirotsu, K.; Ooi, S.; Saito, K. *J. Chem. Soc. Chem. Comm.* **1979,** 707–708.
21. Nishizawa, M.; Saito, K. *Inorg. Chem.* **1980,** *19,* 2284–2288.
22. Zare, K.; Lagrange, J.; Lagrange, P. *Inorg. Chem.* **1979,** *148,* 568–571.

23. Zare, K.; Lagrange, P.; Lagrange, J. *J. Chem. Soc. Dalton Trans.* **1979,** 1372–1376.
24. Crans, D. C.; Shin, P. K. *Inorg. Chem.* **1988,** *27,* 1797–1806.
25. Crans, D. C.; Ehde, P. M.; Shin, P. K.; Pettersson, L. *J. Am. Chem. Soc.* **1991,** *113,* 3728–3736.
26. Crans, D. C.; Shin, P. K. *J. Am. Chem. Soc.* **1994,** *116,* 1305–1315.
27. Theisen, L. A. Ph.D. Thesis, Colorado State University, 1992.
28. Van Geet, A. L. *Anal. Chem.* **1970,** *42,* 679–680.
29. Becker, E. D. *High Resolution NMR Theory and Chemical Applications,* 2nd ed.; Academic: San Diego, CA, 1980.
30. Amos, L. W.; Sawyer, D. T. *Inorg. Chem.* **1972,** *11,* 2692–2697.
31. Yin-Zhuang, Z.; Xiang-Lin, J.; Shun-Cheng, L. *J. Struct. Chem.* **1993,** *12,* 48–51.
32. Crans, D. C.; Rithner, C. D.; Theisen, L. A. *J. Am. Chem. Soc.* **1990,** *112,* 2901–2908.
33. Day, V. W.; Klemperer, W. G.; Yagasaki, A. *Chem. Lett.* **1990,** 1267–1270.
34. Fuchs, J.; Mahjour, S.; Pickardt, J. *Angew. Chem. Int. Ed. Engl.* **1976,** *15,* 374–375.
35. Crans, D. C.; Willging, E. M.; Butler, S. K. *J. Am. Chem. Soc.* **1990,** *112,* 427–432.
36. Crans, D. C. *Comments Inorg. Chem.* **1994,** *16,* 1–33.
37. Sandstrom, J. *Dynamic NMR Spectroscopy;* Academic: New York, 1982.
38. Perrin, C. L.; Dwyer, T. J. *Chem. Rev.* **1990,** *90,* 935–967.
39. Crans, D. C.; Holst, H.; Rehder, D. *Inorg. Chem.* **1995,** *34,* 2524–2534.
40. King-Morris, M. J.; Serianni, A. S. *J. Am. Chem. Soc.* **1987,** *109,* 3501–3507.
41. Bellon, S. F.; Chen, D.; Johnston, E. R. *J. Magn. Reson.* **1987,** *73,* 168–173.
42. Engler, R. E.; Johnston, E. R.; Wade, C. G. *J. Magn. Reson.* **1988,** *77,* 377–381.
43. Bulliman, B. T.; Kuchel, P. W.; Chapman, B. E. *J. Magn. Reson.* **1989,** *82,* 131–138.
44. Derome, A. E. *Modern NMR Techniques for Chemistry Research*; Pergamon Press: Oxford, England, 1987.
45. Shin, P. K. Ph.D. Thesis, Colorado State University, 1993.
46. Green, M. L. H.; Wong, L.-L.; Sella, A. *Organometallics* **1992,** *11,* 2660–2668.
47. Abel, E. W.; Caston, T. P. J.; Orrell, K. G.; Sik, V.; Stephenson, D. *J. Magn. Reson.* **1986,** *70,* 34–53.
48. Kuchel, P. W.; Bulliman, B. T.; Chapman, B. E.; Mendy, G. L. *J. Magn. Reson.* **1988,** *76,* 136–142.
49. Weiss, G. H.; Ferretti, J. A. *J. Magn. Reson.* **1983,** *55,* 397–407.
50. Akitt, J. W.; McDonald, W. S. *J. Magn. Reson.* **1984,** *58,* 401–412.

RECEIVED for review July 19, 1993. ACCEPTED revised manuscript January 24, 1994.

12

Modeling Vanadium Bromoperoxidase

Oxidation of Halides by Peroxovanadium(V) Complexes

Alison Butler and Melissa J. Clague

Department of Chemistry, University of California, Santa Barbara, CA 93106

Functional mimics of the enzyme vanadium bromoperoxidase are the focus of this chapter, and the study of these mimics helps to address the mechanism of vanadium bromoperoxidase. A brief review of vanadium bromoperoxidase is also included. Vanadium bromoperoxidase catalyzes the oxidation of halides (chloride, bromide, and iodide) by hydrogen peroxide. The oxidized halogen species can halogenate appropriate organic substrates or oxidize a second equivalent of hydrogen peroxide, producing dioxygen. The functional mimics include cis-*(dioxo)vanadium(V) and vanadium(V)-liganded complexes (ligands: hydroxyphenylsalicylideneamine, salicylidene–amino acid Schiff bases, citric acid, and others). Some structural considerations concerning model compounds and vanadium bromoperoxidase are discussed.*

THE FIRST TWO VANADIUM ENZYMES, vanadium bromoperoxidase and vanadium nitrogenase, have been discovered in the last decade. Significant progress is also occurring in unraveling the biological role of vanadium in other systems, such as ascidians, phosphate metabolism, the insulin mimetic effect, etc. The reader is referred to the recent book, *Vanadium in Biological Systems*, edited by N. D. Chasteen for a summary of the recent progress in this area (*1*).

The focus of this chapter will be on the mechanistic chemistry of vanadium compounds that catalyze the oxidation of bromide by hydrogen peroxide—that is, functional mimics of vanadium bromoperoxidase. The reader is referred to references 2–8 for more comprehensive reviews of the enzyme.

0065–2393/95/0246–0329$08.18/0

Vanadium bromoperoxidase (V-BrPO) is a nonheme haloperoxidase found primarily in marine algae (*2–8*) although it has been isolated from a lichen (*9*) and very recently from a terrestrial fungus (*10*). The vanadium bromoperoxidases are acidic proteins (*11, 12*) with similar amino acid composition (*13*), subunit molecular weight (65,000), and charge (pI 4–5). The structure of V-BrPO is not known, although V-BrPO from *Ascophyllum nodosum* has been crystallized (*14*). Each subunit contains one active-site vanadium(V), although as isolated, V-BrPO is often deficient in vanadium content (*11, 15, 16*). A full complement of active-site vanadium(V) can be readily obtained by addition of vanadate (*2–8*). Vanadium can also be easily removed, producing the inactive apoprotein (*2–8*). The apo derivative is stable and can be fully reconstituted by addition of vanadate.

The ligands that coordinate the active-site vanadium(V) are not known. The extended X-ray absorption fine structure (EXAFS) analysis is suggestive of a distorted octahedral vanadium(V) coordinated by a single terminal oxide ligand at 1.61 Å, three unknown light-atom donors (O or N) at about 1.72 Å, and two nitrogen donors at 2.11 Å (Figure 1) (*17*). Moreover, electron spin echo results of the V(IV)-BrPO derivative suggest V(IV) coordination by a single terminal oxide at 1.63 Å, and five light-atom ligands (O or N). Three are at 1.91 Å, and two are at 2.11 Å (*17*).

Reactivity of Vanadium Bromoperoxidase

V-BrPO catalyzes halogenation reactions (*16, 19, 20*) and the halide-assisted disproportionation of hydrogen peroxide, producing dioxygen (*21*). In the first step, the enzyme catalyzes the oxidation of the halide by hydrogen peroxide producing an intermediate that is a two-electron oxidized halogen species. In the case of bromide oxidation, the formation of hypobromous acid, bromine, tribromide, or an enzyme-bound Br^+ ion is consistent with the in vitro reactivity of the enzyme. The exact nature of the oxidized halogen intermediate as enzyme-bound or released will not be discussed here (*see* references 2–4). The oxidized intermediate can halogenate an appropriate organic substrate or react

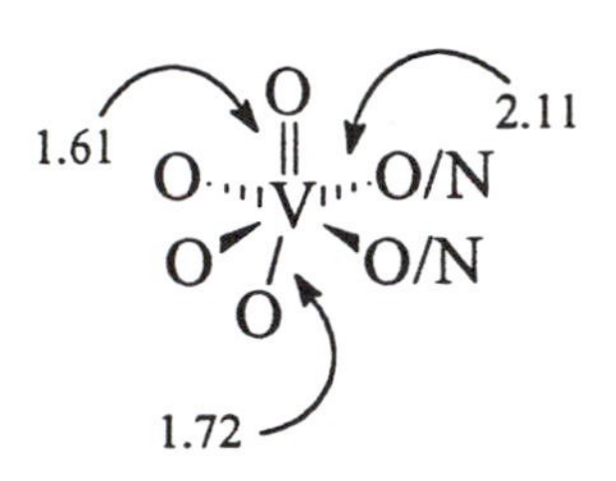

Figure 1. Proposed structure of the active site of vanadium bromoperoxidase based on EXAFS data (17).

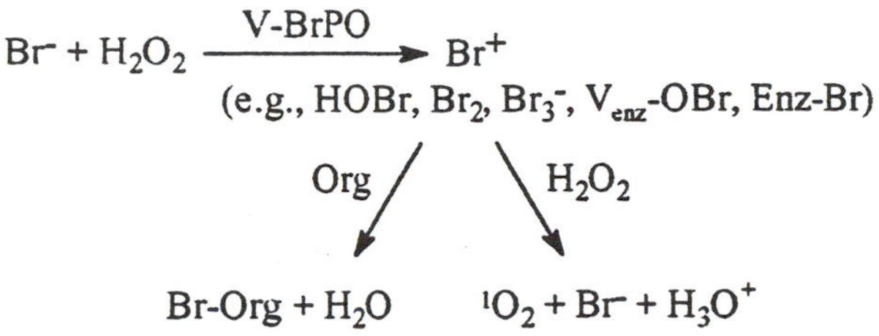

$$Br^- + H_2O_2 \xrightarrow{\text{V-BrPO}} Br^+$$

(e.g., HOBr, Br_2, Br_3^-, V_{enz}-OBr, Enz-Br)

Org ↙ ↘ H_2O_2

$Br\text{-}Org + H_2O$ $\quad\quad$ $^1O_2 + Br^- + H_3O^+$

Scheme 1. Scheme for vanadium bromoperoxidase, showing the formation of an oxidized intermediate which can react by two pathways to give products.

with another equivalent of hydrogen peroxide, forming dioxygen, as depicted in Scheme 1 for bromide (*21, 22*). In the case of bromide, the dioxygen formed is in the singlet excited state ($^1O_2$;$^1\Delta_g$) (*23*). Singlet oxygen formation with chloride is also expected (*24, 25*), however, singlet oxygen formation with iodide is not expected because iodide quenches singlet oxygen.

The oxidation of bromide by hydrogen peroxide is a two-electron process that results in electrophilic bromination of organic substrates as opposed to radical bromination. This reactivity was established from the products of bromination of 2,3-dimethoxytoluene in which ring-substituted bromo-2,3-dimethoxytoluene is the sole product (*26*). The product of bromination by bromine radical, 2,3-dimethoxybenzyl bromide (*27*), was not observed.

Haloperoxidase activity is usually determined spectrophotometrically by the bromination or chlorination of 2-chloro-5,5-dimethyl-1,3-cyclohexanedione (monochlorodimedone; MCD) using hydrogen peroxide as the oxidant of the halide (*28*) (Figure 2). This assay is convenient because of the large change in extinction coefficients between the enolate and the product. However, its use is constrained by the need for conditions that stabilize the enolate. In the early work on V-BrPO, the

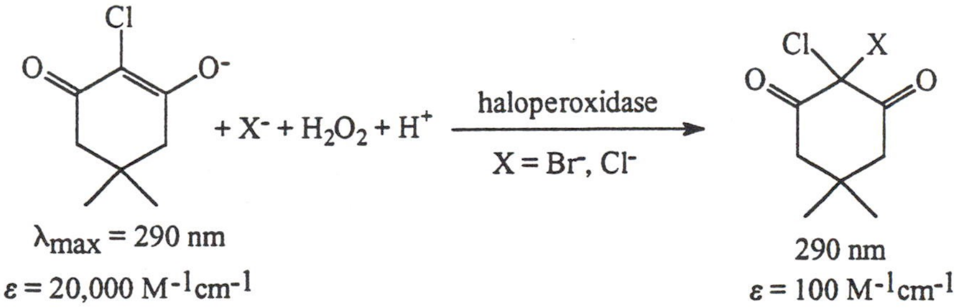

$$\text{MCD enolate} + X^- + H_2O_2 + H^+ \xrightarrow[X = Br^-, Cl^-]{\text{haloperoxidase}} \text{2-chloro-2-X-5,5-dimethyl-1,3-cyclohexanedione}$$

λ_{max} = 290 nm, ε = 20,000 $M^{-1}cm^{-1}$ $\quad\quad$ 290 nm, ε = 100 $M^{-1}cm^{-1}$

Figure 2. MCD assay for haloperoxidase activity.

oxidation of iodide by hydrogen peroxide (*19*), forming triiodide (I_3^-) was used to detect and quantify haloperoxidase activity. This assay, like the MCD assay, permits spectrophotometric analysis. I_3^- can be followed spectrophotometrically at 353 nm (ϵ = 26,400 $M^{-1}cm^{-1}$). Unfortunately, the nonenzymatic oxidation of iodide by hydrogen peroxide can be a significant competing side reaction especially at lower pH; thus the iodoperoxidase activity is not as desirable a method for the determination of haloperoxidase activity as the MCD assay.

The specific chloroperoxidase, bromoperoxidase, and iodoperoxidase activities differ substantially and depend on pH and the concentrations of halide and hydrogen peroxide (*2*). In general, the specific activity for halide oxidation increases in the order of chloride, bromide, and iodide. The pH for maximum specific haloperoxidase activity generally decreases in the order of iodide, bromide, and chloride, but direct comparisons are difficult because the pH maximum can be shifted over several pH units by varying the ratio of halide to hydrogen peroxide and because both halide and hydrogen peroxide can inhibit the enzyme under certain conditions.

The role of vanadium in V-BrPO is a topic of much current interest. Vanadium could function as an electron-transfer catalyst of halide oxidation by hydrogen peroxide or as a Lewis acid catalyst, remaining in the 5+ oxidation state. In an electron-transfer role, one might think that vanadium could cycle between V(V) and V(III). In such a scheme V(V) could be reduced by bromide to form V(III) and HOBr; V(III) could be reoxidized to V(V) by hydrogen peroxide. However, we have found that incubation of V(V)-BrPO with bromide and MCD does not result in stoichiometric bromination of MCD (*29*). On the other hand, certain vanadium(V) complexes have been shown to be reduced by bromide to V(III) (*30*) (*see* following Section). An electron-transfer cycle between V(V) and V(IV) seems unlikely because an electron spin resonance (ESR) signal is not observed under V-BrPO turnover conditions (i.e., in the presence of Br^- and H_2O_2) and because the nonradical nature of the bromination reactions is not consistent with one-electron reduction of V(V) [Of course formation of V(IV) and a protein radical anion could be consistent with both the absence of an ESR signal and the two-electron oxidation of bromide: however, halogenation is not observed in the absence of hydrogen peroxide] (*29*). A Lewis acid role for the vanadium(V) is consistent with the observations although direct observation of the vanadium site in V-BrPO during turnover has not been feasible to date. A small change in the UV spectrum of V(V)-BrPO is observed on addition of hydrogen peroxide; on subsequent addition of bromide, the original spectrum of V(V)-BrPO is restored (*31*). This finding is consistent with initial formation of a vanadium-peroxide species and subsequent oxidation of bromide, reforming V(V)-BrPO. Vanadium(V) peroxide complexes have been well-characterized and are known to be good oxidants, including oxidants of halides (*32, 33*) (*see* next section). Although a short-lived reduced oxidation state cannot be discounted in the oxidation of halides by peroxovanadium(V) species, the vanadium product is in the 5+ oxidation state.

Biomimics of V-BrPO

Haloperoxidases can be readily assayed by the halogenation of MCD (*28*). As discussed previously, the efficacy of this method relies on aqueous conditions to stabilize the enolate form of the β-diketone. Deviation from these conditions in search of active model complexes requires other assays. First, the bromination of 1,3,5-trimethoxybenzene (TMB) yields 2-bromo-1,3,5-trimethoxybenzene (BrTMB) and 2,4-dibromo-1,3,5-trimethoxybenzene (Figure 3) (*33*). Because bromine deactivates aromatic rings, complete monobromination is observed before the second bromination event. Therefore, the use of excess TMB insures conversion only to BrTMB and not to multiply brominated products. This conversion is readily followed by GC analysis. It can also be followed spectrophotometrically at 266 nm, but in some cases, strong absorbances by the vanadium peroxo species interfere. A second method to follow the oxidation of halide by hydrogen peroxide is the evolution of dioxygen, which is formed in the halide-assisted disproportionation of hydrogen peroxide. Dioxygen can be measured by an oxygen electrode under a wide range of conditions. High acid concentrations and alcohol–water mixtures are tolerated, but not neat organic solvents. Finally, direct observation of Br_3^- is also possible in a variety of solvents, as it is in aqueous solution (λ_{max} 267 nm; $\epsilon = 36{,}100\ M^{-1}\ cm^{-1}$), to permit direct spectrophotometric kinetic measurements. Attention must be paid to the equilibria and side-reactions of Br_3^-.

The Reactivity of *cis*-VO_2^+ in Acidic Aqueous Solution. The first functional mimic of vanadium bromoperoxidase was *cis*-

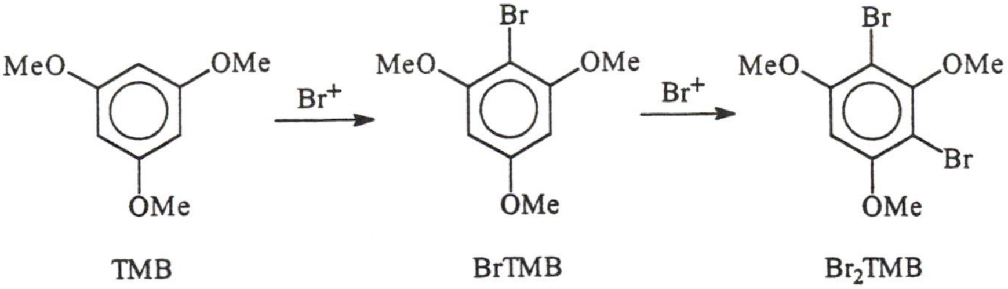

MeO OMe OMe TMB —Br⁺→ Br MeO OMe OMe BrTMB —Br⁺→ Br MeO OMe Br OMe Br_2TMB

Figure 3. Bromination of TMB is an alternative assay for haloperoxidase activity.

(dioxo)vanadium(V) in acidic aqueous solution (*32*, *33*). The proposed mechanism (shown in Scheme 2) involves the coordination of hydrogen peroxide by VO_2^+ to give the corresponding peroxo and diperoxo species in ratios governed by the equilibria shown (*32*).

$$VO_2^+ + H_2O_2 \rightleftarrows VO(O_2)^+ + H_2O$$

$$VO(O_2)^+ + H_2O_2 \rightleftarrows VO(O_2)_2^- + 2H^-$$

The subsequent reaction of the peroxo complexes with bromide gives oxidized bromine species (*see* Scheme 2). HOBr, Br_2, and Br_3^- rapidly equilibrate, and Br_3^- is the predominant spectrophotometrically observed intermediate (λ_{max} 267 nm; $\epsilon = 36{,}100\ M^{-1}\ cm^{-1}$) in the absence of an organic substrate. Tribromide is stabilized with respect to HOBr, Br_2 and decomposition products by high bromide and acid concentrations (*34*). HOBr is reduced by excess hydrogen peroxide to yield bromide, water, and dioxygen, of which dioxygen can be measured. In the presence of TMB, the oxidized species is rapidly consumed in the bromination of TMB to BrTMB. Quantitation of BrTMB demonstrates that bromination is stoichiometric with respect to the concentration of H_2O_2 added. Thus TMB is a rapid, quantitative trap for the oxidized bromine species (*33*).

Reactivity of (HPS)VO(OH). A variety of coordination complexes catalyze the oxidation of bromide by hydrogen peroxide. The first and most studied catalyst is the hydroxooxovanadium(V) complex of hydroxyphenylsalicylideneamine (H_2HPS, shown in Figure 4) (*35*). The proposed mechanism is shown in Scheme 3. The catalyst precursor is the crystalline (HPS)VO(OEt)(EtOH) compound, which is readily prepared from $VO(iOPr)_3$ and H_2HPS (*36*) in absolute ethanol. Dissolution of this solid in *N,N*-dimethylformamide (DMF) gives rise to five species: (HPS)VO(OEt), the active catalyst (HPS)VO(OH), and three stereochemically distinct di-

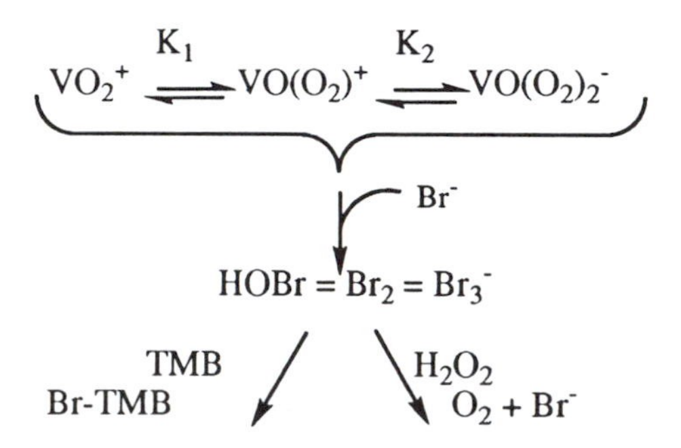

Scheme 2. Bromide oxidation by hydrogen peroxide catalyzed by cis-di-*oxovanadium(V). (See Note Added in Proof).*

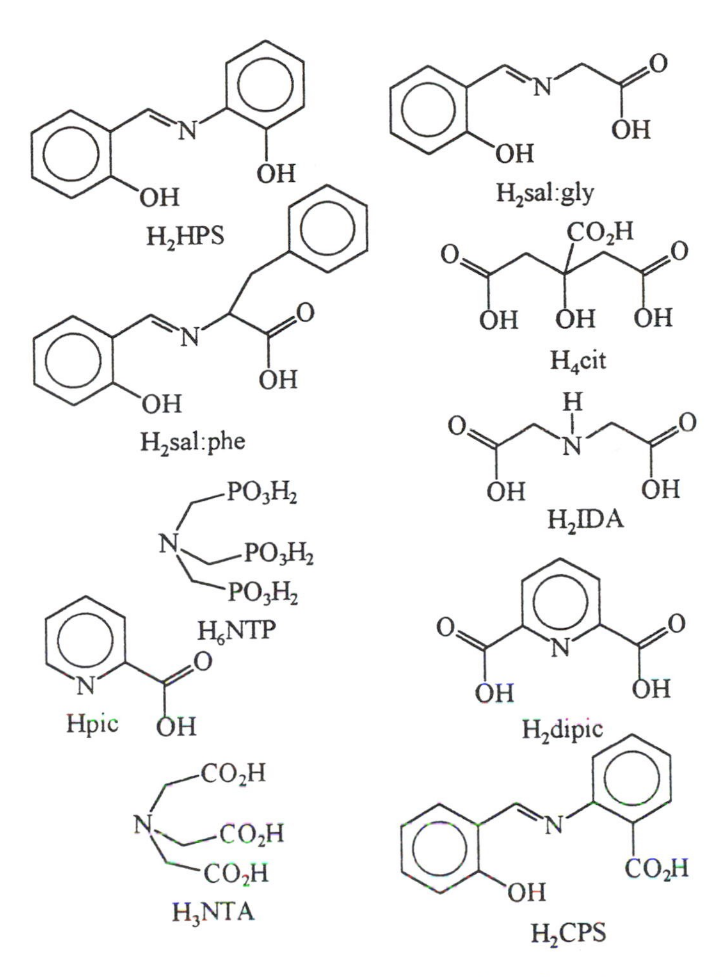

Figure 4. Ligands whose vanadium(V) complexes have been tested for catalysis of bromide oxidation.

mers, $[(HPS)VO]_2O$ (35). Addition of hydrogen peroxide gives rise to another species, $(HPS)VO(O_2)^-$. This peroxo complex is competent to oxidize bromide in the presence of sufficient acid, giving rise to tribromide. The oxidation probably occurs via nucleophilic attack by bromide on the coordinated peroxide or following bromide coordination to the metal. The initial product of oxidation may be vanadium-bound hypobromite, although this species has not been experimentally detected.

The oxidation of bromide can be conveniently and quantitatively measured by the bromination of TMB in DMF solution. Bromination is stoichiometric with the vanadium complex concentration in the absence of added acid (35). The stoichiometric addition of acid results in an

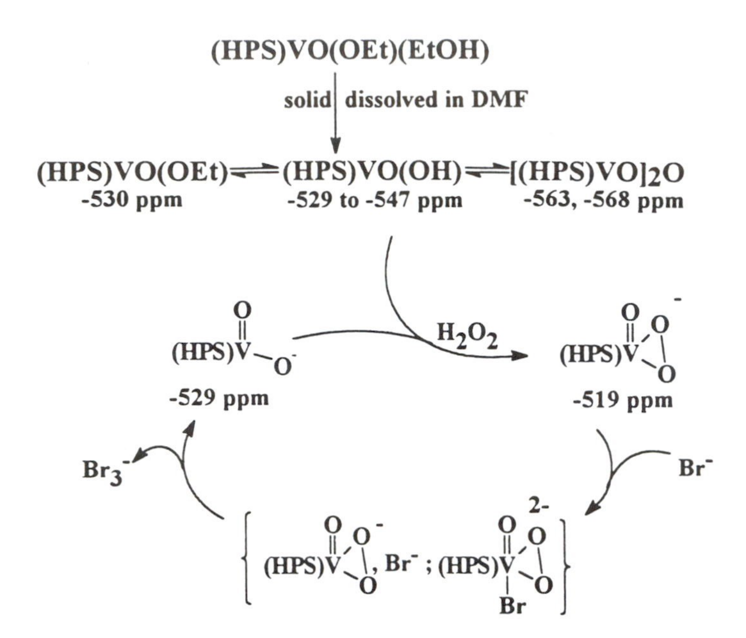

Scheme 3. Proposed mechanism for bromide oxidation by hydrogen peroxide catalyzed by (HPS)VO(OH).

additional equivalent of BrTMB. After the first turnover, one equivalent of acid is required per turnover. These experiments require hydrogen peroxide concentrations in excess of the sum of the vanadium and acid concentrations. When hydrogen peroxide is limiting (i.e., less than the sum of the vanadium and acid concentrations), the formation of BrTMB is quantitative with hydrogen peroxide.

Valuable information about the actual species in solution before and after the oxidation reaction is derived from ^{51}V NMR results. The reactions are too fast under spectroscopically observable conditions (mM vanadium) to permit data collection during the reaction. The observed resonances are assigned to particular species as shown in Scheme 3, but the experiments that facilitated these assignments merit some discussion.

Dissolution of (HPS)VO(OEt)(EtOH) in DMF, as mentioned previously, gives rise to four resonances, which are assigned to five species (Scheme 3). Addition of 0.5 M ethanol converts all the vanadium in solution to a single species with a resonance at −530 ppm. This species is assigned as (HPS)VO(OEt) (35). (Evidence that ethanol binds as ethoxide will be presented.) If methanol is added instead of ethanol, the sole resulting resonance is at −524 ppm, assigned as (HPS)VO(OMe). Addition of ethanol and methanol results in resonances at both −524 and −530 ppm (37). This complex may coordinate a solvent molecule in the axial position trans to the oxo group. Either formulation (five- or six-coordinate) is consistent with the alcohol inhibition described later in this section.

A vanadium complex can be precipitated from a concentrated acetonitrile solution of (HPS)VO(OEt)(EtOH) by the addition of a small amount of water. This complex has an elemental analysis consistent with the formulation $[(HPS)VO]_2O \cdot MeCN$. Its ^{51}V NMR spectrum shows only the three upfield resonances. Addition of ethanol results again in complete conversion to the −530 ppm form (HPS)VO(OEt) (35).

Addition of water to a DMF solution of (HPS)VO(OEt)(EtOH) results in the conversion of all the vanadium to a single species with a resonance at −545 ppm, consistent with the hydrolyzed form (HPS)VO(OH). The position of this resonance is acid- and base-dependent. Addition of NaOH causes this resonance to move incrementally downfield to a position of −529 ppm, which is reached after addition of one equivalent of base. A concomitant decrease in concentrations of the ethoxy and dimeric species is observed. This chemical shift remains constant upon addition of more base. Addition of acid causes this resonance to move upfield to −547 ppm.

This hydrolyzed form, $(HPS)VO(OH)/(HPS)VO_2^-$, is presumably the one to which hydrogen peroxide coordinates, initiating oxidation of bromide. The addition of hydrogen peroxide to a DMF solution of (HPS)VO(OEt)(EtOH) results in a small, additional resonance at −519 ppm. Bulk conversion is effected by the stoichiometric addition of base. In order to assess the protonation state of the hydrogen peroxide that binds, 1 mM (HPS)VO(OEt)(EtOH) was reacted in DMF with excess (4 mM) hydrogen peroxide and 0.5 mM NaOH. The expected resonance at −519 ppm (comprising about half the total signal area) was observed along with resonances at −542, −563, and −568 ppm (Figure 5). Importantly, the position of the acid-dependent resonance shows that the half-equivalent of base exactly neutralizes the protons released upon binding of hydrogen peroxide. Because the initial species is (HPS)VO(OH), the binding of hydrogen peroxide as O_2^{2-} releases one proton (and an equivalent of water). This proton is neutralized by the added base.

$$(HPS)VO(OH) + H_2O_2 \rightleftarrows (HPS)VO(O_2)^- + H_3O^+$$

If hydrogen peroxide were bound as hydroperoxide, HO_2^-, then no protons would be released upon displacement of hydroxide, and the acid-dependent resonance would move downfield as a result of the added base.

A similar experiment was performed to determine whether the ethanol that is bound in the −530 ppm form is ethanol or ethoxide. Addition of ethanol in 1 mM increments to a 1 mM solution of (HPS)VO(OEt)(EtOH) results in an increase in the area of the resonance at −530 ppm (Figure 6). In addition, the acid-dependent resonance

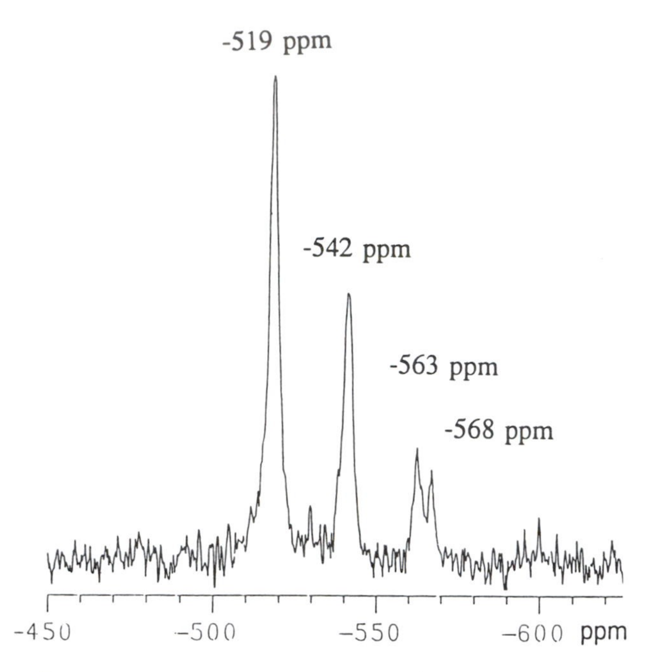

Figure 5. ^{51}V NMR spectrum of 1 mM (HPS)VO(OEt)(EtOH), 0.5 mM NaOH and 4 mM H_2O_2 in DMF.

moves from −542 ppm to −547 ppm, signaling the release of the hydroxyl proton upon ethoxide binding to the vanadium center.

Vanadium(V) without a coordinating organic ligand such as HPS^{2-} is also active as a bromide oxidation catalyst in *wet* DMF (*35*), however, neither sodium vanadate nor ammonium vanadate is soluble in neat, dry DMF (*35*). The mechanism of bromide oxidation appears to be analogous to the one in aqueous solution: peroxide coordinates to the vanadium center, then the peroxovanadium species oxidizes bromide. The rate of bromide oxidation, as measured spectrophotometrically by the appearance of Br_3^- at 272 nm, is linear with vanadium concentration in the range of 1–10 μM, indicating that the vanadium species is monomeric. Data from ^{51}V NMR experiments suggest that the peroxo complex is, in fact, oxo(diperoxo)vanadium(V). The extent of solvation by water or DMF is unknown.

Because unligated V(V) is an effective catalyst for the reaction of hydrogen peroxide and bromide, product studies and spectroscopic characterization are insufficient to establish the identity of the active catalyst. None of the evidence so far presented precludes the possibility that the ligand HPS^{2-} dissociates briefly and that the reaction is catalyzed by a small, spectroscopically undetectable amount of unligated vanadium(V).

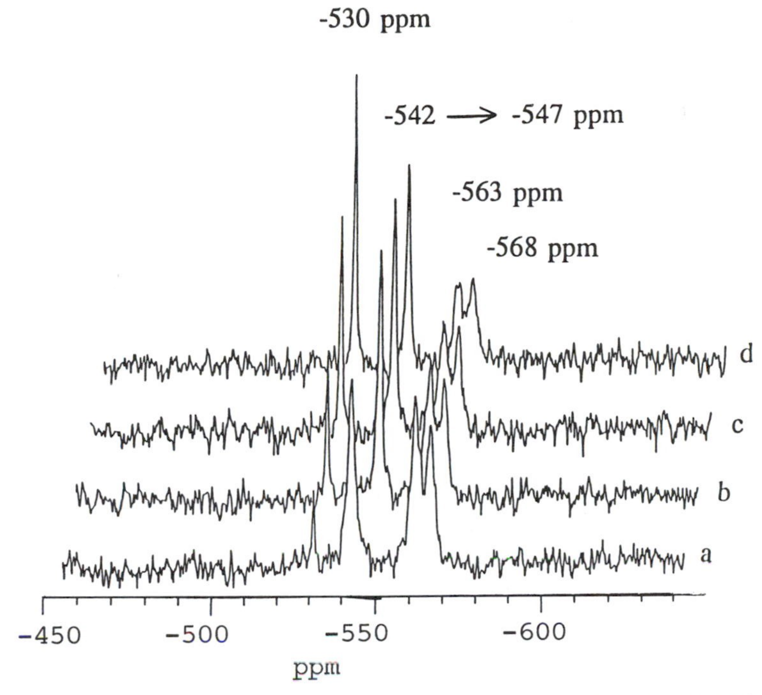

Figure 6. ^{51}*V NMR spectrum of (a) 1 mM (HPS)VO(OEt)(EtOH) in DMF, (b) 1 mM (HPS)VO(OEt)(EtOH) + 1 mM EtOH, (c) 1 mM (HPS)VO(OEt)(EtOH) + 2 mM EtOH, and (d) 1 mM (HPS)VO(OEt)(EtOH) + 3 mM EtOH.*

Fortunately, the equilibrium of the active catalyst (HPS)VO(OH) and another species (HPS)VO(OR) permits a direct probe of the active species. The addition of ethanol or methanol significantly retards the rate of tribromide formation when the catalyst is derived from (HPS)VO(OEt)(EtOH), although the rate is unchanged (within experimental error) when the catalyst is derived from sodium or ammonium vanadate. This disparity is readily rationalized by considering the coordination environment in the two catalysts. In (HPS)VO(OH), only one site is available to bind peroxide, the equatorial site occupied by hydroxide. The presence of ROH in solution results in a competition between peroxide and alkoxide for the sole available site. The equilibrium between (HPS)VO(OR) and (HPS)VO(O_2)$^-$ favors the peroxo species, but addition of 0.13 M methanol (20-fold excess over [H_2O_2]) results in a greater than 2-fold decrease in rate. By contrast, oxovanadium(V) has several available coordination sites, as well as the ability to bind two peroxide moieties. The addition of 0.5 M methanol does not cause any change in the rate of tribromide formation.

The high-field resonances (−563 and −568 ppm) are assigned to μ-O dimers of $(HPS)VO^+$ based on the concentration-dependent variation of their signal areas relative to those of $(HPS)VO(OH)/(HPS)VO_2^-$. The monomer $(HPS)VO^+$ is chiral, so dimerization gives rise to d, l, and meso diastereomers, which contain three vanadium environments. The upfield resonances at −563 and −568 ppm can be fit to three curves (−564, −567, and −568 ppm) in the appropriate ratios (~2:1:1), consistent with dimerization.

The oxidation of bromide can be measured directly by the increase in absorbance at 272 nm (λ_{max} of Br_3^- in DMF). Using catalytic amounts of (HPS)VO(OEt)(EtOH) or $[(HPS)VO]_2O \cdot MeCN$, bromide oxidation is followed under acid-limited conditions. The two equivalents of hydroxide produced upon reduction of hydrogen peroxide are neutralized by acid. Under basic conditions, peroxide can coordinate to vanadium(V), but the oxidation of bromide does not occur. Initial kinetic studies indicate that the reaction proceeds with a partial inverse acid dependence and shows saturation kinetics with respect to hydrogen peroxide and bromide.

Other Complexes. Before we turn to a discussion of other complexes, it is worth making a few general comments about the biomimetic systems studied to date. The model systems are much slower (~10^5-fold) than the enzyme (*33*). All peroxovanadium complexes, whether competent to catalyze bromide oxidation reactions or not, contain η^2-coordinated peroxide (*4*). Little is known about the binding of peroxide in the enzyme (*see* above), but one wonders whether the enhanced reactivity is derived from an alternative binding mode, such as end-on peroxide or hydroperoxide. The rapid enzymatic rate could also arise from the nature or configuration of the ligands to the vanadium ion.

The well-defined chemistry of (HPS)VO(OH) raises questions about other ligands. We have investigated a variety of ligands, which fall into three classes based on the reactivity of their vanadium(V) complexes in catalyzing the oxidation of bromide by hydrogen peroxide in reactions limited by acid. (1) Some ligands stay coordinated to vanadium(V) and produce active catalysts. (2) One ligand, pyridine-2,6-dicarboxylate, remains coordinated but so effectively stabilizes the peroxo complex that no catalysis is observed. (3) Some ligands are displaced from vanadium(V) by hydrogen peroxide under the experimental conditions (0.1 M tetrabutylammonium bromide, mM H_2O_2, in DMF). Our discussion will focus initially on the ligands that remain bound throughout the bromide oxidation cycle and turn briefly, at the end, to ligands that deactivate peroxide or dissociate from vanadium(V).

Vanadium complexes of the first class of ligands catalyze the oxidation of bromide by hydrogen peroxide. These ligands (Figure 4) include two more Schiff bases of salicylaldehyde, H_2sal:gly and H_2sal:phe, as well as citric (H_4cit), and iminodiacetic (H_2IDA).

Complexes of the first two ligands are prepared by reaction of vanadyl sulfate with the ligand, preformed in situ in aqueous ethanol (*38*). Isolated as the vanadium(IV) complexes, they can be oxidized in aerobic methanol and reisolated as the methoxymethanol vanadium(V) derivatives (*39*). Alternatively, the reduced complexes can be dissolved in DMF and allowed to oxidize aerobically before use. Their reactivity parallels that of (HPS)VO(OEt)(EtOH). They have reaction rates for catalyzing bromide oxidation comparable to (HPS)VO(OH) (Table I). Interestingly, the rates of these reactions are quite sensitive to the amount of water. The increase from 0.5% to 1.0% added water causes a 10–20% decrease in rate.

The citrato peroxo complex is formed in situ from equimolar $NaVO_3$ and H_2O_2 with two mole-equivalents of citric acid. The IDA complex is formed analogously.

The catalytic properties of these complexes are similar. The rates show saturation in hydrogen peroxide and a partial inverse dependence on acid concentration. The relative rates are given in Table I.

Pyridine-2,6-dicarboxylic acid (H_2dipic) falls into the second class of ligands: H_2O_2 is deactivated and no catalysis is observed. The dipic^{2-} complex was synthesized by a literature procedure (*40*) and synthesized in situ by addition of equimolar aqueous $NaVO_3$ and H_2dipic. The rate of formation of Br_3^- from H_2O_2 and Br^- in the presence of the dipic^{2-} complex is indistinguishable from the rate in the absence of the vanadium complex. The deactivation of hydrogen peroxide was observed previously in acidic aqueous solution, where oxoperoxo(dipicolinato) vanadium(V) oxidized a (thiolato)Co(III) complex more slowly than free hydrogen peroxide (*41*). Interestingly, the alkyl peroxo adduct of V(V)O(dipic)$^+$ slowly oxidizes organic substrates in acetonitrile solution stoichiometrically (*57*) and catalytically (*42*), although at this point, little is known about the differences in reactivity of halide oxidation by bound peroxide versus bound alkyl peroxide.

Table I. Relative Rates of Bromide Oxidation by Hydrogen Peroxide

Vanadium (V) Complex Ligand	*Rate*
No ligand ($NaVO_3$)	1.0
H_2IDA	1.1
H_2sal:gly	0.9
H_2HPS	0.6
H_4cit	0.6
No catalyst	0.04

Several ligands dissociate from V(V) under the catalytic conditions employed: mM acid and hydrogen peroxide, 100 mM Br^-. Carboxyphenylsalicylideneamine (H_2CPS) is one such ligand. The vanadium complex is synthesized analogously to (HPS)VO(OEt)(EtOH) (*35*). When it is dissolved in DMF, a single resonance is observed in the ^{51}V NMR spectrum at −550 ppm. Because this resonance is unchanged by the addition of excess ethanol, it is assigned by the analogous rationale as for (HPS)VO(OEt) to VO(CPS)(OEt). Upon the addition of hydrogen peroxide, the resonance at −550 ppm disappears and a new one at −585 ppm appears. This position is the same one that is observed when sodium or ammonium vanadate is the vanadium source. No decrease in the rate of bromide oxidation is observed in the presence of alcohol.

Hpic also dissociates from vanadium in the presence of hydrogen peroxide. Vanadium(V) in solution with Hpic and hydrogen peroxide exhibits an equilibrium between the peroxo(picolinato) and diperoxo forms. The presence of peroxo(picolinato)vanadium(V) (−563 ppm) and (diperoxo)vanadium(V) (−585 ppm) upon addition of hydrogen peroxide to a solution of V(V) and Hpic is shown by ^{51}V NMR. Addition of more Hpic to this solution results in the disappearance of the −585 ppm resonance, consistent with an equilibrium between the two species. The rate of bromide oxidation decreases dramatically from the value observed in the absence of any ligand upon addition of stoichiometric Hpic; further decreases in rate are observed upon addition of several equivalents excess Hpic with respect to vanadium. The fact that the rate of bromide oxidation continues to fall as more ligand is added suggests that some reactive species [probably (diperoxo)vanadium(V)] is increasingly complexed by Hpic. The complex $VO(pic)(O_2)_2$ is either unreactive or only slightly reactive. A fit of the observed rates of bromide oxidation gives a value of $3.4 \times 10^5\ M^{-1}$ for the formation of $VO(O_2)_2(pic)^{2-}$ from $VO(O_2)^{2-}$ and Hpic. Similar behavior is observed for the nitrilotriacetic acid (H_3NTA) complex: a 10-fold decrease in rate is observed when the ligand-to-vanadium ratio is changed from 1:1 to 3:1. In addition, displacement of nitrilotris(methylene)triphosphonic acid by hydrogen peroxide was observed. The addition of excess hydrogen peroxide to a solution of equimolar ammonium vanadate, hydrogen peroxide, and H_6NTP results in the conversion of the ^{51}V NMR resonance at −512 ppm for the monoperoxo NTP complex to −585 ppm for diperoxovanadate.

Reactivity of the V(V)–Tetraethylene glycol–HBr–O_2 System. Another system is relevant to this discussion of the oxidation of bromide by hydrogen peroxide. It does not catalyze precisely the same chemistry; instead, the vanadium complex of tetraethylene glycol (H_2teg) catalyzes the aerobic oxidation of bromide (from HBr) in 1,2-dichloroethane (*30*).

From the reaction solution in the absence of oxygen, a vanadium(III) complex, $[V(teg)Br_2]Br$, was isolated and crystallographically characterized. The vanadium sits in a pentagonal bipyramidal site with bromine atoms in the axial positions and the organic ligand defining the equatorial plane. The catalyst is proposed to undergo reduction from V(V) to V(III) by hydrogen bromide followed by aerobic oxidation to V(V) (Scheme 4). This behavior is distinct from the vanadium catalysts discussed previously, where no reduction of vanadium nor oxidation of bromide is observed upon addition of bromide. Vanadium bromoperoxidase also fails to undergo one turnover in the absence of hydrogen peroxide (*29*). Thus, the enzyme does not cycle between V(V) and V(III).

Some Structural Considerations Concerning Model Compounds and Vanadium Bromoperoxidase

The extant literature of vanadium coordination complexes can be used to shed additional light on the vanadium site of vanadium bromoperoxidase. The known structures of vanadium in a variety of coordination environments provide a wealth of structural detail that can be brought to bear on the proposed enzyme structure.

Bond valence sum (BVS) analysis, developed by Brown (*43*) to calculate metal oxidation states in materials such as high-temperature superconductors and zeolites, has recently been shown by Thorp (*44*) to be predictive for metalloenzymes and model compounds. On the basis of crystallographic data, the empirical parameters r_0 and B are determined. These values can then be used to calculate oxidation states from known coordination environments or coordination numbers from known oxidation states and bond lengths. The requisite equations are

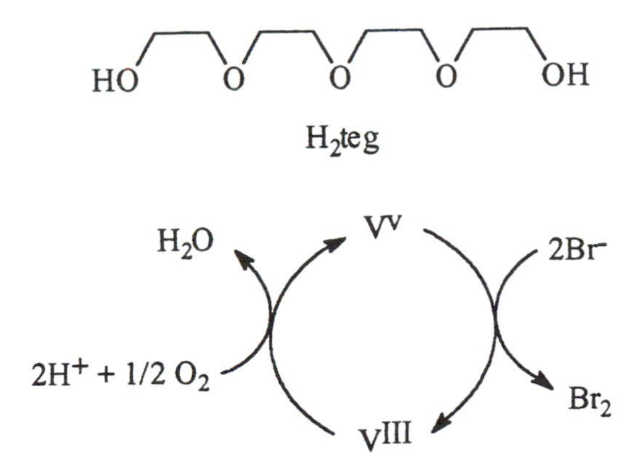

Scheme 4. Structure of H_2teg and proposed mechanism for aerobic bromide oxidation by the $V(III)(H_2teg)^+$ complex.

$$\text{O.S.} = \Sigma s$$

$$s = \exp[(r_0 - r)/B]$$

where O.S. is the oxidation state, and $B = 0.37$. Thorp (*44*) considered some Fe, Mn, Cu, and Zn enzymes and their models. Liu and Thorp (*45*) have now derived new r_0 values for vanadium (and other transition metals) based solely on data from coordination complexes. With these values BVS analysis can be extended to vanadium systems to predict oxidation states (±0.3) for a variety of vanadium(V) complexes in oxygen–nitrogen–halogen environments, including four-, five-, six-, and seven-coordinate complexes.

The r_0 values are listed in Table II. The V(V) values are calculated from V(IV) values using the d-electron correction described by Brown (*43*). Specifically, 0.020 is added to the V(IV) values. The BVS results are tabulated along with the known oxidation state and the coordination environment (Table III). For the most part, the original nomenclature of these complexes has been retained; details are available in the references.

The agreement between the oxidation state calculated from BVS analysis and that determined experimentally is generally quite good (Table III). Even the mixed-valence dimer, entry o, gives good results: intermediate values for each vanadium atom, consistent with the evidence that this dimer is delocalized (not valence-trapped) with a calculated, average oxidation state of 4.5 (*58*). The notable exceptions fall in two categories: the *cis*-(dioxo) complexes (entries c–f) have BVS values significantly lower than 5.0; the vanadium(V) complex of catechol (entry i) is also significantly lower than 5.0. The *cis*-(dioxo) complexes may reflect a decrease in bond order. Because the V=O bond requires the use of a different r_0 from V–O, the calculated BVS value changes with the bond order assigned. In at least one of these *cis*-(dioxo) complexes, hydrogen bonding is proposed to affect the structure significantly (*50*).

Table II. Values Used for BVS Analysis

Bone	r_0 (Å)
V^{3+}–O	1.749
V^{4+}–N	1.875
V^{3+}–N	1.813
V^{5+}=O	1.755
V^{4+}=O	1.735
V^{5+}–O	1.800
V^{4+}–O	1.780
V^{5+}–N	1.895

Table III. Results of BVS Analysis

Compound	*Coordination Sphere*	*BVS*	*Oxidation State*	*Ref.*
Vanadium(V) complexes				
a $[VOCl(OCH_2CH_2O)]_2$	Cl, O_{oxo}, 2RO	4.88	5	47
b (μ-pinacolato)$_2$(VOCl)$_2$	Cl, O_{oxo}, 3RO	4.88	5	48
c $Cs[VO_2(dipic)] \cdot H_2O$	N_{py}, $2O_{oxo}$, $2RCO_2$	4.72	5	49
d *exo*-[VO_2{nap:his}]	N_{im}, $2O_{oxo}$, O_{nap}, RCO_2	4.68	5	50
d2	N_{im}, O_{oxo}, O_{single}, O_{nap}, RCO_2	4.85	5	50
e (n-Bu_4N)[quinolinato$_2VO_2$] $\cdot H_2O$	$2N_{qu}$, $2O_{oxo}$, $2O_{qu}$	4.73	5	51
f $(H_2EDTA)VO_2 \cdot 3H_2O$	$2N_{amine}$, $2O_{oxo}$, $2RCO_2$	4.46	5	52
g VO(HPS)(OEt)(EtOH)	N_{im}, O_{oxo}, O_{ROH}, $2O_{phe}$, RO	5.02	5	35
h VO(sal:ala)(OMe)(MeOH)	N_{im}, O_{oxo}, O_{ROH}, O_{phe}, RO, RCO_2	4.99	5	39
i VO(SALIMH)(CAT)	N_{im}, N_{imid}, O_{oxo}, O_{phe}, $2O_{cat}$	4.5	5	53
j $[HB(Me_2pz)_3]VO(p\text{-Br-pheO})_2$	$3N_{pz}$, O_{oxo}, $2O_{phe}$	4.87	5	54
k $NH_4[VO(O_2)_2(NH_3)]$	N_{amm}, O_{oxo}, $4O_{peroxo}$, $O_{long\ contact}$	5.30	5	55
l $K_2[VO(O_2)(cit)]_2 \cdot 2H_2O$	O_{oxo}, $2O_{peroxo}$, RO, μ-RO, $2RCO_2$	4.94	5	56
m $NH_4[VO(O_2)(H_2O)(dipic)] \cdot xH_2O$	N_{py}, O_{oxo}, O_{aq}, $2O_{peroxo}$, $2RCO_2$	5.18	5	40
n VO(dipic)(OO-t-Bu) $\cdot H_2O$	N_{py}, O_{oxo}, O_{aq}, $2O_{peroxo}$, $2RCO_2$	5.19	5	57
Vanadium(IV/V) complex				
o Na{$[VO(sal:ser)]_2O$}	N_{im}, O_{oxo}, μ-O, O_{phe}, RCO_2	4.65 4.37	4.5	58
Vanadium(IV) complexes				
p $K_2[V(enterobactin)] \cdot 3DMF$	$6O_{cat}$	3.87	4	59
q $Na[VO(hypyb)] \cdot CH_3CN$	N_{py}, $2N_{am}$, O_{oxo}, O_{phe}	4.14	4	60
Vanadium(III) complexes				
r *trans*-$[V(capca)(Cl)_2]$	2Cl, $2N_{py}$, N_{am}, N_{im}	3.11	3	60
s $[V(H_2teg)(Br)_2]Br$	2Br, 2ROH, $3R_2O$	2.95	3	30

NOTE: For ligand identity, *see* references. Abbreviations used in descriptions of coordination sphere: py, pyridine; im, imine; qu, 8-quinolinate; imid, imidazolate; pz, pyrazole; amm, ammine; am, amidate; nap, naphtholate; phe, phenolate; cat, catecholate; aq, aquo.

The fit does improve upon calculating the longer "oxo" bond as a V–O bond (compare entries d and d2).

The case of the vanadium-catechol complex is also interesting. Under certain conditions, vanadium(V) undergoes reduction by catechol to yield the vanadium(IV)-semiquinone. The vanadium in this complex has a ^{51}V NMR signal and no ESR signal, demonstrating its oxidation state of 5+. One wonders whether the low BVS value reflects a simple weakness in the BVS analysis or the "inclination" of catechol to reduce vanadium(V).

As discussed previously, bond lengths are available from extended X-ray absorption fine structure (EXAFS) for the vanadium environment in V-BrPO (*17*). Bond lengths of 1.61, 1.72, and 2.11 Å were found for V–light-atom bonds. However, the number of ligands is never precisely known from EXAFS; typically, the error is ±0.5. In this case, the situation is complicated by a refinement that constrained the coordination numbers to integral values, so the uncertainty in the derived values is not known. A BVS analysis of the enzyme was done using the original values (*46*); we have extended this analysis using the refined values, which distinguish between single and double bonds. The coordination sphere has been formulated as containing one V=O moiety (1.61 Å), three short V–O bonds (1.72 Å), and two long V–O/N bonds (2.11 Å) (Figure 1). Using this ligand set, the oxidations state is calculated to be 6.07 if the long bonds are taken to be V–O, or 6.32 if they are V–N. Thus, this coordination sphere is inconsistent with BVS analysis.

We have considered several other possibilities for vanadium bromoperoxidase. A *cis*-(dioxo) arrangement, even with only two ligands at 1.72 Å, gives an oxidation state by BVS of 6.30, which is also inconsistent with the known valence (5+) of the metal. Other six-coordinate arrangements with the sixth ligand as H_2O at a long distance (2.2–2.3 Å) produce similar results. However, a five-coordinate geometry, with the 2.11 Å bonds as either V–O or V–N, has a calculated valence of 4.83 and 5.08, respectively, consistent with the known oxidation state of 5+.

Conclusion

The V-BrPO-catalyzed oxidation of bromide by hydrogen peroxide is the first step in a series of reactions that results in brominated organic products or the halide-assisted disproportionation of hydrogen peroxide. As studies continue on both the enzymatic and model systems, we can make several comments. The ability to catalyze this reaction seems to be a fairly general property of vanadium(V) complexes with an available coordination site in the equatorial plane in which to bind peroxide. The oxidation rates of the model complexes, with respect to the enzyme, raise the question of what microscopic differences give rise to the high efficiency of vanadium bromoperoxidase. We are investigating the detailed effect of the nature of coordinating groups, active site groups that could function in general acid catalysis, and chelate ring sizes toward a fuller understanding of the factors that determine reactivity.

Note Added in Proof

The kinetics of the VO_{2+}-catalyzed oxidation of bromide have now been investigated. A peroxo vanadium(V) dimer is the oxidant of bromide (Clague and Butler, *J. Am. Chem. Soc.*, 1995, in press).

Acknowledgments

Alison Butler is grateful for support from the National Science Foundation (DMB90–18025) and the National Institutes of Health (GM38130). Melissa J. Clague gratefully acknowledges support from a University of California President's Dissertation Year Fellowship.

References

1. *Vanadium in Biological Systems;* Chasteen, N. D., Ed.; Kluwer Academic: Dordrecht, Netherlands, 1990.
2. Butler, A.; Walker, J. V. *Chem. Rev.* **1993,** *93,* 1937–1944.
3. Butler, A. In *Bioinorganic Catalysis;* Reedijk, J., Ed.; Marcel Dekker: New York, 1992; pp 425–445.
4. Butler, A.; Carrano, C. J. *Coord. Chem. Rev.* **1991,** *109,* 61–105.
5. Wever, R.; Krenn, B. E. In *Vanadium in Biological Systems;* Chasteen, N. D., Ed.; Kluwer Academic: Amsterdam, Netherlands, 1990; pp 81–98.
6. Wever, R.; Kustin, K. *Adv. Inorg. Chem.* **1990,** *35,* 81–115.
7. Rehder, D. *Biometals* **1991,** *5,* 3–12.
8. Rehder, D. *Angew. Chem. Int. Ed.* **1991,** *30,* 148–167.
9. Plat, H.; Krenn, B. E.; Wever, R. *Biochem. J.* **1987,** *248,* 277–279.
10. van Schijndel, J. W. P. M.; Vollenbroek, E. G. M.; Wever, R. *Biochim. Biophys. Acta* **1993,** *1161,* 249–256.
11. Krenn, B. E.; Tromp, M. G. M.; Wever, R. *J. Biol. Chem.* **1989,** *264,* 19287–19292.
12. de Boer, E.; Tromp, M. G. M.; Plat, H.; Krenn, B. E.; Wever, R. *Biochim. Biophys. Acta* **1986,** *872,* 104–115.
13. Wever, R.; Krenn, B. E.; de Boer, E.; Offenberg, H.; Plat, H. *Prog. Clin. Biol. Res.* **1988,** *274,* 477–493.
14. Muller-Fahrnow, A.; Hinrichs, W.; Saenger, W.; Vilter, H. *FEBS Lett.* **1988,** *239,* 292–294.
15. de Boer, E.; van Kooyk, Y.; Tromp, M. G. M.; Plat, H.; Wever, R. *Biochim. Biophys. Acta* **1986,** *869,* 48–53.
16. Soedjak, H. S.; Butler, A. *Biochemistry* **1990,** *29,* 7974–7981.
17. Arber, J. M.; de Boer, E.; Garner, C. D.; Hasnain, S. S.; Wever, R. *Biochemistry* **1989,** *28,* 7968–7973.
18. de Boer, E.; Keijzers, C. P.; Klassen, A. A. K.; Reijerse, E. J.; Collison, D.; Garner, C. D.; Wever, R. *FEBS Lett.* **1988,** *235,* 93–97.
19. Vilter, H. *Phytochemistry* **1984,** *23,* 1387–1390.

20. Wever, R.; Plat, H.; de Boer, E. *Biochim. Biophys. Acta* **1985,** *830,* 181–186.
21. Everett, R. R.; Butler, A. *Inorg. Chem.* **1989,** *28,* 393–395.
22. Everett, R. R.; Soedjak, H. S.; Butler, A. *J. Biol. Chem.* **1990,** *265,* 15671–15679.
23. Everett, R. R.; Kanofsky, J. R.; Butler, A. *J. Biol. Chem.* **1990,** *265,* 4908–4914.
24. Kanofsky, J. R. *J. Biol. Chem.* **1984,** *259,* 5596–5600.
25. Khan, A. U.; Gebauer, P.; Hager, L. P. *Proc. Natl. Acad. Sci. U.S.A.* **1983,** *80,* 5195–5197.
26. Walker, J. V.; Butler, A., unpublished results.
27. Volhardt, K. P. C. *Organic Chemistry;* W. H. Freeman: New York, 1987; p 1090.
28. Hager, L. P.; Morris, D. R.; Brown, F. S.; Eberwein, H. *J. Biol. Chem.* **1966,** *241,* 1769–1777.
29. Soedjak, H. S., Ph.D. Thesis, University of California, Santa Barbara, 1991.
30. Neumann, R.; Assael, I. *J. Am. Chem. Soc.* **1989,** *111,* 8410–8413.
31. Tromp, M. G. M.; Olafsson, B. E. K.; Wever, R. *Biochim. Biophys. Acta* **1990,** *1040,* 192–198.
32. Secco, F. *Inorg. Chem.* **1980,** *19,* 2722–2725.
33. de la Rosa, R. I.; Clague, M. J.; Butler, A. *J. Am. Chem. Soc.* **1992,** *114,* 760–761.
34. Thompson, R. C. *Adv. Inorg. Bioinorg. Mech.* **1986,** *4,* 65–106.
35. Clague, M. J.; Keder, N. L.; Butler, A. *Inorg. Chem.* **1993,** *32,* 4754–4761.
36. Westland, A. D.; Tarafder, M. T. H. *Inorg. Chem.* **1981,** *20,* 3992.
37. Clague, M. J.; Butler, A., unpublished results.
38. Theriot, L. J.; Carlisle, G. O.; Hu, H. J. *J. Inorg. Nucl. Chem.* **1969,** *31,* 2841–2844.
39. Nakajima, K.; Kojima, M. et al. *Bull. Chem. Soc. Jpn.* **1989,** *62,* 760–767.
40. Drew, R. E.; Einstein, F. W. B. *Inorg. Chem.* **1973,** *12,* 829–835.
41. Ghiron, A. F.; Thompson, R. C. *Inorg. Chem.* **1990,** *29,* 4457–4461.
42. Nakajima, K.; Kojima, M.; Toriumi, K.; Saito, K.; Fujita, J. *Bull. Chem. Soc. Jpn.* **1989,** *62,* 760–767.
43. Brown, I. D.; Altermatt, D. *Acta Crystallogr.* **1985,** *B41,* 244–247.
44. Thorp, H. H. *Inorg. Chem.* **1992,** *31,* 1585–1588.
45. Liu, W.; Thorp, H. H. *Inorg. Chem.* **1993,** *32,* 4102–4105.
46. Caranno, C. J.; Mohan, M.; Holmes, S.; de la Rosa, R.; Butler, A.; Charnock, J.; Garner, C. D. *Inorg. Chem.* **1994,** *33,* 646–655.
47. Crans, D. C.; Felty, R. A.; Anderson, O. P.; Miller, M. M. *Inorg. Chem.* **1993,** *32,* 247–248.
48. Crans, D. C.; Felty, R. A.; Miller, M. M. *J. Am. Chem. Soc.* **1991,** *113,* 265–269.
49. Nuber, B.; Weiss, J.; Wieghardt, K. Z. *Naturforsch.* **1978,** *33B,* 265–267.
50. Vergopoulos, V.; Priebsch, W.; Fritzsche, M.; Rehder, D. *Inorg. Chem.* **1993,** *32,* 1844–1849.
51. Giacomelli, A.; Floriani, C.; Duarte, A. O. D. S.; Chiesi-Villa, A.; Guastini, C. *Inorg. Chem.* **1982,** *21,* 3310–3316.
52. Scheidt, W. R.; Collins, D. M.; Hoard, J. L. *J. Am. Chem. Soc.* **1971,** *93,* 3873–3877.
53. Cornman, C.; Kampf, J.; Pecoraro, V. L. *Inorg. Chem.* **1992,** *31,* 1981–1983.
54. Holmes, S.; Carrano, C. J. *Inorg. Chem.* **1991,** *30,* 1231–1235.

55. Drew, R. E.; Einstein, F. W. B. *Inorg. Chem.* **1972,** *11,* 1079–1083.
56. Djordjevic, C.; Lee, M.; Sinn, E. *Inorg. Chem.* **1989,** *28,* 719.
57. Mimoun, H.; Chaumette, P.; Mignard, M.; Saussine, L. *Nouveau J. Chim.* **1983,** *7,* 467–475.
58. Pessoa, J. C.; Sliva, J. A. L.; Vieira, A. L.; Vilas-Boas, L. et al. *J. Chem. Soc. Dalton Trans.* **1992,** 1745–1748.
59. Karpishin, T. B.; Dewey, T. M.; Raymond, K. *J. Am. Chem. Soc.* **1993,** *115,* 1842–1851.
60. Kabanos, T. A.; Anastasios, D. K.; Papaioannou, A. B.; Terzis, A. *J. Chem. Soc., Chem. Commun.* **1993,** 643–645.

RECEIVED for review July 19, 1993. ACCEPTED revised manuscript December 8, 1993.

13

Magnetic Circular Dichroism Spectroscopy as a Probe

of the Active Site Structures of Iron Chlorin- and Formylporphyrin-Containing Green Heme Proteins

John H. Dawson[1,2], Alma M. Bracete[1], Ann M. Huff[1], Saloumeh Kadkhodayan[1], Chi K. Chang[3], and Masanori Sono[1]

[1]Department of Chemistry and Biochemistry and [2]School of Medicine, University of South Carolina, Columbia, SC 29208
[3]Department of Chemistry, Michigan State University, East Lansing, MI 48824

The magnetic circular dichroism (MCD) properties of the two types of green-colored heme systems, iron chlorins (dihydroporphyrins) and iron porphyrins bearing a formyl substituent, have been further developed. The systems to be investigated include enzymes that naturally contain a green heme prosthetic group, proteins such as myoglobin into which an iron chlorin or iron formyl-porphyrin has been reconstituted and synthetic model complexes. As has previously been established for the MCD of protoheme systems, the spectra of iron chlorins are very sensitive to oxidation, spin and ligation state changes, and relatively insensitive to changes in the solvent or protein environment where the chromophore is located. The green catalase from Escherichia coli *is shown to display MCD spectra that are most consistent with the presence of a tyrosine proximal ligand. On the other hand, bovine spleen myeloperoxidase has MCD spectra that closely resemble those of formyl-substituted hemes and are distinct from those of corresponding iron chlorins. The MCD properties of the two types of green heme chromophores are readily distinguishable. Taken together, the data presented herein reveal MCD spectroscopy to be a promising technique for the structural characterization of green heme systems.*

MAGNETIC CIRCULAR DICHROISM (MCD) spectroscopy has been used extensively to examine the structural and magnetic properties of iron

0065–2393/95/0246–0351/$08.18/0

porphyrins and heme proteins (*1–4*). Two general approaches have been developed for these investigations. The more empirical application takes advantage of the fingerprinting capability of the method to characterize structurally undefined heme centers by spectral comparisons to data for structurally defined iron porphyrins, both in synthetic model complexes and heme protein derivatives. The more sophisticated approach requires the use of ultralow temperatures and high magnetic fields to "saturate" MCD signals in certain cases to study the magnetic properties of the heme iron.

The phenomenon of MCD was first reported in 1845 by Michael Faraday (*5*) and is therefore sometimes called the Faraday effect. Chromophores exhibit MCD spectra because they differentially absorb left and right circularly polarized light in the presence of a magnetic field. As a difference spectral technique, both positive and negative signals are seen. MCD "A" terms result from a lifting of excited state degeneracies in the presence of the magnetic field and are derivative shaped. MCD "C" terms result from the lifting of ground-state degeneracies by the magnetic field, are Gaussian in shape and are temperature-dependent in intensity. MCD "B" terms arise from ground- and excited-state mixing brought on by the magnetic field and are Gaussian in shape but are generally temperature-independent in intensity. Thus it is possible to differentiate A, B, and C terms in an MCD spectrum through an analysis of the shapes and temperature dependence of the components of the spectrum.

MCD Spectroscopy as a Probe of Magnetic Properties

As mentioned previously, the intensities of MCD A and B terms are independent of temperature. MCD C terms, however, display temperature-dependent signal intensities (*6*). C terms arise from the lifting of ground electronic state degeneracies by the magnetic field. There will be a difference in population of the two resolved ground states, and that population difference will become more pronounced as the temperature is lowered. Consequently, the MCD signal intensity will increase as the temperature is lowered until a maximum intensity is reached when the temperature is low enough so that only the lower energy ground state is populated. At this point, the signal is said to have been "saturated," i.e., as the temperature is lowered, the C term signal intensity reaches a maximum. A similar effect is seen upon increasing the magnetic field strength at constant temperature.

Schatz and co-workers (*7*) derived theoretical expressions to describe the saturation of MCD signal intensity. Thomson and co-workers (*6*) have extended these results and extensively applied the approach to study paramagnetic systems including those containing heme iron centers. C terms are always seen in paramagnetic metal complexes. Thomson has shown that the rate of MCD saturation (or magnetization) as observed in a plot of signal intensity vs. magnetic field strength at fixed temperature depends on the magnetic properties of the ground state, i.e., low-spin, high-spin, etc. Thus, it is readily possible to determine the magnetic properties, including *g*-values, of particular sites in metalloproteins rather than having to measure the bulk magnetic susceptibility. Furthermore, because the signal intensities of C terms increase dramatically as the temperature is lowered, weak electronic transitions that are not evident at ambient temperature can be observed. Finally, while the determination of *g*-values can be more accurately accomplished using electron paramagnetic resonance (EPR) spectroscopy in half-integer spin cases, for integer spin systems where EPR signals are hard to observe, the study of MCD signal saturation is the best currently available method for *g*-value measurement (*1, 6*). In addition, Stephens and co-workers have shown that variable-temperature MCD provides a practical method for the determination of zero-field splitting parameters in high-spin ferric porphyrins (*8*).

MCD Spectroscopy as a Probe of Heme Iron Coordination Structure

Because MCD signals can be either positive or negative in sign, considerably more fine structure is seen in MCD spectra than in the corresponding electronic absorption spectra. Furthermore, MCD is a property of the molecular electronic structure of a chromophore, and so the only structural changes that will influence the MCD spectrum are those that modify the electronic structure. Furthermore, the MCD spectrum is relatively insensitive to the environment in which the chromophore is located, whether it is the protein microenvironment for a heme protein center or the solvent for a model complex. Thus, comparisons of the MCD spectra of synthetic heme iron model complexes with those of heme protein active sites are possible and have been shown to be of considerable utility in assigning the coordination structures of the heme protein active sites (*1*).

The use of MCD spectroscopy as a probe of heme iron coordination structure relies on the "fingerprinting" capabilities of the technique. In iron porphyrins, changes in oxidation state and in the nature of axial ligands lead to changes in electronic structure that usually are reflected in the MCD spectrum. The sensitivity to the identity of the axial ligand means that by cataloging the MCD spectra of structurally defined model heme complexes or heme protein derivatives, fingerprints for particular combinations of axial ligands as a function of oxidation state can be established that can subsequently be used to ascertain the identity of the axial ligands in structurally undefined "new" heme proteins.

Two examples of this approach will be presented. In Figure 1, the MCD and electronic absorption spectra of phosphine adducts of ferric cytochrome P-450-CAM and chloroperoxidase are displayed (9). Comparison of the absorption and MCD spectra nicely illustrates the increased fingerprinting power of the MCD method. Whereas the absorption spectra of the two complexes feature three peaks in the 300–700-nm region, the corresponding MCD spectra contain four peaks, four troughs, and six crossover points. It is also clear that the respective MCD and absorption spectra of the two complexes are very similar to each other. The comparison of MCD spectra, however, is more powerful in that it involves a comparison of 14 features (peaks, troughs, and crossover points), all of which match, rather than only the three matching peaks compared in the absorption spectra. This particular study involved a comparison of two different proteins, one of which (P-450) had been quite well-established at the time of the investigation to contain a thiolate (cysteinate) axial ligand. The study concluded that chloroperoxidase also contains a thiolate axial ligand.

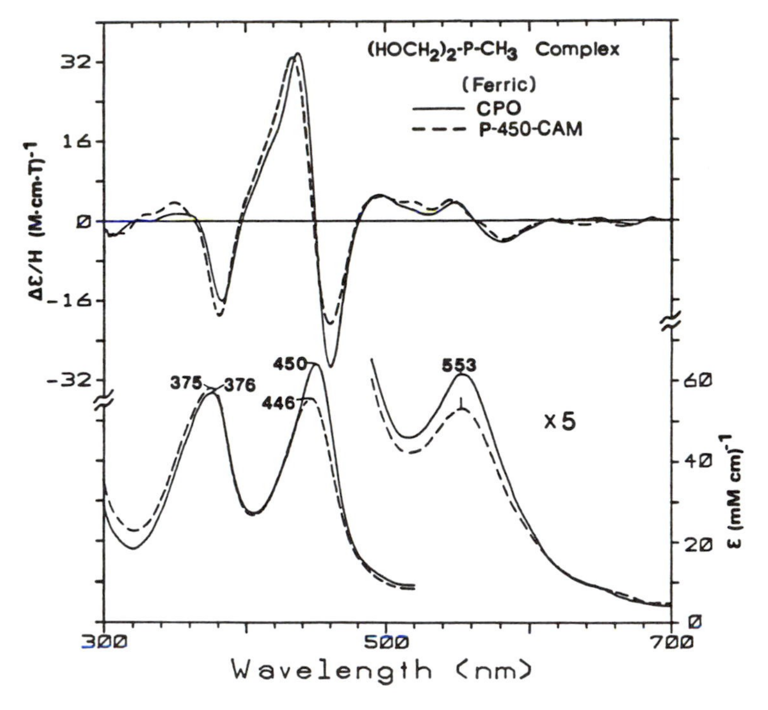

Figure 1. MCD (top) and UV–visible absorption (bottom) spectra at 4 °C of the bis(hydroxymethyl)methylphosphine complexes of ferric chloroperoxidase (CPO) (——) and cytochrome P-450-CAM (- - -). MCD spectra were obtained using a JASCO J-40 spectropolarimeter equipped with a 1.5 T electromagnet. (The data are taken from reference 9.)

An illustration comparing the MCD spectra of a model complex and of the identical ligand combination in a protein is shown in Figure 2. Here, the MCD spectra of ferrous cytochrome b_5 (bis-histidine imidazole ligation) and of bis-imidazole ferrous protoporphyrin IX dimethyl ester are overplotted (*10*). Numerous peaks, troughs, and crossover points are seen that closely match each other throughout the 300–700-nm spectral range examined. Clearly, the fact that the model was studied in a solvent of dichloromethane, whereas the solvent for the cytochrome derivative is water and the chromophore is actually embedded in a polypeptide microenvironment, has little if any effect on the MCD spectra. Once again, when the coordination structures of two heme complexes are the same, the MCD spectra of the two species can be virtually superimposable.

These examples illustrate how MCD spectroscopy has been used to determine the identity of axial ligands in heme proteins through comparison of the MCD spectra of structurally defined systems with those of ill-defined heme derivatives. This approach has been successful in numerous cases involving studies of cytochrome P-450, chloroperoxidase, indoleamine 2,3-dioxygenase, secondary amine mono-oxygenase, hemoprotein H-450, and others (*1, 11–19*). The remainder of this chapter addresses the issue of whether the approach just described that has

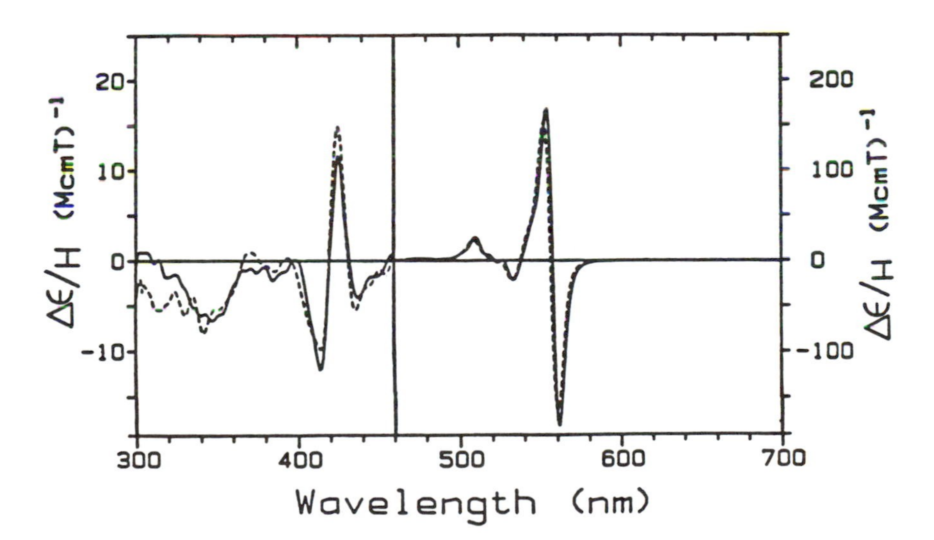

Figure 2. MCD spectra at 25 °C of ferrous cytochrome b$_5$ *(– – –) and bis-imidazole ferrous protoporphyrin IX dimethyl ester in CH_2Cl_2 (——). Note the different scales for the Soret region (300–460 nm) and for the visible region (460–700 nm). MCD spectra were obtained using a JASCO J-40 spectropolarimeter equipped with a 1.5 T electromagnet. (The data are taken from reference 10.)*

worked so well with the more prevalent red-colored protoheme-based systems will also work with modified hemes such as the green-colored iron chlorins and iron formyl-substituted hemes. However, because the exact structure of the porphyrin macrocycle including the presence or absence of peripheral vinyl substituents is not always known, another issue will be addressed first: what effect do vinyl substituents have on the MCD spectra of iron porphyrins?

The Effect of Vinyl Substituents on the MCD Spectra of Porphyrins

One of the variables in the structures of the porphyrins present in heme proteins is the presence or absence of vinyl substituents on the periphery of the macrocycle. For example, *b* hemes have vinyl substituents whereas *c* hemes do not. Because of the sensitivity of such vinyl substituents during synthetic transformations, it has often been desirable to use octa-alkyl porphyrins in model studies of the spectroscopic properties of heme systems. The development of improved methods for the preparation of octa-alkyl porphyrins has likewise increased the availability of such porphyrins for model studies (*20, 21*). To assess the effect that replacement of the two vinyl substituents in protoporphyrin IX with alkyl (ethyl) groups has on the MCD properties of the heme system, an extensive and systematic study of the MCD properties of mesoheme IX-reconstituted myoglobin and horseradish peroxidase in comparison with the spectra of the native protoheme-bound proteins has been carried out (*22*). The structures of these two porphyrins are shown in Figure 3.

The MCD spectra of representative ferric high-spin (the "resting" state, as isolated), ferric low-spin (the azide adduct), ferrous high-spin (the deoxyferrous state), and ferrous low-spin (the NO complex) derivatives of native (protoheme-bound) horseradish peroxidase compared to the parallel mesoheme IX-reconstituted species are displayed in Figure 4 (*22*). Altogether, more than 20 such comparisons were carried out including data obtained with myoglobin derivatives as well as for oxo-ferryl states such as compounds I and II of the peroxidase. The MCD spectra of the mesoheme-reconstituted proteins are consistently blue-shifted by 4–12 nm and are 1.2–2.5-fold more intense in the Soret (300–500 nm) region (*see* Figure 4). Intensity differences in the visible (500–700 nm) region were generally smaller. Importantly, the MCD band patterns seen for the native protein states are also seen for the mesoheme-reconstituted derivatives. These data demonstrate that octa-alkyl porphyrins can be used as models for protoporphyrin IX-containing systems provided the predictable small wavelength and intensity differences are anticipated (*22*). These conclusions provide a basis for the use of octa-alkyl chlorins as models for the protoporphyrin-derived

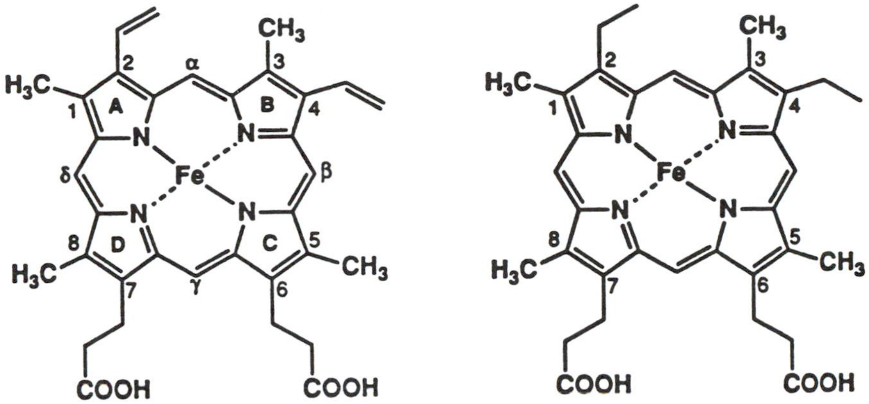

Figure 3. Structures of iron protoporphyrin IX (protoheme IX) (left) and iron mesoporphyrin IX (mesoheme IX) (right).

chlorin prosthetic groups in proteins such as HPII catalase (*see* the next section) and the terminal oxidase from *Escherichia coli*.

Green Hemes: Iron Chlorins and Iron Formyl-Substituted Porphyrins

Iron protoporphyrin IX is the most prevalent heme prosthetic group found in nature. It is present in a large number of heme proteins and enzymes including the most common ones that are involved in oxygen transport and activation, peroxide decomposition, and electron transfer. The macrocycle of protoporphyrin IX has 11 conjugated double bonds; its iron complexes are generally red in color. If 1 or more of the 11 double bonds is saturated, the result is a hydroporphyrin. In the case in which one double bond is reduced, the macrocycle is called a chlorin. Metallochlorins have been found in a number of biological systems such as in chlorophylls, which are magnesium chlorins. Iron chlorins are usually green in color and have been found in several enzymes. *E. coli*, for example, has two enzymes that contain an iron chlorin prosthetic group: the heme *d* terminal oxidase and an alternative catalase called HPII catalase (*23, 24*). Heme *d* and the prosthetic group of HPII catalase are iron chlorins that are structurally derived from protoporphyrin IX. The structure of the latter chlorin, proposed by Timkovich and co-workers (*24*), is shown in Figure 5 along with the structures of octaethylchlorin and of "methyl"chlorin (2,2,4-trimethyldeuterochlorin, a chlorin featuring a *gem*-dimethyl-substituted peripheral carbon) (*25*). The latter two chlorins are used in model studies described herein. A green catalase from *Neurospora crassa* has also been found to contain an iron chlorin

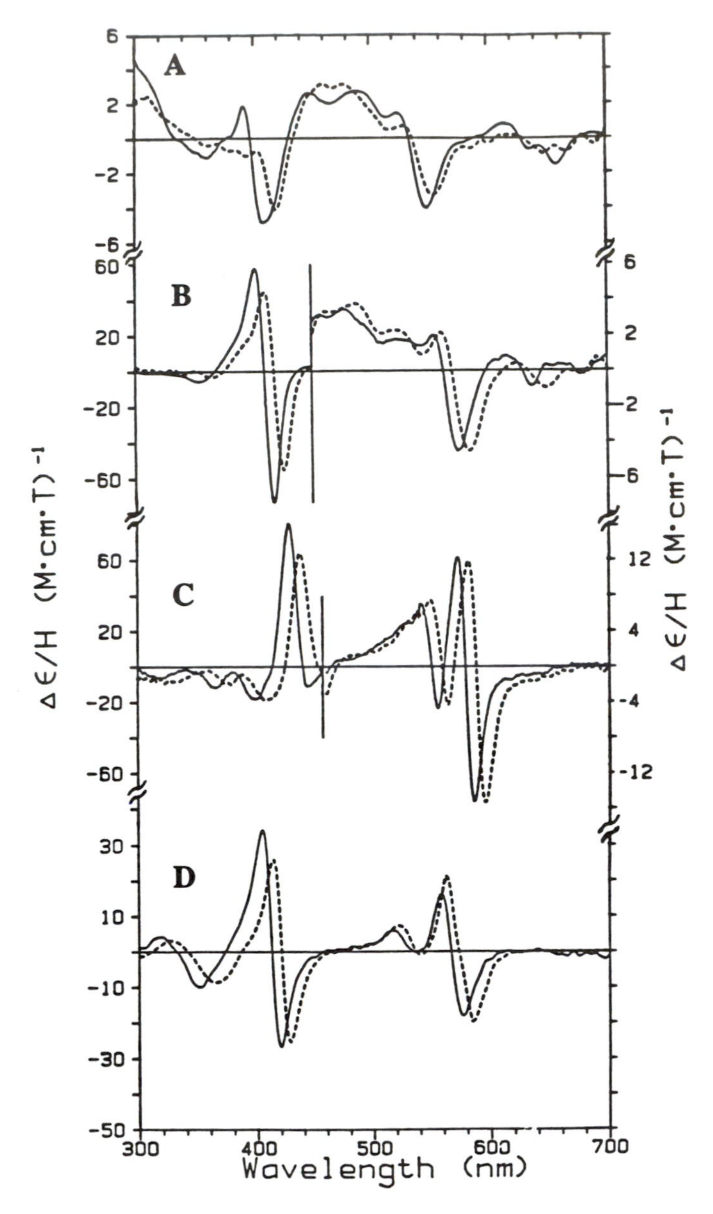

Figure 4. MCD spectra at 4 °C of native (protoheme-bound) (- - -) and mesoheme IX-reconstituted (——) horseradish peroxidase in their ferric high-spin (with no exogenous ligand) (A), ferric low-spin (the azide adduct) (B), ferrous high-spin (deoxy-ferrous) (C), and ferrous low-spin (the NO complex) (D) states. MCD spectra were obtained using a JASCO J-500A spectropolarimeter equipped with a 1.5 T electromagnet. Note *the different scales for the Soret and visible regions for B and C. (The data are taken from reference 22.)*

prosthetic group (*26*). Sulfmyoglobin, an inactive derivative of myoglobin formed by exposure to hydrogen peroxide and inorganic sulfide, contains an iron chlorin-type macrocycle generated by addition of a sulfur atom to one of the pyrrole double bonds of protoporphyrin (*27, 28*).

A second type of green-colored heme is found in the formyl-substituted porphyrins. Myeloperoxidase is a green enzyme that contains

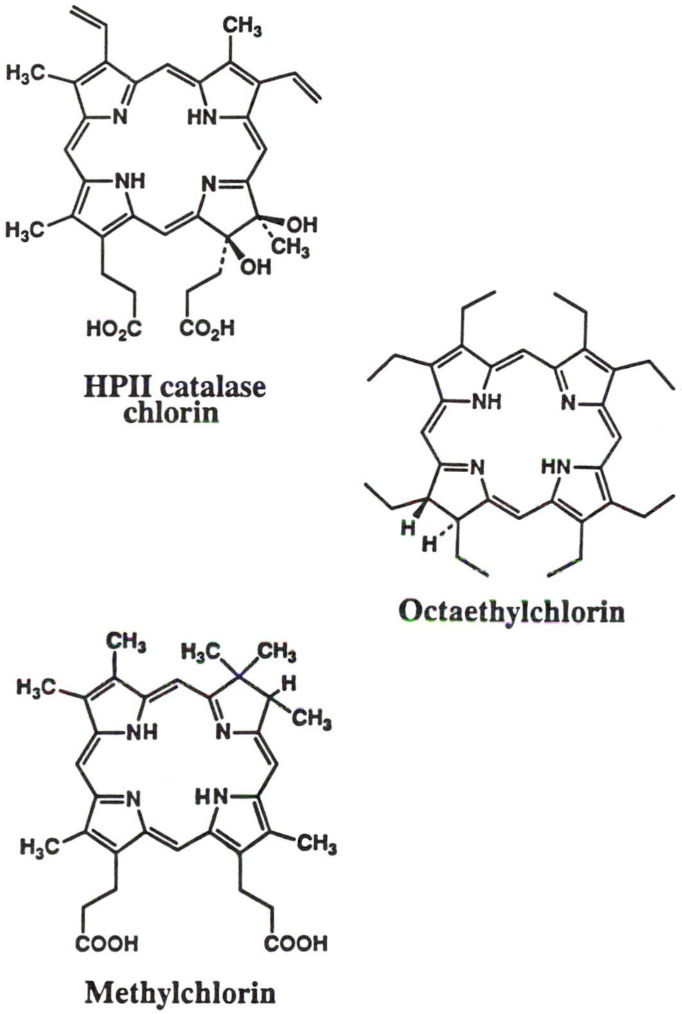

HPII catalase chlorin

Octaethylchlorin

Methylchlorin

Figure 5. Structures of the HPII catalase prosthetic group chlorin (proposed, see Chiu et al. (24)) (top), of octaethylchlorin (the trans isomer) (middle), and of "methylchlorin" (2,2,4-trimethyldeuterochlorin) (bottom). The HPII catalase chlorin has a cis diol structure (24) and is isomeric to heme d *chlorin, which has a trans diol structure.*

a heme-type prosthetic group of unknown structure. Two proposals have been made for the structure of its prosthetic group, an iron chlorin (29–33) or an iron porphyrin with a formyl substituent (*34*). Studies addressing this issue to be described herein have employed *Spirographis* heme (2-formyl-4-vinyldeuteroheme IX, heme *s*) (*35, 36*), the structure of which is shown in Figure 6. The electron-withdrawing formyl substituent in heme *s*, like that present in heme *a*, causes a red shift to occur in the electronic absorption bands of its various iron complexes and causes the color of its solutions to be green. Clearly, the color of a heme protein does not distinguish between the two possibilities described here for green heme prosthetic groups: iron chlorins and iron porphyrins with formyl substituents. The question of whether MCD spectroscopy can distinguish the two types of green hemes will also be answered herein.

MCD Spectroscopy as a Probe of Chlorin Iron Coordination Structure

The use of MCD spectroscopy for the determination of axial ligand identity in iron porphyrin systems was described in previous paragraphs. The success of this approach resulted from the increased fingerprinting power of the method relative to electronic absorption spectroscopy and the relative insensitivity of the technique to the solvent and protein environment, thereby allowing heme sites in proteins to be compared directly to synthetic model systems. The applications of MCD spectroscopy to characterize iron chlorin systems have been rather limited. Until recently, the only study of the MCD properties of iron chlorin complexes was that of Stolzenberg et al. (*37*), which focused on high-spin ferric octaethylchlorin complexes. To address the lack of MCD data on iron chlorin systems, a three-stage effort was initiated involving the investigation of synthetic model complexes; the examination of structurally defined heme proteins such as myoglobin, horseradish peroxidase, and cytochrome b_5 reconstituted with iron chlorin prosthetic groups; and the study of naturally occurring iron chlorin proteins such as those mentioned in preceding paragraphs. To date, the work has focused largely on iron chlorin systems bearing a histidine or other nitrogen-donor axial ligand. The model studies (*38*) have involved the examination of ligand adducts of iron octaethylchlorin, methylchlorin dimethyl ester (Figure 5), and mesochlorin IX dimethyl ester (the dihydroporphyrin derivative of mesoporphyrin; *see* Figure 3). The reconstitution work (*39*) has involved the use of iron methylchlorin and mesochlorin IX.

In Figure 7, the MCD spectra of high- as well as low-spin ferric iron chlorin derivatives are displayed (*38, 39*). The high-spin ferric-fluoride adducts of methylchlorin-reconstituted myoglobin and horseradish peroxidase are compared in Figure 7A. For the ferric low-spin case, the

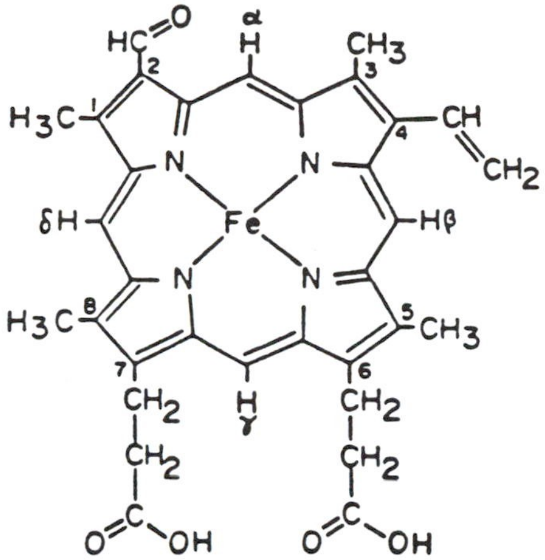

Figure 6. Structure of Spirographis *heme (2-formyl-4-vinyldeuteroheme IX, heme* s*).*

bis-1-methylimidazole adducts of octaethylchlorin and of methylchlorin are compared to that of methylchlorin-reconstituted cytochrome b_5 in Figure 7B. In Figure 8, a similar set of comparisons is made for high- and low-spin ferrous iron chlorins (*38, 39*). The MCD spectrum of the high-spin 1,2-dimethylimidazole adduct of octaethylchlorin is compared in Figure 8A to that of the deoxyferrous state of methylchlorin-reconstituted horseradish peroxidase, presumed to also be five-coordinate imidazole-ligated. In Figure 8B, the spectra of the low-spin ferrous-NO adducts of methylchlorin-reconstituted myoglobin and horseradish peroxidase are compared to that of ferrous (octaethylchlorin)(1-methylimidazole) (NO). In all four cases, ferric high- and low-spin and ferrous high- and low-spin, the MCD spectrum of a particular iron chlorin derivative is essentially independent of the environment, whether the complex is embedded inside a protein or examined in an organic solvent. In addition, the MCD spectra change significantly as a function of oxidation, ligation, and spin state. Thus, the fingerprinting power of MCD appears to be as useful in the study of iron chlorins as has been demonstrated more extensively with iron porphyrins.

The Proximal Ligand to the Chlorin Iron of E. coli *HPII Catalase*

Of the two catalases found in *E. coli*, one has a protoheme prosthetic group whereas the other, the HPII catalase, contains an iron chlorin prosthetic group, the proposed structure of which is displayed in Figure 5 (*24*). Although the nature of the prosthetic group has been established,

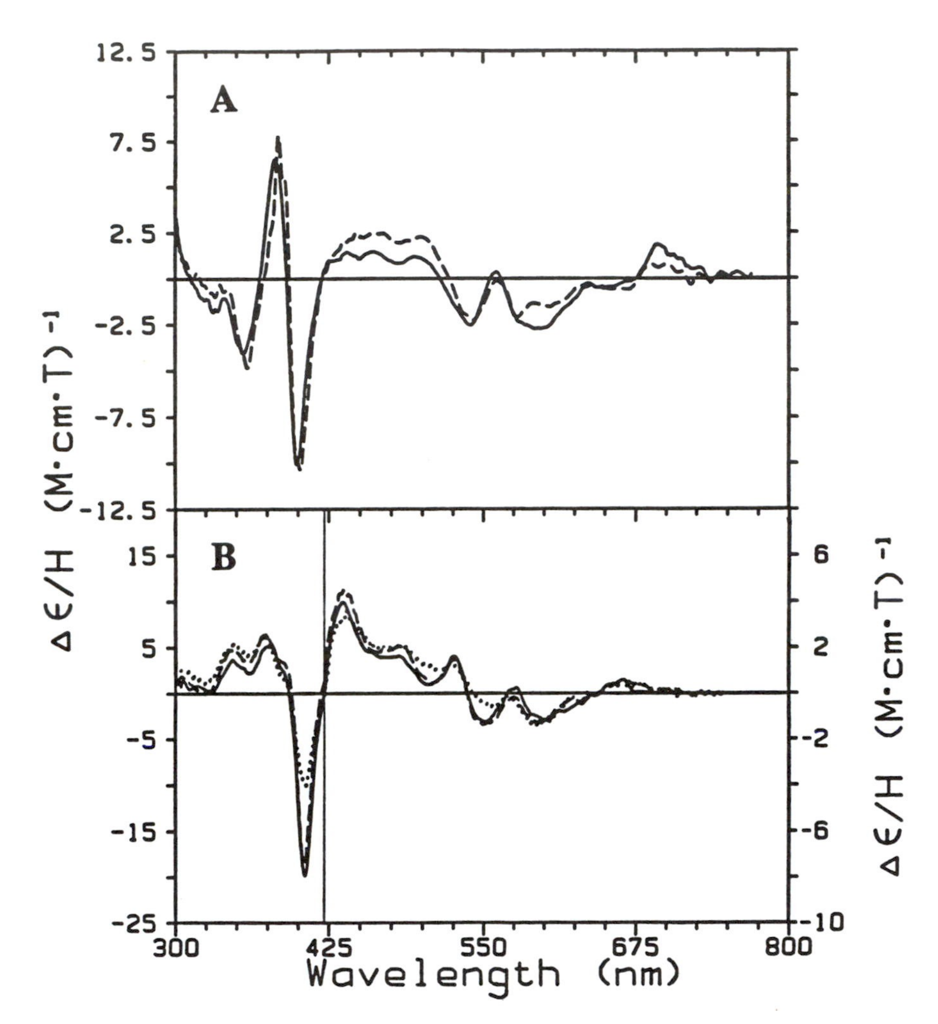

Figure 7. MCD spectra of the (A) high-spin ferric-fluoride adducts of methylchlorin-reconstituted myoglobin (——) and horseradish peroxidase (– – –) and of the (B) low-spin ferric bis-1-methylimidazole adducts of octaethylchlorin (——) and methylchlorin (· · · · ·) overplotted with a spectrum of methylchlorin-reconstituted ferric cytochrome b_5 *(– – –). Spectra of protein samples and of model complexes were measured at 4 and 25 °C, respectively. MCD spectra were obtained using a JASCO J-500A spectropolarimeter equipped with a 1.5 T electromagnet.* Note *the different scales for B for the Soret and visible regions. (The data are taken from reference 39.)*

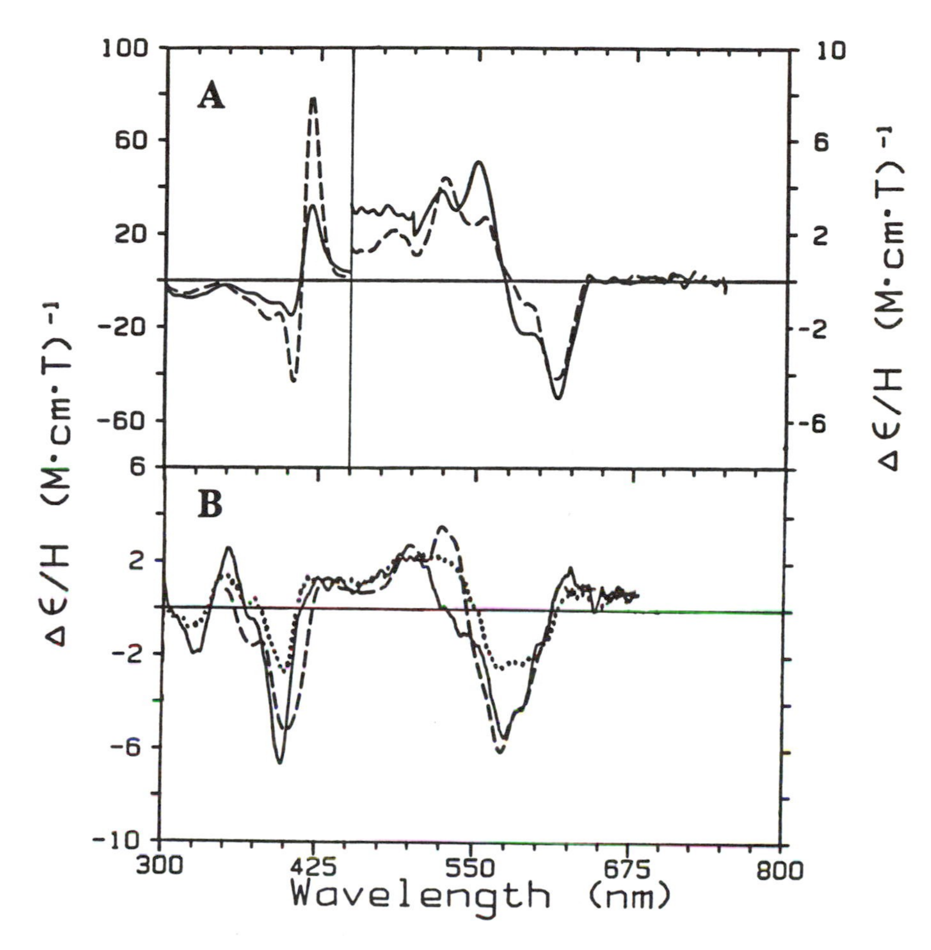

Figure 8. MCD spectra of the (A) deoxyferrous state of methylchlorin-reconstituted horseradish peroxidase (——) and of the high-spin 1,2-dimethylimidazole adduct of octaethylchlorin (- - -) and of the (B) ferrous-NO adducts of methylchlorin-reconstituted myoglobin (——) and horseradish peroxidase (- - -) and of ferrous(octaethylchlorin)(1-methylimidazole)(NO) (· · · · ·). Spectra of protein samples and of model complexes were measured at 4 and 25 °C, respectively. MCD spectra were obtained using a JASCO J-500A spectropolarimeter equipped with a 1.5 T electromagnet. Note *the different scales for A for the Soret and visible regions. (The data are taken from reference 39.)*

the identity of the proximal axial ligand to the chlorin iron is not known with certainty. By using the fingerprinting power of MCD spectroscopy in the manner described in the preceding paragraphs, evidence has been presented favoring the presence of a tyrosine proximal ligand (*40*). To further establish that octaethylchlorin and methylchlorin are appropriate chlorins to use as models for the prosthetic group of HPII catalase, the MCD spectra of the ferrous adducts of all three chlorins with pyridine and CO as axial ligands have been measured (Figure 9). The adduct from the enzyme was generated by extracting the prosthetic group from HPII catalase in the presence of pyridine and CO under reducing conditions. Next, the MCD spectrum of the native ferric state of HPII catalase is compared to that of a five-coordinate phenolate-ligated ferric octaethylchlorin model complex (Figure 10A). The close similarity between the two spectra suggests that the axial ligand in the protein is also a phenolate oxygen donor as would be the case with tyrosine as the proximal ligand. However, the MCD spectrum of the native ferric state of methylchlorin-reconstituted horseradish peroxidase (Figure 10B), presumed to be five-coordinate histidine-ligated, is somewhat similar to those shown in Figure 10A. The MCD spectrum of the related high-spin ferric state with histidine and water axial ligands, namely methylchlorin-reconstituted ferric myoglobin (Figure 10B), is noticeably dif-

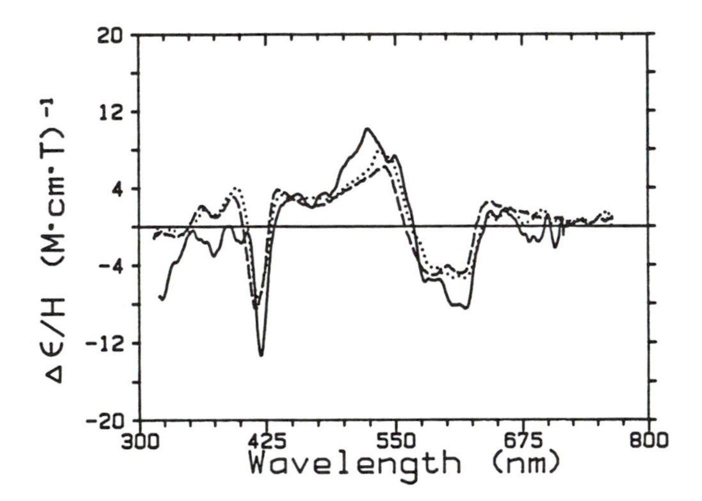

Figure 9. MCD spectra of the ferrous pyridine–CO complexes of the extracted prosthetic group of HPII catalase (——), octaethylchlorin (– – –), and methylchlorin (· · · · ·). Spectra of protein samples and of model complexes were measured at 4 and 25 °C, respectively. MCD spectra were obtained using a JASCO J-500A spectropolarimeter equipped with a 1.5 T electromagnet. The latter two spectra have been red-shifted by 15 nm. (The data are taken from reference 40.)

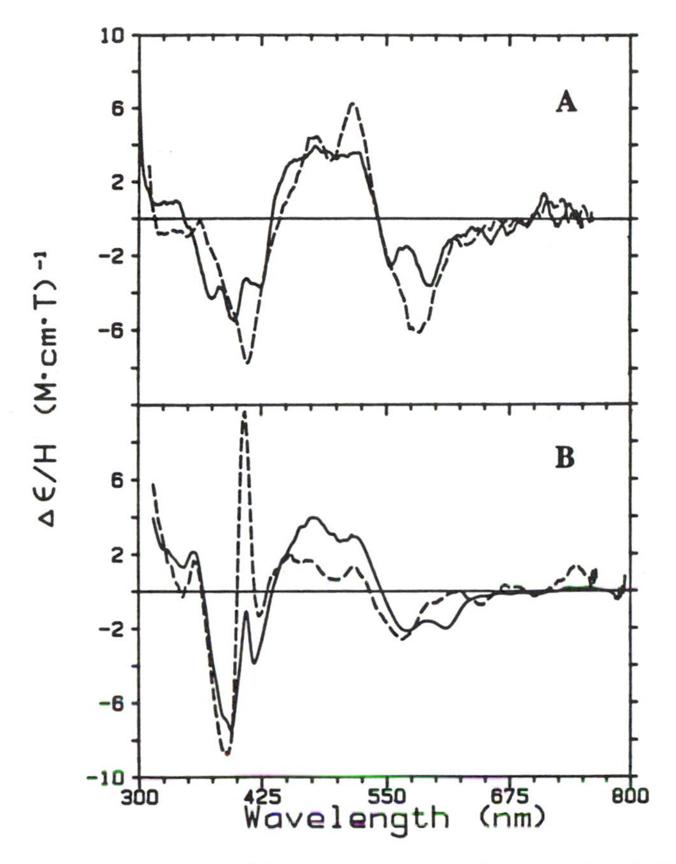

Figure 10. MCD spectra of (A) native HPII catalase (——) and the five-coordinate ferric phenolate adduct of octaethylchlorin (- - -) and of (B) ferric methylchlorin-reconstituted horseradish peroxidase (——) and myoglobin (- - -). Spectra of protein samples and of model complexes were measured at 4 and 25 °C, respectively. MCD spectra were obtained using a JASCO J-500A spectropolarimeter equipped with a 1.5 T electromagnet. Except for the spectrum of native HPII catalase, all MCD spectra have been red-shifted by 10 nm. (The data are taken from reference 40.)

ferent. The data in Figure 10 support the conclusion that the proximal ligand to the chlorin iron of HPII catalase is either tyrosine or histidine.

To rule out the latter possibility, the MCD spectra of the azide and cyanide adducts of ferric HPII catalase have been compared (*40*) to those of the same adducts of methylchlorin-reconstituted myoglobin (Figure 11). Although all four spectra contain a trough at 425 nm, the rest of the spectra of the two myoglobin derivatives are clearly different from those of the corresponding HPII catalase adducts. These data argue against histidine as the proximal ligand in the enzyme. Elimination of

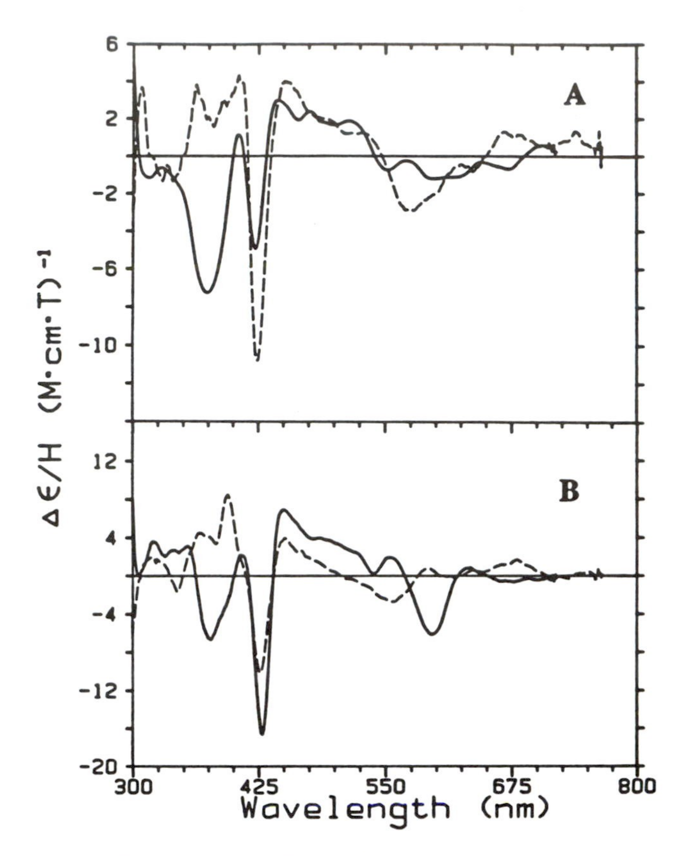

Figure 11. MCD spectra at 4 °C of the (A) ferric azide and (B) cyanide adducts of HPII catalase (——) and of methylchlorin-reconstituted myoglobin (– – –). MCD spectra were obtained using a JASCO J-500A spectropolarimeter equipped with a 1.5 T electromagnet. The spectra of the reconstituted myoglobin derivatives have been red-shifted by 15 nm. (The data are taken from reference 40.)

histidine as a possible proximal ligand leaves tyrosine, or more accurately tyrosinate, as the most likely choice (*40*). This conclusion is consistent with the amino acid sequence of HPII catalase, which contains a tyrosine at the position corresponding to the tyrosine proximal ligand to the heme iron of bovine catalase (*41*). The three-dimensional structure of HPII catalase is being determined (*42*).

The Prosthetic Group of Bovine Spleen Myeloperoxidase

The unusually red-shifted electronic absorption spectra and resulting green color of the various ligation and oxidation state derivatives of myeloperoxidase have been objects of interest for some time. As mentioned in the last section, two proposals have been suggested for the atypical (for a heme protein) green color, either the prosthetic group is an iron chlorin or an iron porphyrin bearing a formyl substituent. The first piece of evidence obtained with MCD spectroscopy (*43*) that addressed this issue is the comparison of the spectra of the pyridine hemochromogens (ferrous bis-pyridine adducts) of the heme from denatured myeloperoxidase compared to those of heme *s* and octaethylchlorin (Figure 12). The spectra of the heme from myeloperoxidase and that of heme *s* are nearly identical and clearly distinguishable from that of the same derivative of octaethylchlorin. A similarly close correspondence is seen between the MCD spectra of the ferric bis-cyanide adducts of the heme from denatured myeloperoxidase and heme *s* shown in Figure 13. The MCD spectrum of the analogous ferric bis-cyanide complex of octaethylchlorin has not been measured.

Several MCD comparisons have been reported involving intact (i.e., nondenatured) myeloperoxidase and heme *s*-reconstituted myoglobin (*43*). The heme *s*-reconstituted myoglobin has been investigated as a model for myeloperoxidase because it has a histidine proximal ligand as

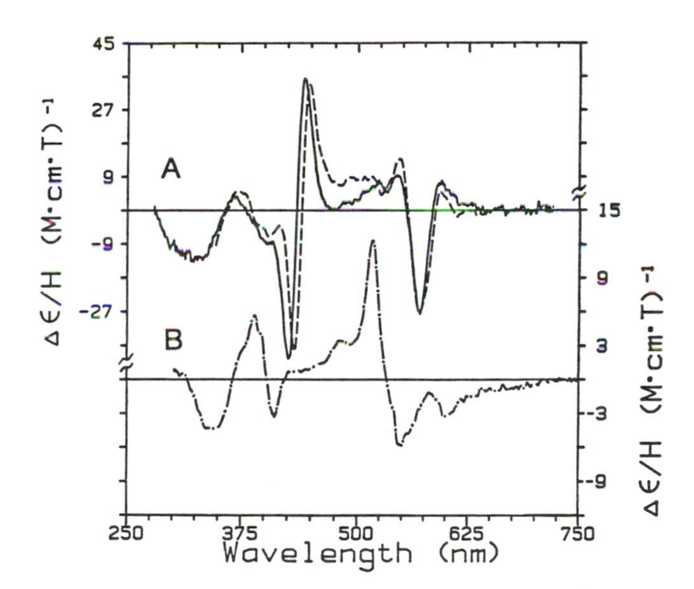

Figure 12. MCD spectra of the pyridine hemochromogens (ferrous bis-pyridine adducts) of alkaline-denatured Spirographis *heme-reconstituted myoglobin (——) and myeloperoxidase (– – –) (both in A) and of iron octaethylchlorin (·–·–·, B). Spectra of protein samples and of model complexes were measured at 4 and 25 °C, respectively. MCD spectra were obtained using a JASCO J-500A spectropolarimeter equipped with a 1.5 T electromagnet. The left- and right-hand side scales apply to A and B, respectively. (The data are taken from reference 43.)*

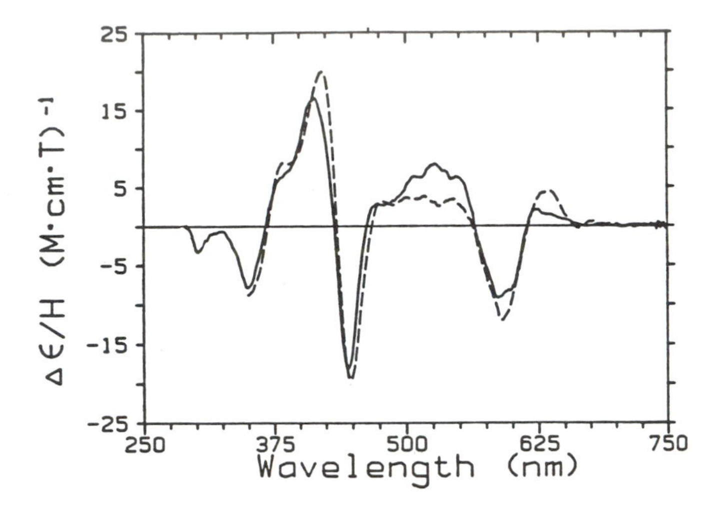

Figure 13. MCD spectra at 4 °C of the ferric bis-cyanide adducts of the prosthetic group from denatured myeloperoxidase (– – –) and of Spirographis *heme (heme* s*) (——) in 6 M guanidine hydrochloride and 50 mM Tris buffer (pH 7.8). MCD spectra were obtained using a JASCO J-500A spectropolarimeter equipped with a 1.5 T electromagnet. (The data are taken from reference 43.)*

has been established for myeloperoxidase. The MCD spectra of parallel derivatives of these two proteins are consistently similar to each other and distinct from the spectra of analogous complexes of octaethylchlorin. An example of this is seen in Figure 14, in which the spectra of the ferrous–CO complexes of myeloperoxidase, heme *s*-reconstituted myoglobin, and Fe(octaethylchlorin)(1-methylimidazole) are compared.

In summary, MCD spectral similarities have been observed between a number of analogous derivatives of myeloperoxidase and heme *s*-reconstituted myoglobin including several not shown in this chapter. This evidence strongly suggests that the heme prosthetic group in myeloperoxidase is an iron porphyrin containing a formyl substituent (*43*). A similar conclusion was reached by Wever et al. (*44*), who observed a carbonyl stretch in the resonance Raman spectrum of the denatured enzyme. The resolution of the published X-ray crystal structure of myeloperoxidase (*45*) does not distinguish between the two proposed explanations for the green color of the enzyme. Finally, in those cases in which parallel derivatives of octaethylchlorin and of heme *s* have been prepared, the MCD spectra of the two types of "green hemes" are readily distinguishable. It therefore seems likely that MCD spectroscopy will be useful in distinguishing the two types of green heme chromophores found in nature.

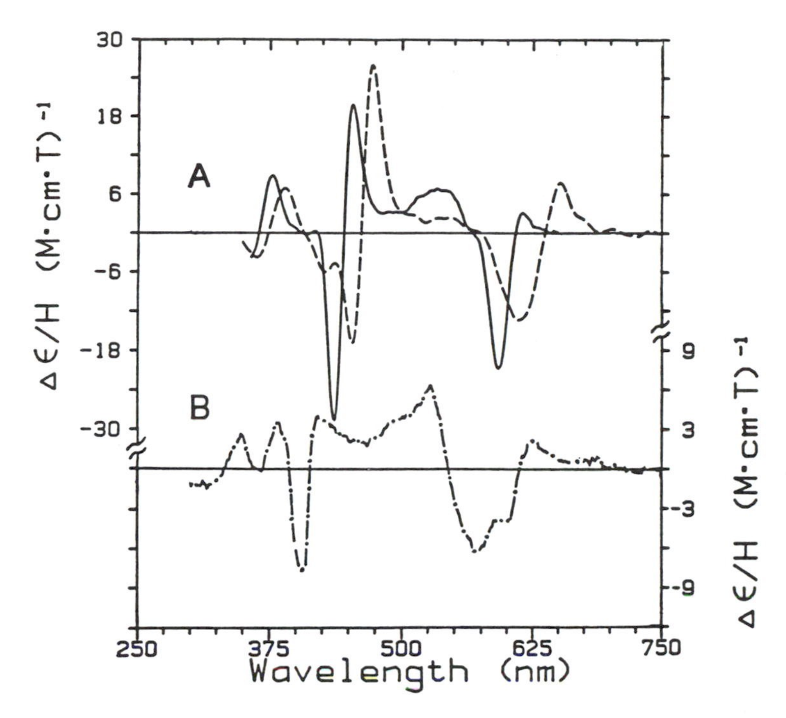

Figure 14. MCD spectra of the ferrous-CO adducts of myeloperoxidase (– – –) and Spirographis *heme (heme* s*)-reconstituted myoglobin (——) (both in A) and of ferrous (octaethylchlorin)(1-methylimidazole)(CO) (·–·–·, B). Spectra of protein samples and of model complexes were measured at 4 and 25 °C, respectively. MCD spectra were obtained using a JASCO J-500A spectropolarimeter equipped with a 1.5 T electromagnet. The left- and right-hand side scales apply to A and B, respectively. (The data are taken from reference 43.)*

Concluding Remarks

MCD spectroscopy is a promising technique for establishing the coordination structures of "green heme" proteins such as those occurring in iron chlorin- and formyl-substituted iron porphyrin-containing systems. As has been well established for the MCD of protoporphyrin IX-based heme systems, the MCD spectra of iron chlorins are quite sensitive to changes in oxidation, spin, and ligation state as well as being insensitive to the solvent or protein environment in which the chromophore is located. By using this approach, it has been shown that the most likely identity of the proximal ligand in the HPII catalase from *E. coli* is tyrosine. Data has also been presented to show that the prosthetic group in a second green heme protein, myeloperoxidase, is not an iron chlorin. Instead, it appears that the protein contains a formyl-substituted iron porphyrin group. Finally, it has been shown that MCD spectroscopy is readily able to distinguish the two types of green heme chromophores. Additional studies are currently underway to further explore the utility of MCD spectroscopy in the structural characterization of green heme systems.

Acknowledgments

We thank Edmund W. Svastits and John J. Rux for developing the computer-based spectroscopic data-handling system; A. Grant Mauk (University of British Columbia), Peter C. Loewen (University of Manitoba), and Masao Ikeda-Saito (Case Western Reserve University) for providing samples of cytochrome b_5, HPII catalase and myeloperoxidase, respectively; and Isam Arafa, Elisabeth T. Kintner, Kevin M. Smith, and Steven H. Strauss for helpful discussions, as well as Chengfeng Zhuang and Caroline M. Zeitler for technical assistance. Support for this research has been provided by National Institutes of Health Grants GM26730 for John H. Dawson and GM34468 for Chi K. Chang. The JASCO J-500 spectropolarimeter and the electromagnet were obtained with grants from NIH (RR-03960) and Research Corporation, respectively.

References

1. Dawson, J. H.; Dooley, D. M. In *Iron Porphyrins;* Part III; Lever, A. B. P.; Gray, H. B., Eds.; VCH: New York, 1989; pp 1–93 and 93–135.
2. Holmquist, B. *Methods Enzymol.* **1986,** *130,* 270–289.
3. Holmquist, B. In *The Porphyrins;* Dolphin, D., Ed.; Academic: New York, 1978; Vol. 3, pp 249–270.
4. Sutherland, J. C.; Holmquist, B. *Annu. Rev. Biophys. Bioeng.* **1980,** *9,* 293–326.
5. *Faraday's Diary;* Vol. IV; G. Bell and Sons: London, 1933.
6. Thomson, A. J.; Johnson, M. K. *Biochem. J.* **1980,** *191,* 411–420.
7. Schatz, P. N.; Mowery, R. L.; Krausz, E. R. *Mol. Phys.* **1978,** *35,* 1535–1557.
8. Browett, W. R.; Fucaloro, A. F.; Morgan, T. V.; Stephens, P. J. *J. Am. Chem. Soc.* **1983,** *105,* 1868–1872.
9. Sono, M.; Dawson, J. H.; Hager, L. P. *Inorg. Chem.* **1985,** *24,* 4339–4343.
10. Svastits, E. W.; Dawson, J. H. *Inorg. Chim. Acta* **1986,** *123,* 83–86.
11. Dawson, J. H.; Sono, M. *Chem. Rev.* **1987,** *8,* 1255–1276.
12. Dawson, J. H.; Andersson, L. A.; Sono, M. *J. Biol. Chem.* **1982,** *257,* 3606–3617.
13. Sono, M.; Dawson, J. H. *Biochim. Biophys. Acta* **1984,** *789,* 170–187.
14. Uchida, K.; Shimizu, T.; Makino, R.; Sakaguchi, K.; Iizuka, T.; Ishimura, Y.; Nozawa, T.; Hatano, M. *J. Biol. Chem.* **1983,** *258,* 2519–2525 and 2526–2533.
15. Alberta, J. A.; Andersson, L. A.; Dawson, J. H. *J. Biol. Chem.* **1989,** *264,* 20467–20473.
16. Svastits, E. W.; Alberta, J. A.; Kim, I.-C.; Dawson, J. H. *Biochem. Biophys. Res. Commun.* **1989,** *165,* 1170–1176.
17. Simpkin, D.; Palmer, G.; Devlin, F. J.; McKenna, M. C.; Jensen, G. M.; Stephens, P. J. *Biochemistry* **1989,** *28,* 8033–8039.
18. Morgan, W. T.; Vickery, L. E. *J. Biol. Chem.* **1978,** *253,* 2940–2945.
19. Siedow, J. N.; Vickery, L. E.; Palmer, G. *Arch. Biochem. Biophys.* **1980,** *203,* 101–107.

20. Wang, C.-B.; Chang, C. K. *Synthesis* **1979,** 548–549.
21. Sessler, J. L.; Mozaffari, A.; Johnson, M. R. *Org. Synth.* **1991,** *70,* 68–78.
22. Dawson, J. H.; Kadkhodayan, S.; Zhuang, C.; Sono, M. *J. Inorg. Biochem.* **1992,** *45,* 179–192.
23. Lorence, R. M.; Gennis, R. B. *J. Biol. Chem.* **1989,** *264,* 7135–7140.
24. Chiu, J. T.; Loewen, P. C.; Switala, J.; Gennis, R. B.; Timkovich, R. *J. Am. Chem. Soc.* **1989,** *111,* 7046–7050.
25. Chang, C. K.; Sotiriou, C. *J. Org. Chem.* **1985,** *50,* 4989–4991.
26. Jacob, G. S.; Orme-Johnson, W. H. *Biochemistry* **1979,** *18,* 2967–2975 and 2975–2980.
27. Berzofsky, J. A.; Peisach, J.; Horecker, B. L. *J. Biol. Chem.* **1972,** *247,* 3783–3791.
28. Chatfield, M. J.; LaMar, G. N.; Kauten, R. J. *Biochemistry* **1987,** *26,* 6939–6950.
29. Sibbett, S. S.; Hurst, J. K. *Biochemistry* **1984,** *23,* 3007–3013.
30. Babcock, G. T.; Ingle, R. T.; Oertling, W. A.; Davis, J. S.; Averill, B. A.; Hulse, C. L.; Stufkens, D. T.; Bolscher, B. G. J. M.; Wever, R. *Biochim. Biophys. Acta* **1985,** *828,* 58–66.
31. Ikeda-Saito, M.; Argade, P. V.; Rousseau, D. I. *FEBS Lett.* **1985,** *84,* 52–55.
32. Eglinton, D. B.; Barber, D.; Thomson, A. J.; Greenwood, C.; Segal, A. W. *Biochim. Biophys. Acta* **1982,** *703,* 187–195.
33. Morell, D. B.; Chang, Y.; Clezy, P. S. *Biochim. Biophys. Acta* **1967,** *136,* 121–130.
34. Newton, N.; Morell, D. B.; Clark, L.; Clezy, P. S. *Biochim. Biophys. Acta* **1965,** *96,* 476–486.
35. Sono, M.; Asakura, T. *J. Biol. Chem.* **1975,** *250,* 5227–5232.
36. Antonini, E.; Brunori, M. *Hemoglobin and Myoglobin in Their Reactions with Ligands;* North-Holland: Amsterdam, Netherlands, 1971; pp 49–50.
37. Stolzenberg, A. M.; Strauss, S. H.; Holm, R. H. *J. Am. Chem. Soc.* **1981,** *103,* 4763–4778.
38. Huff, A. M.; Chang, C. K.; Cooper, D. K.; Smith, K. M.; Dawson, J. H. *Inorg. Chem.* **1993,** *32,* 1460–1466.
39. Bracete, A. M.; Kadkhodayan, S.; Sono, M.; Huff, A. M.; Zhuang, C.; Cooper, D. K.; Smith, K. M.; Chang, C. K.; Dawson, J. H. *Inorg. Chem.* **1994,** *33,* 5042–5049.
40. Dawson, J. H.; Bracete, A. M.; Huff, A. M.; Kadkhodayan, S.; Zeitler, C. M.; Sono, M.; Chang, C. K.; Loewen, P. C. *FEBS Lett.* **1991,** *295,* 123–126.
41. Von Ossowski, I.; Mulvey, M. R.; Leco, P. A.; Borys, A.; Loewen, P. C. *J. Bacteriol.* **1991,** *173,* 514–520.
42. Tormo, J.; Fita, I.; Switala, J.; Loewen, P. C. *J. Mol. Biol.* **1990,** *213,* 219–230.
43. Sono, M.; Bracete, A. M.; Huff, A. M.; Ikeda-Saito, M.; Dawson, J. H. *Proc. Natl. Acad. Sci. U.S.A.* **1991,** *88,* 11148–11152.
44. Wever, L.; Oertling, R. A.; Hoogland, H.; Bolsher, B. G. J. M.; Kim, Y.; Babcock, G. T. *J. Biol. Chem.* **1991,** *266,* 24308–24313.
45. Zheng, J.; Fenna, R. E. *J. Mol. Biol.* **1992,** *226,* 185–207.

RECEIVED for review June 10, 1993. ACCEPTED revised manuscript April 18, 1994.

14

Mechanistic Aspects of the Chemistry of Iron *N*-Alkyl Porphyrins

Charles R. Cornman and Edward P. Zovinka

Department of Chemistry, North Carolina State University, Raleigh, NC 27695–8401

N-*substituted iron porphyrins form upon treatment of heme enzymes with many xenobiotics. The formation of these modified hemes is directly related to the mechanism of their enzymatic reactivity.* N-*alkyl porphyrins may be formed from organometallic iron porphyrin complexes, PFe–R (σ-alkyl, σ-aryl) or PFe = CR_2 (carbene). They are also formed via a branching in the reaction path used in the epoxidation of alkenes. Biomimetic* N-*alkyl porphyrins are competent catalysts for the epoxidation of olefins, and it has been shown that iron* N-*alkylporphyrins can form highly oxidized species such as an iron(IV) ferryl, (N–R P)Fe^{IV}=O, and porphyrin π-radicals at the iron(III) or iron(IV) level of metal oxidation. The* N-*alkylation reaction has been used as a low resolution probe of heme protein active site structure. Modified porphyrins may be used as synthetic catalysts and as models for nonheme and noniron metalloenzymes.*

IN THE EARLY 1970s it was discovered that P-450 cytochromes are irreversibly inhibited during the metabolism of xenobiotics (*1*). The formation of a modified heme prosthetic group is associated with enzyme inhibition and subsequent studies have identified these modified complexes as *N*-alkylated protoporphyrin-IX (*2*). The chemistry of *N*-substituted porphyrins was comprehensively reviewed by Lavallee in 1987 (*3*). Since that time, there have been many significant contributions to this field by several groups. The goal of this chapter is to summarize some of this work as it relates to the mechanism of formation and reactivity of iron *N*-alkyl porphyrins. Biomimetic model complexes have played an important role in elucidating the chemistry of *N*-alkyl hemes in much the same way that synthetic iron tetraarylporphyrins have aided

0065–2393/95/0246–0373$09.98/0

in understanding the role of hemes in oxygen transport (*4*), electron transfer (*5, 6*), and catalysis (*7–10*). This chapter emphasizes the insights obtained from the biomimetic approach to understanding the naturally occurring system (*11*). The abbreviations used for the tetraarylporphyrins are defined in Figure 1.

Formation of N-*Substituted Iron Porphyrins*

N-substituted porphyrins are formed during the metabolism of xenobiotics that include terminal alkenes and alkynes, as well as activated organic molecules such as halocarbons, diazo compounds, and hydrazines. In the synthetic laboratory, *N*-substituted porphyrins are prepared easily via alkylation of a pyrrole nitrogen atom of the porphyrin, followed by metallation. Biomimetic reactions between iron porphyrins, oxidants, and alkenes (or activated carbon sources) may also be used to alkylate the pyrrole nitrogen.

Via Organometallic Intermediates. Metabolic reactions of xenobiotics such as halocarbons, hydrazines, or sydnones result in the formation of *N*-substituted porphyrins. An organometallic complex, in the form of an iron(II)-carbene (for the sydnones and halocarbons) or an iron(III)-σ-alkyl (σ-aryl) (hydrazines), is an isolable intermediate in this process. The novelty of the biological organometallic chemistry has induced a flurry of research activity in this area.

The first carbene complex of an iron porphyrin was prepared and structurally characterized by Mansuy and co-workers (*12, 13*) through the reaction of carbon tetrachloride with (TPP)Fe^{II} under reductive

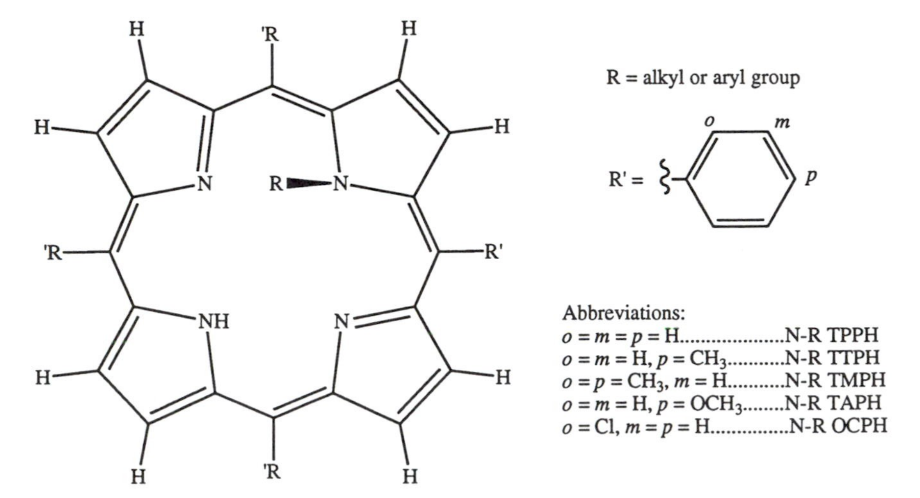

Figure 1. Structures and abbreviations for iron N-*alkyl tetraarylporphyrins.*

conditions. These carbene and σ-alkyl (σ-aryl) complexes are sensitive to the presence of oxidants, and under oxidative conditions they rearrange to form *N*-substituted products. Scheme 1 summarizes the reactions of the halocarbon DDT (1,1-bis(*p*-chlorophenyl)-2,2,2-trichloroethane) with iron(II) tetraarylporphyrin. The intermediacy of the iron–carbene complex **1** (*14*), the oxidative insertion product **2** (*15*), and the *N*-vinylporphyrin **4** (*16*) has been demonstrated by a combination of spectroscopy and X-ray crystallography. The molecular structures of **1**, **2**, and **4** are presented in Figure 2. The reaction used to prepare **2** is formally an oxidatively induced migratory insertion of the carbene into the iron–nitrogen bond to produce the intermediate spin ($S = ^3/_2$) iron(III) product. Addition of acid to the insertion product results in cleavage of the iron–carbon bond and demetallation. The *N*-vinylporphyrin may be metallated with an iron(II) solution.

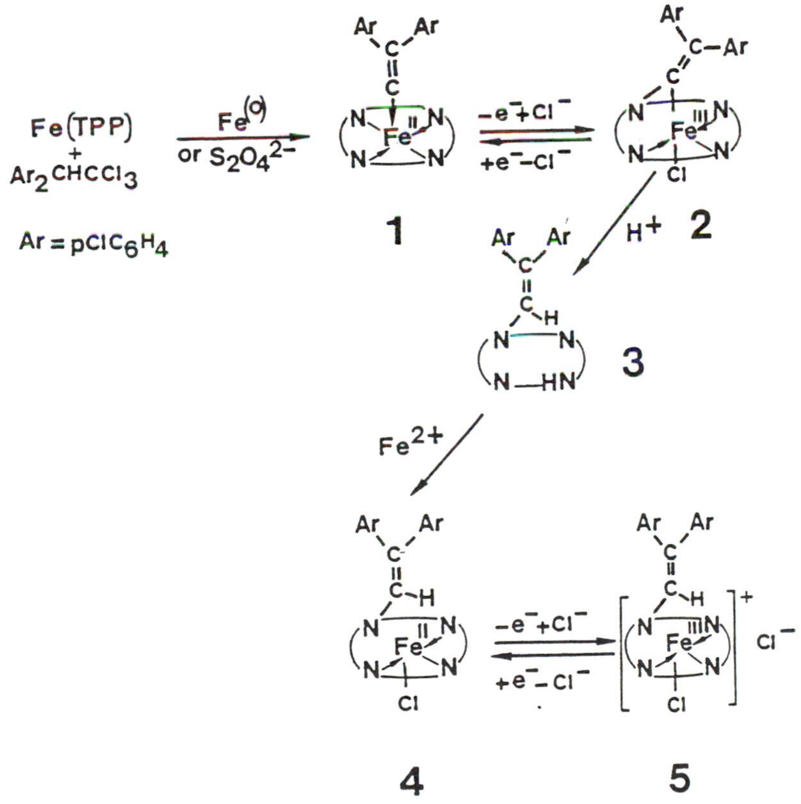

Scheme 1

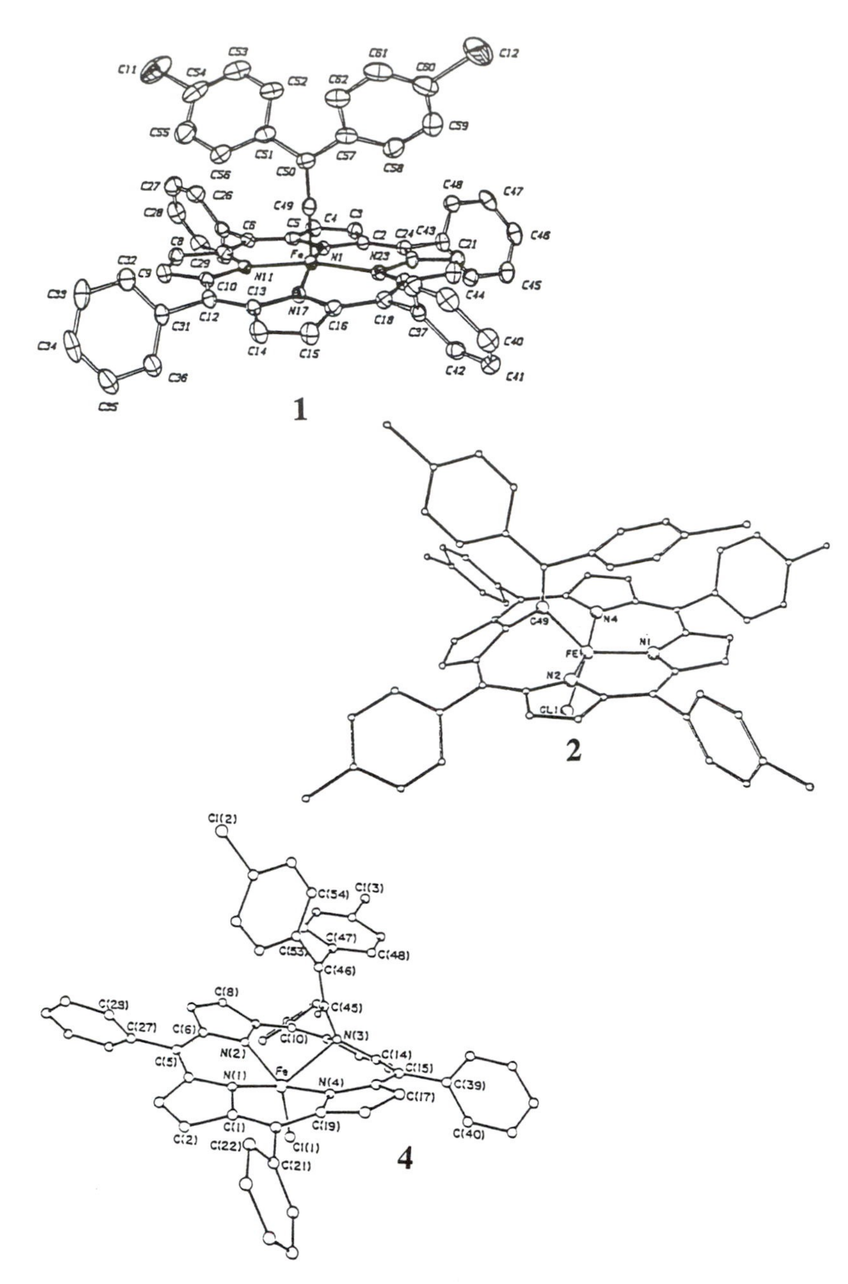

Figure 2. X-ray crystal structures for the proposed intermediates in N-*alkylporphyrin formation from the halocarbon DDT. Top: The carbene complex (TPP)[C=C (*p-ClC_6H_4*)$_2$]Fe (14). Middle: Carbene insertion product (TTP)[(*p-ClC_6H_4*)$_2$C=C]FeCl · 2CH_2Cl_2 (15). Bottom:* N-*alkyl product {*N-*[2,2-bis(*p-*chlorophenyl)vinyl]TPP}FeCl (16).*

Iron porphyrin complexes with axial σ-alkyl and σ-aryl groups have been prepared and fully characterized by several groups (*17, 18*). Addition of a chemical oxidant to (*19, 20*), or electrochemical oxidation of (*21*), the low-spin iron(III)-alkyl (-aryl) porphyrins results in transient formation of an iron(IV) σ-alkyl (σ-aryl) complex that undergoes reductive elimination to give the iron(II) *N*-substituted product as shown in Scheme 2. The iron(IV) intermediate has been directly observed by low temperature ^{1}H NMR spectroscopy (*22*) and spectroelectrochemistry (*21*).

From Iron(III) Tetraarylporphyrins and Alkenes. *N*-alkyl porphyrins are formed via side reactions of the normal catalytic cycle of cytochromes P-450 with terminal alkenes or alkynes. *N*-alkylporphyrins formed from terminal alkenes (with model iron porphyrin catalysts under epoxidation conditions) usually have a covalent bond between the terminal carbon atom of the alkene and a pyrrole nitrogen. The double bond is oxidized selectively to an alcohol at the internal carbon. Mansuy (*23*) showed that, in isolated examples, terminal alkenes can form *N*-alkylated products in which the internal carbon is bound to the nitrogen and the terminal carbon is oxidized to the alcohol. Internal alkenes may also form *N*-alkyl porphyrins (*24, 25*).

The *N*-alkylation reaction represents a bifurcation of the normal alkene epoxidation reaction cycle and, therefore, *N*-alkylation is a "suicide" event that leads to catalytic inhibition in the native system. With synthetic tetraarylporphyrins that mimic the *N*-alkylation reaction, the use of halogen-substituted catalysts that are stable toward oxidative degradation (*26, 27*) provide the most useful model systems because the heme model remains intact for a significantly greater number of turnovers than the partition number. The partition number is the ratio of epoxidation cycles to *N*-alkylation cycles, i.e., *N*-alkyl porphyrins are formed before the heme is oxidatively destroyed.

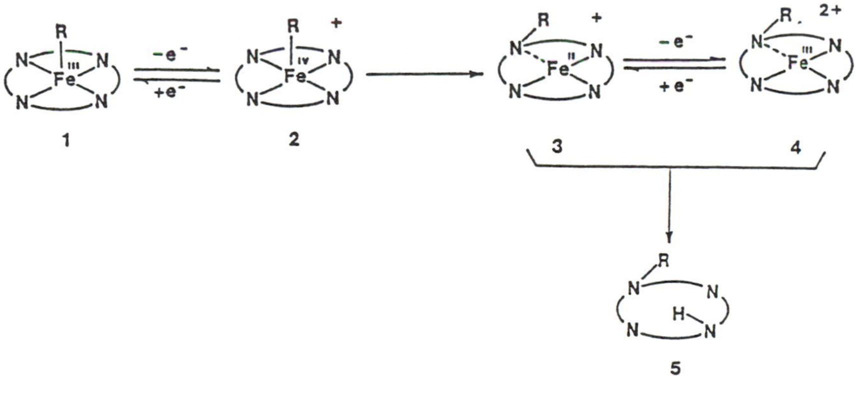

R, N, N, III, Fe, N, N — 1 ⇌ (−e⁻ / +e⁻) — R +, N, N, IV, Fe, N, N — 2 → R +, N, N, II, Fe, N, N — 3 ⇌ (−e⁻ / +e⁻) — R 2+, N, N, III, Fe, N, N — 4 → R, N, N, H, N, N — 5

Scheme 2

Collman and co-workers (*28, 29*) have used [(OCP)Fe^{III}Cl] as the catalyst and pentafluoroiodosylbenzene as the oxidant to determine the partition numbers for a series of alkenes. Iodosylarene oxidants are insoluble in the solvents used and thus the catalytic system was heterogeneous. Kinetic analysis, which took into account contributions from catalysis by the *N*-alkylated heme (*see* subsequent paragraphs), provided partition numbers for the catalytic cycle ranging from 140 to 28,000, depending on the alkene, porphyrin, and oxidant. The partition numbers were reproducible over many measurements including those with different preparations of catalyst and oxidant; however, the rate constants for epoxidation and *N*-alkylation were irreproducible. This irreproducibility was attributed to the heterogeneity of the system. The reproducibility of the partition numbers suggested a common rate limiting step that the authors attributed to formation of an iron-oxo intermediate, not a metallacycle as previously proposed (*30, 31*). The partition numbers are then dependent on the steric and electronic properties of both the iron porphyrin catalyst and the alkene. Because the partition numbers were also dependent on the oxidant, Collman has proposed that the active metalloporphyrin oxidant is different for different oxygen atom sources (ArIO or OCl^-). Given these observations, Collman favors a concerted reaction between the alkene and the ferryl-cation radical oxidant in which the orientation of the alkene determines the product distribution as shown in Figure 3.

Traylor and his collaborators (*32*) have examined the *N*-alkylation reactions for a series of iron tetraarylporphyrins and have reported that *N*-alkyl hemes are reversibly formed during the catalytic cycle of P-450 model systems such as (OCP)Fe^{III}Cl, norbornene, and pentafluoroiodosylbenzene. Mixing these reagents in $CH_2Cl_2/CF_3CH_2OH/H_2O$ (89:10:1) results in a homogeneous solution and formation of a new iron porphyrin complex with a red-shifted Soret band and a modified Q band region as shown in Figure 4. As is evident from the figure, this new species reverts to (OCP)Fe^{III}Cl within several seconds. The intermediate

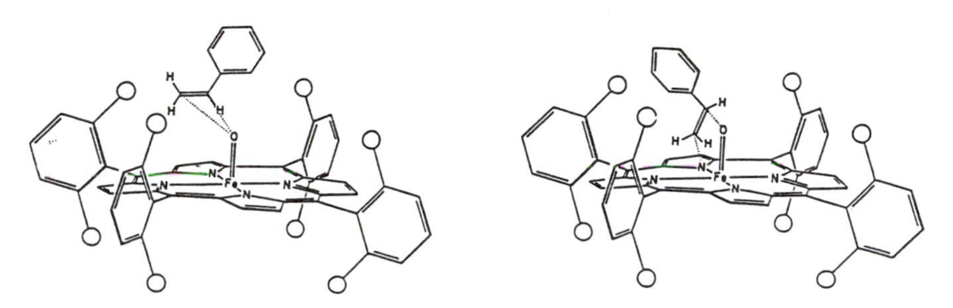

Figure 3. Proposed orientation of alkene relative to the ferryl-porphyrin π-radical for epoxidation (left) or N*-alkylation (right)* (28, 29).

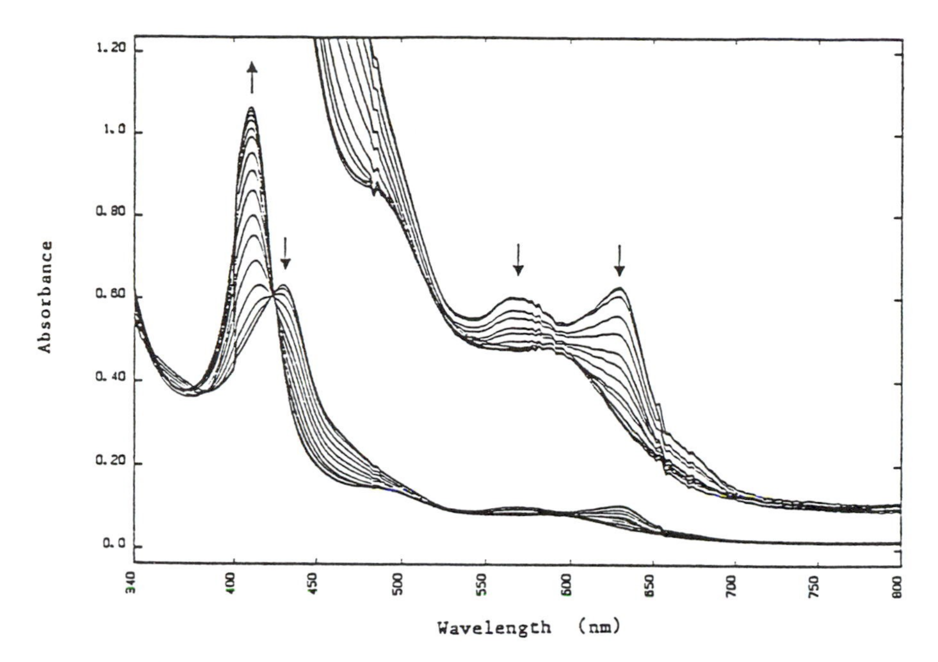

*Figure 4. UV–vis spectra monitoring the decomposition of iron (*N*-norbornyl OCP) transiently generated from the reaction of (OCP)FeCl with norbornene and pentafluoroiodobenzene. The first spectra was taken 3 s after mixing. The first seven spectra are at 7-s intervals and the rest at 14-s intervals. The final spectrum corresponds to the chloride, (OCP)FeCl. (Reproduced from reference 32. Copyright 1987 American Chemical Society.)*

spectrum is very similar to that of an iron *N*-alkylporphyrin isolated from the reaction of (OCP)$Fe^{III}Cl$, PFIB, and 4,4-dimethyl-1-pentene. That the intensity of the Soret band of the starting material is essentially absent at 3 s attests to the nearly complete conversion to the *N*-alkylated derivative. Isosbestic behavior for both the forward and reverse reaction indicates that only two species contribute to the observed spectral changes. The rhombic electron paramagnetic resonance (EPR) spectrum of the transient species, trapped by rapid freezing of the reaction mixture, is similar to that of isolated iron(III) *N*-alkyl porphyrins (*24*). Balch and co-workers (*33*) and Ogoshi and co-workers (*34*) report axial EPR spectra for (*N*-MeTPP)$Fe^{III}Cl^+$ and (*N*-MeOEP)$Fe^{III}Cl^+$, respectively. The larger, chelating *N*-norbornyl alkoxide is probably responsible for the rhombic EPR in the Traylor complex. Mansuy (*40*) reports a rhombic EPR spectrum for an iron(III) *N*-alkyl chelate complex. Kinetic analysis of the formation and decomposition of the transient iron *N*-alkylporphyrin provides rate constants for the forward and reverse reactions of 450 M^{-1} s^{-1} and 0.07 s^{-1}. Because the rate constant for epoxidation was determined to be about 10^5 M^{-1} s^{-1} the *N*-alkylated transient is not an intermediate in the major catalytic cycle. This epoxidation rate is considerably greater than the rates reported by other investigators (about 10^{-4}–10^2 M^{-1} s^{-1}) (*35–37*). Although the authors do not elaborate on why there is such a large discrepancy between their rate constants and those previously reported, it seems likely that the solvent system may play an important role.

Traylor (*38*) has also shown that biomimetic iron *N*-alkylporphyrins themselves are competent catalysts for epoxidation of alkenes with a rate constant of about 10^4 M^{-1} s^{-1}. On the basis of these observations and rearrangement reactions of specific alkenes, Traylor has proposed the reaction sequence outlined in Scheme 3 as representative of the oxidation and *N*-alkylation reactions of the P-450 model systems. In this scheme, the epoxide and the *N*-alkylated heme are derived from a common, electron-transfer intermediate (caged ferrylporphyrin-alkene cation radical). Collman and co-workers (*28, 29*) prefer a concerted mechanism (or a short-lived, acyclic intermediate) for epoxidation and *N*-alkylation reactions. Both authors note that the reactions catalyzed by cytochrome P-450 (and biomimetic reactions) probably can not be ascribed to any single mechanism.

Reactions of Iron N-*Alkylporphyrins*

Steric and Electronic Considerations. Electronically, the *N*-substituted macrocycle is a monoanion in its deprotonated form. For-

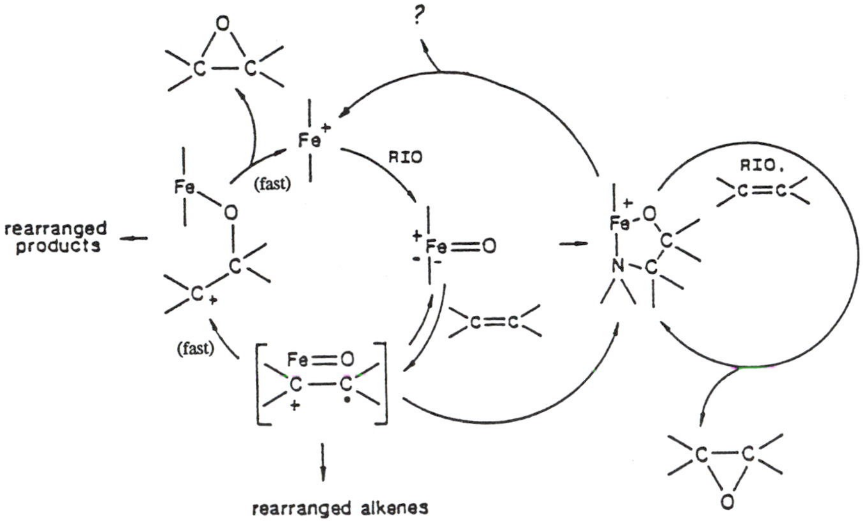

?
(fast)
RIO
RIO.
rearranged products
(fast)
rearranged alkenes

Scheme 3

mally, the porphyrin has two neutral imine donor atoms, one amido anionic donor atom, and one neutral *amine* donor atom. Because of the neutral amine donor, the *N*-substituted macrocycle stabilizes metal ions in lower oxidation states than the corresponding unsubstituted porphyrin. Thus, with ligands such as *N*-MeTPPH, both the iron(II) and iron(III) complexes are stable in air whereas $TPPFe^{II}$ readily oxidizes to iron(III). As reported by Lavallee and associates (*39*), $(N\text{-MeTPP})Fe^{II}Cl$ has a reversible III/II couple at 0.49 V versus SCE (in CH_2Cl_2), whereas $(TPP)Fe^{III}Cl$ has an irreversible reductive wave (due to the loss of the chloride ligand) at −0.29 V under identical conditions.

Three of the nitrogen donor atoms of the deprotonated *N*-alkyl macrocycle are sp^2-hybridized and thus contribute to the planarity of the metal complex. The amine nitrogen donor is sp^3-hybridized, and this sp^3-hybridization forces large deviations of the macrocycle planarity in the metal complex (*40, 41*). These deviations are clearly present in the structure of the cation of $[(N\text{-MeTTP})Fe^{III}Cl]SbCl_6$ as shown in Figure 5A. Two of the pyrrole rings are canted with their nitrogen atoms directed toward the chloro ligand ($<10°$) relative to the mean plane of the three unsubstituted nitrogens. The substituted pyrrole nitrogen and the pyrrole nitrogen *trans* to the substituted pyrrole are directed away from the chloro ligand by 39° and 7°, respectively, relative to the plane defined by the three unsubstituted nitrogen atoms. This change in direction is a graphic example of the ability of the aromatic porphyrin macrocycle to distort to accomodate steric demands. As illustrated in Figure 5B, the *N*-substituent crowds the sixth metal coordination site and forces most metal complexes to be five-coordinate.

Mansuy (*40*) has characterized structurally a six-coordinate iron(II) *N*-alkylporphyrin. Figure 6 presents the molecular structure of this novel complex in which the strained six-coordinate geometry is stabilized by the formation of two five-membered chelate rings. The substituted pyrrole rings are canted a remarkable 50° from the mean equatorial plane as defined by the four pyrrole nitrogens.

Balch and co-workers have been interested in the formation and reactivity of iron *N*-alkylporphyrins. Especially interesting is the catalytic activity, observed by Traylor and co-workers (*32*) (*see* previous paragraphs), of the iron *N*-substituted complexes, an observation that suggests that the iron-*N*-alkylporphyrins can form highly oxidized reaction centers analogous to the compound I state of horseradish peroxidase (HRP), i.e., the ferryl porphyrin π-radical $(Fe^{IV}{=}O)P^{\bullet}$. Although no precedent exists for these highly oxidized iron *N*-alkylporphyrins, the coordination chemistry of iron(II) and iron(III) complexes of *N*-alkylporphyrins parallels that of iron(II) and iron(III) porphyrins. These similarities are summarized in Chart I. The porphyrin macrocycle and the *N*-alkylporphyrin macrocycle both form five-coordinate iron(II) complexes (*42*).

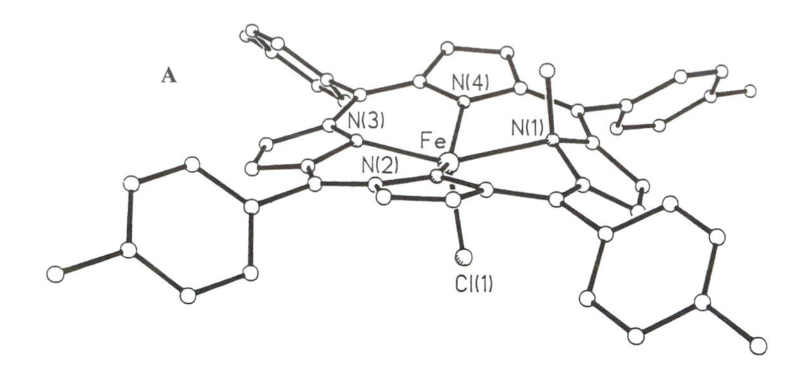

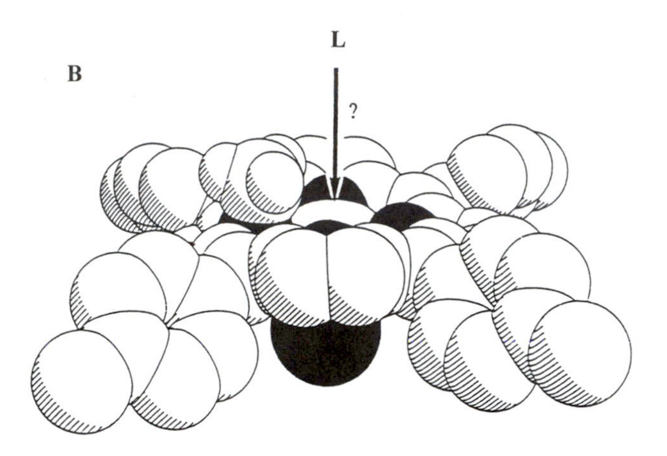

Figure 5. X-ray crystal structure of the cation of [(N-*MeTTP*)Fe^{III}*Cl*]$SbCl_6$ *(top) and a space-filling representation of the same molecule indicating the crowded nature of the sixth metal coordination site (bottom).*

Five-coordinate iron(III) complexes are also known for both macrocycles (*41*), including oxo-bridged "dimers" (*43, 44*). Iron *N*-alkyl complexes that correspond to low-spin six-coordinate iron(III) porphyrins such as $[(TPP)Fe^{III}(Im)_2]^+$, or to highly oxidized iron porphyrins such as $(TPP^{\cdot})Fe^{III}(ClO_4)_2$ (*45*), $TPPFe^{IV}{=}O$ (*7*), and $(TMP^{\cdot})Fe^{IV}{=}O$ (*8*) have only recently been reported, and these are discussed in subsequent paragraphs (*41, 46*).

NMR spectroscopy is uniquely effective in probing the oxidation state, spin state, and ligation state of iron porphyrins (*47, 48*) and iron *N*-alkylporphyrins (*22*). For iron-tetraarylporphyrins the β-pyrrole proton resonance is most indicative of the electronic structure of the metal

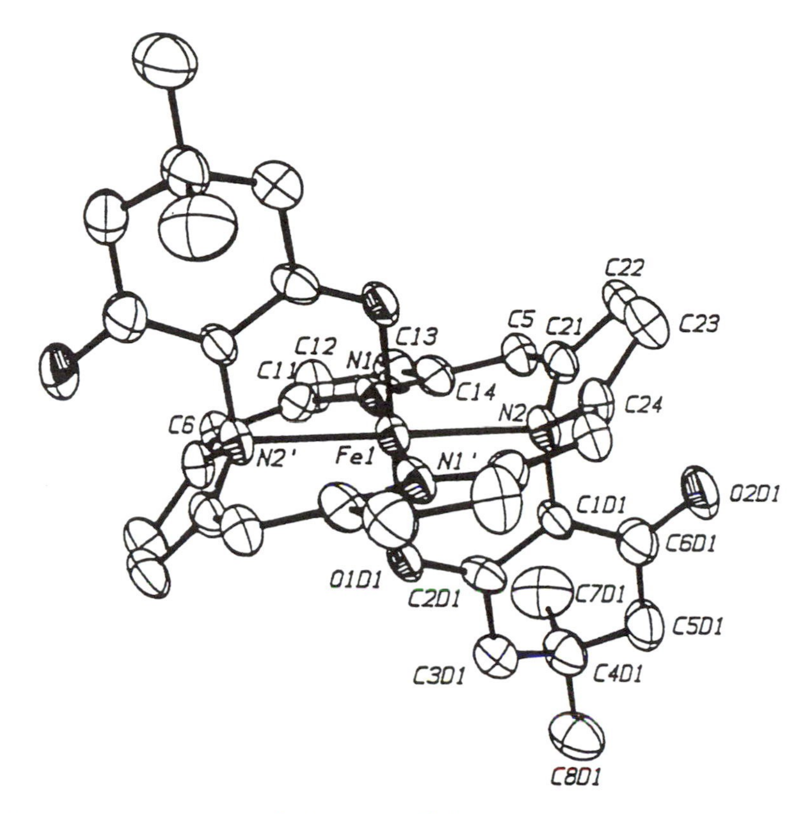

*Figure 6. X-ray crystal structure of the iron(II)-bis-*N*-alkylporphyrin complex incorporating two five-membered metallacycles to stabilize the distorted porphyrin macrocycle* (40).

complex. In Table I the chemical shifts of the β-pyrrole and meso-aryl proton resonances for representative iron porphyrins and iron *N*-alkylporphyrins are presented. The decrease in symmetry from C_4 to C_s associated with *N*-substitution results in four β-pyrrole magnetic environments. As can be seen in Table I, the chemical shifts for these four resonances are, in general, in the same region as the corresponding iron porphyrins. As is apparent for $[(N\text{-MeTTP})Fe^{III}Cl]^+$, three β-pyrrole proton resonances are downfield as expected for an $S = \frac{5}{2}$ iron(III)porphyrin. However, one β-pyrrole proton resonance is upfield of the diamagnetic region. Although this resonance has not been rigorously assigned, it is intuitively appealing to assign this to the protons on the *N*-substituted pyrrole ring. The high-field shift may then be attributed to decreased delocalization of the unpaired spin in the $d_{x^2-y^2}$ through the sp^3 nitrogen atom to the β-carbons of the pyrrole ring (33). The high-spin iron(II) complex $(N\text{-MeTPP})Fe^{II}Cl$ has *two* resonances

Chart I

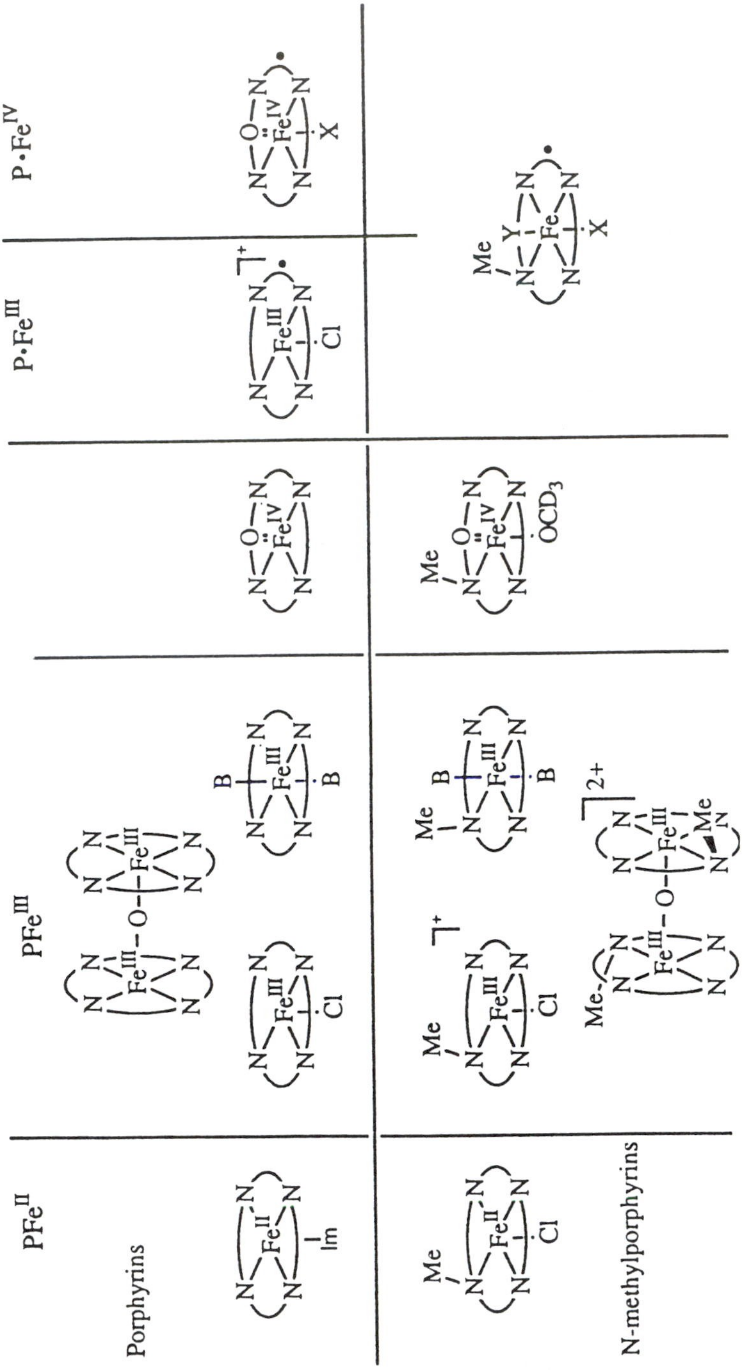

	PFe^{II}	PFe^{III}		$P{\cdot}Fe^{III}$	$P{\cdot}Fe^{IV}$
Porphyrins	N N N N Fe^{II} Im	N N N N Fe^{III}–O–Fe^{III} N N N N; N N N N Fe^{III} Cl; B N N N N Fe^{III} B	N O N N N Fe^{IV}	N N N N Fe^{III} Cl]$^{+}$ •	N O N N N Fe^{IV} X •
N-methylporphyrins	Me N N N N Fe^{II} Cl	Me N N N N Fe^{III} Cl]$^{+}$; Me B N N N N Fe^{III} B; Me- N N N N Fe^{III}–O–Fe^{III} N N N N Me]$^{2+}$	Me N O N N N Fe^{IV} OCD_3	Me N Y N N N Fe X •	

Table I. Representative ^{1}H and ^{2}H NMR Data for Iron Complexes of Porphyrins and *N*-Methylporphyrins

		Chemical Shifts (ppm)						
Compound	*Spin State, S (symmetry)*[a]	*Pyrrole*	*N-Me*	*p-aryl*	*m-aryl*	*o-aryl*	*Temp. (°C)*	*Ref.*
(*N*-MeTTP)$Fe^{II}Cl$	2 (C_S)	62.8, 42.5, −6.3 (2)	193			18.5	−90	34
(*N*-MeTTP)$Fe^{II}OCD_3$	2 (C_S)	47.7, 43.6, −7.6, −20.8	—[b]			13.1	−90	40
(*N*-CD_3d_8-TTP)$Fe^{II}OCH_3$	2 (C_S)	**48, 44, −8, −21**[c]	**252**				−90	40
(*N*-MeTTP)Fe^{II}-(OC_6H_4-*p*-CH_3)	2 (C_S)	43.2, 41.2, −6.1, −16.7	197			12.9	−60	40
[(*N*-MeTTP)$Fe^{III}Cl]^{+1}$	$^5/_2$ (C_S)	128, 92, 79, 2	272				−50	28
[(*N*-MeTTP)$Fe^{III}(CN)_2]^{-1}$	$^1/_2$ (C_S)	0.0, −31.5, −34.2, −56.9	−61				−90	28
(CD_3O)(*N*-MeTTP)-Fe^{IV}=O	1 (C_S)	−1.6, −6.3, −7.6, −17.1	—[d]		9.8, 10.5 (2), 11.0		−90	40
(CH_3O)(*N*-CD_3d_8-TTP)Fe^{IV}=O	1 (C_S)	**−1.6, −6.3, −7.6, −17.1**	**−59**				−90	40
(*N*-MeTTP π radical)-FeLL′	*see* text	−152[e]		125.8, 100.8[f]	58.8, 58.2, 49.7, 47.7	−72.5, −74.3 (2), −76.2	−90	40
(*N*-Med_{28}-TTP π radical)-FeLL′	*see* text			**127.2, 102.3**[f]	**61.1, 51.4**	**−70.9**	−90	40
[*N*-MeT(3,5-$Me_2C_6H_3$)P π radical]FeLL′	*see* text	−154[e]		−71.2, −85.1	−12.4, −13.3, −15.8, −17.8	−63.9, −64.8, −67.4, −69.9	−90	40

Continued on next page

Table I. Continued

Compound	Spin State, S (symmetry)[a]	Chemical Shifts (ppm) Pyrrole	N-Me	p-aryl	m-aryl	o-aryl	Temp. (°C)	Ref.
(TMP)Fe^{IV}=O	1 (C_4)	8.4		2.6[f]	6.4, 6.6	3.3[f]		6a
(*N*-MeIm)(TPP)Fe^{IV}=O	1 (C_S)	5.05		7.9	7.9	9.2	−80	72
(Me_3N)(TPP)Fe^{IV}=O	1 (C_S)	(1.1)					−76	73
(CD_3O)(TMP)Fe^{IV}=O	1 (C_S)	0.1, −4.1		2.7[f]		1.5, (3.0)	−30, (−90)	40
(TMP π radical)Fe^{IV}=O	S = 1 metal $S = {}^1\!/_2$ porph. (C_4)	−27		11.1[f]	68	26, 24[f]	−77	6[g]
[(TPP π radical)Fe^{III}Cl]-$SbCl_6$	S = 1 metal $S = {}^1\!/_2$ porph. (C_{2v})	64.8		−30.5	−14.7	48	25	33
(TTP π radical)Fe^{III}-$(ClO_4)_2$	S = 1 metal $S = {}^1\!/_2$ porph. (D_{4h})	31.6		0.1[f]	31.6	−16.9	25	33
(TMP)$Fe^{IV}(OCH_3)_2$	1 (D_{4h})	−37.5		2.86[f]	7.72	2.4[f]	−78	6[f]
(TTP)Fe^{IV}(Ph)Br	1 (C_{4v})	−60		4.3[f]	9.5	9.8, 10.3	−50	15

[a] Symmetry designations disregard any constraints from axial ligand orientations.
[b] Not observed.
[c] Boldface denotes ^{2}H resonances.
[d] Not observed at −90 °C, −41 ppm at −30 °C.
[e] Other resonances presumed to be in the diamagnetic region.
[f] Methyl resonances.
[g] Intensity twice that of related protons.

upfield of the diamagnetic region. One of these has been assigned, using homonuclear correlation spectroscopy (COSY), to a β-pyrrole proton on the ring *cis* to the substituted pyrrole ring (*49*). Thus, the spin delocalization is significantly modified by deformations of the carbon skeleton of the macrocycle.

Oxidation of the porphyrin π-system of iron(III) (*45*) and iron(IV) tetraarylporphyrins (*8*) leads to large chemical shifts for the *meso*-aryl protons. This is a result of spin density on the *meso*-carbon, which is delocalized into the aryl rings via a π-spin delocalization mechanism (*50*). Accordingly, the chemical shifts of *ortho-*, *meta-*, and *para*-protons have alternating signs. Coupling of the porphyrin spin to the metal spin in both a ferro- and antiferromagnetic manner has been demonstrated (*45*).

Given the similarities in chemical shifts and linewidths, as well as the contributions of symmetry to the appearance of the spectrum, the electronic and molecular structure of new iron complexes of *N*-alkylporphyrins may be ascertained, to a first approximation, from ^{1}H NMR data. Thus for low-spin iron(III) complexes one would expect at least four sharp resonances upfield of the diamagnetic region. Iron(IV) complexes should have at least four resonances upfield of the diamagnetic region. Iron(III) can be differentiated from iron(IV) by measurement of the solution susceptibility (*51*).

Formation of Five- and Six-Coordinate Complexes (*41, 52*). The ability of the five-coordinate *N*-substituted complexes to accomodate a sixth ligand was demonstrated by observing the change in iron spin state from high spin ($S = \frac{5}{2}$) to low spin ($S = \frac{1}{2}$) upon adding a base such as imidazole or cyanide. The stable iron(III) starting material, (*N*-MeTTP)$Fe^{III}Cl^+$, was prepared from the corresponding iron(II) precursor by oxidation with thianthrene radical. This same complex prepared by chlorine oxidation is unstable at ambient temperatures and decomposes via demetallation (*33*). The ^{1}H NMR spectrum of this complex, shown in Figure 7A, indicates that the complex has C_s symmetry in solution. The four pyrrole resonances are distributed across a large chemical shift range with three of the four resonances downfield (156, 108, and 89 ppm) near the shift for nonsubstituted TTPFe^{III}Cl at 183 K. The fourth resonance appears upfield at −2 ppm. There are four *meta*-phenyl resonances and two *para*-methyl resonances, also consistent with the C_s symmetry. The *ortho*-phenyl resonances are very broad and are located in the diamagnetic region. These assignments are consistent with those reported previously for the chlorine oxidation product (*33*).

Addition of 4-methylimidazole to the high-spin starting material yields a low-spin complex that has decreased symmetry. As seen in Figure 7B, *five* β-pyrrole resonances are present in the upfield region and *two* 5-methyl resonances are present downfield. The 5-methyl resonances

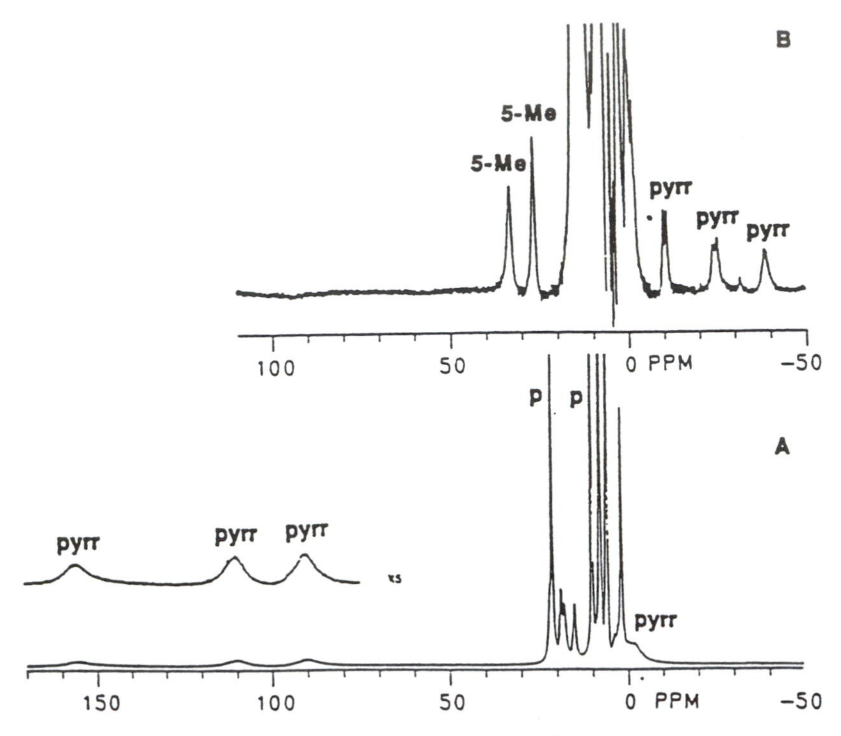

Figure 7. Proton NMR spectra of (N*-MeTTP)Fe*^{III}*Cl*$^{+}$ *(A) and the product from the addition of 4-methylimidazole to* (N*-MeTTP)Fe*^{III}*Cl*$^{+}$*, [(*N*-MeTTP)(5-MeIm)*$_2$*Fe*$^{+}$*] (B). Both spectra acquired at −90 °C in* CD_2Cl_2*. Resonance assignments: pyrr, β-pyrrole-H; p,* para*-methyl-H; 5-Me, 5-methyl-Hs from 5-MeIm axial ligands. (Reproduced from reference 41. Copyright 1990 American Chemical Society.)*

were assigned by their absence in the spectrum of the corresponding imidazole complexes and by comparison to TPPFe^{III}(5-MeIm)$_2$$^{+}$ (53). The presence of two 5-methyl resonances indicates that there are two magnetically inequivalent 5-methylimidazole ligands. This presence can be accomplished if the complex is indeed six-coordinate and if ligands are present at both the proximal and distal coordination sites (methyl-substituted face defined as proximal). By integration, the five upfield resonances correspond to six β-pyrrole protons. The presence of five resonances for six of the eight β-pyrrole protons requires a decrease in symmetry from C_s to C_1. This decrease in symmetry is attributed to hindered rotation of the proximal 5-MeIm ligand. Groves (54) showed that the axial ligands of TMPFe^{III}(2-MeIm)$_2$$^{+}$ undergo hindered rotation due to steric interactions between the 2-methyl substituent of the 2-MeIm ligand and the *ortho*-methyl groups of the TMP macrocycle. However, with less hindered imidazoles or tetraarylporphyrins, free rotation is observed. Under the conditions used, the six-coordinate complexes were unstable above −60 °C.

Cyanide ion may also be used to form a six-coordinate, low-spin complex. The effect of adding 2.6 equivalents of cyanide is shown in Figure 8A. The β-pyrrole resonances were assigned by ^{2}H NMR of [(*N*-CD_3-pyrrole-d_8-TTP)$Fe^{III}Cl]^+$ as shown in Figure 8B. The high-field chemical shifts are consistent with the change in spin state. Because CN^- is a strong field ligand, a change in spin state does not necessarily require addition of a sixth ligand. Titration of the five-coordinate complex with CN^- provides evidence for formation of two intermediates during the reaction. The first is a high-spin complex that has β-pyrrole chemical shifts similar to those of the starting material. This intermediate is reasonably assigned to the five-coordinate ligand exchange product [(N-MeTTP)$Fe^{III}CN]^+$ (structure **1**):

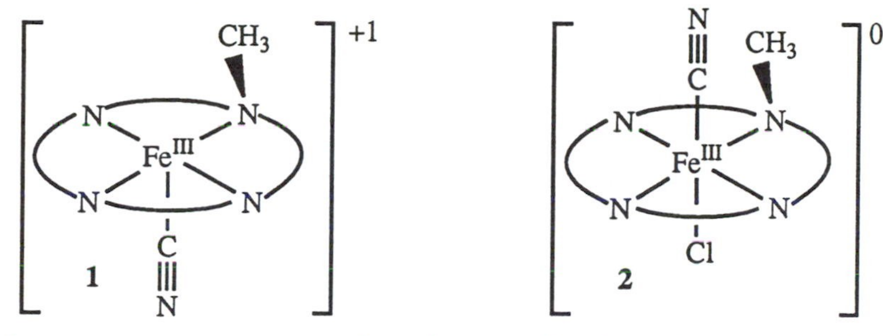

If this assignment is correct, then the coordination of one cyanide ligand is not responsible for the spin-state change. The second intermediate is a low-spin complex that has high-field resonances that are distinct from the product. If the final product is assumed to be the bis-CN complex, it is likely that the cyanide ion is coordinated to the proximal face and the chloride is distal (or possibly absent) as shown in structure **2**. Formation of the six-coordinate complexes is summarized in Scheme 4.

Oxidation of *N*-MeTTPFe^{II}Cl *(46, 52)*. Catalytic alkene oxidation by iron *N*-alkylporphyrins requires that the modified heme center can form an active oxidant, presumably at the HRP compound I level of oxidation. To show that iron *N*-alkyl porphyrins could form highly oxidized complexes, these reactive species were generated by chemical oxidation and examined by NMR spectroscopy. Reaction of the (*N*-MeTTP)Fe^{II}Cl with chlorine or bromine at low temperatures results in formation of the corresponding iron(III)–halide complex. Addition of ethyl- or *t*-butyl-hydroperoxide, or iodosylbenzene, to a solution of *N*-MeTTPFe^{II}Cl at low temperatures has no effect on the ^{1}H NMR spectrum. However, addition of *m*-chloroperoxybenzoic acid (*m*-CPBA) results in the formation of iron(III) and iron(IV) products as well as porphyrin radical compounds that retain the *N*-substituent.

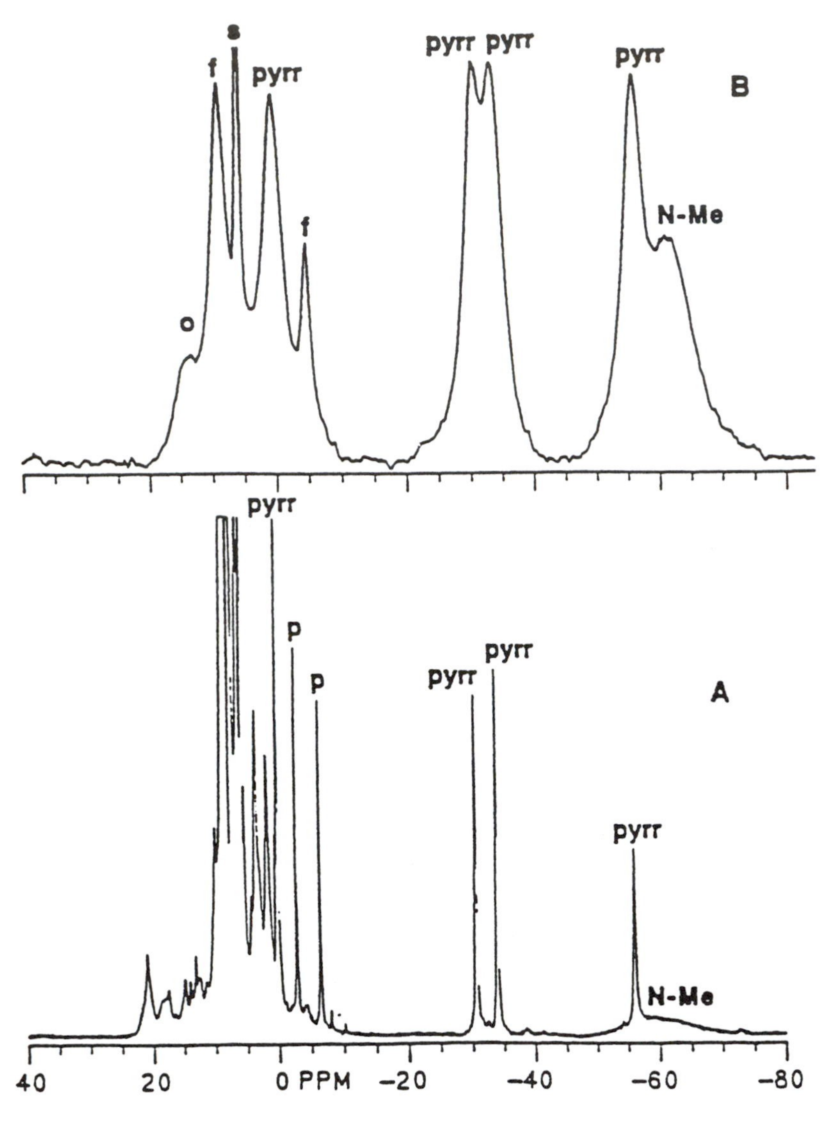

Figure 8. *^{1}H and ^{2}H NMR (A and B, respectively) of the product from the addition of cyanide to* (N-MeTTP)*$Fe^{III}Cl^+$ at −90 °C in CD_2Cl_2. Resonance assignments: pyrr, β-pyrrole-H; p,* para-*methyl-H; f, β-pyrrole-H of* N-*MeTTPH; o, β-pyrrole-H of $(TTPFe^{III})_2O$. (Reproduced from reference 41. Copyright 1990 American Chemical Society.)*

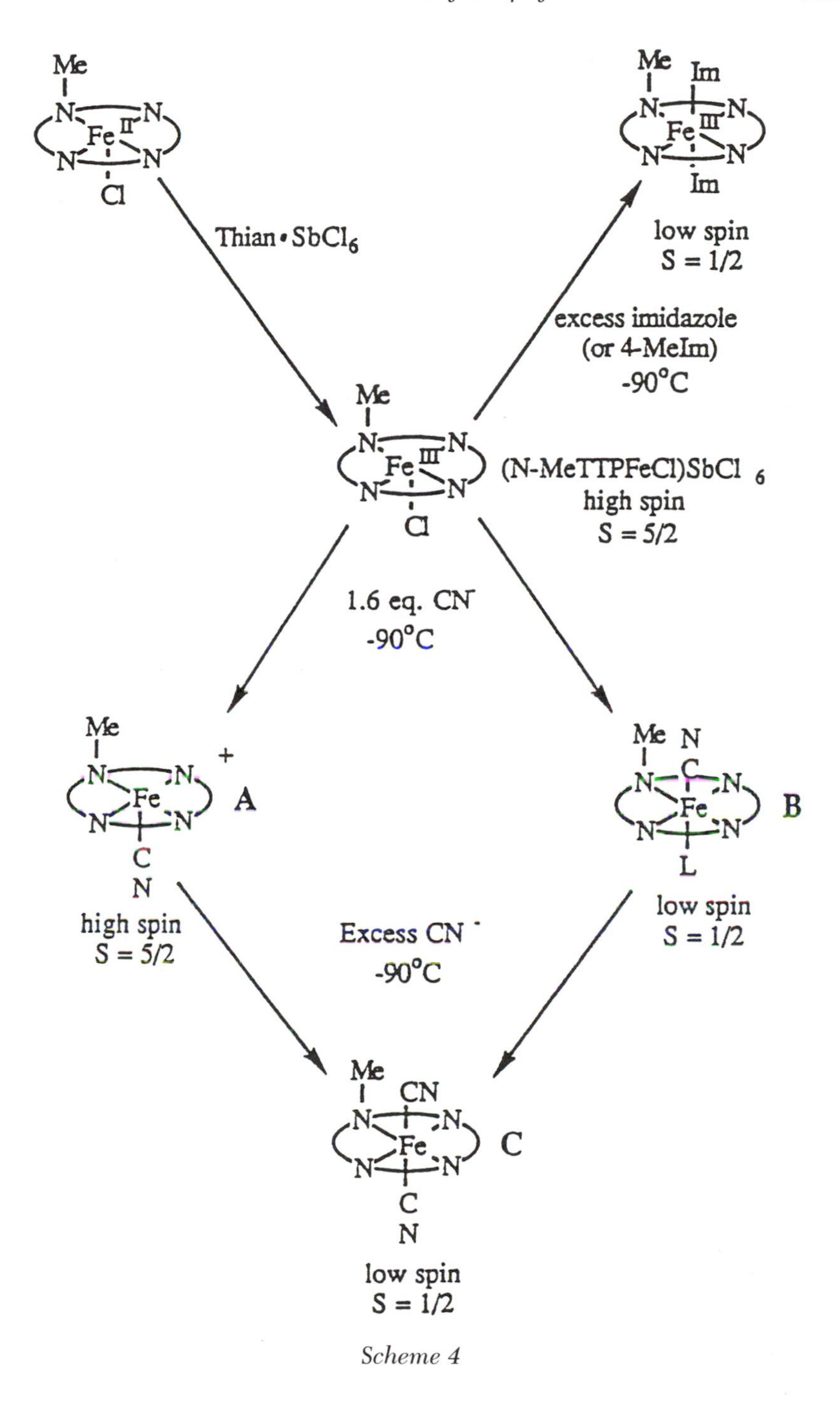
Me
N N
Fe II
N N
Cl
Thian • SbCl6
Me
Im
Fe III
Im
low spin
S = 1/2
excess imidazole
(or 4-MeIm)
-90°C
Me
Fe III
Cl
(N-MeTTPFeCl)SbCl 6
high spin
S = 5/2
1.6 eq. CN⁻
-90°C
Me
+
Fe
A
C
N
high spin
S = 5/2
Me N
C
Fe
B
L
low spin
S = 1/2
Excess CN⁻
-90°C
Me
CN
Fe
C
C
N
low spin
S = 1/2

Scheme 4

Axial Ligand Exchange. *m*-CPBA oxidation of (*N*-MeTTP)$Fe^{II}Cl$ is sensitive to solvent and the presence of methoxide. Treatment of (*N*-MeTTP)$Fe^{II}Cl$, with sodium methoxide-d_3 in a solvent mixture of dichloromethane-d_2/methanol-d_4 (4:1, vol/vol) at −90 °C results in formation of the ligand exchange product (*N*-MeTTP)$Fe^{II}OCD_3$, as shown in Figure 9. Trace A shows the ^1H NMR spectrum of (*N*-MeTTP)$Fe^{II}Cl$, whereas trace B shows the effect of adding five equivalents of sodium methoxide-d_3. The four pyrrole resonances and the *N*-methyl substituent for (*N*-CD_3-d_8-TTP)$Fe^{II}OCH_3$ have been unambiguously identified by ^2H NMR as shown in inset B′. The magnetic susceptibility of (*N*-MeTTP)$Fe^{II}OCD_3$ at −90 °C, as determined by the Evans' method (*51*), is 5.0(2) μ_B. This is similar to the value of 4.9(2) μ_B measured for (*N*-MeTTP)$Fe^{II}Cl$ in a control experiment. In the basic reaction medium,

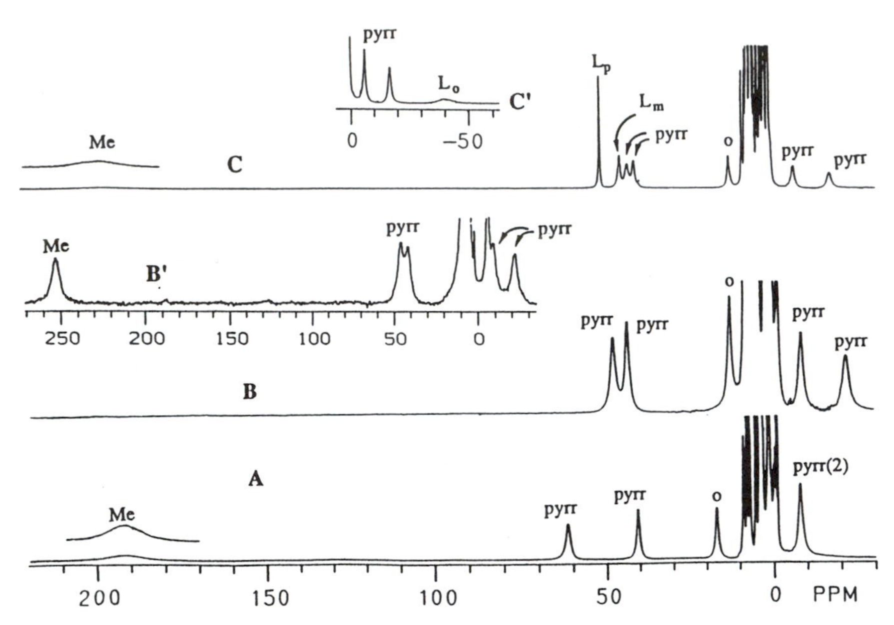

*Figure 9. Trace A: The ^1H NMR spectrum from a solution of (*N*-MeTTP)$Fe^{II}Cl$ in CD_2Cl_2/CD_3OD (4:1, vol/vol) at −90 °C. Trace B: The ^1H NMR spectrum from the same solution after the addition of five equivalents of $NaOCD_3$ at −90 °C. Trace B′: The ^2H NMR spectrum from a similar solution as that in trace B prepared from the deuterated complex (*N*-CD_3 d_8-TTP)$Fe^{II}Cl$. Trace C: The ^1H NMR spectrum of a solution of (*N*-MeTTP)$Fe^{II}Cl$ and five equivalents of sodium *p*-cresolate at −60 °C in toluene-d_8/methanol-d_4 (9:1, vol/vol). Trace C′: The upfield region of trace C. Resonance assignments: Me, the *N*-methyl protons; pyrr, the pyrrole protons; o, the ortho protons of the *p*-tolyl substituents; L_o, L_m, L_p, the ortho, meta, and para resonances of the axial cresolato ligand. (Reproduced from reference 46. Copyright 1992 American Chemical Society.)*

(*N*-MeTTP)$Fe^{II}OCD_3$ is unstable to warming. Above −90 °C it gradually decomposes to form *N*-MeTTPH through the loss of iron.

Other ligand exchange reactions of (*N*-MeTTP)$Fe^{II}Cl$ have been examined. Addition of sodium *p*-cresolate to *N*-MeTTP$Fe^{II}Cl$ in toluene-d_8 yields the corresponding cresolato complex. In contrast to the methoxide complex, the axial ligand protons are observable for the cresolato ligand. The spectrum and the important peak assignments for the cresolato complex are presented in Figures 9C and 9C′. Addition of 15 equivalents of sodium *m*-chlorobenzoate to *N*-MeTTP$Fe^{II}Cl$ yields spectral evidence for only limited exchange. Addition of an excess of *m*-chlorobenzoic acid to *N*-MeTTP$Fe^{II}Cl$ results in no observable ligand exchange.

Oxidation of (*N*-MeTTP)$Fe^{II}(OCD_3)$ to Form an Fe(IV) Complex *(46, 52)*. Addition of 1.1 equivalents of *m*-chloroperoxybenzoic acid to a sample of (*N*-MeTTP)$Fe^{II}(OCD_3)$ (Figure 10A) in the presence of excess sodium methoxide at −90 °C yields trace B of Figure 10. Four new pyrrole resonances are readily observed in the region from 0 to −20 ppm. Three resonances in the 10–12-ppm range are assigned the meta protons of the *p*-tolyl groups on the basis of their intensities, multiplicities, and line widths. The *N*-methyl resonance is not observable at −90 °C because of its linewidth, but it does become detectable when the sample is warmed. Attempts to oxidize *N*-MeTTP$Fe^{II}OCD_3$ at higher temperatures results in considerable demetallation, yielding *N*-MeTTPH; however, once the product is formed at −90 °C, it is stable to warming and can be observed up to −20 °C, at which it suffers only slow decomposition.

Because of the thermal stability of the *m*-CPBA oxidation product, it has been possible to monitor the ^1H NMR spectrum throughout the temperature range −90–0 °C. Plots of chemical shifts versus 1/T are linear as expected for a paramagnetic substance and have intercepts at the expected diamagnetic chemical shift values. The magnetic susceptibility for the new species, (Evans' technique) is 2.9(3) μ_B over the temperature range −90° to −20 °C, consistent with an $S = 1$ iron(IV) spin system. The formation of this iron(IV) complex in reproducible fashion requires the presence of a substantial excess of sodium methoxide, which appears to provide an axial ligand. This complex has been tentatively formulated as (*N*-MeTTP)$Fe^{IV}LL'$, where L may be an oxo ligand derived from the peracid and L′ is methoxide or *m*-chlorobenzoate. Alternatively, L may be methoxide or *m*-chlorobenzoate.

(*N*-MeTTP)$Fe^{IV}LL'$ undergoes reduction when treated with tertiary phosphines. Addition of a large excess of dimethylphenylphosphine to a solution of *N*-MeTTP$Fe^{IV}LL'$ results in conversion of the spectrum

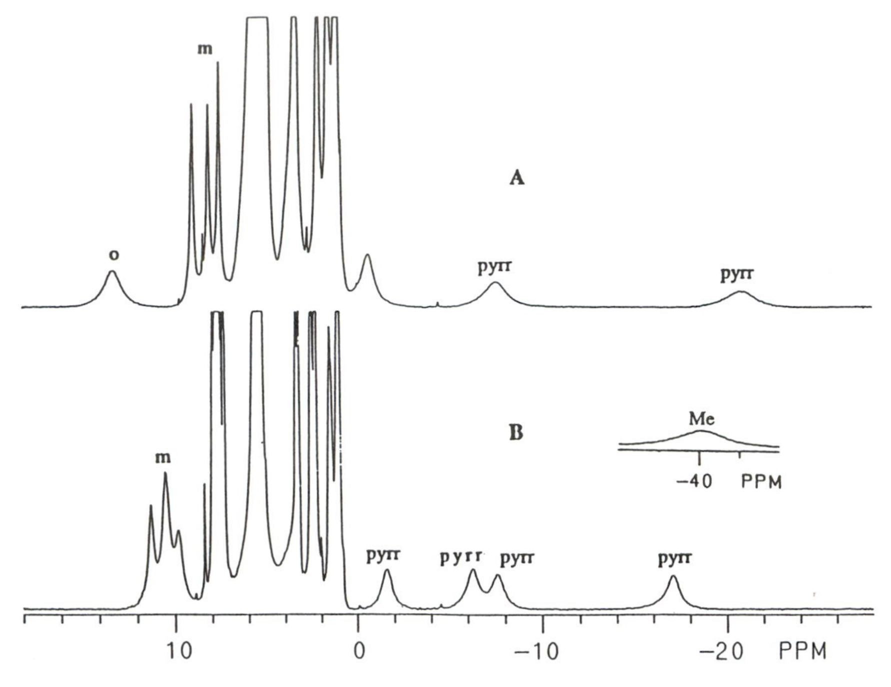

Figure 10. Upfield region of the ^{1}H NMR spectrum of (N-MeTTP)$Fe^{II}OCD_3$ (A) and the full ^{1}H NMR spectrum (B) observed upon adding 1.1 equivalents of m-CPBA to the solution used to obtain trace A. Both spectra acquired at −90 °C. Resonance assignments as in Figures 7–9. (Reproduced from reference 52. Copyright 1990.)

seen in Figure 11A to the spectrum of *N*-MeTTP$Fe^{II}OCD_3$ shown in Figure 11C. Addition of the weaker reductant triphenylphosphine also results in the two-electron reduction of (*N*-MeTTP)Fe^{IV}LL′; however, in this case the reaction is much slower and *N*-MeTTPFe^{IV}LL′ conproportionates with *N*-MeTTP$Fe^{II}OCD_3$ to form the iron(III) complex (*N*-MeTTPFe^{III}X)$^+$, where X is Cl^- or OCD_3^-. With both phosphines the only phosphine product is the corresponding phosphine oxide as observed by ^{31}P NMR. Apparently, (*N*-MeTTP)Fe^{IV}LL′ is not competent for the oxidation of alkenes. This fact is not surprising because the corresponding ferryl porphyrins (at the compound II level of oxidation) are also poor oxidants for alkene epoxidation (7).

Formation of an Iron Complex of an *N*-alkyl Porphyrin π-Radical *(46, 52)*. In the absence of sodium methoxide, the reaction between (*N*-MeTTP)Fe^{II}Cl and *m*-CPBA takes two different paths. When 1.1 equivalents of the peroxy acid are added, the Fe^{II} complex is converted into a high-spin, five-coordinate Fe^{III} complex. However,

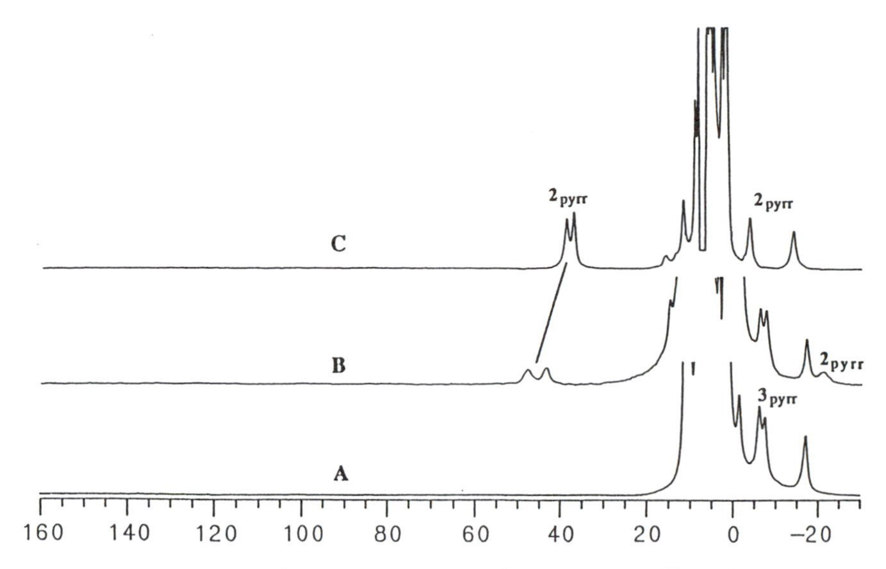

Figure 11. Trace A: ^1H NMR spectrum of (N-MeTTP)Fe^{IV}LL′, denoted **3**. *Trace B: Same solution as in trace A 30 min after the addition of dimethylphenylphosphine. Trace C: The same solution as in trace B after warming to −30 °C. The final spectrum in trace C is for (N-MeTTP)$Fe^{II}OCD_3$, denoted* **2**. *β-pyrrole resonances are labeled "pyrr." (Reproduced with permission from reference 52. Copyright 1990.)*

when 5–10 equivalents of peroxy acid are added, a new species with a remarkable ^1H NMR spectrum, shown in Figure 12A, is produced. The new species that is formed has proton resonances from 130 to −160 ppm. The individual resonances have been assigned through labeling experiments and consideration of relative intensities as shown in Figures 12B and 12C. This complex can only be observed over the limited temperature range −99 ° to −75 °C; above this range it undergoes decomposition. A plot of the chemical shifts versus l/T, shown in Figure 13, reveals that all resonances exhibit linear behavior over this small temperature range consistent with a simple paramagnetic complex. However, the extrapolated shifts at infinite temperature deviate considerably from the anticipated diamagnetic chemical shifts. These deviations indicate that the magnetism is the result of two interacting spin systems.

The large paramagnetic contribution to the chemical shifts for the *meso*-aryl protons is consistent with the presence of a porphyrin π-radical species (*45*). The shift pattern for these signals (*o*-H upfield, *m*-H and *p*-CH_3 downfield) is similar to that observed for (TPP$^{\bullet}$)$Fe^{III}(ClO_4)_2$ (*o*-H, *p*-H upfield, *m*-H downfield), which exhibits ferromagnetic coupling between the porphyrin π-radical spin and the metal spin (*45*). The small

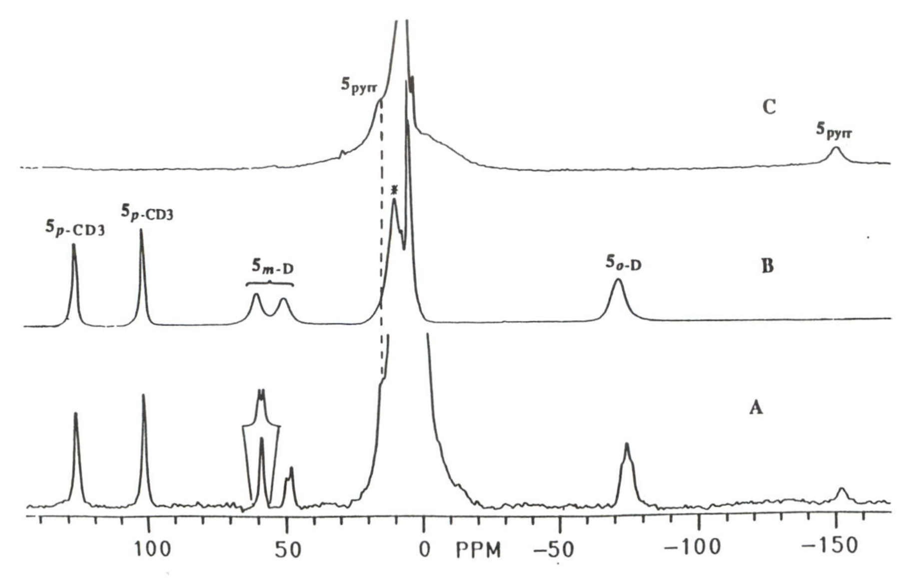

Figure 12. Trace A: ^1H NMR spectrum of the porphyrin π-radical generated from (N-*MeTTP)Fe*^{II}*Cl and excess* m-*CPBA (denoted* **5** *in this figure) at −90 °C in CD_2Cl_2. Trace B: ^2H NMR spectrum of* **5** *selectively deuterated at the* meso-*phenyl positions. Trace C: ^2H NMR spectrum of* **5** *selectively deuterated at the β-pyrrole positions. (Reproduced with permission from reference 52. Copyright 1990.)*

paramagnetic shift for three of the four β-pyrrole proton resonances is consistent with an empty $d_{x^2-y^2}$ orbital. Thus the iron may be intermediate-spin iron(II), low- or intermediate-spin iron(III), or iron(IV). The unique high-field pyrrole resonance is obviously due to a novel electronic structure. The presence of iron(III) and iron(IV) impurities in all preparations of the radical species has precluded more rigorous magnetic studies.

The chemical redox reactions of (*N*-MeTTP)Fe^{II}Cl are summarized in Scheme 5. In this scheme, the iron(IV) complex (*N*-MeTTP)Fe^{IV}LL′ (labeled **3**) has been formulated as a ferryl, Fe^{IV}=O, complex with the oxo ligand proximal or distal to the *N*-substituent. Alternatively, this complex may be formulated as the bis-methoxide, [(*N*-MeTTP)-Fe^{IV}(OCD$_3$)$_2$]$^+$, by analogy to (TMP)Fe^{IV}(OCD$_3$)$_2$, which has been reported by Groves and co-workers (*8*). The average β-pyrrole resonance for (*N*-MeTTP)Fe^{IV}LL′ is −6.9 ppm at −80 °C. This value is much more downfield than the −37.5-ppm chemical shift reported for the β-pyrrole resonance of (TMP)Fe^{IV}(OCD$_3$)$_2$ at −78 °C. The β-pyrrole chemical shifts for ferryl porphyrin complexes range from −5 to +9 ppm depending on axial ligation (*see* Table I). Thus the

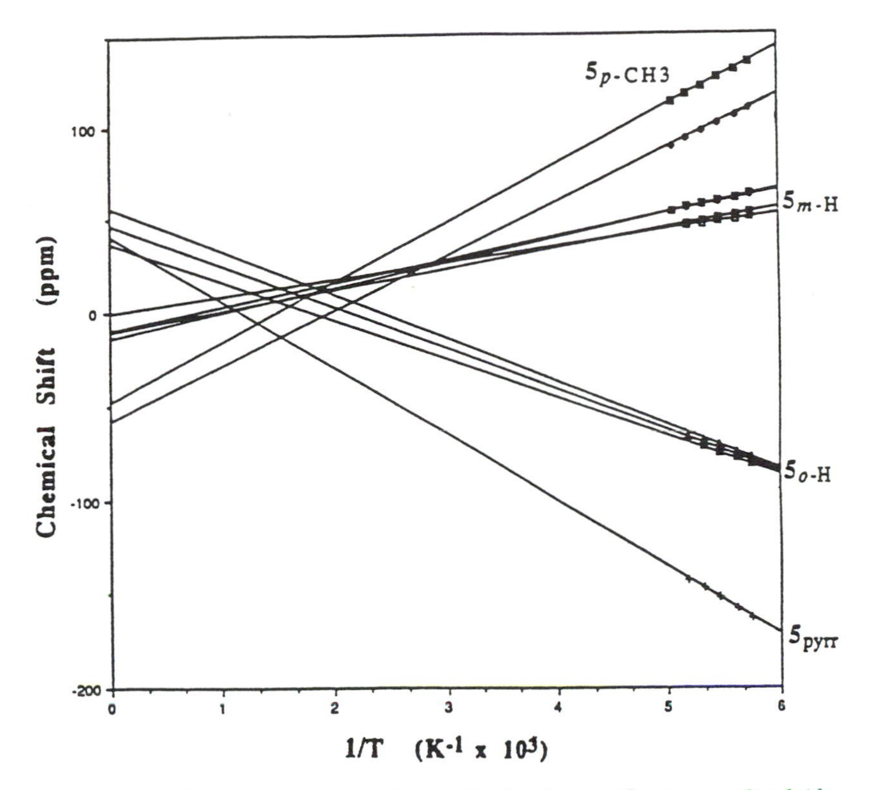

Figure 13. Plot of chemical shift vs. 1/T for the porphyrin π-radical (denoted **5** *in figure) generated from (N-MeTTP)Fe*^{II}*Cl and excess* m-*CPBA. (Reproduced from reference 46. Copyright 1992 American Chemical Society.)*

ferryl complex, (*N*-MeTTP)(Fe^{IV}=O)L′ is the preferred formulation for (*N*-MeTTP)Fe^{IV}LL′.

The metal oxidation state of the porphyrin π-radical complex is unknown although it is nearly certain that the metal $d_{x^2-y^2}$ orbital is unoccupied. This unoccupied orbital is also suggested based on the symmetry arguments first proposed by Reed and Marchon (*45*). In C_s symmetry, the metal orbitals should have the symmetries and relative energies shown in Figure 14. In all possible metal spin states, the d_{xz} and d_{yz} orbitals (a in C_s symmetry) have at least one unpaired electron. Because the two spins are ferromagnetically coupled on the basis of the signs of the chemical shifts for the *meso*-tolyl substituents (*45*), the porphyrin radical spin must be in an orbital of orthogonal a′ symmetry. The observed ferromagnetism thus rules out having unpaired spin in the d_{z^2} and the $d_{x^2-y^2}$ orbital because these also have a′ symmetry and would lead to antiferromagnetic coupling between the unpaired electrons of

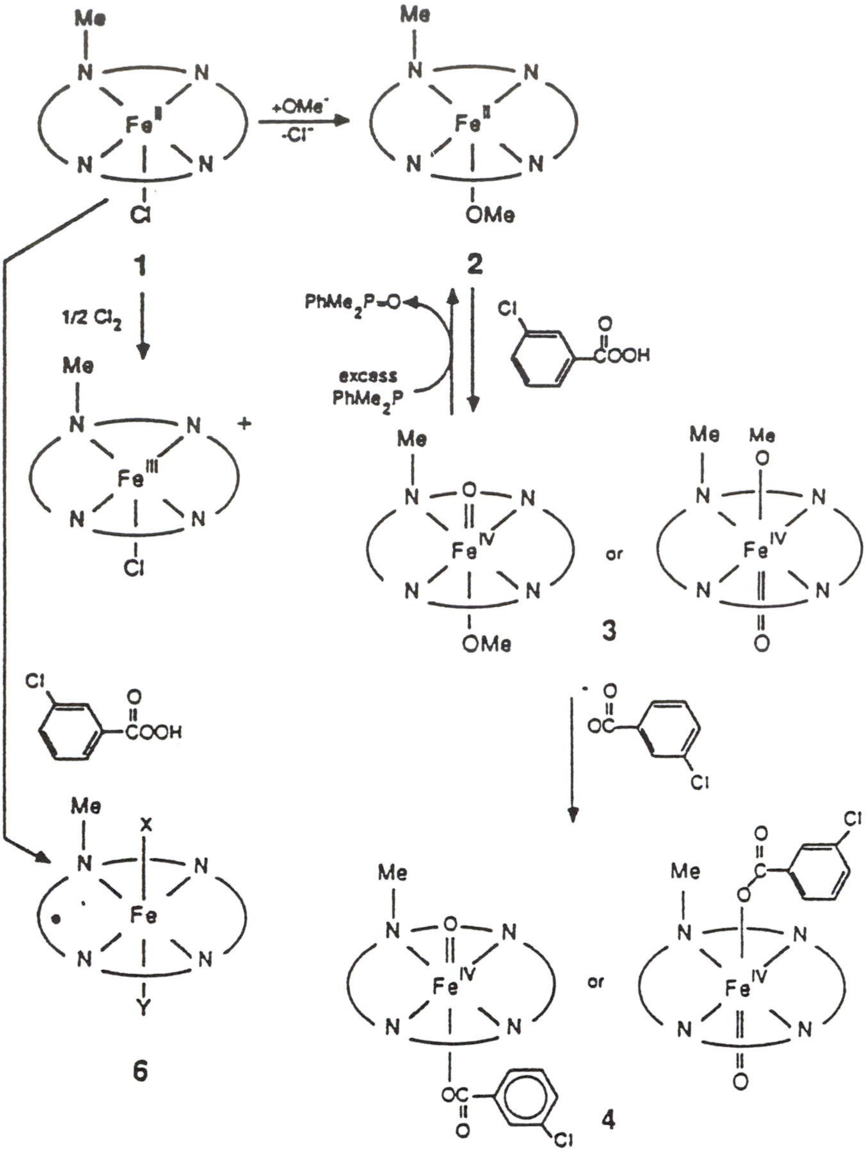

Scheme 5

the metal and the radical. Therefore, diamagnetic low-spin iron(II), intermediate-spin iron(II), and iron(III), and high-spin iron(II) and iron (III) are unlikely electronic configurations for the π radical complex.

Low-spin iron(III) and low-spin iron(IV) are the remaining possibilities for the metal oxidation and spin state. The thermal instability of

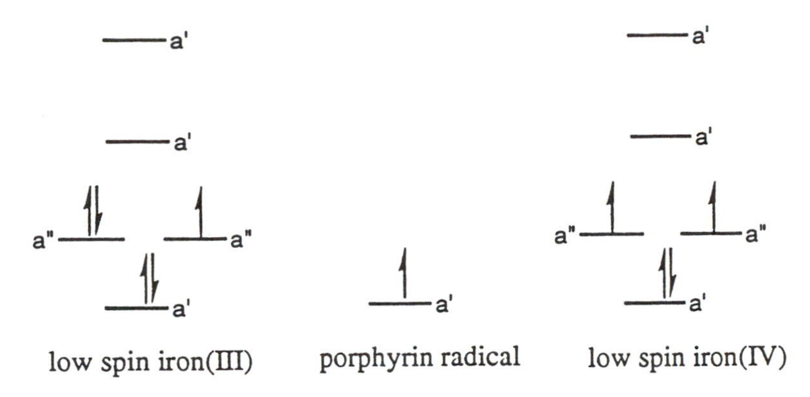

Figure 14. Expected symmetries and the relative energies of the metal d orbitals in the C_S point group for the porphyrin π-radical generated from (N-MeTTP)Fe^{II}Cl and excess m-CPBA. Note that the d_{xz} and d_{yz} are close in energy but not degenerate.

the radical complex may indicate that it is a very highly oxidized complex, possibly at the HRP compound 1 level of oxidation. Alternatively, this instability may arise from the large excess of oxidant needed to form the radical. Lavallee (*26*) has reported two one-electron oxidation waves for the electrochemical oxidation of (*N*-MeTPP)Fe^{II}Cl and has attributed the second oxidative wave (at 1.51 V vs. SCE in CH_2Cl_2) to a ligand-centered oxidation by comparison to the corresponding zinc complex. Thus it seems that iron(III) complexes of *N*-alkylporphyrin π-radicals are electrochemically generated. In the absence of methoxide ion, which is known to stabilize high-valent iron porphyrins (*48*), a low-spin iron(III) complex may be favored. A low-spin iron(III)-porphyrin π-radical has precedence in the complex $[(TAP^{\cdot})Fe^{III}(Im)_2]^{2+}$ (*55*).

If this radical is not at the HRP compound 1 level of oxidation, then the catalysis observed by Traylor and co-workers may require a different mechanism, possibly an iodosyl arene complex of the iron *N*-alkylporphyrin. No reactions were observed between iodosylbenzene and (*N*-MeTTP)Fe^{II}Cl. The lack of reactions does not preclude reactions under the reaction conditions used in the previous works.

Future Prospects

N-Arylporphyrin Formation as a Probe of Protein Structure. Addition of phenyldiazene to myoglobin (*56*), cytochromes P-450 (*57*), and the novel peroxidase from *Caldariomyces fumago* (*58*), results in the formation of four regioisomers of *N*-phenylprotoporphyrin-IX. These have been separated by HPLC and unambiguously assigned by NMR spectroscopy. The degree of substitution at rings A, B, C, or

D of protoporphyrin-IX depends on the steric constraints of the substrate binding pocket. Ortiz de Montellano (*59*) has used this selectivity to probe the active site structure of several heme enzymes. The structure of phenyl-cyt P-450_{cam} has been determined by X-ray crystallography and indicates that *N*-phenyl heme formation is an accurate, low-resolution probe of active site structure.

Catalysis. Zinc *N*-methyl- and *N*-phenyltetraarylporphyrins with thiolate or alkoxide as an axial ligand have been examined as catalysts for polymerization reactions that are also catalyzed by aluminum tetraphenylporphyrins (*60–64*). These studies have been undertaken to obtain polymers of uniform molecular weight to be used in polymer design. Unlike the aluminum porphyrins, the *N*-substituent protects the back side of the active site from coordination of monomers, which may then cause changes in the reactivity of the growing polymers. Polymerizations of both epoxides and episulfides have yielded living and immortal polymers in the molecular weight range up to 26,600 in the case of propylene sulfide.

Given the observation of catalysis of alkene epoxidation by iron *N*-alkyl porphyrins, it is likely that these complexes may yield synthetically useful catalysts (*32, 65*). The possibility of chiral induction by using either a chiral *N*-alkyl group (*66*) or a chiral macrocycle such as *N*-Me etioporphyrin (*67*) is an area that should prove fruitful in the near future.

Models for Nonporphyrin Metalloproteins. Because the *N*-alkyl macrocycle is electronically dissimilar to the porphyrin macrocycle (amine vs. imine donor), the observation of highly oxidized iron complexes forms an important bridge between heme enzymes and nonheme iron enzymes such as methane monooxygenase. The nature of the active oxidizing agent in methane monooxygenase and other nonheme iron oxygenases is an area of intense current interest (*68*). Other modified porphyrinic macrocycles such as thiaporphyrins (*69*), oxophlorins, and oxoporphyrins (*70, 71*) may also yield good models for the electronic properties of metalloenzymes in nonporphyrin environments. The rigorous, multidisciplinary characterization of many metalloporphyrin complexes allows extrapolation of the properties of the metalloporphyrin to the properties of the nonporphyrin system via complexes of the modified porphyrinic macrocycle.

Acknowledgments

We thank Alan Balch and Lechoslaw Latos-Grazynski for helpful discussions and inspiration. The preparation of this manuscript was financially supported by the Department of Chemistry at North Carolina State University.

References

1. Ortiz de Montellano, P. R.; Mico, B. A. *Mol. Pharmacol.* **1980,** *18*, 128–135, and references therein.
2. Ortiz de Montellano, P. R.; Reich, N. O. In *Cytochrome P-450: Structure, Mechanism, and Biochemistry;* Ortiz de Montellano, P. R., Ed.; Plenum: New York, 1986; p 273.
3. Lavalee, D. K. *The Chemistry and Biochemistry of* N-*Substituted Porphyrins;* VCH: New York, 1987.
4. Collman, J. P.; Brauman, J. I.; Rose, E.; Suslick, K. S. *Proc. Natl. Acad. Sci. U.S.A.* **1978,** *75*, 1053–1055.
5. Zhang, H.; Simonis, U.; Walker, F. A. *J. Am. Chem. Soc.* **1990,** *112*, 6124–6126.
6. Safo, M. K.; Gupta, G. P.; Watson, C. T.; Simonis, U.; Walker, F. A.; Scheidt, W. R. *J. Am. Chem. Soc.* **1992,** *114*, 7066–7075.
7. Balch, A. L.; Chan, Y.-W.; Cheng, R.-J.; La Mar, G. N.; Latos-Grazynski, L.; Renner, M. W. *J. Am. Chem. Soc.* **1984,** *106*, 7779–7785.
8. Groves, J. T.; Haushalter, R. C.; Nakamura, M.; Nemo, T. E.; Evans, B. J. *J. Am. Chem. Soc.* **1981,** *103*, 2884–2886.
9. Balch, A. L.; Latos-Grazinski, L.; Renner, M. W. *J. Am. Chem. Soc.* **1985,** *107*, 2983.
10. Groves, J. T.; Quinn, R.; McMurray, T. J.; Nakamura, M.; Lang, G.; Boso, B. *J. Am. Chem. Soc.* **1985,** *107*, 354.
11. Ibers, J. A.; Holm, R. H. *Science (Washington, D.C.)* **1980,** *209*, 223–233.
12. Mansuy, D.; Lange, M.; Chottard, J. C.; Guerrin, P.; Morliere, P.; Brault, D. *J. Chem. Soc. Chem. Commun.* **1977,** 648–649.
13. Mansuy, D.; Lange, M.; Chottard, J. C.; Bartoli, J. F.; Chevrier, B.; Weiss, R. *Angew. Chem. Int. Ed. Engl.* **1978,** *17*, 781–782.
14. Mansuy, D.; Battioni, J.-P.; Lavalee, D. K.; Fischer, D. K.; Weiss, R. *Inorg. Chem.* **1988,** *27*, 1052–1056.
15. Olmstead, M. M.; Cheng, R.-J.; Balch, A. L. *Inorg. Chem.* **1982,** *21*, 4143–4148.
16. Balch, A. L.; Chan, Y.-W.; Olmstead, M. M.; Renner, M. W. *J. Org. Chem.* **1986,** *51*, 4651–4656.
17. Ogoshi, H.; Sugimoto, H.; Yoshida, Z.; Kobayashi, H.; Sakai, H.; Maeda, Y. *J. Organomet. Chem.* **1982,** *234*, 185–195.
18. Cocolios P.; Laviron, E.; Guilard, R. *J. Organomet. Chem.* **1982,** *228*, C39–C42.
19. de Montellano, P. R.; Kunze, K. L.; Augusto, O. *J. Am. Chem. Soc.* **1982,** *104*, 3545–3546.
20. Mansuy, D.; Battioni, J.-P.; Dupre, D.; Sartori, E. *J. Am. Chem. Soc.* **1982,** *104*, 6159–6161.
21. Lancon, D.; Cocolios, P.; Guillard, R.; Kadish, K. M. *J. Am. Chem. Soc.* **1984,** *106*, 4472–4478.
22. Balch, A. L.; Renner, M. W. *J. Am. Chem. Soc.* **1986,** *108*, 2603–2608.
23. Artaud, I.; Devocelle, L.; Battioni, J. P.; Girault, J.-P.; Mansuy, D. *J. Am. Chem. Soc.* **1987,** *109*, 3732–3738.
24. Mashiko, T.; Dolphin, D.; Nakano, T.; Traylor, T. G. *J. Am. Chem. Soc.* **1985,** *107*, 3735–3736.
25. Collman, J. P.; Hampton, P. D.; Brauman, J. I. *J. Am. Chem. Soc.* **1986,** *108*, 7861–7862.
26. Traylor, T. G.; Tsuchiya, S. *Inorg. Chem.* **1987,** *26*, 1338–1339.

27. Traylor, P. S.; Dolphin, D.; Traylor, T. G. *J. Chem. Soc. Chem Commun.* **1984,** 279–280.
28. Collman, J. P.; Hampton, P. D.; Brauman, J. I. *J. Am. Chem. Soc.* **1990,** *112,* 2977–2986.
29. Collman, J. P.; Hampton, P. D.; Brauman, J. I. *J. Am. Chem. Soc.* **1990,** *112,* 2986–2998.
30. Collman, J. P.; Kodadek, T.; Raybuck, S. A.; Brauman, J. I.; Papazian, L. M. *J. Am. Chem. Soc.* **1985,** *107,* 4343–4345.
31. Collman, J. P.; Kodadek, T.; Brauman, J. I. *J. Am. Chem. Soc.* **1986,** *108,* 2588–2594.
32. Traylor, T. G.; Nakano, T.; Miksztal, A. R.; Dunlap, B. E. *J. Am. Chem. Soc.* **1987,** *109,* 3625–3632.
33. Balch, A. L.; La Mar, G. N.; Latos-Grazynski, L.; Renner, M. W. *Inorg. Chem.* **1985,** *24,* 2432–2436.
34. Ogoshi, H.; Kitamura, S.; Toi, H.; Aoyama, Y. *Chem. Lett.* **1982,** 495–498.
35. Groves, J. T.; Watanabe, Y. *J. Am. Chem. Soc.* **1986,** *108,* 507–508.
36. Collman, J. P.; Kodadek, T.; Raybuck, S. A.; Brauman, J. I.; Papazian, L. M. *J. Am. Chem. Soc.* **1985,** *107,* 4343–4345.
37. Dicken, C. M.; Woon, T. C.; Bruice, T. C. *J. Am. Chem. Soc.* **1986,** *108,* 1636–1643.
38. Traylor, T. G.; Miksztal, A. R. *J. Am. Chem. Soc.* **1987,** *109,* 2770–2774.
39. Anderson, O. P.; Kopelove, A. B.; Lavallee, D. K. *Inorg. Chem.* **1980,** *19,* 2101–2107.
40. Battioni, J.-P.; Artaud, I.; Dupre, D.; Leduc, P.; Akhrem, I.; Mansuy, D.; Fischer, J.; Weiss, R.; Morgenstern-Badarau, I. *J. Am. Chem. Soc.* **1986,** *108,* 5598–5607.
41. Balch, A. L.; Cornman, C. R.; Latos-Grazynski, L.; Olmstead, M. M. *J. Am. Chem. Soc.* **1990,** *112,* 7552–7558.
42. Goff, H.; La Mar, G. N. *J. Am. Chem. Soc.* **1977,** *99,* 6599–6606.
43. Wyslouch, A.; Latos-Grazynski, L.; Grzeszczuk, M.; Drabent, K.; Bartczak, T. *J. Chem. Soc. Chem. Commun.* **1988,** 1377–1378.
44. Bartczak, T.; Latos-Grazynski, L.; Wyslouch, A. *Inorg. Chim. Acta* **1990,** *171,* 205–212, and references therein.
45. Gans, P.; Buisson, G.; Duee, E.; Marchon, J.-C.; Erler, B. S.; Scholz, W. F.; Reed, C. A. *J. Am. Chem. Soc.* **1986,** *108,* 1223–1234.
46. Balch, A. L.; Cornman, C. R.; Latos-Grazynski, L.; Renner, M. W. *J. Am. Chem. Soc.* **1992,** *114,* 2230–2237.
47. La Mar, G. N.; Walker, F. A. In *The Porphyrins;* Dolphin, D., Ed.; Academic: New York, 1979; Vol. 4, p 61.
48. Bertini, I.; Luchinat, C. *NMR of Paramagnetic Molecules in Biological Systems;* Benjamin Cummings: Menlo Park, CA, 1986.
49. Keating, K. A.; de Ropp, J. S.; La Mar, G. N.; Balch, A. L.; Shiau, F.-Y., Smith, K. M. *Inorg. Chem.* **1991,** *30,* 3258–3263.
50. Horrocks, W. DeW. In *NMR of Paramagnetic Compounds;* La Mar, G. N.; Horrocks, W. DeW.; Holm, R. H., Eds.; Academic: New York, 1973; Chapter 4.
51. Evans, D. F. *J. Chem. Soc.* **1959,** 2003–2005.
52. Cornman, C. R. Ph.D. Thesis, University of California, Davis, 1990.
53. Satterlee, J. D.; La Mar, G. N. *J. Am. Chem. Soc.* **1976,** *98,* 2804–2808.
54. Nakamura, M.; Groves, J. T. *Tetrahedron* **1988,** *44,* 3225–3230.
55. Goff, H. M.; Phillippi, M. A. *J. Am. Chem. Soc.* **1983,** *105,* 7567–7571.

56. (a) Ringe, D.; Petsko, G. A.; Kerr, D. E.; Ortiz de Montellano, P. R. *Biochemistry* **1984,** *23,* 2–4. (b) Swanson, B. A.; Ortiz de Montellano, P. R. *J. Am. Chem. Soc.* **1991,** *113,* 8146–8153.
57. Tuck, S. F.; Peterson, J. A.; Ortiz de Montellano, P. R. *J. Biol. Chem.* **1992,** *267,* 5614–5620.
58. Samokyszyn, V. M.; Ortiz de Montellano, P. R. *Biochemistry* **1991,** *30,* 11646–11653.
59. Raag, R.; Swanson, B. A.; Poulos, T. L.; Ortiz de Montellano, P. R. *Biochemistry* **1990,** *29,* 8119–8126.
60. Aida, T.; Kawaguchi, K.; Inoue, S. *Macromolecules* **1990,** *23,* 3887–3892.
61. Inoue, S. *Polym. Prepr. (Am. Chem. Soc., Div. Polym. Chem.)* **1990,** *31,* 70–71.
62. Watanabe, Y.; Aida, T.; Inoue, S. *Macromolecules* **1990,** *23,* 2612–2617.
63. Watanabe, Y.; Yasuda, T.; Aida, T.; Inoue, S. *Macromolecules* **1992,** *25,* 1396–1400.
64. Watanabe, Y.; Aida, T.; Inoue, S. *Macrolmolecules* **1991,** *24,* 3970–3972.
65. Traylor, T. G.; Miksztal, A. R. *J. Am. Chem. Soc.* **1989,** *111,* 7443–7448.
66. Inoue, S.; Aida, T.; Konishi, K. *J. Mol. Catal.* **1992,** *74,* 121–129.
67. Kubo, H.; Aida, T.; Inoue, S.; Okamoto, Y. *J. Chem. Soc., Chem. Commun.* **1988,** *15,* 1015–1017.
68. Liu, K. E.; Johnson, C. C.; Newcomb, M.; Lippard, S. J. *J. Am. Chem. Soc.* **1993,** *115,* 939–947.
69. Chmielewski, P. J.; Latos-Grazynski, L. *Inorg. Chem.* **1992,** *31,* 5231–5235.
70. Balch, A. L.; Noll, B. C.; Reid, S. M.; Zovinka, E. P. *J. Am. Chem. Soc.* **1993,** *115,* 2531–2532.
71. Balch, A. L.; Latos-Grazynski, L.; Noll, B. C.; Olmstead, M. M.; Szterenberg, L.; Safari, N. *J. Am. Chem. Soc.* **1993,** *115,* 1422–1429.
72. Chin, D. H.; Balch, A. L.; La Mar, G. N. *J. Am. Chem. Soc.* **1980,** *102,* 1446–1448.
73. Shin, K.; Goff, H. M. *J. Am. Chem. Soc.* **1987,** *109,* 3140–3142.

RECEIVED for review June 10, 1993. ACCEPTED revised manuscript January 21, 1994.

15

Mechanisms of DNA Cleavage by High-Valent Metal Complexes

Gregory A. Neyhart, William A. Kalsbeck, Thomas W. Welch, Neena Grover, and H. Holden Thorp*

Department of Chemistry, University of North Carolina, Chapel Hill, NC 27599–3290

The mechanisms of DNA cleavage are reviewed for a number of different types of metal-based cleavage agents. Mechanisms are discussed both from the point of view of the metal complex and from the point of view of the DNA. Recent work on oxoruthenium(IV) complexes is discussed in detail. In particular, recent binding studies using a new luminescence method give binding affinities of relatively high precision. The affinity, viscometry, helical unwinding, thermal denaturation, and hypochromicity results for a complex of dipyridophenazine are consistent with binding by intercalation. The kinetics of oxidation of Ru(IV)O complexes can be studied in a straightforward manner, and a complete kinetic model is presented. Recent results show that lowering the oxidation potential of the Ru(IV)O oxidant leads to new specificities in DNA cleavage. This result is discussed in light of the kinetic model and partitioning between dissociation and oxidation.

THE DEVELOPMENT OF FUNCTIONALITIES THAT CLEAVE DNA is of prime importance in cancer chemotherapy and design of synthetic restriction enzymes (*1, 2*) and most importantly, in probing the complex tertiary structure of nucleic acids (*3, 4*). In recent years, transition-metal complexes have been shown to be effective cleaving agents that operate by a variety of mechanisms (*3, 4*). These mechanisms involve base oxidation by photosensitized singlet oxygen (*5*) and H-atom abstraction from the sugar functionality (*6*). The H-atom abstraction may occur through reaction of a photoactivated ligand (*4*), generation of hydroxyl radical (*3*), or formation of a reactive metal-oxo species (*6–8*). The metal-oxo species

* Corresponding author

0065–2393/95/0246–0405$08.90/0

formation mechanism is attractive because inorganic synthesis can be used to prepare metal-oxo species that are particularly suited to the DNA cleavage reaction.

The pathways of sugar oxidation involve abstraction of H-atoms bound to secondary or tertiary carbons (*6*). Many transition-metal oxidants will perform these abstractions on simple organic substrates such as secondary alcohols and hydrocarbons (*9–11*), and many of these reagents should therefore be capable of DNA cleavage. Because detailed information on available mechanistic pathways in sugar oxidation is now available (*6, 12*), there is a new opportunity for inorganic chemists to design reagents that are specifically tailored to the appropriate pathway.

Mechanisms of Cleavage Agents

In this section, cleavage reactions will be discussed from the point of view of the metal complex. To cleave DNA, an oxidant must be generated. Numerous redox reactions of metal complexes have been used to generate appropriate DNA oxidants (*4*). The simplest comes from Fenton chemistry, in which peroxide is used to generate hydroxy radicals from Fe^{2+} (*3*):

$$Fe^{2+} + H_2O_2 \rightarrow Fe^{3+} + OH^- + \cdot OH \quad (1)$$

When this reaction is performed in the vicinity of the DNA helix, the hydroxy radicals that are generated abstract hydrogen atoms from DNA sugars, leading to strand scission. Using $Fe(EDTA)^{2-}$, which does not bind to DNA, as a delivery agent allows the hydroxy radicals to be delivered in a nonspecific fashion, allowing for footprinting of DNA–protein interactions and structural imaging of nucleic acids based on solvent accessibility of individual sites (*3, 13*). Because hydroxy radical is a diffusible species, these reagents generate diffuse reaction patterns in sequencing gels.

The natural product bleomycin mediates strand scission in the presence of iron and oxygen by a related mechanism (*6, 12, 14*); however, no diffusible intermediate is generated. Electrochemical and related studies showed that the mechanism of iron bleomycin (Fe-BLM) involves first reduction of Fe(III)-BLM, reaction of Fe(II)-BLM with O_2, and reduction of the Fe(II)-BLM-O_2 adduct to "activated Fe-BLM" (*15*).

$$\text{Fe(III)-BLM} + e^- \rightarrow \text{Fe(II)-BLM} \quad (2)$$

$$\text{Fe(II)-BLM} + O_2 \rightarrow O_2^{-}\text{-Fe(III)-BLM} \quad (3)$$

$$O_2^{-}\text{-Fe(III)-BLM} + e^- \rightarrow \text{"activated Fe-BLM"} \quad (4)$$

The precise nature of activated Fe-BLM is still a subject of debate; however, both $Fe(IV)O^{2+}$ and $Fe(V)O^{3+}$ species have been suggested (*15*). In any case, activated Fe-BLM is capable of oxidizing many organic substrates, including ribose and deoxyribose in nucleic acids (*6*).

Copper-peroxide chemistry is also capable of generating a potent oxidant (*16, 17*). Addition of H_2O_2 to $Cu(phen)_{2+}$ generates a novel oxidant according to equation 5:

$$Cu(phen)_2^+ + H_2O_2 \rightarrow Cu(II)(phen)_2(\cdot OH)^{2+} + OH^- \qquad (5)$$

The oxidant can be formulated either as a hydroxy radical bound to Cu(II) or as a hydroxide ligand on Cu(III). In either case, the activated complex is capable of strand scission, and the cleavage pattern suggests that the oxidant is nondiffusible and therefore bound to the metal (*17*). Cleavage by this reagent involves sugar oxidation, as discussed in subsequent paragraphs.

Work in our laboratory has shown that oxoruthenium(IV) and hydroxoruthenium(III) complexes are capable of DNA cleavage (*7*). These reagents can be generated by oxidation of complexes based on $Ru(tpy)(bpy)OH_2^{2+}$, either chemically or electrochemically:

$$Ru^{II}(tpy)(bpy)OH_2^{2+} \rightleftharpoons Ru^{III}(tpy)(bpy)OH^{2+} + e^- \qquad (6)$$

$$Ru^{III}(tpy)(bpy)OH^{2+} \rightleftharpoons Ru^{IV}(tpy)(bpy)O^{2+} + e^- \qquad (7)$$

These reagents can be used either catalytically, by application of a potential of 0.8 V to a solution of $Ru(tpy)(bpy)OH_2^{2+}$ or stoichiometrically, by isolation of the $Ru(tpy)(bpy)O^{2+}$ or $Ru(tpy)(bpy)OH^{2+}$ complexes and addition of these reagents to a DNA solution. The chemistry of these complexes in DNA will be discussed in detail in this chapter.

Photoreactions can also be used to induce strand scission. Upon photolysis, polypyridyl complexes of Ru(II) sensitize the formation of singlet oxygen, which is capable of oxidizing nucleic acid bases (*5*). Following piperidine treatment, base oxidation leads to strand scission (*18*). Uranyl salts are also capable of DNA cleavage via sugar oxidation upon irradiation (*19*). By far the most studied and widely applied cleavage photoreactions involve polypyridyl and other diimine complexes of rhodium(III) (*4*). Upon UV irradiation, these complexes induce DNA strand scission by sugar oxidation. When a complex of phenanthrenequinone diimine (phi) is used, hydrolysis occurs during the cleavage reaction to produce a diaqua rhodium complex and the free phi ligand (*20, 21*). The precise mechanism at the metal center is still unknown; however, irradiation into the ligand π-π^* bands is required for reaction, thereby implicating hydrogen abstraction into the ligand as the primary mechanism. Work with these complexes has shown that tuning the shape of the ligand environment can lead to stereo- and shape-selective cleavage (*20*).

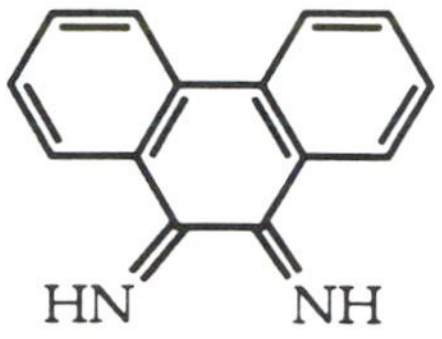

HN NH

phi

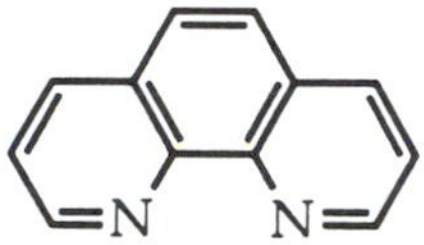

N N

phen

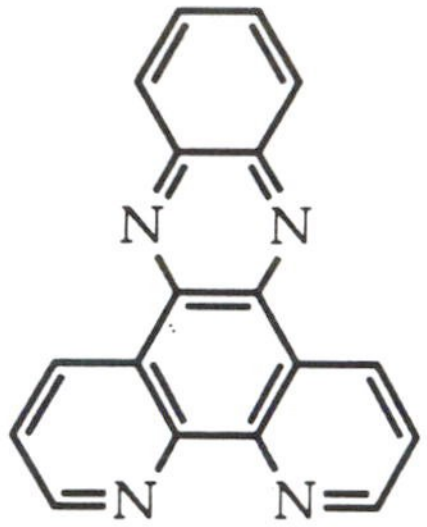

N N

N N

dppz

A photoanalogue of $Fe(EDTA)^{2-}$ has also been reported (*22*). The complex $Pt_2(pop)_4^{4-}$ (pop = $P_2O_5H_2^{2-}$) has a high-energy excited state ($Pt_2(pop)_4^{4-*}$) that is capable of abstracting hydrogen atoms from organic substrates to generate a hydridoplatinum(III) complex and organic radicals (*23*):

$$Pt_2(pop)_4^{4-*} + R\text{-}H \rightarrow Pt_2H(pop)_4^{4-} + R\cdot \qquad (8)$$

When $Pt_2(pop)_4^{4-}$ is irradiated at 450 nm in the presence of DNA, strand scission occurs, and scavenger studies show that no diffusible intermediates are involved (*22*). Thus, scission must involve abstraction of hydrogen atoms from the sugar functionalities by the metal complex itself. Because $Pt_2(pop)_4^{4-}$ is an anion, binding of the complex to DNA does not occur. This presents the possibility that $Pt_2(pop)_4^{4-}$ may be a particularly sensitive probe of solvent accessibility and hence nucleic acid structure.

Complexes of nickel have been shown to induce DNA damage by base oxidation in the presence of $KHSO_5$ (*8*). These macrocyclic complexes of Ni(II) apparently covalently bind to guanine bases prior to oxidation. Treatment with $KHSO_5$ leads to formation of a Ni(III) species that oxidizes guanine. Upon base treatment, strand scission is observed. Because covalent binding must occur prior to oxidation, only guanines present in single-stranded residues are oxidized. This feature has been used to demonstrate selective cleavage of DNA at guanines in bulges, mismatches, ends, and hairpin loops (*24*). Guanines present in double-stranded regions are not oxidized.

Mechanisms of Strand Scission

In this section we discuss cleavage mechanisms from the point of view of the nucleic acid. The accepted mechanism for the oxidation of DNA by activated Fe-bleomycin (Fe-BLM) is shown in Scheme 1 (*6, 14*). The initial step involves abstraction of the 4′-hydrogen in the minor groove to form the radical species. The generated radical may undergo subsequent oxidation to form the tertiary carbocation, which reacts with water to generate the hydroxylated species. This oxidation immediately liberates free base, and after base treatment, strand scission occurs to provide a phosphate terminus and a modified terminus, as shown in Scheme 1. Alternatively, the radical may react with oxygen to form a hydroperoxy species, which goes on without base treatment to liberate base propenal, a phosphate terminus, and a phosphoglycolate terminus. Partitioning between the oxygen-dependent base propenal pathway and the oxygen-independent free base pathway is controlled by the oxygen tension in solution.

The ability to incorporate deuterium or tritium selectively into appropriate positions on nucleotides has had a tremendous impact on mechanistic studies on sugar oxidation. In particular, Stubbe and coworkers have shown that when thymidine deuterated at the 4′ position is incorporated into restriction fragments, sequence-selective isotope effects are observed for cleavage by Fe-BLM (*12*). These isotope effects are essentially equal for the radical and carbocation pathways (Scheme 1). The magnitude of these isotope effects ($k_H/k_D = 2.5–4$) is consistent with the magnitude of the analogous tritium isotope effects ($k_H/k_T = 7–11$). These isotope effects are determined from the net amount of cleavage observed at the labeled nucleotide, not from a time-resolved study of the true cleavage rate constant.

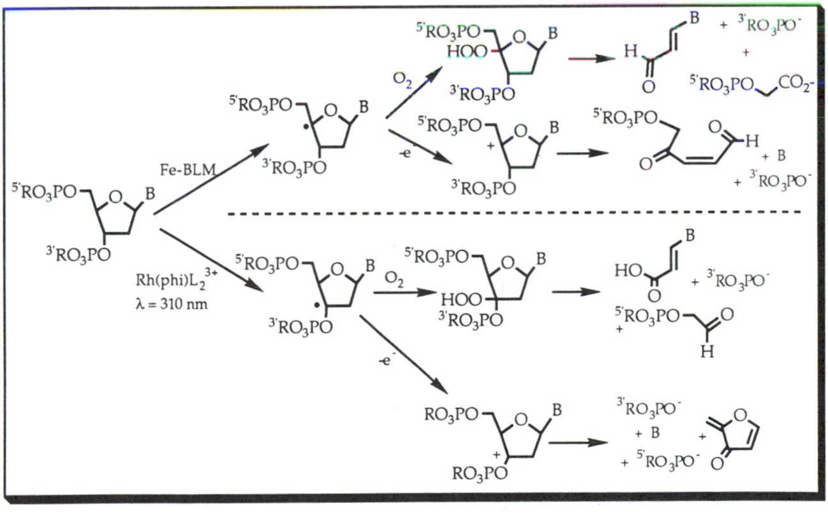

Scheme 1

Barton et al. (*21*) have shown that complexes based on $RhL_2(phi)^{3+}$ cleave DNA by abstraction of the 3′-hydrogen in the major groove. This reaction proceeds as shown in Scheme 1, with an analogous oxidation of the 3′ position to yield a radical that can react with oxygen to form base propenoic acid, a phosphate terminus, and a phosphoglycaldehyde terminus. Alternatively, the radical can be oxidized to a carbocation that reacts with water to yield free base, two phosphate termini, and a furenone product. To date, the furenone product has not been detected. Mechanistic studies show that when the metal complex has a very high binding affinity, only the carbocation (free base) pathway is observed, regardless of the oxygen tension in solution. However, complexes that bind less strongly produce substantial amounts of base propenoic acid upon irradiation in the presence of oxygen.

In addition to chemistry at the 4′ position, Hecht and co-workers (*25*) have demonstrated that when oligonucleotides contain a ribo- or ara-sugar, these nucleotides are cleaved by Fe-BLM partly by chemistry occurring at the 1′ position (Scheme 2). The oxygen-dependent pathway results in the formation of a 1′-hydroperoxy sugar, which upon treatment with 1,2-diaminobenzene gives a product that is readily detected by high-performance liquid chromatography (HPLC). With an ara-sugar, oxidation at 1′ accounts for as much as 58% of the total oxidation. Oxidation of DNA by $Cu(phen)_2^+$ and H_2O_2 yields free base and a product that can be assigned as 5-methylene-2,5-*H*-furanone (5-MF) (*16*). This product has been suggested to arise from oxidation by the copper complex at the 1′ position in an oxygen-independent reaction (Scheme 2).

Oxidation of DNA by nickel macrocycles and $KHSO_5$ occurs via base oxidation (*8*). It has been proposed that the guanine oxidation leads to the formation of 8-oxoguanine, which is known to promote strand scission upon base treatment. It has been reported that one-electron oxidation of guanine leads to a radical cation that reacts with water to

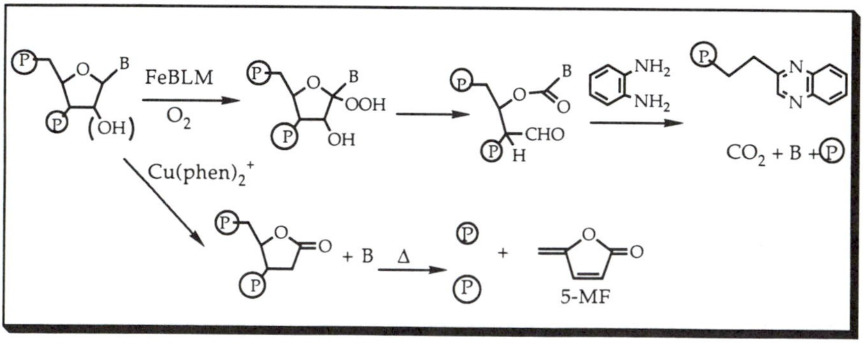

Scheme 2

form 8-oxoguanine (*26*). Labeling studies show that the oxo group in the 8-oxoguanine formed by this reaction comes from solvent water.

Binding of Metal Complexes to DNA

Any cleavage reaction is governed by the binding of the metal complex to DNA. Cleavage reactions of $Fe(EDTA)^{2-}$ and $Pt_2(pop)_4{}^{4-}$ by design do not involve binding of the cleavage agent to DNA (*3, 13, 22*). Thus, any analysis of the cleavage mechanism must incorporate this fact. In most cases, however, cleavage is mediated by cationic complexes that bind to DNA. In the case of FeBLM, sugars and bithiazoles in the bleomycin molecule control binding and hence, selectivity (*27*). With small metal complexes, however, the binding model is less complex, although large selectivities are still observed (*4*).

For simple octahedral metal complexes containing polypyridyl or phi ligands, a number of binding modes have been discussed (*28–31*). The first involves simple electrostatic binding and is observed for complexes with ligands such as bipyridine, that do not have extended planar surfaces (*28*). When complexes contain large, flat ligands, such as phi or dppz, binding of the metal complex via intercalation has been proposed (*28–36*). Intercalation involves unwinding of the DNA to create room for the flat ligand to insert in between two base pairs. Complexes of dppz have been shown to have high binding affinities, to unwind DNA, and to exhibit photophysical properties consistent with intercalation (*30–33*). Results from our laboratory on dppz intercalation are discussed in subsequent paragraphs. Complexes of phi also unwind DNA and exhibit high binding affinities (*21, 28*) and have been shown by using two-dimensional NMR to intercalate specifically into oligonucleotides (*34*). A number of binding modes for octahedral phen complexes have been proposed; however, the current results discussed in subsequent paragraphs and elsewhere are consistent with a model in which some fraction of the phen complexes are partially intercalated (*29, 35, 36*).

The advantage of an intercalative binding mode is that it provides a basis from which to begin devising a mechanistic scheme (*34, 37*). For example, intercalative binding is thought to occur in the major groove, which suggests certain sugar hydrogens that might be more accessible to the metal complex (*21*). Thus, the intercalator serves to anchor the metal complex into the major groove. On the other hand, binding that is primarily electrostatic in nature occurs primarily in the minor groove, because the close proximity of phosphate groups creates a higher density of negative charge (*38*).

In devising a mechanistic scheme for DNA cleavage by a particular metal complex, it is important to consider the binding equilibrium that must occur prior to cleavage. For example, consider cleavage by a complex that is added in the activated form to a solution of DNA. This complex must first bind to DNA before cleavage can occur:

$$M_{ox}^{n+} + DNA \rightleftharpoons M_{ox}^{n+} \cdot DNA \tag{9}$$

$$M_{ox}^{n+} \cdot DNA \rightarrow M_{red}^{n+} \cdot DNA \tag{10}$$

An important question involves the relative rates of binding, dissociation, and cleavage. Dramatically different cleavage efficiencies may be observed depending on the relative rates of each of these processes. Also, if an excess of oxidized complex is present in solution, dissociation of the reduced form may be required before continued oxidation of DNA can occur. This may profoundly affect the oxidation efficiency. If selectivity is desired, the complex should be able to bind without oxidation always occurring, so that cleavage only occurs in certain sites. In these cases, the rate of oxidation should be similar to the dissociation rate. Successfully developing a kinetic model for DNA cleavage requires a detailed, quantitative understanding of the binding equilibrium (equation 10). These issues will be discussed in detail.

DNA Cleavage by Oxoruthenium(IV)

Most of our early work in this area centered on the complexes $Ru(tpy)(L)OH_2^{2+}$, where L = bpy, phen, and dppz (Figure 1, tpy = 2,2′,2″-terpyridine) (*7, 30, 33*). These complexes cleave DNA upon oxidation to $Ru(tpy)(L)OH^{2+}$ and $Ru(tpy)(L)O^{2+}$, which can be performed chemically or electrochemically. We have a number of other complexes in our laboratory with similar properties (*39*); however, we will begin here with a discussion of the binding and kinetics of the simple, achiral tpy complexes, which have provided a basis for the design of cleavage agents with other desirable properties.

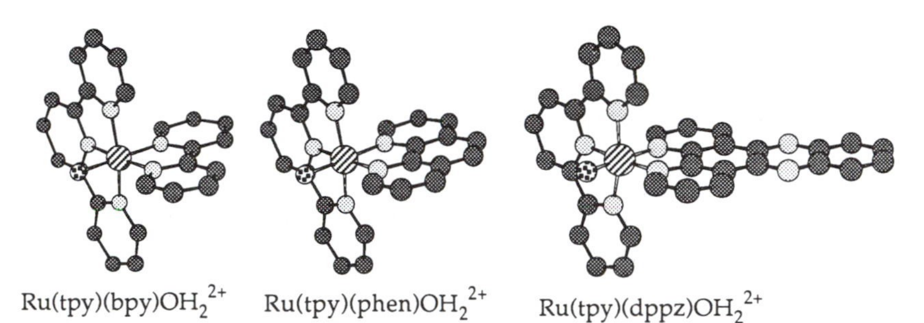

Figure 1. Structures of $Ru(tpy)(L)OH_2^{2+}$ complexes. (Reprinted from reference 30. Copyright 1993 American Chemical Society.)

Binding Studies. As discussed in the preceding paragraph, an understanding of the mechanism of any DNA cleavage agent must also involve a complete understanding of the binding of the complex to DNA. Furthermore, in cleavage reactions of oxoruthenium(IV), the binding equilibrium and kinetics play a very important role in the overall cleavage mechanism. One of the most challenging and important aspects of studying the binding of metal complexes to DNA is obtaining a quantitatively reliable value for the binding constant of the metal complex. Difficulties in determining these binding constants arise because often the changes in the absorption spectrum of the metal complex upon binding are small compared to analogous changes in the spectra of organic molecules (*40*). This occurs because only a portion of the metal complex may interact with the nucleic acid, and the effect of binding on the absorbance would thereby be smaller than with porphyrins or organic molecules in which the entire chromophore interacts directly with the biomolecule (*41*). In addition, the absolute magnitudes of extinction coefficients for metal-based chromophores are often much lower than those of organic species that bind to DNA.

Difficulties in measuring binding constants for metal complexes arise both when binding constants are quite high ($>10^5$ M^{-1}) and very low (≤ 1000 M^{-1}), because of the lower extinction coefficients and hypochromocities (*28, 33, 42*). We developed a new technique for measuring the binding constants of metal complexes to DNA that does not depend on the absorbance properties of the metal complex of interest. This method involves the complex $Pt_2(pop)_4^{4-}$, which has a high-energy (55 kcal/mol) excited state that is readily quenched by electron and energy transfer by a numerous metal complexes (*23*), many of which have been studied with regard to their DNA-binding properties (*43*). The method relies on the quenching of the strong emission of the excited-state of $Pt_2(pop)_4^{4-}$ by the metal complex with and without DNA. In the absence of DNA, the emission is quenched very efficiently. Because metal complexes that bind to DNA are generally cations, the quenching is efficient partly because of ion pairing between the $Pt_2(pop)_4^{4-}$ tetraanion and the cationic metal complex. In the presence of DNA, the quenching is slowed down by as much as two orders of magnitude, because the metal complex of interest binds to the DNA polyanion from which $Pt_2(pop)_4^{4-}$ is electrostatically repelled.

By using the difference in quenching rate constant brought about by binding of the metal complex to DNA, binding constants for the metal complex quenchers can be determined. Because of the resolution provided by the large difference in rate constants, particularly small ($<10^3$ M^{-1}) and quite large ($>10^4$ M^{-1}) binding constants can be determined, provided the appropriate equations are used to fit the data. Shown in Figure 2 is the binding isotherm for $Ru(tpy)(dppz)OH_2^{2+}$, which is fit to equation 11:

$$\frac{(k_a - k_f)}{(k_b - k_f)} = \frac{b - \left(b^2 - \frac{2K^2C_t[\mathrm{DNA}]}{s}\right)^{1/2}}{2KC_t} \tag{11a}$$

$$b = 1 + KC_t + \frac{K[\mathrm{DNA}]}{2s} \tag{11b}$$

where k_a is the observed quenching rate constant, k_f is the quenching rate constant in the absence of DNA, k_b is the quenching rate constant in the presence of excess DNA, C_t is the total concentration of metal complex, [DNA] is the concentration of DNA in nucleotide phosphate, s is the site size of the metal complex is base pairs, and K is the binding constant. For $Ru(tpy)(dppz)OH_2^{2+}$, the binding constant determined is 730,000 M^{-1} and the site size is 2, consistent with nearest-neighbor exclusion (*44*). Attempts to quantitate the binding affinity of this complex and other metallointercalators by equilibrium dialysis provided only an

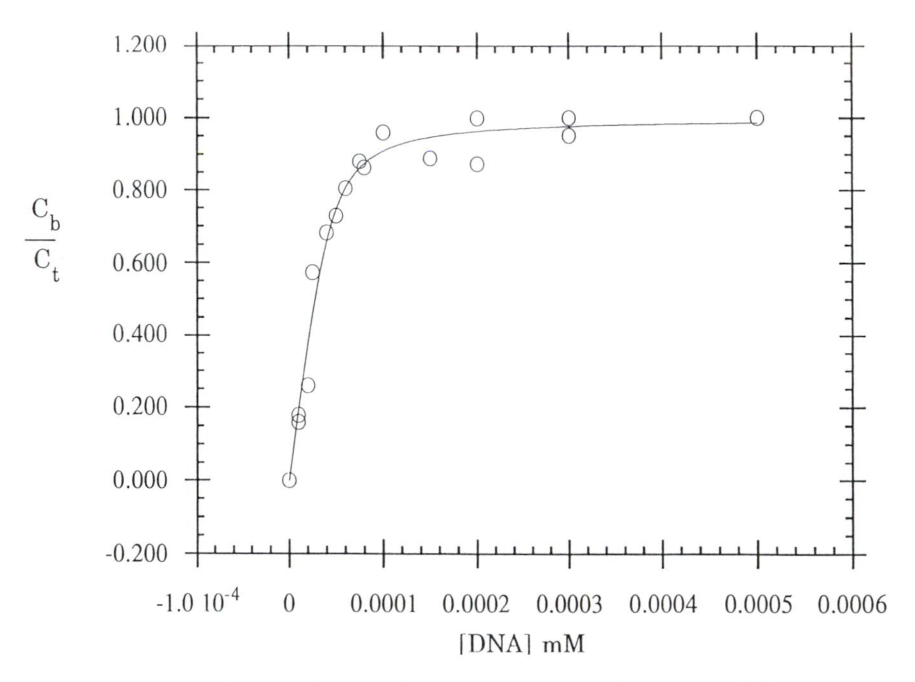

Figure 2. Binding isotherm of $Ru(tpy)(dppz)OH_2^{2+}$ measured by using $Pt_2(pop)_4^{4-}$ quenching. (Reproduced from reference 56. Copyright 1993 American Chemical Society.)

order-of-magnitude estimate (*21, 32, 33*). Binding constants for the $Ru(tpy)(L)OH_2^{2+}$ complexes are shown in Table I.

In addition to quantitating the binding affinity, we performed thermal denaturation, viscometry, and absorption hypochromicity studies on the $Ru(tpy)(L)OH_2^{2+}$ complexes. In thermal denaturation experiments (Table I), the L = bpy complex shows a modest value of ΔT_m, consistent with perhaps simple electrostatic binding of the complex to the DNA (*44,45*). The low binding affinity of 660 M^{-1} is also consistent with simple electrostatic binding. When L = dppz, however, the value for ΔT_m is strikingly high, actually higher than that of the organic intercalator ethidium bromide (*44*). A higher value of ΔT_m is probably due in part to the dicationic charge of $Ru(tpy)(dppz)OH_2^{2+}$ compared to the ethidium monocation. The L = dppz complex also shows an extremely high ($\Delta T_m > 33°$) affinity for poly(dA)·poly(dT), whereas the ΔT_m values of the bpy and phen complexes for this polymer are similar to those for calf thymus DNA. Recent NMR studies of oligonucleotides show that related complexes appear to have higher affinities for AT regions (*38, 46, 47*). Absorption hypochromicity studies are similar, with a small red shift of the MLCT band upon binding of the bpy complex, but a sizable red shift upon binding of the dppz complex to DNA. In terms of thermal denaturation, binding affinity, and absorbance hypochromicity, the phen complex gives results intermediate between those of the bpy and dppz derivatives.

The ability of metal complexes to unwind DNA has been put forth as an important criterion for proving an intercalative binding mode and has been observed with other complexes of phen, dppz, and phi (*21, 28, 30, 33*). The enzyme topoisomerase can be used to determine if small molecules unwind DNA, according to published procedures (*28*). We find that by using this assay, $Ru(tpy)(dppz)OH_2^{2+}$ unwinds DNA by 17°, which is consistent with intercalative binding.

Another convincing test of intercalation comes from viscometry studies that test the ability of a small molecule to lengthen DNA, which occurs when base pairs separate to accommodate intercalators (*41, 44, 48*). The slope of the viscometry plot gives the amount of lengthening per metal complex. The bpy complex does not appear by viscometry to lengthen DNA; however, $Ru(tpy)(dppz)OH_2^{2+}$ gives results in the viscometry experiment that are quantitatively identical to those for ethidium bromide. As can be seen in Figure 3, the ratio of the length of DNA in the presence of $Ru(tpy)(dppz)OH_2^{2+}$ to the length of DNA in the absence of the complex increases linearly with the concentration of metal complex, with a slope of 1.13. Shown in Figure 3 are two separate titrations for $Ru(tpy)(dppz)OH_2^{2+}$ (● and ▲), one of which is over a smaller range of metal complex/DNA ratios (●). Both experiments clearly give the same slope within experimental error. The same behavior is observed

Table I. Binding Parameters of Polypyridyl Complexes to Calf Thymus DNA

Complex	ΔT_m *(°C)*[a]	*Slope*[b]	$\Delta\lambda_{max}$*(nm) (%H)*[c]	*Relative Rate*[d]	ϕ[e]	K_B (M^{-1})[f]
$Ru(tpy)(bpy)OH_2^{2+}$	4.2 ± 0.5	—	4.1 ± 0.7 (2.3%)	Fast (1.0)	—	660
$Ru(tpy)(phen)OH_2^{2+}$	7.2 ± 0.4	—	7.3 ± 0.7 (8.0%)	Fast (1.6)	—	3,700
$Ru(tpy)(dppz)OH_2^{2+}$	14.1 ± 0.8	1.13	8.7 ± 0.5 (9.6%)	Slow (0.12)	17°	700,000
Ethidium bromide	13.0[g]	1.14	39 (26%)[h]	Slow[i]	26°[i]	

[a] Determined in 5.0 mM Tris-HCl buffer, 10:1 DNA/metal complex.

[b] Slope of the plot of L/L° vs. the metal complex/DNA ratio between metal/DNA ratios of 0 to 0.15.

[c] $H = (A_{free} - A_{DNA})/A_{free}$ in 50 mM phosphate buffer, 350:1 DNA/metal complex.

[d] Ratio of k_2 at 10:1 DNA/metal complex for each complex relative to $Ru(tpy)(bpy)O^{2+}$. *See* text and Table II.

[e] Unwinding angle measured using topoisomerase; error ±2°.

[f] Measured by emission titration at 50 mM ionic strength; error ±30%.

[g] Reference 44.

[h] Reference 40.

[i] Reference 51.

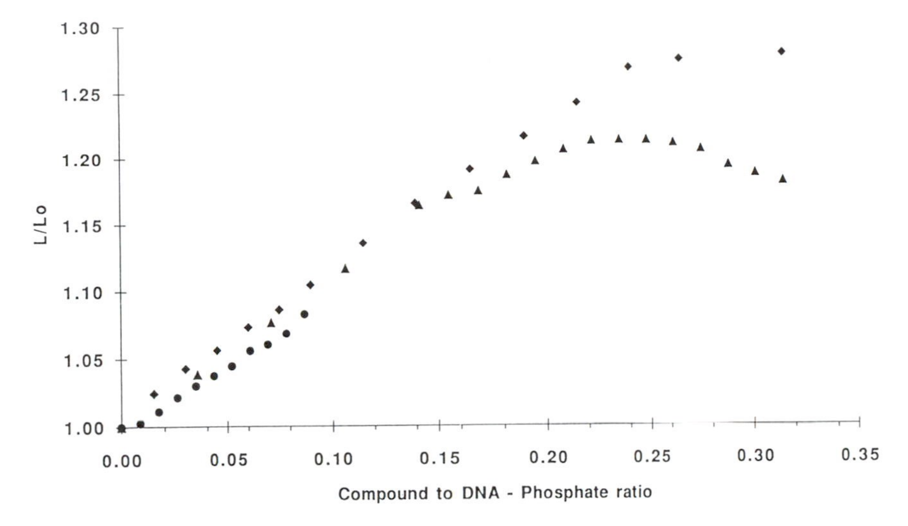

Figure 3. Viscometric titrations with sonicated calf thymus DNA (5 mM ionic strength) for ethidium bromide (♦) and Ru(tpy)(dppz)OH_2^{2+} (●) and (▲). (Reproduced from reference 30. Copyright 1993 American Chemical Society.)

with ethidium bromide (♦), which gives an identical slope of 1.14. The values of L/L° for both ethidium and Ru(tpy)(dppz)OH_2^{2+} plateau at a small molecule/DNA ratio of 0.2–0.25, which is indicative of neighbor-exclusion intercalation in both cases (*44*, *48*), with saturation occurring once metal complexes are bound approximately every two base pairs. Saturation at this metal/DNA ratio is consistent with the site size of 2 determined in Figure 2. Recent luminescence experiments on $Ru(phen)_2(dppz)^{2+}$ are consistent with these findings (*31*).

Because metal complexes that bind to DNA are cations, part of the binding affinities arises from simple electrostatic interactions. Others have suggested that this is the primary force responsible for binding (29). However, the role of buffer cation concentration in determining the binding constant of cations to DNA can be quantitated by using polyelectrolyte theory (*49*). To date, a similar study has not been conducted for a higher-affinity metal complex, such as a complex of dppz or phi, because of the difficulty in accurately determining large ($\geq 10^6$ M^{-1}) binding constants for metal complexes. Using our method of $Pt_2(pop)_4^{4-*}$ quenching, we can determine these large binding constants with sufficient precision to permit such a study.

The binding constant for Ru(tpy)(dppz)OH_2^{2+} was determined by using our method as a function of buffer ionic strength. As the ionic strength is decreased, the binding constant for the complex increases dramatically, up to 27 x 10^6 M^{-1} at 5 mM ionic strength from 0.73 x 10^6

M^{-1} at 50 mM ionic strength. The dependence of the binding constant on the concentration of the monovalent buffer cation is shown in Figure 4. The plot shown gives log *K* versus the logarithm of the monovalent (Na^+ or K^+) cation concentration. The linear dependence observed is predicted from polyelectrolyte theory, which states that the logarithm of *K* should depend on buffer cation concentration as (*49*):

$$\ln K_{obs} = \ln K°_t + Z\xi^{-1}[\ln(\gamma \pm \delta)] + Z\,\Psi(\ln[M^+]) \qquad (12)$$

where K_{obs} is the measured binding concentration at a monovalent cation concentration of $[M^+]$, Z is the charge on the metal complex, $K°_t$ is the "thermodynamic" binding constant, $\gamma_{\pm}$ is the mean activity coefficient at cation concentration $[M^+]$, and $\xi = 4.2$ and $\delta = 0.56$ for calf thymus DNA. The parameter Ψ is the number of counterions associated with each DNA phosphate, which is 0.88 for calf thymus DNA (*49*). The magnitude of $K°_t$ represents the contribution to binding from nonelectrostatic forces.

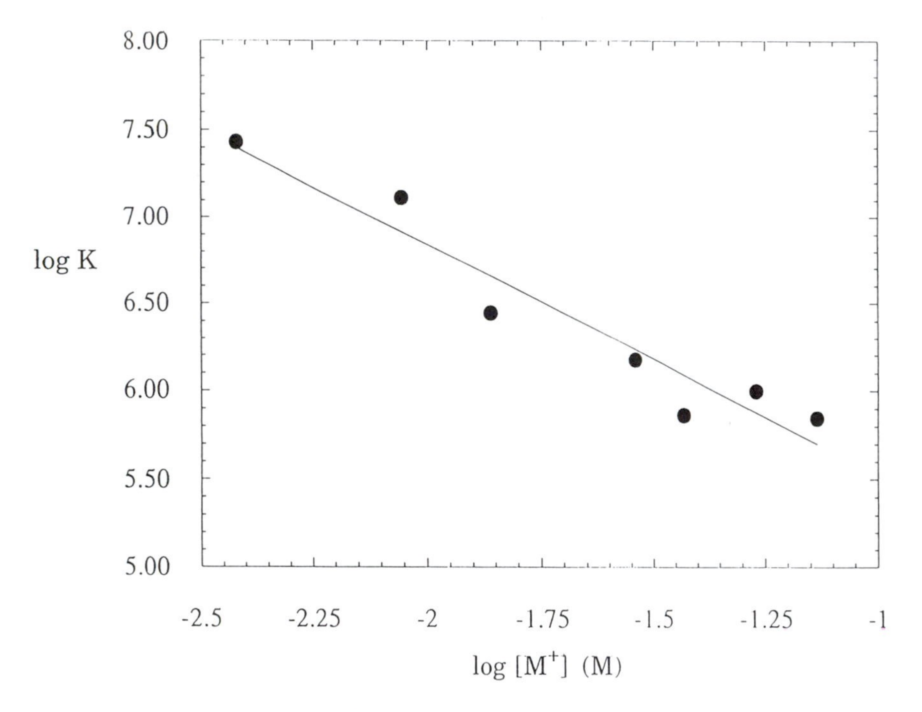

Figure 4. Dependence of the observed binding constant of Ru(tpy)-(dppz)OH_2^{2+} on the concentration of the monovalent buffer cation. Solid line is the least-squares fit to equation 12. (Reproduced from reference 56. Copyright 1993 American Chemical Society.)

The results of fitting equation 12 to the data in Figure 4 are given in Table II. The slope of the plot provides an experimental measure of $Z\Psi$. Our value of $Z\Psi = -1.34$ for $Ru(tpy)(dppz)OH_2^{2+}$ is somewhat lower than the theoretical value; however, Chaires et al. (29) have determined nearly identical values for Δ- and Λ-$Ru(phen)_3^{2+}$. Slightly lower values of $Z\Psi$ can be attributed to changes in DNA hydration or coupled anion release from the metal complex upon binding. Results from the analyses of $Ru(phen)_3^{2+}$ and ethidium binding by Chaires et al. are given for comparison in Table II.

Also given in Table II are values for $K°_t$ and $\Delta G°_t$ for binding of $Ru(tpy)(dppz)OH_2^{2+}$ to DNA. These quantities represent the binding energetics in the absence of electrostatic forces. The values for $Ru(tpy)(dppz)OH_2^{2+}$ are clearly much larger than those for $Ru(phen)_3^{2+}$ and approach quite closely the values for ethidium. Strikingly, $\Delta G°_t$ is only 0.6 kcal/mol greater for ethidium than for $Ru(tpy)(dppz)OH_2^{2+}$.

The confirmation that the binding of these metal complexes to DNA is governed by equation 12 provides insight into the interplay of electrostatic and other forces in controlling the binding affinities. In general, the $Z\xi^{-1}[\ln(\gamma\pm\delta)]$ term is small compared to the other two, so the binding affinity is basically a sum of the electrostatic $Z\Psi$ term and the $K°_t$ term. For complexes that bind solely by electrostatics, $\ln K_{obs}$ should be essentially the same as the $Z\Psi$ term. If $Z\Psi \sim -1.3$, as it is in $Ru(phen)_3^{2+}$ and $Ru(tpy)(dppz)OH_2^{2+}$, it is straightforward to calculate that this is indeed the case for $Ru(bpy)_2^{2+}$ and $Ru(tpy)(bpy)OH_2^{2+}$. This indicates that at 50 mM salt concentration, electrostatic forces account for 10^2 M^{-1} in binding affinity (for dications), and the remainder of the binding affinity arises from nonelectrostatic forces. On going from bpy complexes to dppz complexes, $\Delta G°_t$ increases from essentially zero to about 6 kcal/mol. This 6 kcal/mol is responsible for the difference in binding affinity between 10^2 M^{-1} for bpy complexes and 10^6 M^{-1} for dppz complexes. In the case of dppz complexes that are known to be metallointercalators,

Table II. Comparative Energetics of Binding to DNA

Compound	$K_{obs}/10^4$ M^{-1} [a]	$Z\Psi$	$K°_t/10^4$ M^{-1}	$\Delta G°_t$ *(kcal/mol)* [b]
$Ru(tpy)(dppz)OH_2^{2+}$	73	1.32	2.7 ± 1.1	−5.9 ± 0.3
Ethidium[c]	49.4	0.75	6.1	−6.5
Δ-$Ru(phen)_3^{2+}$ [c]	0.97	1.38	0.02	−3.1
Λ-$Ru(phen)_3^{2+}$ [c]	1.07	1.24	0.03	−3.4

[a] Measured in μ = 50 mM phosphate buffer.

[b] Determined from $\Delta G°_t = -RT \ln K°_t$.

[c] Reference 29.

it is clear that intercalation is responsible for the increased binding affinity. For phen complexes, which have been shown to exhibit $\Delta G°_t$ values of about 3 kcal/mol (29), the contribution to the binding affinity arising from nonelectrostatic forces is not as high as that for the metallointercalators, but it is readily measurable, and the binding affinity is certainly greater than the 10^2 M^{-1} expected for a complex that binds solely by electrostatics.

Kinetic Studies. We studied the present complexes in their oxidized forms, Ru(tpy)(L)O^{2+}, because these complexes are effective DNA cleavage agents (*7, 30, 33, 50*). Oxidation of calf thymus DNA by Ru(tpy)(bpy)O^{2+} leads to release of all four nucleic acid bases, as shown in Figure 5. The cleavage is nonspecific, leading to release of approximately equal amounts of adenine, thymine, guanine, and cytosine. The observation of base release implicates sugar oxidation as the primary cleavage mechanism (*6, 14, 16*).

One of the interesting features of the cleavage reactions of these complexes is that the kinetics can be followed by using optical spec-

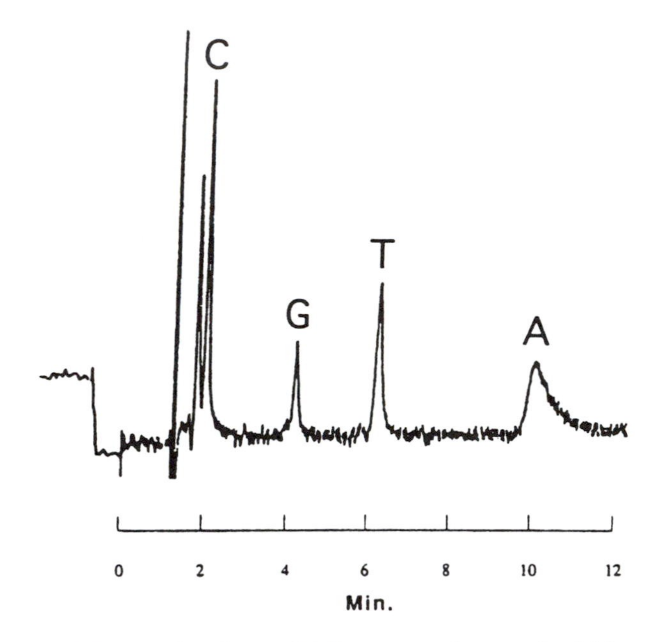

Figure 5. HPLC results obtained following oxidation of calf thymus DNA (1.0 mM) by Ru(tpy)(bpy)O^{2+} (0.05 mM). Chromatograms were run on a Rainin Microsorb-MV "Short-One" C_{18} column with 0.1 M ammonium formate buffer (pH 7) at a flow rate of 1.0 mL/min. Peaks are labeled as adenine (A), thymine (T), guanine (G), and cytosine (C). (Reproduced from reference 30. Copyright 1993 American Chemical Society.)

troscopy (7). We will now present kinetics results taken at an isosbestic point at which only the disappearance of $Ru(IV)O^{2+}$ is monitored (9). The decay curves for the disappearance of $Ru(IV)O^{2+}$ show two distinct reaction phases (Figure 6). The first phase is rapid and occurs within the time of conventional mixing. We determined that the kinetics of this early "burst" phase can be investigated by stopped-flow spectrophotometry under the appropriate conditions, and these experiments are currently underway in our laboratory. For $Ru(tpy)(bpy)O^{2+}$, the contribution of the early "burst" phase is a function of R (= [DNA-phosphate]/[Ru(IV)]), with the burst fraction being much larger at high R. For $Ru(tpy)(phen)O^{2+}$, the contribution is higher than for $Ru(tpy)(bpy)O^{2+}$ at the same R, and also increases dramatically with increasing R. For $Ru(tpy)(dppz)O^{2+}$, the contribution is much larger than that for either of the other two complexes at low R (Table III).

We interpret the trend in the burst fraction in terms of the model shown in Scheme 3. The results can be accounted for if the rate of binding of $Ru(IV)O^{2+}$ (k_{on}) is faster than the rate of oxidation (k_1), which is in turn faster than the dissociation rate, k_{off}. The rate-determining step in the first phase is therefore the zero-order oxidation of DNA by bound $Ru(IV)O^{2+}$. Thus, the burst fraction represents the amount of metal complex that is bound at time zero, which is higher when the binding constant of the metal complex is increased or R is increased.

The second phase of the reaction occurs over a longer time period and can be analyzed by fitting to a biexponential decay. The rates are given in Table III. Interestingly, the rates for a single complex do not vary outside of experimental error as a function of either R or the absolute concentrations of metal complex or DNA. In addition, the rates are the same within experimental error for both $Ru(tpy)(bpy)O^{2+}$ and $Ru(tpy)(phen)O^{2+}$. However, the rates are an order of magnitude slower for $Ru(tpy)(dppz)O^{2+}$ than for the other two complexes.

The results for the second phase can also be interpreted in terms of the model shown in Scheme 3. Following the k_1 step, excess $Ru(IV)O^{2+}$ is present in solution, and $Ru(II)OH_2^{2+}$ is bound to DNA. For continued oxidation of DNA to occur, inactive $Ru(II)OH_2^{2+}$ must dissociate before another active $Ru(IV)O^{2+}$ complex can bind. Both oxidation states must have approximately the same binding constant, because they have the same charge and their structures differ only by two protons. We know from the analysis of the first phase that dissociation (k_{off}) is slower than oxidation (k_1) or binding (k_{on}). Thus, the rate-limiting step becomes the dissociation of the reduced ruthenium complex so that another $Ru(IV)O^{2+}$ can bind. Because dissociation is a zero-order process, we would expect the rate to be concentration-independent, as is observed.

An intercalator, such as the dppz complex, would be expected to exhibit slower dissociation kinetics than the bpy complex, which binds

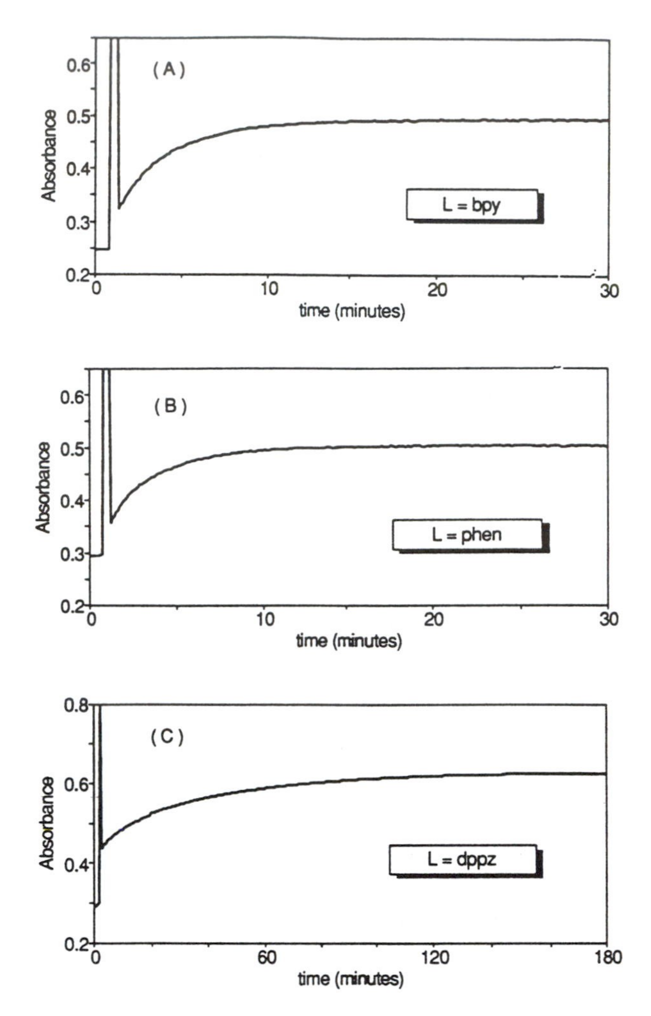

Figure 6. Absorbance vs. time curves for oxidation of DNA by Ru(IV)O^{2+} species in 50 mM phosphate buffer, pH 7, containing (A) 1.2 mM DNA and 0.12 mM Ru(tpy)(bpy)O^{2+}, (B) 1.1 mM DNA and 0.11 mM Ru(tpy)(phen)O^{2+}, and (C) 1.3 mM DNA and 0.12 mM Ru(tpy)(dppz)O^{2+}. (Reproduced from reference 30. Copyright 1993 American Chemical Society.)

electrostatically. Classical intercalators, such as ethidium bromide, have been shown to exhibit exchange kinetics approximately an order of magnitude slower than related surface binding molecules, because structural rearrangements of the DNA occur upon binding and dissociation (*38, 51*). Gross structural changes in the DNA are apparently not required for binding and dissociation of the phen complex, because its kinetics are similar to those of the bpy complex.

Table III. Kinetics Results for Oxidation of DNA by $Ru(tpy)(L)O^{2+}$

L	*[Ru(IV)]*[a]	*[DNA]*[b]	*R*[c]	$k_2 \times 10^3$[d]	$k_2' \times 10^3$[d]	*Burst Fraction*[e]
bpy	0.12	1.2	10	11 ± 6	3.2 ± 1.5	0.10 ± 0.03
	0.40	4.0	10	13 ± 6	5.2 ± 2.5	0.29 ± 0.03
	0.12	5.3	45	16 ± 7	5.5 ± 2.5	0.53 ± 0.05
phen	0.11	1.1	10	20 ± 3	5.2 ± 0.9	0.15 ± 0.04
	0.37	3.8	10	19 ± 7	5.3 ± 0.9	0.15 ± 0.09
	0.11	4.9	45	20 ± 10	6.5 ± 10	0.76 ± 0.03
dppz	0.12	1.3	11	2.4 ± 0.9	0.38 ± 0.12	0.47 ± 0.07

[a] Concentration in mM.
[b] Concentration in nucleotide phosphate, mM.
[c] [DNA-phosphate]/[Ru(IV)].
[d] Rate constants in s^{-1}.
[e] Contribution of the "burst" (k_1) to the overall decay.

The apparent biexponential kinetics for the k_2 phase must arise because of a variety of distinct binding sites that would be expected to exist in calf thymus DNA. The two rate constants then reflect the best fit for what is no doubt a complex ensemble of sequence- and structure-dependent binding sites, each with its own innate affinity for the metal complexes and reactivity toward the $Ru(IV)O^{2+}$ oxidant.

Cleavage Selectivity. There has been significant effort directed toward the selective cleavage of DNA by metal complexes (*4, 20, 34, 37*). Significant progress has been made in achieving selectivity by altering the binding specificity of the complex by appropriate tailoring of the ligands, leading to shape-selective cleavage (*20*). In contrast, little

$$Ru(IV)O^{2+} + DNA \underset{k_{off}}{\overset{k_{on}}{\rightleftharpoons}} Ru(IV)O^{2+}\bullet DNA \quad (K_B)$$

$$Ru(IV)O^{2+}\bullet DNA \xrightarrow{k_1} Ru(II)OH_2^{2+}\bullet DNA$$

$$Ru(II)OH_2^{2+}\bullet DNA \underset{k_{on}}{\overset{k_2\,(k_{off})}{\rightleftharpoons}} Ru(II)OH_2^{2+} + DNA \quad (K_B)$$

Scheme 3

effort has been directed toward achieving selectivity by altering the reactivity of the metal complex. Burrows and Rokita (*8*, *24*) have developed a nickel complex that, in the presence of persulfate, can oxidize guanine bases, leading to guanine-selective cleavage in single-stranded DNA. We have been interested in determining whether altering the reactivity of complexes that cleave DNA via sugar oxidation could lead to an increase in the specificity of oxidation of double-stranded DNA.

The results in Figure 5 show that oxidation of DNA by $Ru(tpy)(bpy)O^{2+}$ proceeds via abstraction of sugar hydrogen atoms, resulting in release of free nucleic acid bases. Because significant amounts of all four bases are released, $Ru(tpy)(bpy)O^{2+}$ apparently has no specificity with regard to the base to which the oxidized sugar is attached, at least via the pathway that leads to base release. Quantitation of the released bases shows that sugar oxidation accounts for 10% of the total ruthenium concentration. Identical results are observed for oxidation of calf thymus DNA by $Ru(bpy)_2(py)O^{2+}$. These complexes are known to undergo self-reduction, especially at relatively high concentration (*52*). The increased local concentration of the metal complexes upon binding to DNA may trigger this self-reduction, which could account for the remainder of the oxidizing equivalents. Self-inactivation is also known for Fe-bleomycin, and partitioning between cleavage and nonproductive reactions such as self-inactivation has been invoked to explain cleavage specificity by Fe-bleomycin (*12*).

We have studied the complex $Ru(bpy)_2(EtG)OH_2^{2+}$ (EtG = 9-ethylguanine), as a model for covalent binding reactions. This complex can be oxidized to the reactive Ru(IV)O form, $Ru(bpy)_2(EtG)O^{2+}$, at a potential of $E_{1/2}$(IV/II) = 0.43 V versus SSCE, which is 210 mV lower than that required for $Ru(tpy)(bpy)O^{2+}$. The $Ru(bpy)_2(EtG)O^{2+}$ complex is significantly less reactive than $Ru(tpy)(bpy)O^{2+}$; for example, $Ru(tpy)(bpy)O^{2+}$ is an efficient oxidant of 2-propanol(*9*) whereas solutions of $Ru(bpy)_2(EtG)O^{2+}$ are stable in the presence of 2-propanol. Analysis by HPLC of DNA solutions oxidized by $Ru(bpy)_2(EtG)O^{2+}$ provides the results shown in Figure 7 (*13*). Surprisingly, the only nucleic acid base released is thymine (*53*). Quantitation of the thymine released reveals a 12% efficiency for sugar oxidation. Thus, for both $Ru(bpy)_2(EtG)O^{2+}$ and $Ru(tpy)(bpy)O^{2+}$, the yield of sugar oxidation is the same, but only thymidine sugars are oxidized by $Ru(tpy)(bpy)O^{2+}$, whereas all four sugars are oxidized by $Ru(bpy)_2(EtG)O^{2+}$. Other oxidation pathways that do not lead to base release may or may not be specific; however, the base release pathway detected by our current HPLC analysis certainly is specific for thymidine sugars.

To rule out specific recognition by the ethylguanine ligand as the mechanism of thymidine specificity, the complex $Ru(bpy)_2(dmap)OH_2^{2+}$ was prepared (dmap = 4-dimethylaminopyridine). The spectral and

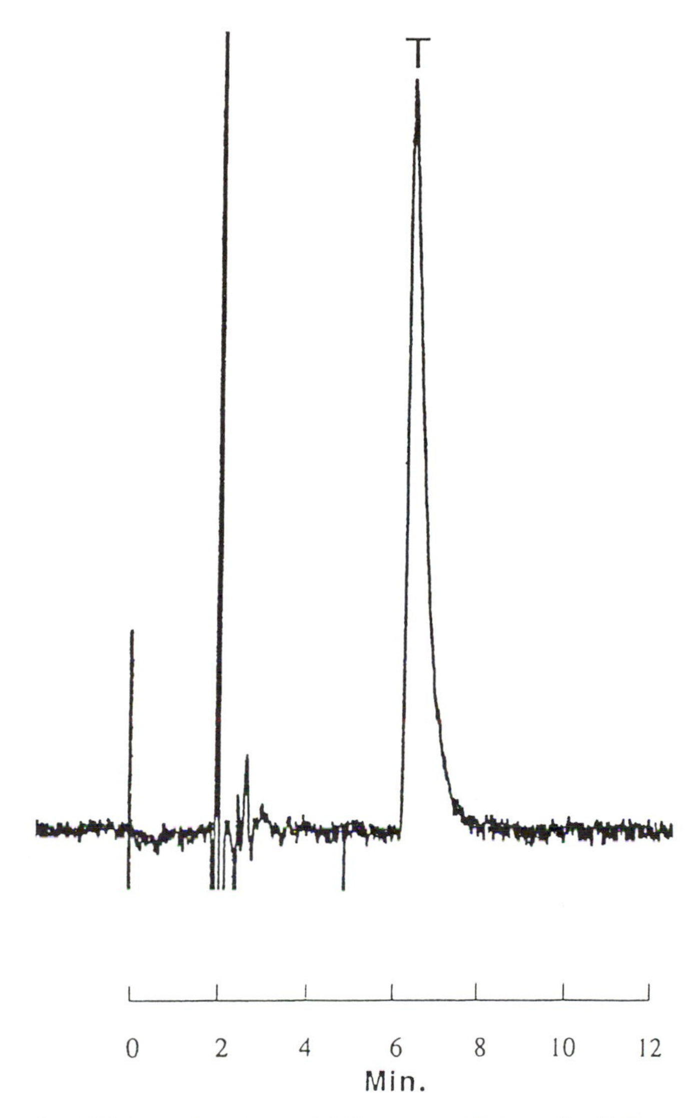

Figure 7. HPLC results obtained following oxidation of calf thymus DNA by $Ru(bpy)_2(EtG)O^{2+}$.

electrochemical properties are identical to those of the thymidine-specific complex $Ru(bpy)_2(EtG)O^{2+}$, but the structure differs from that of the nonspecific complex $Ru(bpy)_2(py)O^{2+}$ by only the addition of a single dimethylamino group, which would not be expected to exhibit special DNA-recognition properties. Oxidation reactions using $Ru(bpy)_2(dmap)O^{2+}$ lead to the release of only thymine, as seen with

$Ru(bpy)_2(EtG)O^{2+}$. No base release was detected upon reaction with DNA of $Os(tpy)(bpy)O^{2+}$ for which $E_{1/2}$(IV/III) = 0.41 V (*54*).

The results summarized in Table IV show strikingly that simple alteration of the reactivity of the M(IV)O oxidant dramatically alters the specificity of DNA damage with regard to sugar oxidation leading to base release. The binding affinity of $Ru(bpy)_2(EtG)OH_2{}^{2+}$ (K_B = 2300 M^{-1}) is modest and is between those of $Ru(tpy)(bpy)OH_2{}^{2+}$ (K_B = 660 M^{-1}) and $Ru(tpy)(phen)OH_2{}^{2+}$ (K_B = 3700 M^{-1}); oxidation of calf thymus DNA by $Ru(tpy)(phen)O^{2+}$($E_{1/2}$(IV/III) = 0.61 V) gives identical release of all four bases. Addition of a single dimethylamino group to the pyridine ligand of $Ru(bpy)_2(py)O^{2+}$ converts the complex from a nonspecific oxidant to a specific oxidant. Thus, it seems certain that the lower oxidation potential is the source of the specificity. We are exploring two mechanisms for the thymidine specificity. First, the sugar oxidation may occur at the 1′-position (*6*, *16*, *25*), at which the reactivity would be strongly influenced by the nature of the coordinated base. Second, the structure of thymidine sugars may permit close approach of the oxo group to sugar hydrogens, and the low oxidation potential may necessitate a close approach in order for oxidation to occur.

The kinetics of oxidation by complexes that exhibit specificity should be different from the kinetics of oxidation by $Ru(tpy)(bpy)O^{2+}$. What is desired for specificity is a model in which dissociation (k_{off}, Scheme 3) can compete with oxidation (k_1). Because the selective complexes are weaker oxidants, it follows that k_1 would be slower, allowing k_{off} to compete with oxidation, leading to the standard saturation (Michaelis-Menton) kinetics model (*55*). The ability of dissociation to compete with oxidation demonstrates that binding can occur without oxidation in the case of $Ru(bpy)_2(EtG)O^{2+}$. In the $Ru(tpy)(bpy)O^{2+}$ case, the metal complex is reduced every time binding occurs. Clearly, the chances of selective oxidation are much higher when every binding event does not necessarily lead to DNA oxidation. Thus, only binding sites that are particularly reactive are prone to oxidation.

Conclusions. These studies show how a thorough understanding of the fundamentals of the oxidation process can lead to new DNA ox-

Table IV. Influence of Redox Potential on Base Release from DNA Oxidation by $M(IV)O^{2+}$ Complexes

Complex	$E_{1/2}$ *V vs. SSCE*	*Bases Released*
$Ru(tpy)(bpy)O^{2+}$	0.64	A, T, G, C
$Ru(bpy)_2(py)O^{2+}$	0.52	A, T, G, C
$Ru(bpy)_2(EtG)O^{2+}$	0.43	T
$Ru(bpy)_2(dmap)O^{2+}$	0.43	T
$Os(tpy)(bpy)O^{2+}$	0.41	none

idants with special properties. We discussed a new method that allows binding affinities of metal complexes to be quantitated with unprecedented precision (*56*). A special property of the oxoruthenium(IV) system is the ability to study the kinetics using optical spectroscopy. By using the binding affinities and a knowledge of the binding mode afforded by polyelectrolyte theory and other experiments, the complete kinetic model for DNA oxidation can be developed. From this kinetic model, the prediction can be made that the $Ru(tpy)(bpy)O^{2+}$ complexes are too reactive to be specific, because every time the complex binds to DNA, oxidation occurs. Indeed, less reactive complexes show a surprising specificity for oxidation of thymidine sugars, at least via the pathway that leads to base release. Preliminary studies of the kinetics of DNA oxidation by these complexes are consistent with a model in which dissociation can compete with oxidation, leading to a greater chance of specific oxidation.

Acknowledgments

H. Holden Thorp thanks the National Science Foundation, David and Lucile Packard Foundation, Camille and Henry Dreyfus Foundation, North Carolina Biotechnology Center, and the Burroughs-Wellcome Company for financial support.

References

1. Barton, J. K. *Science (Washington, D.C.)* **1986,** *233,* 727.
2. Hecht, S. M. *Acc. Chem. Res.* **1986,** *19,* 83.
3. Burkhoff, A. M.; Tullius, T. D. *Nature (London)* **1988,** *331,* 455.
4. Barton, J. K.; Pyle, A. M. *Prog. Inorg. Chem.* **1990,** *38,* 413.
5. Fleisher, M. B.; Waterman, K. C.; Turro N. J.; Barton J. K. *Inorg. Chem.* **1986,** *25,* 3549.
6. Stubbe, J.; Kozarich, J. W. *Chem. Rev.* **1987,** *87,* 1107.
7. Grover, N.; Thorp, H. H. *J. Am. Chem. Soc.* **1991,** *113,* 7030.
8. Chen, X.; Burrows, C. J.; Rokita, S. E. *J. Am. Chem. Soc.* **1991,** *113,* 5884.
9. Thompson, M. S.; Meyer, T. J. *J. Am. Chem. Soc.* **1982,** *104,* 4106.
10. Meyer, T. J. *J. Electrochem. Soc.* **1984,** *131,* 221C.
11. Thompson, M. S.; DeGiovani, W. F.; Moyer, B. A.; Meyer, T. J. *J. Org. Chem.* **1984,** *49,* 4972.
12. Worth, L., Jr.; Frank, B. L.; Christner, D. F.; Absalon, M. J.; Stubbe, J.; Kozarich, J. W. *Biochemistry* **1993,** *32,* 2601.
13. Tullius, T. D. *Trends Biochem. Sci.* **1987,** *12,* 297.
14. Sugiyama, H.; Xu, C.; Murugesan, N.; Hecht, S. M.; van der Marel, G. A.; van Boom, J. H. *Biochemistry* **1988,** *27,* 58.
15. Sam, J. W.; Peisach, J. *Biochemistry* **1993,** *32,* 4698.
16. Goyne, T. E.; Sigman, D. S. *J. Am. Chem. Soc.* **1987,** *109,* 2846.
17. Sigman, D. S.; Bruice, T. W.; Mazumder, A.; Sutton, C. L. *Acc. Chem. Res.* **1993,** *26,* 98.
18. Mei, H.-Y.; Barton, J. K. *Proc. Natl. Acad. Sci. U.S.A.* **1988,** *85,* 1339.

19. Nielsen, P. E.; Hiort, C.; Sönnichsen, S. H.; Burhardt, O.; Dahl, O.; Nordèn, B. *J. Am. Chem. Soc.* **1992,** *114,* 4967.
20. Pyle, A. M.; Morii, T.; Barton, J. K. *J. Am. Chem. Soc.* **1990,** *112,* 9432.
21. Sitlani, A.; Long, E. C.; Pyle, A. M.; Barton, J. K. *J. Am. Chem. Soc.* **1992,** *114,* 2303.
22. Kalsbeck, W. A.; Grover, N.; Thorp, H. H. *Angew. Chem. Int. Ed. Engl.* **1991,** *30,* 1517.
23. Roundhill, D. M.; Gray, H. B.; Che, C.-M. *Acc. Chem. Res.* **1989,** *22,* 55.
24. Chen, X.; Burrows, C. J.; Rokita, S. E. *J. Am. Chem. Soc.* **1992,** *114,* 322.
25. Duff, R. J.; de Vroom, E.; Geluk, A.; Hecht, S. M.; van der Marel, G. A.; van Boom, J. H. *J. Am. Chem. Soc.* **1993,** *115,* 3350.
26. King, P. A.; Anderson, V. E.; Edwards, J. O.; Gustafson, G.; Plumb, R. C.; Suggs, J. W. *J. Am. Chem. Soc.* **1992,** *114,* 5430.
27. Hamamichi, N.; Natrajan, A.; Hecht, S. M. *J. Am. Chem. Soc.* **1992,** *114,* 6278.
28. Pyle, A. M.; Rehmann, J. P.; Meshoyrer, R.; Kumar, C. V.; Turro, N. J.; Barton, J. K. *J. Am. Chem. Soc.* **1989,** *111,* 3051.
29. Satyanarayana, S.; Dabrowiak, J. C.; Chaires, J. B. *Biochemistry* **1992,** *31,* 9319.
30. Neyhart, G. A.; Grover, N.; Smith, S. R.; Kalsbeck, W. A.; Fairley, T. A.; Cory, M.; Thorp, H. H. *J. Am. Chem. Soc.* **1993,** *115,* 4423.
31. Hiort, C.; Lincoln, P.; Nordén, B. *J. Am. Chem. Soc.* **1993,** *115,* 3448.
32. Friedman, A. E.; Chambron, J. C.; Sauvage, J. P.; Turro, N. J.; Barton, J. K. *J. Am. Chem. Soc.* **1990,** *112,* 4960.
33. Gupta, N.; Grover, N.; Neyhart, G. A.; Liang, W.; Singh, P.; Thorp, H. H. *Angew. Chem. Int. Ed. Engl.* **1992,** *31,* 1048.
34. David, S. S.; Barton, J. K. *J. Am. Chem. Soc.* **1993,** *115,* 2984.
35. Kumar, C. V.; Barton, J. K.; Turro, N. J. *J. Am. Chem. Soc.* **1985,** *105,* 5518.
36. Barton, J. K.; Danishefsky, A. T.; Goldberg, J. M.; Kumar, C. V.; Turro, N. J. *J. Am. Chem. Soc.* **1986,** *108,* 2081.
37. Krotz, A. H.; Kuo, L. Y.; Shields, T. P.; Barton, J. K. *J. Am. Chem. Soc.* **1993,** *115,* 3877.
38. Eriksson, M.; Leijon, M.; Hiort, C.; Nordén, B.; Graslund, A. *J. Am. Chem. Soc.* **1992,** *114,* 4933.
39. Gupta, N.; Grover, N.; Neyhart, G. A.; Singh, P.; Thorp, H. H. *Inorg. Chem.* **1993,** *32,* 310.
40. Waring, M. J. *J. Mol. Biol.* **1965,** *13,* 269.
41. Marzilli, L. G.; Petho, G.; Lin, M.; Kim, M. S.; Dixon, D. W. *J. Am. Chem. Soc.* **1992,** *114,* 7575.
42. Smith, S. R.; Neyhart, G. A.; Kalsbeck, W. A.; Thorp, H. H. *New J. Chem.* **1994,** *18,* 397.
43. Barton, J. K.; Raphael, A. L. *J. Am. Chem. Soc.* **1984,** *106,* 2466.
44. Cory, M.; McKee, D. D.; Kagan, J.; Henry, D. W.; Miller, J. A. *J. Am. Chem. Soc.* **1985,** *107,* 2528.
45. Kelly, J. M.; Tossi, A. B.; McConnell, D. J.; OhUigin, C. *Nucleic Acids Res.* **1985,** *13,* 6017.
46. Rehmann, J. P.; Barton, J. K. *Biochemistry* **1990,** *29,* 1701.
47. Rehmann, J. P.; Barton, J. K. *Biochemistry* **1990,** *29,* 1710.
48. Frederick, C. A.; Williams, L. D.; Uhgetto, G.; Van Der Marel, G. A.; Van Boom, J. H.; Rich, A. *Biochemistry* **1990,** *29,* 2538.
49. Record, M. T., Jr.; Anderson, C. F.; Lohman, T. M. *Q. Rev. Biophys.* **1978,** *11,* 103.

50. Grover, N.; Gupta, N.; Singh, P.; Thorp, H. H. *Inorg. Chem.* **1992,** *31,* 2014.
51. Leupin, W.; Feigon, J.; Denny, W. A.; Kearns, D. R. *Biophys. Chem.* **1985,** *22,* 299.
52. Roecker, L.; Kutner, W.; Gilbert, J. A.; Simmons, M.; Murray, R. W.; Meyer, T. J. *Inorg. Chem.* **1985,** *24,* 3784.
53. Welch, T. W.; Neyhart, G. A.; Ciftan, S. A.; Goll, J. G.; Thorp, H. H. *J. Am. Chem. Soc.* **1993,** *115,* 9311.
54. Takeuchi, K. J.; Thompson, M. S.; Pipes, D. W.; Meyer, T. J. *Inorg. Chem.* **1984,** *23,* 1845.
55. Fersht, A. *Enzyme Structure and Mechanism;* W. H. Freeman: New York, 1983.
56. Kalsbeck, W. A.; Thorp, H. H. *J. Am. Chem. Soc.* **1993,** *115,* 7146.

RECEIVED for review July 19, 1993. ACCEPTED revised manuscript December 21, 1993.

16

Metal Ion Macrocyclic Complexes as Artificial Ribonucleases

Janet R. Morrow, Kimberly A. Kolasa, Shahid Amin, and K. O. Aileen Chin

State University of New York at Buffalo, Department of Chemistry, Buffalo, NY 14214

Macrocyclic ligands that form thermodynamically stable or kinetically inert complexes with metal ions are used in the design of metal complexes that behave as artificial ribonucleases. Artificial ribonucleases are compounds that cleave RNA by transesterification of the phosphate diester linkages. Zn(II) complexes of tetraaza and triaza macrocycles promote RNA cleavage, but cleavage rates are modest. Hexadentate Schiff base macrocyclic complexes of the trivalent lanthanides promote rapid cleavage of RNA oligomers. New ligands that encapsulate the trivalent lanthanides have been constructed from the addition of four pendent ligating groups to the 1,4,7,10-tetraazacyclododecane macrocycle. Several of the new lanthanide complexes are resistant to metal ion release and show promise as artificial ribonucleases. The mechanism of RNA cleavage by metal complexes and the effect of RNA structure on cleavage are discussed.

THE DESIGN OF INORGANIC COMPOUNDS that may be useful as therapeutic or diagnostic agents is a topic of great interest in bioinorganic chemistry. Because free metal ions may be highly toxic, many pharmaceutical applications require the use of strong chelates. Macrocyclic ligands are useful where strong chelates are required. Pharmaceutical applications where macrocycles have been used include the design of magnetic resonance imaging agents (*1–3*) and the construction of metal complex–antibody conjugates for use as new radiopharmaceuticals (*4–6*). The utility of macrocyclic ligands in controlling the reactivity of metal ions and in forming highly stable metal complexes make them ideal for use in the design of new metallodrugs.

0065–2393/95/0246–0431$08.00/0

The use of macrocycles as ligands may be an important strategy in the design of metal complexes for RNA cleavage (artificial ribonucleases). Inertness to metal ion release under physiological conditions and a high degree of thermodynamic stability are important properties that must be considered if metal ions are to be used for cleavage. A motivation for the study of metal complexes that catalyze RNA cleavage lies in the development of sequence-specific cleaving agents for RNA. The attachment of a complex that catalyzes RNA cleavage to an oligonucleotide that is complementary in sequence to a message RNA (an antisense oligonucleotide) may be one method to promote sequence-specific cleavage of RNA. Antisense oligonucleotides bearing cleaving groups may be more effective in promoting translation arrest than are other types of antisense oligonucleotides (*7*).

The RNA cleavage reaction of interest involves transesterification of the phosphate esters of RNA and is analogous to the first step of the reaction catalyzed by enzymes such as RNase A. Small molecules that catalyze this type of cleavage reaction are more likely to be useful for therapeutic applications because the reaction is specific for RNA. However, one of the difficulties in the use of molecules that catalyze RNA cleavage by transesterification is the slow rate of the reaction. For example, slow rates of cleavage are observed for cleaving agents that contain no metal ions including phosphate ester receptors (*8*), polyamines (*9*), or peptides (*10, 11*). In contrast, metal complexes have been shown to promote RNA cleavage rapidly at 37 °C and neutral pH (*12–14*). In this chapter we discuss the design of macrocyclic complexes of metal ions for RNA cleavage and the mechanism of metal ion promoted RNA cleavage.

Experimental Details

The free base form of cyclen(1,4,7,10-tetraazacyclododecane) was generated by passing the tetrahydrochloride salt (Parish Chemicals or Strem Chemicals) through a Dowex 1X8-200 anion exchange column (30 × 2.5 cm, hydroxide form). Acetonitrile was dried over CaH_2. Reagent grade absolute alcohols were employed. Milli-Q purified water was used for kinetic experiments. The ligand 1,4,7,10-tetrakis(2-hydroxyethyl)-1,4,7,10-tetraazacyclododecane (THED) was synthesized as reported previously (*15*). The 4-nitrophenylphosphate ester of propylene glycol was synthesized (*16*), and kinetics studies were performed as described previously (*17*).

An Orion research digital ion analyzer 510 equipped with a temperature compensation probe was used for all pH measurements. A Hewlett-Packard diode array 8452A spectrophotometer with a thermostatted cell compartment was employed for UV-vis spectra and for kinetic measurements. All ^{1}H,^{13}C NMR spectra were recorded by use of a Varian 400 XL spectrophotometer. Elemental analyses were performed by E and R Microanalytical Laboratories. A VG 70-SE mass spectrometer with fast atom bombardment was used.

1,4,7,10-tetrakis(2-carbamoylethyl)-1,4,7,10-tetraazacyclododecane (TCEC). Cyclen (0.200 g, 1.16 mmol) and acrylamide (0.397 g, 5.58 mmol) were dissolved in methanol (2 mL) and the solution was heated under a nitrogen atmosphere in an oil bath maintained at 75 °C for 50 h. A few drops of methanol were added periodically to redissolve the gelatinous mixture over the course of the reaction. The solution was cooled and the gelatinous solid was dissolved in 1 mL of methanol. Diethylether (1 mL) was added dropwise. A crystalline solid formed after several hours at room temperature. The solid was collected by filtration, washed with chloroform, and dried in vacuo at 50 °C: yield, 87%; melting point, 174–177 °C; fast atom bombardment mass spectroscopy (FABMS) m/e: 457 (ligand + H).

1,4,7,10-tetrakis(carbamoylmethyl)-1,4,7,10-tetraazacyclododecane (TCMC). Cyclen (500 mg), bromoacetamide (1.8095 g), and triethylamine (2 mL) were refluxed in anhydrous ethanol (35 mL) for 4 h. The white solid that precipitated from solution was recrystallized from an ethanol/water mixture: yield, 61%. Analysis calculated for $C_{16}H_{32}N_8O_4$: C, 48.00; H, 8.00; N, 28.00. Found C, 47.86; H, 7.90; N, 27.83. FABMS m/e: 401.3 (TCMC ligand + H^+).

Lanthanide Complexes. Typically, $Ln(SO_3CF_3)_3$ (0.55 mmol, Ln = Eu or La) was refluxed under nitrogen in a mixture of 60 mL of dry acetonitrile and 6.5 mL of trimethylorthoformate. Alternately, ethanol was used as a solvent. Upon dissolution of the salt, the ligand (TCMC, THED, or TCEC) (0.55 mmol) was added in a minimum amount of anhydrous methanol or ethanol. The mixture was refluxed for 3.5 h. The solution was concentrated in vacuo and methylene chloride or hexanes was added until cloudiness was observed. Yields were typically ≥35%. Analytical data (C,H,N analysis) for the complexes were satisfactory. $La(TCEC)(CF_3SO_3)_3(CH_3CN)$: FABMS m/e: 893 (complex-CF_3SO_3). $La(TCMC)(CF_3SO_3)_3(CH_3CH_2OH)$: FABMS m/e: 837.0 (complex-CF_3SO_3). $Eu(TCMC)(CF_3SO_3)_3$: analytical data satisfactory. $Eu(THED)(CF_3SO_3)_3$: Analysis calculated for $C_{19}H_{36}N_4O_{13}F_9Eu$: C, 24.10; H, 3.80; N, 5.91. Found C, 23.82; H, 3.82; N, 5.79. FABMS m/e: 798. $La(THED)(CF_3SO_3)_3$: FABMS m/e 785 (complex-CF_3SO_3).

Kinetics. The rate of dissociation of La^{3+} or Eu^{3+} from the macrocyclic complex was monitored at 37 °C in the presence of Cu^{2+} by following the increase in absorbance characteristic of the Cu(II) macrocyclic complex. Beer's law plots with varying concentrations of the Cu(II) complex (0.100–1.00 mM) gave an extinction coefficient of 6830 M^{-1} cm^{-1} (312 nm, THED), 6,500 M^{-1} cm^{-1} (304 nm, TCEC), and 5500 M^{-1} cm^{-1} (312 nm, TCMC). Solutions for experiments to monitor the rate of dissociation of Ln^{3+} from the macrocyclic complex contained the complex (0.1 mM), 10 mM Mes buffer, pH 6.0, with different concentrations of $CuCl_2$. The Cu^{2+} concentration was varied from 0.1 to 1.0 mM for certain experiments. In most experiments the Cu^{2+} concentration was maintained in 10-fold excess to complex.

Transfer-RNA Experiments. Transfer RNA^{phe} was purchased from Sigma and used as received. The RNA was 3′-end labeled with cytidine bisphosphate by use of T_4 RNA ligase as described previously (18). The labeled tRNA was purified by polyacrylamide gel electrophoresis. The bands were cut out and the RNA was eluted at room temperature overnight and recovered by ethanol precipitation. All standard precautions were taken to avoid ribonuclease contamination. All solutions were autoclaved and gloves were worn for all manipulations. Recrystallization of the europium complex $[Eu(L^1)](OAc)_2Cl$ from chloroform was performed multiple times. This had no effect on the europium promoted RNA cleavage reactions.

Complexes of the Transition Metals and Zn(II)

Several complexes of the transition metals and Zn(II) promote transesterification of RNA (*12, 13, 19, 20*). A few of these complexes have macrocyclic ligands. $Zn(CR)^{2+}$, Zn-*N*-methyl-$(CR)^{2+}$, $Zn(cyclam)^{2+}$, $Zn([12]aneN_3)^{2+}$, and $Zn([9]aneN_3)^{2+}$ promote RNA transesterification although rates are modest (CR = 2,12-dimethyl-3,7,11,17-tetraazabicyclo[11.3.1]heptadeca-1(17)2,11,13,15-pentaene; *N*-methyl-CR = 7-(*N*-methyl)-2,12-dimethyl-3,7,11,17-tetraazabicyclo[11.3.1]heptadeca-1(17)2,11,13,15-pentaene; $[12]aneN_3$ = 1,5,9-triazacyclododecane; $[9]aneN_3$ = 1,4,7-triazacyclononane). $Zn(CR)^{2+}$ (0.160 mM) promotes 70% cleavage of the RNA oligomer A_{12}–A_{18} over a 20-h period (*12*). Dinucleotide cleavage is accelerated by $Zn(cyclam)^{2+}$, $Zn([9]aneN_3)^{2+}$, and $Zn([12]aneN_3)^{2+}$ at 64 °C. Cleavage was too slow to observe at 37 °C (*19*). Cleavage of dinucleotides with metal complexes is generally much more difficult than is cleavage of longer oligomers of RNA.

Strongly chelating ligands may serve to maintain the metal ion in solution in an active form. The initial rate of cleavage of dinucleotides by $Zn([9]aneN_3)^{2+}$ is slower than cleavage by $Zn(NO_3)_2$. However, over time a greater extent of cleavage is observed with the metal complex than with the Zn(II) salt. The metal complex showed catalytic turnover, whereas the Zn(II) salt precipitates from solution (*19*). For a Zn(II) complex, the addition of nitrogen donor ligands may reduce the number of coordination sites for catalysis but may favorably modify the Lewis acidity of the Zn(II) center. Using the pK_a of the metal-bound water of the Zn(II) complexes as a measure of their Lewis acidity, one would predict that the triazamacrocyclic complexes would promote RNA cleavage more rapidly than would $Zn(cyclam)^{2+}$ (*19*). The following order of catalytic efficiency is observed for dinucleotide cleavage: $(Zn([9]aneN_3)^{2+} \approx (Zn[12]aneN_3)^{2+} > Zn(cyclam)^{2+}$.

For transition metals and Zn(II), how might metal complexes be designed to promote rapid cleavage of RNA? One approach is the functionalization of ligands to participate in catalysis. The *N*-methyl-CR ligand was modified to contain a basic group for bifunctional catalysis (*21*). A Zn(II) complex of one of the modified macrocycles was shown to accelerate cyclization of the RNA model substrate **1** (**1** = 4-nitrophenylphosphate ester of propylene glycol) 20-fold more rapidly than the Zn(II) complex of *N*-methyl-CR.

$$\mathbf{1} \longrightarrow \text{cyclic phosphate} + 4\text{-}NO_2PhO^- + H^+$$

Trivalent Lanthanides in RNA Cleavage

Several kinetic studies of phosphate diester transesterification by metal ions indicate that lanthanide ions are among the most efficient promoters (*17, 22, 23*). The trivalent lanthanides are good Lewis acids and have flexible coordination geometries and a high cationic charge. Inasmuch as lanthanide hydroxides form readily at near neutral pH, the overall charge of the complex will decrease. A metal hydroxide ligand, however, may serve as a general base catalyst. The lanthanides are considered to be oxophilic metal ions and bind well to phosphate diesters. Oxophilicity is an important property for an artificial nuclease, as it is desirable that the metal ion bind to the phosphate ester in preference to binding a nitrogenous base of RNA. Thus, strong coordination of a metal complex to one or more of the nitrogenous bases of RNA may inhibit metal complex promoted transesterification.

Ligands for the lanthanides must strongly chelate lanthanide ions but not inactivate them as catalysts. Coordination sites must be available for catalysis and the metal ion should retain a high degree of Lewis acidity. An overall positive charge on the complex may aid in catalysis, as discussed in subsequent paragraphs. Macrocyclic complexes of the lanthanides abound (*24*). Few lanthanide macrocyclic complexes, however, are inert to metal ion release in water. For example, perhaps the most well-known class of lanthanide macrocyclic compounds are the crown ethers. Crown ether complexes are synthesized under anhydrous conditions and are known to hydrolyze in water.

When we began to study lanthanide macrocyclic complexes for RNA cleavage, we began our search with complexes that were effective for another biomedical application: magnetic resonance imaging (MRI). MRI agents must be stable under physiological conditions and must also have at least one coordination site available for coordination to water (*1*). Properties of artificial ribonucleases may be similar. Lanthanide complexes that will efficiently promote RNA cleavage will require available coordination sites for catalysis and must be inert to metal ion release or have large formation constants.

Hexadentate Schiff Base Macrocyclic Complexes

The hexadentate Schiff base complexes [$Ln(L^1)^{3+}$, $Ln(L^2)^{3+}$] have been developed by several groups (*2, 25, 26*). Applications for these complexes include their use as fluorescent agents and as MRI agents. The neutral Schiff base ligand imparts a +3 charge to the complex at neutral pH (*27*). Six-coordination sites are occupied by the nitrogen donors, leaving three to four coordination sites for water or counter ions. In the solid state, the macrocycle is not planar, but is best described as bowl- or butterfly-shaped. ^{13}C and ^{1}H NMR studies indicate a higher degree of symmetry in solution. Luminescence decay measurements in H_2O and in D_2O indicate approximately three bound water molecules for $Eu(L^1)^{3+}$ at neutral pH (*28*).

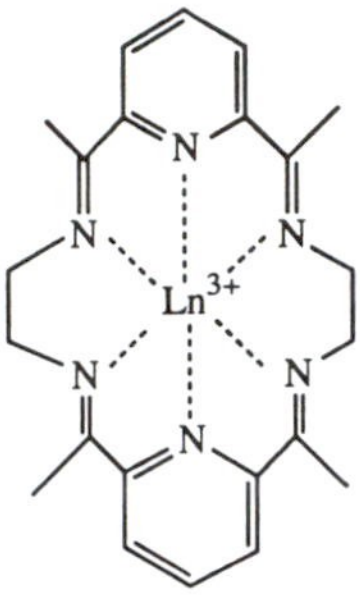

$Ln(L^1)^{+3}$ 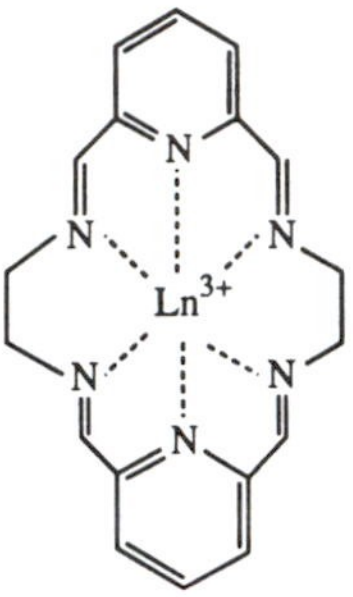 $Ln(L^2)^{+3}$

Several lanthanide complexes of L^1 promote rapid cleavage of oligomers or dinucleotides of RNA (*14*). The Gd(III), Eu(III), Tb(III), and La(III) complexes all promote greater than 70% cleavage of A_{12}–A_{18} after 4 h at 37 °C, pH 7.00. Pseudo-first-order rate constants for the cleavage of ApUp by 0.490 mM $Eu(L^1)^{3+}$ or of A_{12}–A_{18} by 0.160 mM $Eu(L^1)^{3+}$ are 0.14 and 1.5 h^{-1}, respectively. Catalytic turnover is observed for the cleavage of a dinucleotide in the presence of the europium complex.

RNA Structure

Most substrates examined to date are flexible fragments of RNA. How might RNA structure modulate the rate of RNA cleavage by artificial ribonucleases? Certainly in the cleavage of transfer RNA (tRNA) by metal ions, RNA structure has a dramatic effect on the site of cleavage (*29*). RNA structure dictates the highly specific cleavage observed in self-cleaving RNAs that require metal ions (*30*). For metal complexes there is little information on whether transesterification catalysts that readily cleave single-stranded RNA are able to cleave RNA with a large degree of secondary and tertiary structure. In addition, before a metal complex is attached to an oligodeoxynucleotide for sequence-specific RNA cleavage, it would be useful to know whether the metal complex is able to cleave a DNA–RNA hybrid. Attachment of the metal complex cleaver to the middle of the oligodeoxynucleotide would necessitate cleavage of a double-stranded DNA–RNA hybrid. Information on the effect of RNA structure on cleavage rates may have a significant effect on the design of oligonucleotide–artificial ribonuclease conjugates.

To probe the effect of RNA structure on cleavage by $Eu(L^1)^{3+}$, $tRNA^{phe}$ (yeast) 3′-end labeled with a ^{32}P label was incubated with the europium complex. Fragments from cleavage reactions were resolved by high-resolution gel electrophoresis as shown in Figure 1. Cleavage of $tRNA^{phe}$ by $Eu(L^1)^{3+}$ occurred at sites that are quite distinct from those observed for $Eu(CH_3CO_2)_3$ (*31, 32*) as shown in a recent publication (*33*). $La(L^1)^{3+}$ gave an identical pattern (data not shown). Bands co-migrated with those produced by alkaline hydrolysis or by digestion with RNase T_1. Over long time periods, the $Eu(L^1)^{3+}$ complex induced cleavage at nearly every nucleotide to produce a ladder of cleavage sites.

A 20-base oligodeoxynucleotide complementary to A_{38} to G_{57} of $tRNA^{phe}$ was annealed to t-RNA^{phe} by heating the oligodeoxynucleotide with the tRNA to 65 °C followed by cooling to 0 °C. Sites in the RNA sequence complementary to the DNA strand were protected from cleavage by $Eu(L^1)^{3+}$ with the exception of sites at the ends of the hybrid where fraying probably occurs. Longer incubation times led to cleavage at nearly every nucleotide of the tRNA with the exception of those protected by the oligodeoxynucleotide. The sequence is as follows:

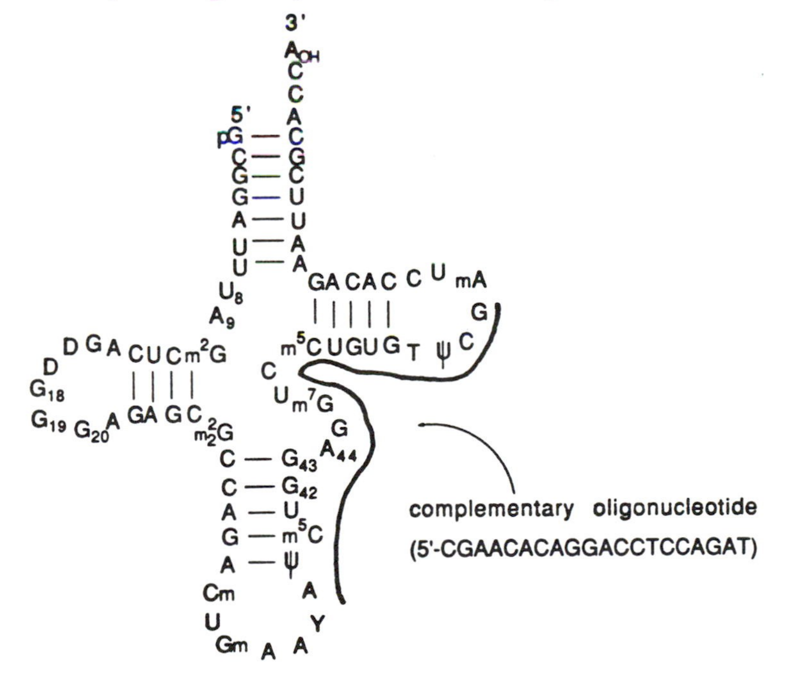

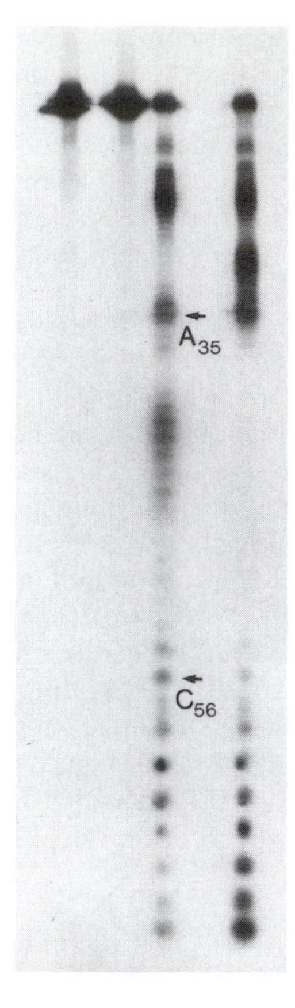

Figure 1. Autoradiogram showing the effect of annealing a complementary oligodeoxynucleotide (complementary to RNA bases A_{38} to G_{57}) to $tRNA^{phe}$ followed by treatment with $Eu(L^1)^{3+}$. L^1 is 2,7,13,18-tetramethyl-3,6,14,17,-23,24-hexaazatricyclo[17.3.1.1]tetracosa-1(23),2,6,8,10,12(24),13,17,19,-21-decane). Autoradiograms are of 8 M urea denaturing polyacrylamide sequencing gels of $tRNA^{phe}$ labeled with ^{32}P at the 3′ end. Approximately 1×10^5 cpm of labeled $tRNA^{phe}$ was loaded onto the gel for each sample. Cold $tRNA^{phe}$ was added to give a total concentration of tRNA of 20 μM (1.3 mM nucleotide). Metal complex concentrations were 1 mM and HEPES buffer was 0.4 M. Reactions were run at pH 7.86, 37 °C for the times indicated. Lane 1: control, 5 h; lane 2: control and oligonucleotide (20 μM), 5 h; lane 3: $Eu(L^1)^{3+}$, 5 h; and lane 4: $Eu(L^1)^{3+}$ and oligonucleotide (20 μM), 5 h. (Reproduced from reference 33. Copyright 1993 American Chemical Society.)

These results indicate that RNA in a DNA–RNA hybrid is protected from cleavage by $Eu(L^1)^{3+}$ under conditions where nearly all other sites in $tRNA^{phe}$ are cleaved. We cannot rule out poorer binding of the metal complex to the hybrid than to RNA alone. There is ample precedence for binding of Eu(III) ions to double-stranded nucleic acids (*34*). Eu(III) is also known to bind well to pockets in highly structured RNAs such as tRNA (*31*). It is not known how the europium complex will bind to these different structures.

Earlier studies suggest that the conformation of RNA will probably play a major role in phosphate ester transesterification reactions. The flexibility of RNA in the DNA–RNA hybrid will be limited in comparison to single-stranded RNA. Studies have shown that RNA cleavage with ethylene diamine as a catalyst is inhibited for RNA in a triple helix relative to single-stranded RNA (*35*). More recent work (*36*) demonstrated that polyvinylpyrrolydone promotes hydrolysis of single-stranded oligoribonucleotides but not those in double-stranded form. One explanation for this behavior is based on the orientation of the 3′-hydroxyl group and the phosphate diester in an RNA double helix; nucleophilic attack of the 3′-hydroxyl and displacement of the 5′-hydroxyl cannot occur by an in-line displacement mechanism (*37*). Most DNA–RNA double helices are structurally similar to RNA–RNA double helices (*38*) and geometric constraints may also be similar. Further study is underway in the author's laboratory to probe the effect of structure on cleavage.

Decomposition of Lanthanide(III) Schiff Base Macrocycles

The macrocyclic ligand in the Schiff base macrocyclic complexes $Ln(L^1)^{3+}$ slowly hydrolyzes in water. Decomposition products of the lanthanide(III) complexes of L^1 and L^2 macrocycles as determined by use of ^1H NMR are pyridine dicarboxaldehyde or diacetylpyridine, respectively, and ethylenediamine. For the $Ln(L^1)^{3+}$ complexes, it is not known whether the coordinated macrocycle is initially hydrolyzed, followed by metal ion dissociation, or whether dissociation of the lanthanide ion is followed by hydrolysis of the free macrocycle. The first pathway, hydrolysis of the Schiff base complex, may occur through the formation of a carbinolamine, followed by expulsion of the amine and cleavage of the C–N bond; carbinolamine complexes are intermediates in the hydrolysis of imine bonds for certain Schiff base ligands (*39*). Carbinolamine complexes (*40*) of L^2 have been reported for all lanthanides from Nd(III) to Lu(III) [with the exception of Eu(III)]. The second pathway, dissociation of the lanthanide followed by hydrolysis of the free ligand, has not been ruled out. Testing this possibility is made more difficult by the fact that the free L^1 ligand has not been isolated. Ligands such as L^1 are generally susceptible to hydrolysis and difficult to isolate. However, if it were to occur, dissociation of Ln^{3+} from the macrocycle is probably not reversible. Incubation of $La(L^1)^{3+}$ with excess Ce^{3+} or Y^{3+} for several hours does not result in incorporation of the Ce^{3+} or Y^{3+} label into the macrocycle (*26, 28*).

The L^1 complexes of the middle lanthanides Gd(III), Eu(III), and Tb(III) decompose less rapidly at pH 7.4, 37 °C than do the L^1 complexes of La(III) or Lu(III) (*14*). The fit of the lanthanide ion into the macrocycle may be important here. Certainly, the macrocycle fit will vary for La^{3+} (116 pM) compared to Lu^{3+} (97.7 pM) (*41*). A recent study using luminescence measurements suggests a greater lability of the $Eu(L^1)^{3+}$ complex than previously reported (*28*). Detection of the $Eu(DPTA)^-$ complex produced upon addition of diethylenetriaminepentaacetic acid (DTPA) to $Eu(L^1)^{3+}$ indicates that the complex decomposes approximately 12% in 48 h at 37 °C, pH 7.4. It is noteworthy that solutions of $Eu(L^1)^{3+}$ contain two different species (*28*). One of them, possibly a hydroxy-bridged dimer, is present in greater amounts at high concentrations of $Eu(L^1)^{3+}$.

Encapsulated Lanthanide Ions

The most thermodynamically stable and kinetically inert complexes of the trivalent lanthanides are those of the ligand DOTA (1,4,7,10-tetraazacyclododecane-1,4,7,10-tetraacetate) (*42, 43*). Our search for lanthanide macrocyclic complexes that would remain intact for longer time periods led us to examine derivatives of DOTA. There are two potential difficulties with the use of DOTA complexes of the trivalent lanthanides for RNA cleavage. First, the overall negative charge on the complex is not conducive to anion binding; for example, $Gd(DOTA)^-$ does not bind hydroxide well (*44*). Second, DOTA complexes of the middle lanthanides Eu(III) and Gd(III) have only one available coordination site for catalysis. The previous lanthanide complexes that we used, e.g., $Eu(L^1)^{3+}$, were good catalysts and had at least two available coordination sites.

A neutral ligand was prepared by replacement of the negatively charged acetate groups with neutral amide or hydroxyalkyl groups. Three of these ligands are shown in Figure 2. All lanthanide complexes were characterized fully by elemental analysis and mass spectrometry (*see* Experimental Details) and by ^1H and ^{13}C NMR. In addition $La(TCMC)(CF_3SO_3)_3 \cdot (EtOH)$ (*45*) and $La(TCEC)(CF_3SO_3)_3$ (*46*) have been characterized by single-crystal X-ray diffraction studies. A diagram of the $[La(TCEC)]^{3+}$ cation is shown in Figure 3. ^1H and ^{13}C NMR data in d^3-acetonitrile or d^4-methanol support lanthanide chelation of the ligands through all four nitrogen donors and all four oxygen donors in a manner similar to that observed in the solid-state structures of the

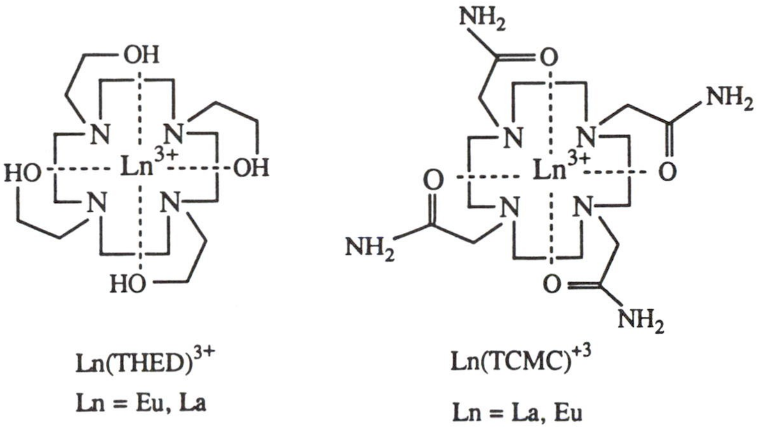

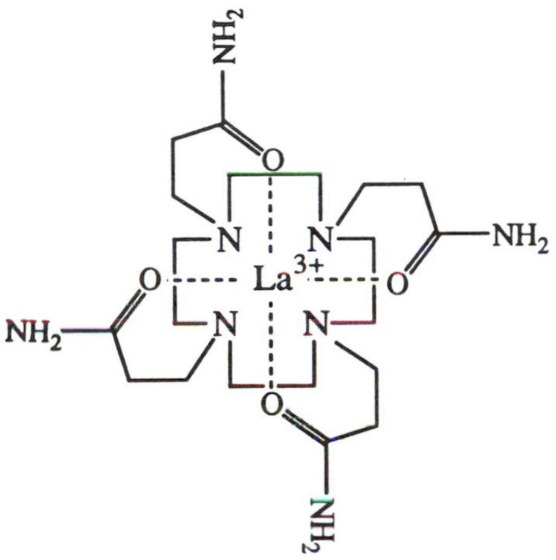

$Ln(THED)^{3+}$
Ln = Eu, La

$Ln(TCMC)^{+3}$
Ln = La, Eu

$La(TCEC)^{+3}$

Figure 2. Encapsulated lanthanide complexes.

complexes (*45, 46*). Variable temperature studies for all lanthanum complexes indicated a large degree of ligand rigidity at temperatures below 0 °C, a property also a characteristic of $Ln(DOTA)^-$ complexes (*42, 47*).

The ORTEP of the cation of $La(TCEC)^{3+}$ is shown in Figure 3 (*46*). The $[La(TCEC)]^{3+}$ cation (Figure 3) consists of an encapsulated eight-coordinate lanthanum(III) ion (*46*). The primary coordination polyhedron of the lanthanum cation can be described as a distorted square antiprism. The amide substituents are arranged in a clockwise fashion around the lanthanum ion. Lanthanum–nitrogen bond distances are as follows: La(1)–N(1) = 2.727(6), La(1)–N(2) = 2.711(6), La(1)–N(3) = 2.710(7), and La(1)–N(4) = 2.724(6). Lanthanum–oxygen bond dis-

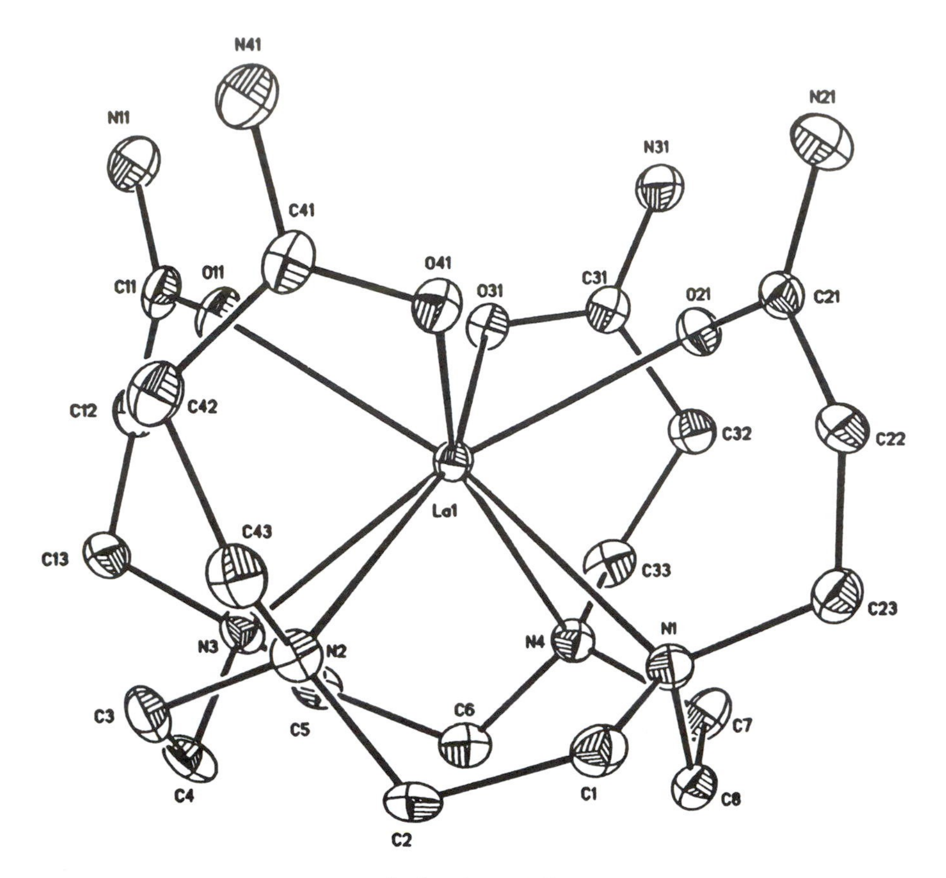

Figure 3. The $[La(TCEC)]^{3+}$ cation (46).

tances are as follows: La(1)–O(11) = 2.390(5), La(1)–O(21) = 2.434(5), La(1)–O(31) = 2.411(5), and La(1)–O(41) = 2.455(5) [average La–N = 2.718(±0.009), La–O = 2.423(±0.032)]. Comparison of the two sets of bond lengths indicates that the La–N bond lengths are longer than would be predicted from their ionic radii (*48*), while the La–O bond lengths are shorter than would be predicted from their ionic radii. The trend of short La–O bond distances and long La–N bond distances is also found in structures of macrocyclic polyaminocarboxylate complexes of europium (*49*) and gadolinium (*50*). The La^{3+} cation, which cannot fit into the cavity of the small 12-membered ring [*trans* N to N distance is 4.356 (±0.003)], is found above the ring with the amide groups folding over to encapsulate the ion.

NMR studies indicated that dissociation of the encapsulated lanthanum complexes varied dramatically with the ligand. The $La(TCEC)^{3+}$ complex contains pendent groups that form a six-membered ring. This complex undergoes decomposition within minutes in D_2O at 37 °C, initial pH 6.5. Resonances attributed to free ligand rapidly appear. The pendent amide group of the TCMC ligand differs from that of the TCEC ligand by a single methylene group, yet the lanthanum(III) complex of the TCMC ligand is inert to metal ion release; the majority of the complex was intact after 4 days at 37 °C, neutral pH. In contrast, dissociation of $La(THED)^{3+}$ is extensive over a period of a day.

Several of the complexes in Figure 2 were examined further for their resistance to dissociation. The europium complexes $Eu(THED)^{3+}$ and $Eu(TCMC)^{3+}$ were more difficult to study quantitatively by ^1H NMR because of their broad ^1H resonances. Decomposition was monitored by use of a UV–vis assay. Excess Cu^{2+} was added to solutions containing the lanthanide macrocycles. The Cu^{2+} ion served the dual purpose of trapping the free macrocycle and as an indicator to monitor the amount of macrocycle that had dissociated. All Cu(II) macrocyclic complexes gave an absorbance peak in the UV–vis spectrum that was characteristic of the Cu(II) macrocycle complex. For all macrocycles, Cu^{2+} was an effective trap; formation of the Cu(II) macrocyclic complex went to completion in the presence of 0.10 mM La^{3+} or 0.10 mM Eu^{3+}, 0.10 mM ligand and excess Cu^{2+} (1.0 mM). The increase in the concentration of Cu(II) macrocycle complex over time is a measure of the inertness of the lanthanide complex to dissociation. For the $La(THED)^{3+}$ complex, the reaction rate (*51*) was independent of the concentration of Cu^{2+}, consistent with the following mechanism:

$$\begin{aligned} La(THED)^{3+} &\rightarrow THED + La^{3+} \quad (k_1) \\ THED + Cu^{2+} &\rightarrow Cu(THED)^{2+} \quad (\text{rapid}) \end{aligned} \tag{1}$$

For $La(THED)^{3+}$ and $Eu(THED)^{3+}$, rate constants (k_1) of 9.2 (± 0.5) $\times 10^{-6}$ s^{-1} and 7.1 (± 0.4) $\times 10^{-7}$ s^{-1} corresponding to half-lives of 21 h and 11 days, respectively, were determined at pH 6.0, 37 °C. The $La(TCMC)^{3+}$ complex had a half-life of 6.7 days ($k_1 = 1.2 \times 10^{-6}$ s^{-1}) at pH 6.0, 37 °C. $Eu(TCMC)^{3+}$ showed <1% decomposition at pH 6.0, 37 °C after 6 weeks. That the $Eu(TCMC)^{3+}$ complex appears to be highly inert to metal ion release is not surprising given its similarity to $Gd(DOTA)^-$. The half-life for the dissociation of Gd^{3+} from DOTA at pH 5.0 is approximately 200 days (*43*).

Several of the encapsulated lanthanide complexes promote transesterification of *1* (Table I). The $Eu(TCMC)^{3+}$ complex alone did not promote cyclization of *1* over a 1-h period with 1.00 mM europium complex. Preliminary studies with $tRNA^{phe}$ indicate substantial RNA cleavage by $Eu(THED)^{3+}$ after 1 h at 37 °C, pH 7.4.

Mechanism of RNA Cleavage

Further studies are underway to characterize the solution properties of the encapsulated macrocyclic complexes of the lanthanides. One of the

Table I. Apparent Second-Order Rate Constants for the Transesterification of *1* by Lanthanide(III) Complexes at 37 °C, pH 7.4

Complex	k_2 *($M^{-1}s^{-1}$)*
$La(L^1)^{3+a}$	0.046
$Eu(L^1)^{3+}$	0.12
$La(TCMC)^{3+}$	0.016
$Eu(THED)^{3+}$	0.058

[a] pH 6.85. All reactions contained 0.10 M NaCl, 0.010 M HEPES buffer. [*1*] = 5 × 10^{-5} M to 1 × 10^{-4} M, [complex] = 2 × 10^{-4} to 1 × 10^{-3} M.

properties that is of interest is the number of coordination sites that are available for binding small molecules. Most of the better catalysts such as $Eu(L^1)^{3+}$ or Co(III) amine complexes (*9*) have at least two available adjacent coordination sites. One of them may be occupied by a hydroxide ligand, and the other may be available for binding the phosphate ester. A metal hydroxide may participate in the reaction by acting as a general base to deprotonate the 2′-hydroxyl group (*12, 19, 20*). Evidence for such a mechanism arises from the pH profile observed for RNA transesterification by metal complexes (*19*). Three possible pathways are shown in Scheme 1. It is clear that the metal does not participate purely as a general base (a). Organic bases with similar pK_as show much lower activity than observed for metal complexes with similar pK_as. Pathways b and c with the metal hydroxide participating as a base or hydroxide acting as a base, respectively, are kinetically indistinguishable.

How many open coordination sites are optimal to produce catalytically active lanthanide complexes? In the solid state (*45*), the

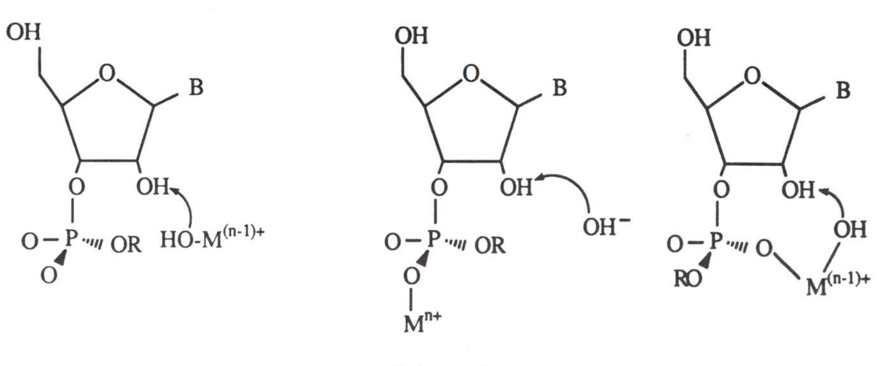

Scheme 1

$La(TCMC)^{3+}$ complex has two coordination sites occupied by ligands other than the macrocycle. $Eu(TCMC)^{3+}$, similar to $Eu(DOTA)^{-}$, has only one available coordination site (*52*). Thus for the TCMC ligand, the larger lanthanides that have more open coordination sites may be catalytically active, whereas the smaller, heavier lanthanides may lack activity. On the basis of its similarity to $Eu(DOTA)^{-}$, the THED complex of Eu(III) probably has only one available coordination site. Why then is this complex active? One possible explanation is that the active catalyst is produced upon dissociation of one of the hydroxyethyl groups. Another possibility is the participation of an hydroxyethyl group as a general base catalyst. Further studies are need to delineate these possibilities. Studies are in progress to determine how many water molecules are bound to the europium complexes in solution and to determine the pK_a of the bound water molecules.

The overall charge on the complex may be important in catalyst design. It is well-known that in reactions where a metal ion acts as a Lewis acid, the addition of anionic ligands to the metal ion catalyst may decrease the rate of the reaction (*53*). For phosphate ester substitution reactions involving anionic phosphate esters, charge neutralization of the phosphate diester by the catalyst may be especially important (*54*). In support of this hypothesis, polyaminocarboxylate ligands such as EDTA do not form complexes with La(III) that are active in cleavage (*14*). The lanthanide complexes discussed here that are active transesterification catalysts have neutral ligands. Conductivity measurements indicate an overall +3 charge on $La(L^1)^{3+}$ (*27*), $Eu(THED)^{3+}$, and $La(TCMC)^{3+}$ (*45*, *51*).

Acknowledgment

We thank the National Institutes of Health (GM46539) for support of this work.

References

1. Lauffer, R. B. *Chem. Rev.* **1987,** *87*, 901–927.
2. Smith, P. H.; Brainard, J. R.; Morris, D. E.; Jarvinen, G. D.; Ryan, R. R. *J. Am. Chem. Soc.* **1989,** *111*, 7437–7443.
3. Dischino, D. D.; Delaney, E. J.; Emswiler, J. E.; Gaughan, G. T.; Prasad, J. S.; Srivastava, S. K.; Tweedle, M. F. *Inorg. Chem.* **1991,** *30*, 1265–1269.
4. Moi, M. K.; Meares, C. F.; DeNardo, S. J. *J. Am. Chem. Soc.* **1988,** *110*, 6266–6267.
5. McMurray, T. J.; Brechbiel, M.; Kumar, K.; Gansow, O. A. *Bioconjugate Chem.* **1992,** *3*, 108–117.
6. Morphy, J. R.; Parker, D.; Kataky, R.; Harrison, A.; Eaton, M. A. W.; Millican, A.; Phipps, A.; Walker, C. *J. Chem. Soc. Chem. Commun.* **1989,** 792–794.
7. Stein, C. A.; Cohen, J. S. *Cancer Res.* **1988,** *48*, 2659–2688.
8. Smith, J.; Ariga, K.; Anslyn, E. V. *J. Am. Chem. Soc.* **1993,** *115*, 362–364.

9. Yoshinari, K.; Yamazaki, K.; Komiyama, M. *J. Am. Chem. Soc.* **1991,** *113,* 5899–5901.
10. Barbier, B.; Brack, A. *J. Am. Chem. Soc.* **1988,** *110,* 6880–6882.
11. Tung, C.-H.; Wei, Z.; Leibowitz, M. J.; Stein, S. *Proc. Natl. Acad. Sci. U.S.A.* **1992,** *89,* 7114–7118.
12. Stern, M. K.; Bashkin, J. K.; Sall, E. D. *J. Am. Chem. Soc.* **1990,** *112,* 5357–5359.
13. Modak, A. S.; Gard, J. K.; Merriman, M. C.; Winkeler, K. A.; Bashkin, J. K.; Stern, M. K. *J. Am. Chem. Soc.* **1991,** *113,* 283–291.
14. Morrow, J. R.; Buttrey, L. A.; Shelton, V. M.; Berback, K. A. *J. Am. Chem. Soc.* **1992,** *114,* 1903–1905.
15. Madeyski, C. M.; Michael, J. P.; Hancock, R. D. *Inorg. Chem.* **1984,** *23,* 1487–1489.
16. Brown, D. M.; Usher, D. A. *J. Chem. Soc.* **1965,** 6558–6564.
17. Morrow, J. R.; Buttrey, L. A.; Berback. K. A. *Inorg. Chem.* **1992,** *31,* 16–20.
18. England, T. E.; Uhlenbeck, O. C. *Nature (London)* **1978,** *275,* 560–561.
19. Shelton, V. M.; Morrow, J. R. *Inorg. Chem.* **1991,** *30,* 4295–4299.
20. Matsumoto, Y.; Komiyama, M. *J. Chem. Soc. Chem. Commun.* **1990,** 1050–1051.
21. Breslow, R.; Berger, D.; Huang, D.-L. *J. Am. Chem. Soc.* **1990,** *112,* 3686–3687.
22. Komiyama, M.; Matsumura, K.; Matsumoto, Y. *J. Chem. Soc. Chem. Commun.* **1992,** 640–641.
23. Breslow, R.; Huang, D.-L. *Proc. Natl. Acad. Sci. USA* **1991,** *88,* 4080–4083.
24. Bunzli, J.-C. G.; Wessner, D. *Coord. Chem. Rev.* **1984,** *60,* 191–253.
25. De Cola, L.; Smailes, D. L.; Vallarino, L. M. *Inorg. Chem.* **1986,** *25,* 1729–1732.
26. Arif, A. M.; Backer-Dirks, J. D. J.; Gray, C. J.; Hart, F. A.; Hursthouse, M. B. *J. Chem. Soc. Dalton Trans.* **1987,** 1665–1673.
27. Hay, R. W.; McGovan, N. *J. Chem. Soc. Chem. Commun.* **1990,** 714–715.
28. Bruno, J.; Herr, B. R.; Horrocks, W. DeW., Jr. *Inorg. Chem.* **1993,** *32,* 756–762.
29. Brown, R. S.; Dewan, J. C.; Klug, A. *Biochemistry* **1985,** *24,* 4785–4801.
30. Dahm, S. C.; Uhlenbeck, O. C. *Biochemistry* **1991,** *30,* 9464–9469.
31. Rordorf, B. F.; Kearns, D. R. *Biopolymers* **1976,** *15,* 1491–1504.
32. Marciniec, T.; Ciesiolka, J.; Wrzesinski, J.; Krzyzosiak, W. J. *FEBS Lett.* **1989,** *243,* 293–298.
33. Kolasa, K. A.; Morrow, J. R.; Sharma, A. P. *Inorg. Chem.* **1993,** *32,* 3983–3984.
34. Topal, M. D.; Fresco, J. R. *Biochemistry* **1980,** *19,* 5531–5537.
35. Usher, D. A.; McHale, A. H. *Proc. Natl. Acad. Sci. U.S.A.* **1976,** *73,* 1149–1153.
36. Kierzek, R. *Nucleic Acids Res.* **1992,** *20,* 5073–5077.
37. Usher, D. A. *Nature New Biol.* **1972,** *235,* 207–208.
38. Saenger, W. *Principles of Nucleic Acid Structure;* Springer-Verlag: New York, 1984; pp 277–279.
39. Suh, J.; Min, D. W. *J. Org. Chem.* **1991,** *56,* 5710–5712.
40. Abid, K. K.; Fenton, D. E. *Inorg. Chim. Acta* **1984,** *95,* 119–125.
41. *Lanthanide Probes in Life, Chemical and Earth Sciences;* Bunzli, J.-C. G.; Choppin, G. R., Eds.; Elsevier: New York, 1989; Chapter 1.
42. Desreux, J. F. *Inorg. Chem.* **1980,** *19,* 1319–1324.

43. Wang, X.; Jin, T.; Comblin, V.; Lopez-Mut, A.; Merciny, E.; Desreux, J. F. *Inorg. Chem.* **1992,** *31,* 1095–1099.
44. Bryden, C. C.; Reilley, C. N.; Desreux, J. F. *Anal. Chem.* **1981,** *53,* 1418–1425.
45. Amin, S.; Morrow, J. R.; Lake, C. H.; Churchill, M. R. *Angew. Chem. Int. Ed. Engl.* **1994,** *33,* 773–775.
46. Morrow, J. R.; Amin, S.; Lake, C. H.; Churchill, M. R. *Inorg. Chem.* **1993,** *32,* 4566–4572.
47. Aime, S.; Botta, M.; Ermondi, G. *Inorg. Chem.* **1992,** *31,* 4291–4299.
48. *International Tables for X-ray Crystallography;* Kynoch Press: Birmingham, England, 1974; Vol. 4, pp 99–101.
49. Spirlet, M. R.; Rebizant, J.; Desreux, J. F.; Longcin, M.-F. *Inorg. Chem.* **1984,** *23,* 359–363.
50. Aime, S.; Anelli, P. L.; Botta, M.; Fedeli, F.; Spiller, M. *Inorg. Chem.* **1992,** *31,* 2422–2428.
51. Morrow, J. R.; Chin, K. O. A. *Inorg. Chem.* **1993,** *32,* 3357–3361.
52. Amin, S.; Voss, D. A.; Horrocks, W. DeW.; Lake, C. H.; Churchill, M. R.; Morrow, J. R. *Inorg. Chem.* **1995,** *34,* 3294–3300.
53. Bender, M. L. *Mechanisms of Homogeneous Catalysis from Protons to Proteins;* John Wiley and Sons: New York, 1971; Chapter 8.
54. Kirby, A. J.; Younas, M. *J. Chem. Soc. B* **1970,** 1172–1182.

RECEIVED for review June 10, 1993. ACCEPTED revised manuscript December 20, 1993.

17

Metallointercalators as Probes of the DNA π-way

Michelle R. Arkin[1], Yonchu Jenkins[1], Catherine J. Murphy[1,3], Nicholas J. Turro[2], and Jacqueline K. Barton[1]*

[1] Division of Chemistry and Chemical Engineering, Beckman Institute, California Institute of Technology, Pasadena, CA 91125
[2] Department of Chemistry, Columbia University, New York, NY 10027

This chapter describes efforts in our laboratory to characterize the role of double helical DNA in catalyzing electron-transfer reactions. Using intercalating metal complexes as donor and acceptor, we have shown that the luminescence of $[Ru(phen)_2(dppz)]^{2+*}$ is efficiently quenched by $[Rh(phi)_2(phen)]^{3+}$ in the presence of B-form DNA. Covalent attachment of these metal complexes to either ends of a short duplex leads to complete quenching of luminescence over a separation distance between intercalated donor and acceptor of >40 Å. These results with metallointercalators point to the π-stacked array of heterocyclic DNA base pairs as an effective intervening medium for long-range electron transfer and provides a new approach in applying the DNA helical polymer as a "molecular wire."

Double helical DNA is a water-soluble polymer that contains an electronically well-coupled stack of aromatic heterocyclic base pairs. This review describes efforts in our laboratory to characterize electron-transfer reactions between transition metal complexes bound by intercalation within the π-stack of DNA. Much information is available concerning the structure, synthesis, and methods of characterization of this polymer. Also, research in our laboratories has been directed toward describing the photophysical and photochemical properties of metal complexes bound to DNA. Using these metal complexes to probe the DNA π-way, we are now in a position to ask: Is DNA a molecular wire?

* Corresponding author
[3] Current address: Department of Chemistry and Biochemistry, University of South Carolina, Columbia, SC 29208

0065–2393/95/0246–0449/$08.18/0

Electron-transfer chemistry has been the focus of substantial research over the past 40 years and understanding it is fundamental to elucidating electron-transport processes in biology and in developing artificial photosynthetic systems and electroactive sensors (*1–4*). Experiments in many laboratories have focused on measurements of electron-transfer rates between metal centers over long distances in proteins or protein pairs as a function of distance, driving force, and the intervening medium (*5–8*). Model complexes have also been prepared to explore how different structural and electronic factors may mediate electron-transfer reactions (*9–13*), and theories exploring optimal pathways for electron transfer have sought to reconcile experimental studies (*14–17*). Among the many ideas put forth concerning how the medium may serve to modulate or direct electron transfer has been the notion that stacked aromatic heterocyclic moieties might serve as "π-ways" through which electron-transfer reactions might be promoted efficiently. Few experimental measurements of electron transfer through π-stacked arrays have been accomplished, however (*18–22*).

Irrespective of its biological function, the DNA double helix may be described as a prototype π-stacked column and therefore a novel medium through which to examine electron-transfer reactions. The double helix is a polymer containing a relatively rigid, electronically coupled column of stacked base pairs within a water-soluble polyanion, the sugar-phosphate backbone. The electronic coupling within the column is reflected in the extensive hypochromicity of the stacked double helix compared to the random coil, and it is this stacking interaction that accounts substantially for the stabilization of the helical form (*23*).

Theoretical studies have proposed the importance of charge transfer in nucleic acids for some time (*24–26*), but only recently has DNA been examined experimentally as a medium for electron-transfer reactions (*27–30*). One impetus for such study has come from experiments with radiation-damaged DNA that show a link between DNA-mediated electron transfer and nucleic acid-based disease (*31–33*). For example, electron trapping experiments on DNA subjected to γ-rays at low temperature have suggested that radical species can migrate up to 100 base pairs away from the initial site of damage (*33*). Pulse radiolysis studies of the antitumor drug daunorubicin intercalated into DNA reveal that this electron mobility is comparable to that found in conducting polymers (*34*). Such dissipation of charge may actually be a mechanism by which redox damage to DNA at localized sites is avoided.

Researchers have also studied DNA-mediated electron transfer by using donor–acceptor pairs that bind DNA noncovalently (*27–30*). Early work in our laboratory used cationic tris(phenanthroline) metal complexes as donor–acceptor pairs (*29, 30*). These complexes, shown in Figure 1, associate with DNA through two modes, (i) intercalation and

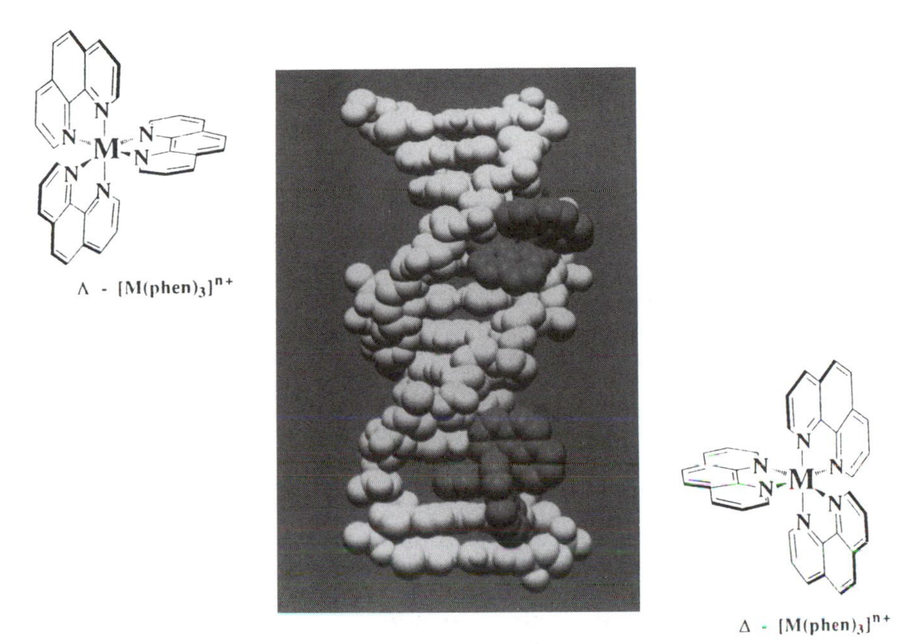

Figure 1. Binding to DNA by enantiomers of tris(phenanthroline) metal complexes. The computer graphic representation (center) depicts our model for noncovalent binding to right-handed double-helical DNA by the Δ- (right) and Λ- (left) isomers. Δ-$[M(phen)_3]^{n+}$ is shown bound to the lower half of the helix through intercalation in the major groove. In this binding mode, preferred for the Δ-isomer, one ligand is inserted partially and stacked between the DNA base pairs. Λ-$[M(phen)_3]^{n+}$, shown bound to the upper half of the DNA helix, is illustrated bound against the minor groove through a hydrophobically stabilized surface- or groove-bound interaction; for this surface-bound mode, we find enantioselectivity favoring the Λ-isomer.

(ii) surface or groove binding, with an overall binding constant of 10^3 M^{-1} (*35–38*). In these experiments, the donor was photoexcited $[Ru(phen)_3]^{2+}$, whereas the acceptors were $[M(phen)_3]^{3+}$, where M = Rh(III), Co(III), or Cr(III). The rates of photoinduced electron transfer are increased by an order of magnitude in the presence of DNA, and quenching rates are dependent on the mode of interaction of each complex with the DNA helix. The enhancements in rate of electron transfer were attributed to a combination of (i) long-range electron transfer through the DNA medium, (ii) an increase in local concentration of donors and acceptors bound to DNA, and (iii) facilitated diffusion along the helix. The rapid equilibration between binding modes and positions of donors and acceptors, however, made it difficult to evaluate the relative importance of each factor.

To probe more effectively the role of the DNA π-way in mediating electron-transfer reactions, we now focus on avid metallointercalators with binding constants for intercalation of $\geq 10^7$ M^{-1}. The strong preference of these molecules to intercalate rather than groove-bind clarifies the relationship of the donor and acceptor to the DNA medium.

Figure 2 displays the donor $[Ru(phen)_2(dppz)]^{2+}$ (dppz = dipyrido-phenazine) and the acceptor $[Rh(phi)_2(phen)]^{3+}$ (phi = 9,10-diaminophenanthrene) that we are using in these studies. The donor, photoexcited (*) $[Ru(phen)_2(dppz)]^{2+}$, shows no luminescence in aqueous solution, but glows intensely when the complex binds to DNA (Figure 3). Similar to the parent complex $[Ru(phen)_3]^{2+}$, the absorption spectrum of this complex is characterized by a metal-to-ligand charge-transfer band, and studies of $[Ru(bpy)_2(dppz)]^{2+}$ in the absence of DNA have shown that charge transfer is directed onto the phenazine ring (*38–40*). The luminescence quencher $[Rh(phi)_2(phen)]^{3+}$ is also pictured in Figure 2. Rhodium(III) complexes containing phi are known to bind tightly to nucleic acids via intercalation of this ligand (*41–43*), and the lowest energy absorption bands of these complexes result from transitions centered on the phi (*44, 45*). We were intrigued by the possibility that intercalation by the rhodium and ruthenium complexes could afford easy access to the π-way, where the stacked bases might readily accept and direct an electron from the donor to the intercalated acceptor.

$Ru(phen)_2dppz^{2+}$: A Molecular Light Switch

The photoluminescence of dipyridophenazine complexes of ruthenium(II) in the presence and absence of DNA has been well-characterized (*38–40, 46–52*). Excitation of the dppz complexes with visible light (440 nm) leads to localized charge transfer from the metal center (*39, 40*). In aqueous solution, the emission resulting from the metal-to-ligand charge-transfer excited state is deactivated via nonradiative energy transfer

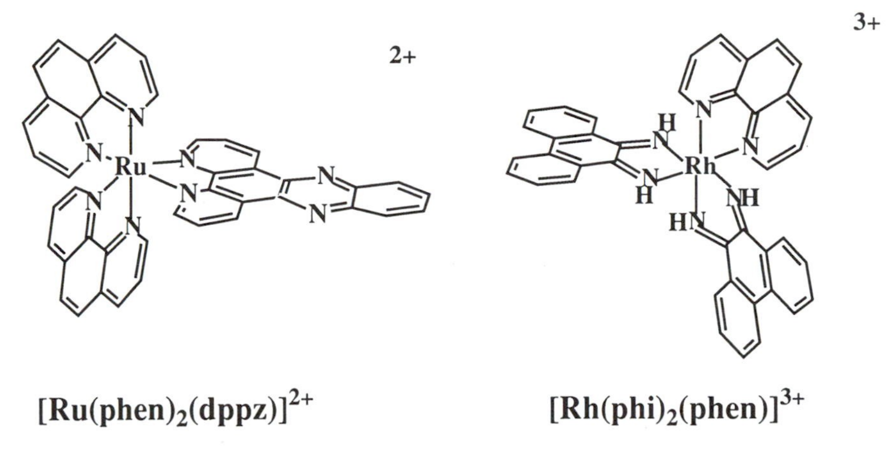

Figure 2. Structures of donor and acceptor metallointercalators. The photoexcited donor $[Ru(phen)_2(dppz)]^{2+}$ is shown on the left, and the acceptor $[Rh(phi)_2(phen)]^{3+}$ is pictured on the right.

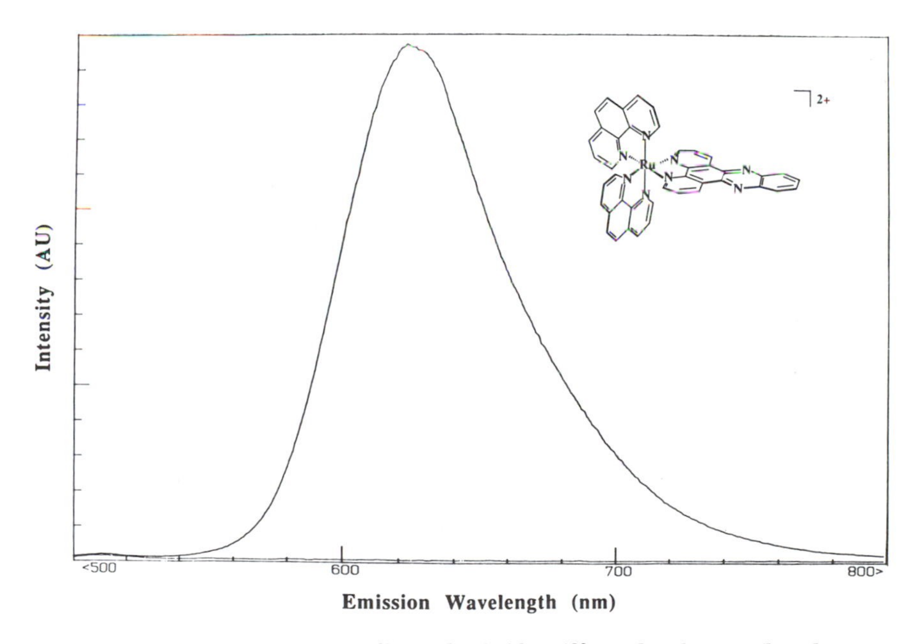

Figure 3. Emission spectra of $[Ru(phen)_2(dppz)]^{2+}$ in the absence (baseline) and presence of calf thymus DNA.

from the phenazine nitrogens to solvent water molecules (*46, 48*). When the complex intercalates into double-stranded DNA, the stacked bases protect the phenazine nitrogens from water, and $[Ru(phen)_2(dppz)]^{2+}$ photoluminesces brightly. Evidence for intercalation is provided by DNA unwinding studies, emission titrations, and luminescence depolarization experiments (*38, 46–51*). Shown in Figure 3 are the emission spectra of the complex in the absence and presence of calf thymus DNA. In the absence of DNA, no detectable emission is evident. Upon addition of double-stranded DNA, the complex luminesces intensely, with an emission enhancement upon binding to DNA of $>10^3$.

Table I shows examples of the steady-state and time-resolved emission characteristics of $[Ru(phen)_2(dppz)]^{2+}$ upon binding to various DNAs. The time-resolved luminescence of DNA-bound Ru(II) is characterized by a biexponential decay, consistent with the presence of at least two binding modes for the complex (*47, 48*). Previous photophysical studies conducted with tris(phenanthroline)ruthenium(II) also showed biexponential decays in emission and led to the proposal of two noncovalent binding modes for the complex: (i) a surface-bound mode in which the ancillary ligands of the metal complex rest against the minor groove of DNA and (ii) an intercalative stacking mode in which one of the ligands inserts partially between adjacent base pairs in the double helix (*36, 37*). In contrast, quenching studies using both cationic quenchers such as $[Ru(NH_3)_6]^{3+}$ and anionic quenchers such as $[Fe(CN)_6]^{4-}$ have indicated that for the dppz complex both binding modes

Table I. Luminescent Parameters for Photoexcited $[Ru(phen)_2(dppz)]^{2+}$ Bound to Nucleic Acids of Varying Conformations

Nucleic Acid	τ *(ns)*[a,b]	*%*[c]	λ_{max} *(nm)*	Φ[a,b,d]
Calf thymus DNA	770	40	617	0.039
	120	60		
Z-form poly[d(GC)] · poly[d(GC)]	270	60	608	0.025
	70	40		
poly[r(AU)] · poly[r(AU)]	490	20	620	0.004
	80	80		
poly(dT) · poly(dA) · poly(dT)	530	60	621	0.061
	170	40		

NOTE: All steady-state and time-resolved measurements were taken at 20 °C using instrumentation described in reference 28.

[a] Error was estimated to be ±10% for both steady-state and time-resolved measurements.

[b] Samples used in steady-state and time-resolved measurements contained 10 μM ruthenium complex/100 μM nucleotides.

[c] Lifetime ratios were calculated from the magnitudes of the pre-exponential factors produced by the program used in the deconvolution of the time-resolved data.

[d] Quantum yields, Φ, were determined relative to $[Ru(bpy)_3]^{2+}$ (Φ = 0.042) (*48*).

are intercalative in nature (*48, 53*). Additional studies have shown that for dppz complexes both emissive components also maintain polarization (*46*), and thus both lifetimes arise from species that are rigidly held on the time scale of the emission. For tris(phenanthroline) metal complexes, the Δ-isomer was found to favor the intercalative mode and the Λ-isomer, the surface-bound mode. Recent experiments have shown that $[Ru(phen)_2(dppz)]^{2+}$ also displays enantioselectivity in emission, with highest luminescence observed by the Δ-isomer on binding right-handed DNA. Here too the data are consistent with two families of photoluminescent species for each enantiomer that bind by intercalation (*51*). Two specific intercalative modes have been proposed on the basis of photophysical studies of $[Ru(phen)_2(dppz)]^{2+}$ derivatives: (i) a perpendicular mode in which the dppz ligand intercalates from the major groove such that the long axis of the metal complex lies along the dyad axis and (ii) a side-on mode in which the long axis of the dppz lies more closely to the long axis of the base pairs (*47*). Recent NMR results lend further support to these models.

As also described in Table I, the luminescent parameters for the metal complex bound to different conformations of DNA can be correlated with the accessibility of the phenazine ligand to water (*48*). This correlation is most clearly illustrated in the examples of A-form poly[r(AU)]·poly[r(AU)] and the triple helix poly(dT)·poly(dA)·poly(dT). In A-form nucleic acids, the base pairs are pushed back toward the periphery of the major groove, creating a major groove that is both very deep and very narrow (*54*). The shape of this cavity likely hinders the intercalation of the dppz ligand, as was found with tris(phenanthroline)complexes of ruthenium(II). This relatively poor protection results in short excited-state lifetimes and correspondingly low luminescent intensities. Intercalation into the triplex, on the other hand, results in an interaction in which the base triples adjacent to the intercalating ligand completely surround the phenazine nitrogens, resulting in greater protection from water and therefore longer luminescent lifetimes and higher luminescent intensities.

The sensitive emission properties of $[Ru(phen)_2(dppz)]^{2+}$ and its derivatives make these complexes ideal electron donors in the study of DNA-mediated electron transfer. Because luminescence is due to intercalated species, our photophysical studies will probe only those complexes bound to DNA. The steady-state and time-resolved luminescence assists also in characterizing novel metal/DNA assemblies.

Phi Complexes of Rhodium(III): Intercalators and Photocleavage Agents

Phi complexes of rhodium(III) bind avidly to DNA through intercalation (*41–45, 55–60*). ^1H NMR results (*43*) on Δ-$[Rh(phen)_2(phi)]^{3+}$ bound to a hexamer oligonucleotide offer specific evidence for intercalation of the phi ligand and support earlier spectroscopic and helical unwinding studies. Measurements of this complex intercalated into a short oligonucleotide show preferential shifts of phi protons compared to those on the ancillary phenanthroline ligands. Importantly, two-dimensional nuclear Overhauser effect spectroscopy (NOESY) experiments indicate a selective loss of the intramolecular NOE between the central base and the adjacent sugar of the hexamer, providing compelling evidence for intercalation of the rhodium(III) complex at that base step. These NOESY experiments also indicate intermolecular NOEs between the rhodium complex and protons in the DNA major groove.

Rhodium complexes have proven to be particularly useful because these complexes promote strand breaks in DNA and RNA upon photoactivation (*41, 42*). Analysis of the DNA-derived products of the photocleavage reaction are consistent with abstraction of the C3′ hydrogen atom from the nucleotide in the 5′ position of the intercalation site. Because cleavage occurs directly at the base step of intercalation, these complexes have been very effective as probes of higher-order structures in nucleic acids and as high-resolution DNA photofootprinting reagents (*55–58*). The product analysis for photocleavage, consistent with the NMR results, demonstrates that the complexes intercalate from the major groove.

Much of the work in our laboratory has been directed toward tuning the recognition properties of phi complexes of rhodium for different nucleic acid sites by altering the ancillary ligands (*42, 59–61*). Figure 4 illustrates some of the complexes that we have prepared that differ substantially with respect to DNA recognition characteristics. The recognition of these octahedral complexes is governed by the ensemble of noncovalent interactions between the metal complex and the nucleic acid site. Such interactions arise from (i) the complementarity of the three-dimensional shapes of the metal complex and its site and (ii) the positioning of ligand functionalities for hydrogen bonding and van der Waals contacts to functional groups in the DNA major groove. Δ-$[Rh(phen)_2(phi)]^{3+}$, for instance, binds preferentially at base steps with a propeller twisted and opened major groove, because only at such open sites are steric clashes of the phen protons with the bases relieved. $[Rh(phi)_2(bpy)]^{3+}$, on the other hand, contains a phi ligand in one of the ancillary positions; in this complex the ancillary phi is pulled away from the helix and steric clash with protons in the major groove is avoided. Hence $[Rh(phi)_2(bpy)]^{3+}$ is essentially sequence-neutral in its interactions with B-form DNA, making this complex a high-resolution photofootprinting agent. For example, $[Rh(phi)_2(bpy)]^{3+}$ has been used to map the association of *Eco*RI with DNA, because specific binding of the protein inhibits intercalation of the metal complex and therefore eliminates

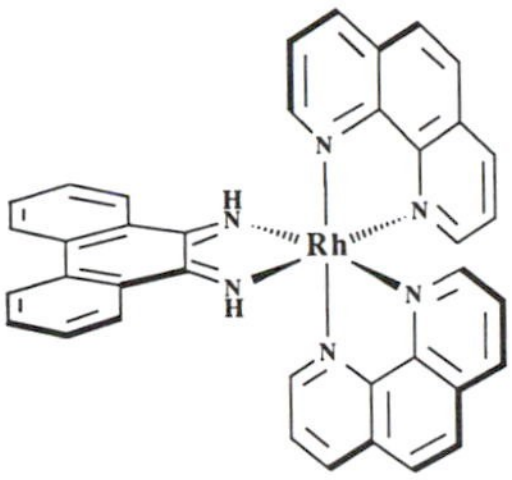

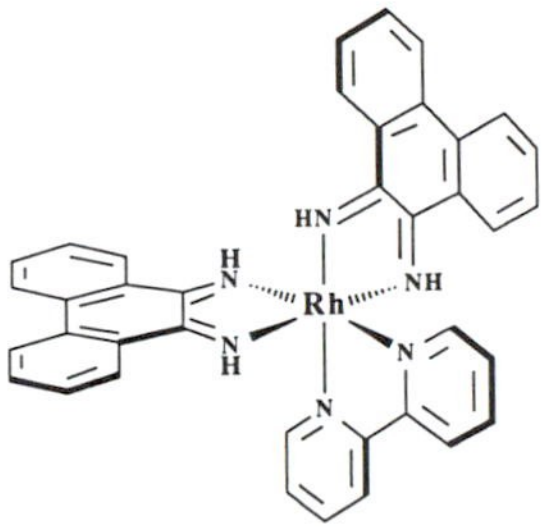

$[Rh(phen)_2(phi)]^{3+}$ $[Rh(phi)_2(bpy)]^{3+}$

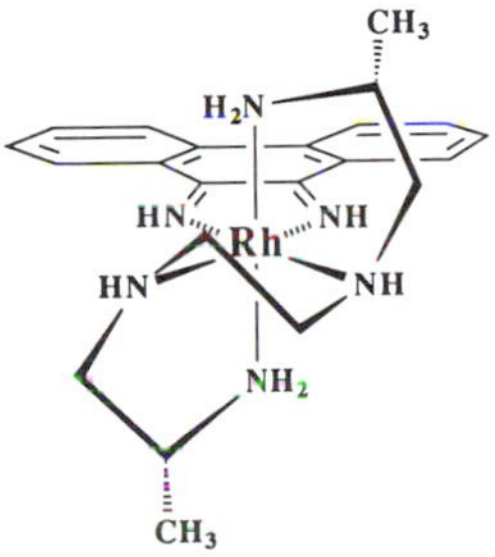

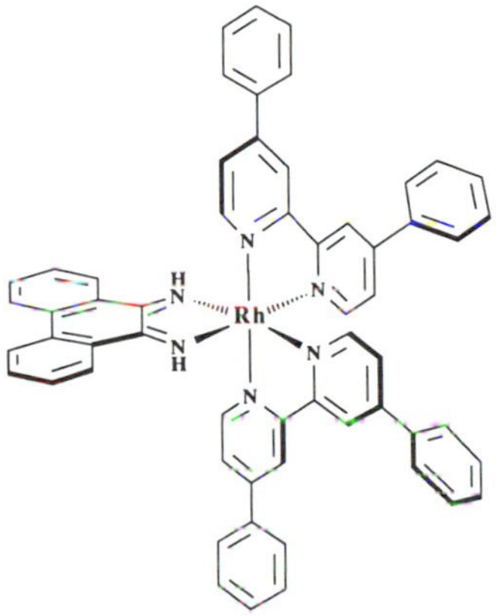

Δ,α-(R,R)-$[Rh(Me_2trien)(phi)]^{3+}$ [Rh(4,4'-dimethylbpy)$_2$(phi)]$^{3+}$

Figure 4. Phi complexes of rhodium that recognize DNA with differing site selectivity. Clockwise, from upper left: $[Rh(phen)_2phi]^{3+}$ *recognizes 5'-pyr-pyr-pur-pur-3' sequences, characterized by an open major groove (22).* $[Rh(phi)_2(bpy)]^{3+}$ *binds and cleaves B-form DNA without sequence selectivity, making it a high-resolution photofootprinting reagent (37). Δ-*[Rh(4,4'-dimethylbpy)$_2$(phi)]$^{3+}$ *recognizes the palindromic sequence 5'-CTCTAGAG-3' and displays striking enantioselectivity (40). Δ,α,-(R,R)* $[Rh(Me_2trien)(phi)]^{3+}$ *recognizes 5'-TGCA-3' sequences through a combination of van der Waals interactions involving the methyl groups on the ligand and hydrogen bond donation by the axial amines (38).*

photocleavage at the protein's binding site (*58*). The highest degree of site-selectivity attained by a rhodium(III) complex to date has been with the bulky complex $[Rh(4,4'\text{-diphenylbpy})_2(phi)]^{3+}$. This complex recognizes the 8 base-pair sequence 5'-CTCTAGAG-3' with a specificity and binding strength that rivals DNA-binding proteins (*61*).

$[Rh(phi)_2(phen)]^{3+}$ is a particularly suitable luminescence quencher for our investigations of electron-transfer reactions on DNA. Its electronic properties are favorable for electron transfer, and this rhodium complex is primarily sequence neutral, so that nearly random binding of the donor and acceptor is expected. Moreover, the photocleavage reaction actually allows us to identify the positions of binding of the acceptor to the DNA double helix.

Electron-Transfer Reactions Between Metal Complexes in the Presence of DNA

When $[Rh(phi)_2(phen)]^{3+}$ is titrated into a solution containing $[Ru(phen)_2(dppz)]^{2+}$ and B-form DNA, the photoinduced luminescence of the ruthenium(II) complex is quenched dramatically (*53*). In these experiments, luminescence is monitored by laser flash as quencher is added. Data are then plotted in Stern–Volmer format, where the ratio of initial intensity/intensity (I_0/I) is given as a function of quencher concentration $[Q]$. The degree of lifetime quenching can also be described by plotting the inverse of the lifetime (τ_0/τ) versus $[Q]$. Normally, when chromophore and quencher interact bimolecularly, Stern-Volmer graphs are linear with $[Q]$ and the slope for τ_0/τ is the same as that for I_0/I.

What is actually observed, however, is that $[Rh(phi)_2(phen)]^{3+}$ intercalated into DNA quenches the intensity of $[Ru(phen)_2(dppz)]^{2+*}$ much more effectively than it quenches the two lifetimes, as summarized in Table II. This effect is most pronounced when the DNA helix is a short oligonucleotide. The direct comparison of quenching in the absence of DNA cannot be accomplished because the ruthenium(II) complex does not luminesce in aqueous solution; however, electron transfer from $[Ru(phen)_3]^{2+*}$ to $[Rh(phi)_2(phen)]^{3+}$ in buffered solution provides a control with the same thermodynamic driving force (*40*).

The solution-phase quenching of $[Ru(phen)_3]^{2+*}$ luminescence is minimal at these concentrations, supporting the notion that the remarkably efficient quenching of $[Ru(phen)_2(dppz)]^{2+*}$ luminescence by rhodium(III) is catalyzed by DNA.

To test specifically the role of the DNA π-way, we monitored the quenching of $[Ru(phen)_2(dppz)]^{2+*}$ in DNA by hexa(amine)ruthenium (III). $[Ru(NH_3)_6]^{3+}$ is an effective oxidative quencher of the luminescence of ruthenium(II) polypyridyl complexes (*62*) and binds to DNA by electrostatic and hydrogen bonding interactions (*63*). The resulting Stern–

Table II. Luminescence Quenching of Ru^{2+} Donors by M^{3+} Acceptors

Donor	*Acceptor*	*Medium*	τ_o/τ *at 50 μM Quencher*[a]	I_o/I *at 50 μM Quencher*[b]	*Curvature*[c]
$[Ru(phen)_2dppz]^{2+}$	$[Rh(phi)_2phen]^{3+}$	calf thymus DNA in buffer	1.3	3.5	upward-curving
$[Ru(phen)_2dppz]^{2+}$	$[Rh(phi)_2phen]^{3+}$	28-mer oligonucleotide in buffer	2	11	upward-curving
$[Ru(phen)_2dppz]^{2+}$	$[Ru(NH_3)_6]^{3+}$	calf thymus DNA in buffer	1.3	1.5	linear
$[Ru(bpy)_3]^{2+}$	$[Rh(phi)_2phen]^{3+}$	calf thymus DNA in buffer	1.1	1.3	linear
$[Ru(phen)_2dppz]^{2+}$	$[Rh(phi)_2phen]^{3+}$	ethanol	1.1	1.3	linear
$[Ru(phen)_3]^{2+}$	$[Rh(phi)_2phen]^{3+}$	buffer	1.05	1.2	linear

NOTE: All measurements were taken at ambient temperature using instrumentation described in reference 48.

[a] Lifetimes were determined by fitting time-resolved data to a biexponential decay using a computer fitting program.

[b] Steady-state luminescence intensities were determined by integrating time-resolved data, using a computer fitting program.

[c] Refers to the shape of Stern–Volmer plot of initial intensity/intensity at [Q].

Volmer plots, also described in Table II, are linear and the loss of luminescent intensity is found to mirror the reduction in luminescent lifetimes. These kinetics reflect dynamic quenching in which the donor and acceptor molecules are brought together by molecular diffusion, which occurs on a time scale comparable to the inherent luminescence decay (*64, 65*). Thus, the results of quenching of $[Ru(phen)_2(dppz)]^{2+*}$ luminescence by $[Ru(NH_3)_6]^{3+}$ in the presence of DNA are consistent with a quenching mechanism in which $[Ru(NH_3)_6]^{3+}$ is a diffusible species. This normal Stern–Volmer behavior differs significantly from that observed when intercalated $[Rh(phi)_2(phen)]^{3+}$ is the quencher, when steady-state Stern–Volmer plots are nonlinear and upward-curving, and the steady-state quenching far exceeds the reduction in luminescent lifetime (Table II).

The large loss of intensity and small loss in the lifetimes of $[Ru(phen)_2(dppz)]^{2+*}$ luminescence in the presence of an intercalated quencher is, instead, consistent with a "static" mechanism of quenching, one which occurs faster than the diffusion of these rigidly bound complexes. There are two models that are often put forth to describe this phenomenon. The "sphere of action" model for static quenching requires that quenchers within a critical distance of the excited molecule will quench the excited state on a time scale that is shorter than diffusion (*64–66*). In the second model, the complexes simply interact in the ground state, precluding population of the emissive excited state. In our system no evidence for ground-state complex formation between these cationic species has been found. Indeed DNA photocleavage assays, in which the position of the rhodium may be monitored on the DNA helix in the presence and absence of ruthenium, have indicated that the two complexes bind independently and are situated randomly on the double helix (*53*). Therefore, luminescence quenching of the intercalated complexes likely requires a fast, long-range electronic interaction. Because this static quenching is found only when both donor and acceptor are intercalated, we propose that electronic communication is mediated by the DNA π-way.

Energy Transfer or Electron Transfer?

Excited-state quenching arises in general because of energy transfer or electron transfer or some mixture thereof. There are several reasons why the quenching in this system may most reasonably be attributed to a long-range electron-transfer reaction rather than energy transfer. With a driving force of −0.8 V, electron transfer is thermodynamically favored (*53*). (The reduction of $[Rh(phi)_2phen]^{3+}$ in *N,N*-dimethylformamide is quasireversible, with a reduction potential (E^o) of +0.01 V vs. NHE. The E_{00} of photoexcited $[Ru(phen)_2dppz]^{2+*}$ is 2.4 V and the ground state potential is −1.6 V. The driving force is calculated by the equation $E°(*D/D^+) = E°(D/D^+) + E_{00}(*D) + E°(A/A^-)$.) In quenching studies of other ruthenium(II) polypyridyl complexes by rhodium(III) polypyridyl complexes, researchers have demonstrated that electron transfer is the dominant mechanism of luminescence quenching (*67–71*). In addition, although Förster energy transfer is known to occur over the distances proposed (*23*), this mechanism requires spectral overlap between the absorbance band of the acceptor and the emission band of the photoexcited energy donor; ruthenium(II)* emission, with a maximum at 617 nm, does not overlap the lowest energy absorbance of the rhodium(III) complex, with a maximum at 360 nm. In recent transient absorption spectroscopic measurements, we have, furthermore, identified the Ru(III) electron transfer intermediate. Luminescence quenching of $Ru(DMP)_2dppz^{2+}$ (DMP = dimethylphen) excited state by Δ-$Rh(phi)_2bpy^{3+}$ bound to DNA yields a long-lived (>1 μs) transient intermediate, whose intensity parallels the fraction of luminescence quenching; the wavelength dependence, additionally, is consistent with the Ru(III) species (*72*). We cannot rule out the possibility that some quenching proceeds by Dexter energy transfer (*64, 73–75*), but it is notable that this exchange energy-transfer mechanism is itself a form of electron transfer. Thus, thermodynamics, literature precedence, and the direct spectroscopic identification of the intermediate all support the proposition that luminescence quenching between $[Ru(phen)_2(dppz)]^{2+*}$ and $[Rh(phi)_2(phen)]^{3+}$ proceeds by an electron-transfer reaction.

Electron-Transfer Reactions Between Metal Complexes Bound Covalently to DNA

To investigate in more detail the effects of the DNA medium on long-range electron transfer between intercalated species, we have designed a system in which the donor and acceptor metal complexes are tethered to the 5′-termini of a 15-base pair DNA duplex (*76*). Attachment of one metal complex to each end of the oligonucleotide duplex through a flexible linker allows the formation of a well-defined electron-transfer assembly with donor and acceptor bound at distinct positions with a discrete distance of separation. The assembly is shown schematically in Figure 5. Covalent attachment of each metal complex to the 5′-terminus of an oligonucleotide also permits two companion experiments, represented in Figure 5, which are useful in characterizing the electron-transfer chemistry. Hybridization of a ruthenated oligonucleotide to its unmodified complement permits measurements of luminescence in the absence of rhodium quencher and thus offers a means to characterize the intercalated species. Furthermore, photocleavage reactions on the rhodium-modified oligonucleotide hybridized to its unmodified complement permits a measurement of the position of intercalation on the duplex, because photoactivated cleavage can be used to mark the site of binding by these rhodium complexes. Modeling studies have suggested that our tether is sufficiently flexible to permit intercalation two base pairs from the end of the helix.

When the ruthenium-modified oligomer is annealed to its unmetallated complement, the metal complex intercalates and intense luminescence is observed (*77*). By contrast, the ruthenium-modified oligonucleotide alone or in the presence of noncomplementary single-stranded DNA displays little luminescence. These results are consistent with previous studies; luminescence is observed in aqueous solution only when the stacked bases of a DNA helix provide a platform for intercalation of the dppz ligand.

Table III shows that the luminescent lifetimes and the relative luminescent intensities for the covalently bound duplex and its noncovalent analogue are similar. As with $[Ru(phen)_2(dppz)]^{2+}$, a biexponential decay in emission is observed for the ruthenated oligonucleotide hybridized to its complement. A small shift in the wavelength of maximum emission is also observed compared to the noncovalent complex. This shift likely reflects the sensitivity in emission to the stacking of the oriented dppz ligand; a dependence of the maximum emission wavelength on base

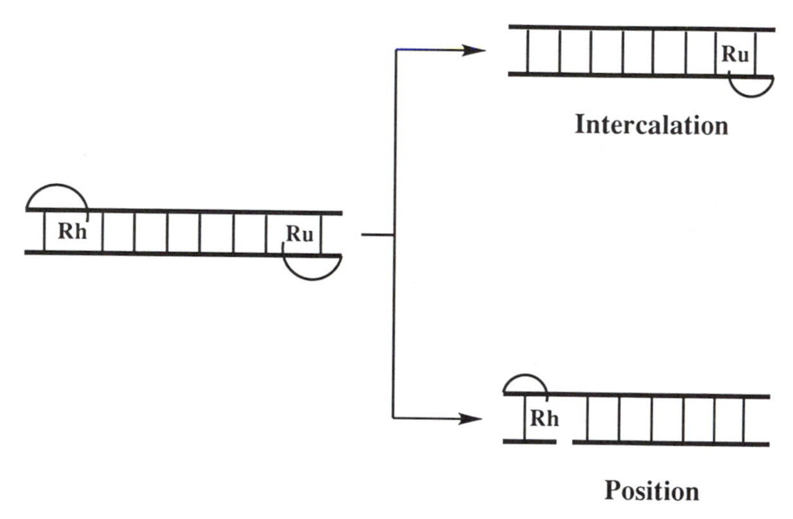

Figure 5. Schematic drawing of intramolecular, covalently bound intercalators on an oligonucleotide. The luminescent properties of the ruthenium-modified duplex provide information about the mode of intercalation; photocleavage of the oligonucleotide by covalently bound rhodium provides a determination of the position(s) of intercalation.

composition has been observed with noncovalently bound complexes (*48*; Table I).

The luminescence of the hybridized $[Ru(phen)_2(dppz)]^{2+}$ derivative may be used to characterize the molecular assembly (*77*). Dilution experiments show that intercalation is intramolecular at concentrations ≤5 mM duplex; addition of unmodified duplex to the covalently bound duplex results in ≤5% change in the luminescence. The results of experiments performed on duplexes containing mismatches in various positions along the duplex are also consistent with intramolecular intercalation. In this series, luminescence is higher for mismatches near the ruthenated end of the oligomer, where the ruthenium complex can intercalate intramolecularly and stabilize the mismatched site.

Luminescence titrations further demonstrate that the ruthenated duplex behaves as a 15 mer bearing one intercalator (*76*). As free $[Ru(phen)_2(dppz)]^{2+}$ is added to a solution of unmetallated 15 mer duplex, the luminescence increases linearly until the emission reaches saturation at about three equivalents of ruthenium(II) per duplex, consistent with competitive binding of $[Ru(phen)_2(dppz)]^{2+}$ to the 15-mer duplex and an average binding site size of a little more than four base pairs. When the analogous experiment is conducted with the ruthenated duplex, saturation of luminescence occurs after almost two equivalents of $[Ru(phen)_2(dppz)]^{2+}$ are added. This comparison indicates that the covalently bound ruthenium(II) complex is not displaced by additional intercalators.

Table III. Excited-State Lifetimes and Integrated Intensities for Covalently Bound Duplex and Its Noncovalent Analogue

Sample	τ *(ns)*[a,b]	%[c]	λ_{max} *(nm)*	Φ[a,d,e]
Ru-duplex	500	60	598	0.0071
	110	40		
Ru + duplex	420	35	610	0.0063
	90	65		

NOTE: All steady-state and time-resolved measurements were taken at 20 °C using instrumentation described in reference 50. Ru-duplex refers to 5′-[Ru(phen′)$_2$(dppz)]-AGTGCCAAGCTTGCA-3′ annealed to its complement.

[a] Error was estimated to be ±10% for both steady-state and time-resolved measurements.

[b] Samples were used in time-resolved measurements contained either 5 μM of the covalently bound duplex or 5 μM $[Ru(phen)_2(dppz)]^{2+}$ and 5 μM of the analogous 15-mer duplex.

[c] Lifetime ratios were calculated from the magnitudes of the pre-exponential factors produced by the program used in the deconvolution of the time-resolved data.

[d] Quantum yields, Φ, were determined relative to $[Ru(bpy)_3]^{2+}$ (Φ = 0.042).

[e] Samples used in steady-state measurements contained either 4 μM of the covalently bound duplex or 4 μM $[Ru(phen)_2(dppz)]^{2+}$ and 4 μM of the analogous 15-mer duplex.

Hybridization of the [Rh(phi)$_2$(phen)]-modified oligonucleotide with its unmodified complement permits the position of intercalation on the helix to be determined, because photoactivation of phi complexes of rhodium promotes strand cleavage at the site of intercalation (*41*). The complementary strand is radioactively labeled at its 5′-end and annealed to the rhodium-modified strand. Photocleavage followed by gel electrophoresis shows that the covalently bound rhodium complex cleaves with high specificity at sites 2 and 3 from the 3′-terminus of the ^{32}P-labeled strand (*76*; Figure 6). In contrast, DNA cleavage by free [Rh(phi)$_2$(phen)]$^{3+}$ yields reaction at all positions on the oligomer. This result indicates the positioning of the covalently bound intercalator with similar probability one or two base pairs in from the 5′-end of the modified strand. The specificity of the rhodium(III) complex also argues that intercalation of these covalently attached complexes is largely intramolecular. Because the linker arm is the same for both metal complexes, one may deduce that both the rhodium(III) and the ruthenium(II) complexes are able to intercalate one and two bases from the 5′-sites of attachment. Because the stacking distance between π-systems is 3.4 Å (*54*), hybridization of ruthenated 15 mer hybridized to its rhodium-modified complement creates an assembly in which the intercalated ligands of the two metal complexes are separated by at least 41 Å (the most probable distance of separation is 44 Å). Figure 7 illustrates the assembly and the functionalized metal complexes used in its construction.

What is the level of intramolecular luminescent quenching observed when the metallated DNA strands are annealed to each other? In this system with donors and acceptors separated by ≥41 Å we had expected to find a small but significant level of quenching. What we actually observe is the complete quenching of luminescence, a result that perhaps is not surprising given the static quenching experiments on DNA with noncovalently bound species. On the basis of this quenching, laser flash photolysis studies put a lower limit on the rate of electron transfer of 3×10^9 s^{-1} through the π-stack (*76*).

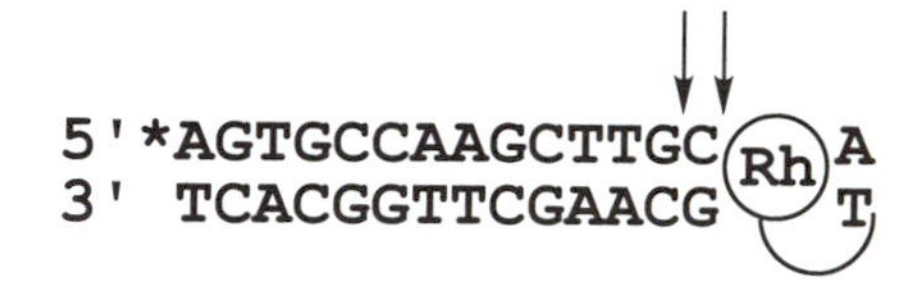

Figure 6. Sequence of a 15-mer oligonucleotide bearing covalently bound rhodium(III) complex, hybridized to its ^{32}P-labeled complement. Arrows point to the sites of photocleavage by the metal complex, establishing that it is intercalated either adjacent to the first (as shown) or second base steps from the site of covalent attachment.

Table IV compares this result to the steady-state luminescence of related species. For example, the addition of the doubly modified duplex to the ruthenium-modified duplex does not quench the luminescence from the ruthenium-modified duplex. The lack of luminescence demonstrates the absence of any adventitious quenchers in the rhodium(III) sample. Addition of an equimolar amount of rhodium-modified duplex to ruthenium-modified duplex also does not promote significant quenching of the ruthenium duplex, consistent with the quenching being substantially intramolecular at these concentrations. These studies complement the photocleavage experiments (Figure 6), from which we estimate less than 15% intermolecular interaction at these concentrations.

It is also useful to consider the luminescence from metallated oligonucleotides in the presence of noncovalent metallointercalator. Adding one equivalent of free $[Ru(phen)_2(dppz)]^{2+}$ to the ruthenium-modified duplex doubles the intensity in luminescence, consistent with independent intercalation by the two species. As described earlier, steady-state luminescence reaches saturation at approximately three times the luminescence of the ruthenium-modified duplex when two equivalents of $[Ru(phen)_2(dppz)]^{2+}$ have been added. It is not surprising, then, that addition of a stoichiometric amount of $[Rh(phi)_2(phen)]^{3+}$ to the ruthenium-modified duplex leads to substantial but not complete quenching of the ruthenium emission. Statistically, some duplexes will accommodate two rhodium(III) complexes, leaving a few ruthenium-modified duplexes unoccupied and therefore unquenched. Thus, complete quenching is observed only when the acceptor is covalently bound to the same duplex as the donor.

An analogue of $[Ru(phen)_3]^{2+}$ has also been linked to the smaller oligonucleotide 5′-CTATTAGC-3′ and $[Rh(phen)_3]^{3+}$ has been linked to

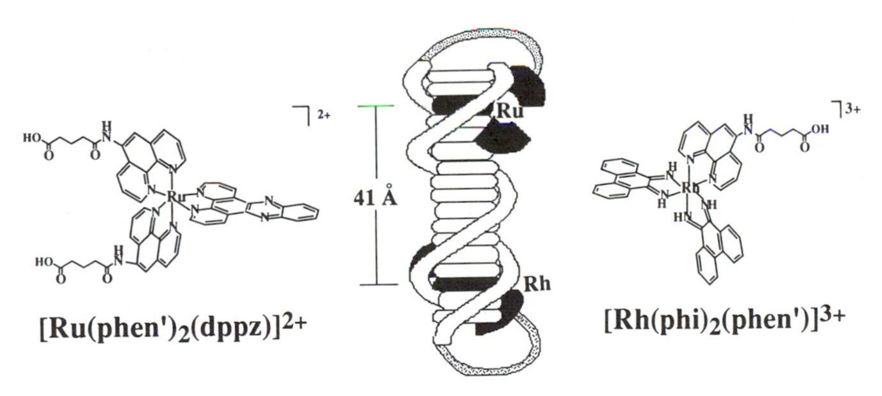

Figure 7. Structures of covalently bound donor and acceptor metal complexes with schematic picture of the doubly metallated 15-mer duplex. The closest metal-to-metal separation is 41 Å.

Table IV. Luminescent Intensities of Covalent Metal–DNA Intercalation Complexes

Sample[a]	*Relative Luminescence Intensity*[b]
Ru-duplex	1.00
Ru-duplex-Rh	0.00
$\frac{1}{2}$ Ru-duplex + $\frac{1}{2}$ Ru-duplex-Rh	0.57
$\frac{1}{2}$ Ru-duplex + $\frac{1}{2}$ duplex-Rh	0.43
Ru-duplex + Ru(II)	2.78
Ru-duplex + Rh(III)	0.08

[a] Ru-duplex refers to 5′-[Ru(phen′)$_2$(dppz)]-AGTGCCAAGCTTGCA-3′ annealed to its complement; Ru-duplex-Rh refers to 5′-[Ru(phen′)$_2$(dppz)]-AGTGCCAAGCTTGCA-3′ annealed to 5′-[Rh(phi)$_2$(phen′)]-TGCAAGCTTGGCACT-3′; Ru(II) and Rh(III) refer to [Ru(phen′)$_2$(dppz)]$^{2+}$ and [Rh(phi)$_2$(phen′)]$^{3+}$, respectively.

[b] All spectra were taken on an SLM Aminco 8000 spectrofluorimeter. Intensities were integrated from 500 to 800 nm and are normalized to the luminescence intensity of Ru-duplex. The uncertainty of integrated intensities is ±10% (*76*).

its complement (*76*). As described in the preceding paragraphs, [Ru(phen)$_3$]$^{2+}$ luminesces in the presence of DNA with two lifetimes, a short one comparable to free ruthenium and an increased excited-state lifetime that arises from the intercalatively bound species. When the [Ru(phen)$_3$]-labeled oligonucleotide is hybridized to its complement, however, no long-lived luminescence is observed, indicating the absence of intercalation by the covalently attached ruthenium complex. Also, although noncovalently bound [Ru(phen)$_3$]$^{2+}$ is protected from quenching by [Fe(CN)$_6$]$^{4-}$, the double helix offers no protection to the covalently bound ruthenium(II) complex from ferrocyanide quenching. Importantly, when this [Ru(phen)$_3$]-modified oligonucleotide is hybridized to the [Rh(phen)$_3$]-modified complement, there is no quenching of luminescence. Taking these results with those for the avid intercalators, we conclude that intercalation is required for rapid electron transfer to occur. Thus, electron transfer appears to proceed much more efficiently through noncovalent π interactions than along a covalent σ framework.

The results for covalently bound analogues of [Ru(phen)$_2$(dppz)]$^{2+}$ and [Rh(phi)$_2$(phen)]$^{3+}$ intercalated into a 15-mer oligonucleotide therefore demonstrate that photoinduced electron transfer between intercalators can occur rapidly over >40 Å through a DNA helix over a pathway consisting of π-stacked base pairs. The DNA π-stack may be considered a remarkably effective medium for electronic coupling of intercalated species.

Conclusion

The experiments described in this chapter point to the DNA π-stack as an effective intervening medium for long-range electron transfer. The results suggest more generally that a π-stack may provide a substantially preferred pathway for electron-transfer reactions, a pathway that should be considered in the study of biological transport and in the design of new artificial sensors. These experiments hopefully also underscore the utility of DNA as a polymer in exploring these and other long-range reactions. The DNA oligonucleotide offers synthetic flexibility and a range of tools to characterize assemblies that are formed. The results described here mark the beginning of our exploration of how the DNA double helix mediates electron-transfer chemistry and of how metallointercalators may be used to probe this chemistry.

Acknowledgments

We are grateful to the efforts of our collaborators and co-workers, as noted in the individual references. We acknowledge in particular the contributions of C. V. Kumar, who first discovered the remarkable efficiency of DNA in promoting reactions between transition metal complexes. We thank also Jay Winkler, who has provided expert technical assistance in the Beckman Institute laser laboratory. In addition we thank the National Institutes of Health, the National Science Foundation, the Air Force Office of Scientific Research, the Ralph M. Parsons Foundation, and the National Foundation for Cancer Research for financial support.

References

1. Marcus, R. A.; Sutin, N. *Biochim. Biophys. Acta* **1985,** *811,* 265.
2. Bowler B. E.; Raphael, A. L; Gray, H. B. *Prog. Inorg. Chem.* **1990,** *38,* 259.
3. Meyer T. J. *Acc. Chem. Res.* **1989,** *22,* 163.
4. *Energy Resources through Photochemistry and Catalysis;* Gratzel, M., Ed.; Academic: New York, 1983.
5. Moser, C. C.; Keske, J. M.; Warncke, K.; Farid, R. S.; Dutton, P. L. *Nature (London)* **1992,** *355,* 796.
6. Wuttke, D. S.; Bjerrum, M. J.; Winkler, J. R.; Gray, H. B. *Science (Washington, D.C.)* **1992,** *256,* 1007.
7. McLendon, G. *Acc. Chem. Res.* **1988,** *21,* 160.
8. Hoffman, B. M.; Natan, M. J.; Nocek, J. M.; Wallin, S. A. *Struct. Bond.* **1991,** *75,* 85.
9. Closs, G. L.; Calcaterra, L. T.; Green, N. J.; Penfield, K. W.; Miller, J. R. *J. Phys. Chem.* **1986,** *90,* 3673.
10. Joran, A. D.; Leland, B. A.; Felker, P. M.; Zewail, A. H.; Hopfield, J. J.; Dervan, P. B. *Nature (London)* **1987,** *327,* 508.
11. Meier, M. S.; Fox, M. A.; Miller, J. R. *J. Org. Chem.* **1991,** *56,* 5380.
12. Closs, G. L.; Miller, J. R. *Science (Washington, D.C.)* **1988,** *240,* 440.
13. Isied, S. S.; Vassilian, A. *J. Am. Chem. Soc.* **1984,** *106,* 1726.

14. Beratan, D. N.; Onuchic, J. N. In *Electron Transfer in Inorganic, Organic, and Biological Systems;* Bolton, J. R.; Mataga, N.; McLendon, G. L., Eds.; Advances in Chemistry 228; American Chemical Society: Washington, DC, 1991; p 71.
15. Ratner, M. A. *J. Phys. Chem.* **1990,** *94,* 4876.
16. Naleway, C. A.; Curtiss, L. A.; Miller, J. R. *J. Phys. Chem.* **1991,** *95,* 8434.
17. Risser, S. M.; Beratan, D. N.; Mead, T. J. *J. Am. Chem. Soc.* **1993,** *115,* 2508.
18. Marks, T. J. *Science (Washington, D.C.)* **1985,** *227,* 881.
19. Tanaka, M.; Yoshida, H.; Ogasawara, M. *J. Phys. Chem.* **1991,** *95,* 955.
20. Schouten, P. G.; Warman, J. M.; deHaas, M. P.; Fox, M. A.; Pan, H.-L. *Nature (London)* **1991,** *353,* 736.
21. Markovitsi, D.; Bengs, H.; Ringsdorf, H. *J. Chem. Soc. Faraday Trans. 1* **1992,** *88,* 1275.
22. Gaudiello, J. G.; Kellogg, G. E.; Tetrick, S. M.; Marks, T. J. *J. Am. Chem. Soc.* **1989,** *111,* 5259.
23. Cantor, C. R.; Schimmel, P. R. *Biophysical Chemistry;* W. H. Freeman and Company: New York, 1980.
24. Otto, P.; Clementi, E.; Ladik; J. *J. Chem. Phys.* **1983,** *78,* 4547.
25. Dee, D.; Baur, M. E. *J. Chem. Phys.* **1974,** *60,* 541.
26. Risser, S. M.; Beratan, D. J.; Meade, T. J. *J. Am. Chem. Soc.* **1993,** *115,* 2508.
27. Fromherz, P.; Rieger, B. *J. Am. Chem. Soc.* **1986,** *108,* 5361.
28. Brun, A. M.; Harriman, A. *J. Am. Chem. Soc.* **1992,** *114,* 3656.
29. Barton, J. K.; Kumar, C. V.; Turro, N. J. *J. Am. Chem. Soc.* **1986,** *108,* 6391.
30. Purugganan, M. D.; Kumar, C. V; Turro, N. J.; Barton, J. K. *Science (Washington, D.C.)* **1988,** *241,* 1645.
31. Miller, J. H.; Swenberg, C. E. *Can. J. Phys.* **1990,** *68,* 962.
32. Baverstock, K. F.; Cundall, B. *Radiat. Phys. Chem.* **1988,** *32,* 553.
33. Cullis, P. M.; McClymont, J. D., Symons, M. C. R. *J. Chem. Soc. Faraday Trans. 1* **1990,** *86,* 591.
34. Houee-Levin, C.; Gardes-Albert, M.; Rouscilles, A.; Ferradini, C.; Hickel, B. *Biochemistry* **1991,** *30,* 8216.
35. Barton, J. K.; Goldberg, J. M.; Kumar, C. V.; Turro, N. J. *J. Am. Chem. Soc.* **1986,** *108,* 2081.
36. Kumar, C. V.; Barton, J. K.; Turro, N. J. *J. Am. Chem. Soc.* **1985,** *107,* 5518.
37. Rehmann, J. P.; Barton, J. K. *Biochemistry* **1990,** *29,* 1701.
38. Friedman, A. E.; Chambron, J.-C.; Sauvage, J.-P.; Turro, N. J.; Barton, J. K. *J. Am. Chem. Soc.* **1990,** *112,* 4960.
39. Chambron, J.-C.; Sauvage, J.-P.; Amouyal, E.; Koffi, P. *New J. Chem.* **1985,** *9,* 527.
40. Amouyal, E.; Homsi, A.; Chambron, J.-C.; Sauvage, J.-P. *J. Chem. Soc. Dalton Trans.* **1990,** *1841,* 1990.
41. Sitlani, A.; Long, E. C.; Pyle, A. M.; Barton, J. K. *J. Am. Chem. Soc.* **1992,** *14,* 2303.
42. Pyle, A. M.; Long, E. C.; Barton, J. K. *J. Am. Chem. Soc.* **1989,** *111,* 4520.
43. David, S. S.; Barton, J. K. *J. Am. Chem. Soc.* **1993,** *115,* 2984.
44. Pyle, A. M.; Chiang, M. Y.; Barton, J. K. *Inorg. Chem.* **1990,** *29,* 4487.
45. Krotz, A. H.; Kuo, L. Y.; Barton, J. K. *Inorg. Chem.* **1993,** *32,* 5963.
46. Friedman, A. E.; Kumar, C. V.; Turro, N. J.; Barton, J. K. *Nucleic Acids Res.* **1991,** *19,* 2595.
47. (a) Hartshorn, R. M.; Barton, J. K. *J. Am. Chem. Soc.* **1992,** *114,* 5925. (b) Dupureur, C. M.; Barton, J. K.; *J. Am. Chem. Soc.* **1994,** *116,* 10286.

48. Jenkins, Y.; Friedman, A. E.; Turro, N. J.; Barton, J. K. *Biochemistry* **1992,** *31,* 10811.
49. Jenkins, Y.; Barton, J. K. *Proc. SPIE—Int. Soc. Opt. Eng.* **1993,** *1885,* 129.
50. Gupta, N.; Grover, N.; Neyhart, G. A.; Singh, P.; Thorp, H. H. *Inorg. Chem.* **1993,** *32,* 310.
51. Hiort, C.; Lincoln, P.; Norden, B. *J. Am. Chem. Soc.* **1993,** *115,* 3448.
52. Fees, J.; Kaim, W.; Moscherosch, M.; Matheis, W.; Klima, J.; Krejcik, M.; Zalis, S. *Inorg. Chem.* **1993,** *32,* 166.
53. Murphy, C. J.; Arkin, M. R.; Ghatlia, N. D.; Bossmann, S.; Turro, N. J.; Barton, J. K. *Proc. Natl. Acad. Sci. U.S.A.* **1994,** *91,* 5315.
54. Saenger, W. *Principles of Nucleic Acid Structure;* Springer-Verlag: New York, 1984.
55. Pyle, A. M.; Barton, J. K. *Prog. Inorg. Chem.* **1990,** *38,* 259.
56. Chow, C. S.; Barton, J. K. *Methods Enzymol.* **1992,** *212,* 219.
57. Chow, C. S.; Behlen, L. S.; Uhlenbeck, O. C.; Barton, J. K. *Biochemistry* **1992,** *31,* 972.
58. Uchida, K.; Pyle, A. M.; Morii, T.; Barton, J. K. *Nucleic Acids Res.* **1989,** *17,* 10259.
59. Krotz, A. H.; Kuo, L. Y.; Shields, T. P.; Barton, J. K. *J. Am. Chem. Soc.* **1993,** *115,* 3876.
60. Krotz, A. H.; Hudson, B. P.; Barton, J. K. *J. Am. Chem. Soc.* **1993,** *115,* 12577.
61. Sitlani, A.; Dupureur, C.; Barton, J. K. *J. Am. Chem. Soc.* **1993,** *115,* 12589.
62. Navon, G.; Sutin, N. *Inorg Chem* **1974,** *13,* 2159.
63. Ho, P. S.; Frederick, C.; Saal, D.; Wang, A. H.-J.; Rich, A. *J. Biomol. Struct. Dynam.* **1987,** *4,* 521.
64. Wagner, P. J. In *Creation and Detection of the Excited State;* Lamola, A. A., Ed.; Marcel Dekker: New York, 1971; Chapter 4, p 173.
65. Turro, N. J. *Modern Molecular Photochemistry;* Benjamin-Cummings: Menlo Park, CA, 1978.
66. Laws, W. R.; Contino, P. B. *Methods Enzymol.* **1992,** *210,* 448.
67. Chan, S.-F.; Chou, M.; Creutz, C.; Matsubara, T.; Sutin, N. *J. Am. Chem. Soc.* **1981,** *103,* 369.
68. Creutz, C.; Keller, A. D.; Sutin, N.; Zipp, A. P. *J. Am. Chem. Soc.* **1982,** *104,* 3618.
69. Kalyanasundaram, K.; Gratzel, M.; Nazeeruddin, M. K. *J. Phys. Chem.* **1992,** *96,* 5865.
70. Nozaki, K.; Ohno, T.; Haga, M. *J. Phys. Chem.* **1992,** *96,* 10880.
71. Furue, M.; Hirata, M.; Kinoshita, S.; Kushida, T.; Kamachi, M. *Chem. Lett.* **1990,** 206.
72. Stemp, E.; Arkin, M. R.; Barton, J. K. *J. Am. Chem. Soc.,* in press.
73. Dexter, D. L. *J. Chem. Phys.* **1953,** *21,* 836.
74. Endicott, J. F. *Acc. Chem. Res.* **1988,** *21,* 59.
75. Closs, G. L; Johnson, M. D.; Miller, J. R.; Piotrowiak, P. *J. Am. Chem. Soc.* **1989,** *111,* 3751.
76. Murphy, C. J.; Arkin, M. R.; Jenkins, Y.; Ghatlia, N. D.; Bossman, S.; Turro, N. J.; Barton, J. K. *Science (Washington, D.C.)* **1993,** *262,* 1025.
77. Jenkins Y.; Barton, J. K. *J. Am. Chem. Soc.* **1992,** *114,* 8737.

RECEIVED for review August 23, 1993. ACCEPTED revised manuscript December 21, 1993.

18

Donor–Acceptor Electronic Coupling in Ruthenium-Modified Heme Proteins

Danilo R. Casimiro[1], David N. Beratan[2], José Nelson Onuchic[3], Jay R. Winkler[1], and Harry B. Gray[1]

[1] Beckman Institute, California Institute of Technology, Pasadena, CA 91125
[2] Department of Chemistry, University of Pittsburgh, Pittsburgh, PA 15260
[3] Department of Physics, University of California, San Diego, La Jolla, CA 92093

The rates of electron transfer (ET) in six Ru-modified cytochrome c derivatives were analyzed in terms of four theoretical models describing donor–acceptor electronic coupling. The simplest model, which treats the protein as a homogeneous medium, fails to describe the variations in ET rates with changes in donor–acceptor separation. The three other models explicitly account for the inhomogeneity of the polypeptide matrix and are more successful in describing the electronic couplings. Calculations of relative coupling strengths give results within an order of magnitude of experimentally determined values for cytochrome c. *The homogeneous-medium model is more successful in describing ET in Ru-modified myoglobin, and two of the inhomogeneous-medium models suggest that multiple pathways are important in mediating the electronic coupling.*

MANY BIOENERGETIC AND BIOSYNTHETIC PROCESSES involve electron-transfer (ET) steps in which an electron tunnels several angstroms from donor to acceptor through protein (*1*). The problem of understanding the detailed mechanisms of these processes is being addressed experimentally through studies of ET in synthetic model complexes (*2–4*), chemically modified metalloproteins (*5*), and protein complexes (*6–8*). Theoretical efforts are aimed at describing the electronic coupling between distant redox sites (*9–21*) and clarifying the importance of protein conformational dynamics in regulating ET (*22*). A particular emphasis

0065–2393/95/0246–0471$08.00/0

of our work is to identify the role of the protein matrix in mediating biological electron transfer.

The weak electronic coupling between distant donor and acceptor sites leads to long-range ET rates (k_{ET}) that are well-described by a nonadiabatic formulation (*23*):

$$k_{ET} = \frac{2\pi}{\hbar} |T_{DA}|^2(FC) \tag{1}$$

The rate is proportional to an electronic coupling factor, $|T_{DA}|^2$, and a Franck–Condon factor, FC, that arises from nuclear motion coupled to the ET process. The FC factor describes the trade-off between reaction free energy and the nuclear reorganization energy. At the optimum driving force, FC is unity, and rates reach a maximum value (k_{max}), limited by $|T_{DA}|^2$. The simplest description of T_{DA} treats the medium between donor and acceptor as a one-dimensional square tunneling barrier (1DSB). As such, the rate is predicted to drop exponentially with $R - R_0$, the tunneling distance (*24, 25*):

$$k_{ET}{}^{1DSB} = \frac{2\pi}{\hbar} |T_{DA}(R_0)|^2 \exp[-\beta(R - R_0)](FC) \tag{2}$$

Accounting for the role of protein-mediated coupling in the 1DSB models amounts to assigning a barrier height for electron tunneling. Early estimates of the electronic coupling decay constants by Hopfield (β = 1.4 Å^{-1}) (*24*) and Jortner (β = 2.6 Å^{-1}) (*25*) stimulated numerous experiments on small molecules and proteins of varied bridge structure. Both proteins and hydrocarbon bridges of known structure were used to link donors and acceptors. Homologous series of organic compounds displayed different β values. Improved theoretical analysis showed that much of this diversity could arise from bridge orbital-energy and symmetry effects. Dutton and co-workers (*26, 27*) suggested that the uniform 1DSB model for electron tunneling with a single exponential decay constant (eq 2) qualitatively describes a broad range of natural and synthetic ET systems. The value of β derived for biological ET systems was 1.4 Å^{-1}. This treatment provides a reliable zero-order estimate of coupling strengths for many protein ET systems over a wide range of donor–acceptor separations. However, some discrepancies may be due to specific structural aspects of the proteins involved.

A 1DSB is a crude approximation to a polypeptide matrix. A more sophisticated model can be developed by decomposing the protein medium between two redox sites into smaller subunits (*28–31*). The coupling between the termini in a bridge that is composed of identical repeating units drops by a simple multiplicative factor (ϵ) as the chain is lengthened. This value of ϵ depends on the energy of the tunneling electron as well as the composition of the repeat unit in the bridge. For simplicity, the decay can be divided into components associated with each bonded and nonbonded contact within the repeat unit.

The convenient and practical "decay per unit" strategy can be readily extended to aperiodic protein structures by differentiating among the many types of interactions present in a protein structure. Tunneling is much more efficient (the electronic coupling decays more slowly) through bonded orbitals than through space, because the potential barrier is effectively lower. In proteins, the covalently bonded path between donor and acceptor can be extremely long compared to the direct through-space distance. As a first approximation, we distinguish only between covalent bonds, hydrogen bonds, and nonbonded through-space contact (subscripts C, H, and S, respectively). If a single pathway dominates the matrix element (9):

$$T_{DA} \propto \Pi_i \, \epsilon_C(i) \, \Pi_j \, \epsilon_H(j) \, \Pi_k \, \epsilon_S(k) \tag{3}$$

Just as the ϵ values are highly "renormalized" parameters in the periodic chain calculation, these decay factors are also renormalized for proteins. The values of the parameters take into account, in an average way, high order contributions to the coupling, a distribution in bond energies and interactions, and local quantum interference effects. Any combination of covalent, hydrogen-bonded, and through-space contacts between the donor and acceptor sites defines a specific physical tunneling pathway.

We estimate relative tunneling matrix elements by finding the single physical pathway that maximizes the product in eq 3 using the parameters (*9*, *30*, *32*):

$$\epsilon_H = \epsilon_C^{\,2} \exp[-1.7(R - 2R_C)] \tag{4a}$$

$$\epsilon_S = \sigma_S \epsilon_C \exp[-1.7(R - R_C)] \simeq 0.5\epsilon_C \exp[-1.7(R - R_C)] \tag{4b}$$

The constant chosen for ϵ_C is usually 0.6, and for σ_S (the average through-space orientation factor) the constant is 0.5. These parameters are based on experiments, where available, and on theory where experiments are preliminary or sparse. The nature of the dominant paths does not change drastically as the parameters are varied over a physically reasonable range. The most critical parameter of the model is the value of ϵ_C relative to the exponential decay of through-space interactions.

Using the decay factors in eqs 3 and 4, one can search for the lowest order contributions to T_{DA}. In this way, a macromolecule electronic structure problem is reduced to the relatively simple task of finding the minimum distance in a graph. The minimum-distance problem is well-known and can be solved in a reasonable amount of time. Thus, a given protein structure defines the connectivity and decay factors between neighboring sites according to the prescription in eq 3, and the optimum coupling pathways between any two points are readily calculated.

The 1DSB model for protein ET predicts that ln k_{ET} should scale linearly with distance. In the tunneling pathway (TP) model, rates should scale with the product of decay factors in eq 3. Thus, if pathways between donor and acceptor consist of a small number of strong links (covalent or hydrogen-bonded connections), the couplings and maximum rates will be large. If, however, a donor–acceptor pair has circuitous bonded pathways (i.e., a large number of covalent or hydrogen-bonded links) or weak direct paths (i.e., long or numerous through-space contacts), the maximum rate will be anomalously slow. These qualitative expectations can be quantified by calculating the effective tunneling pathway length, $\sigma_1 = [\ln \Pi_i \epsilon(i)/\ln \epsilon_C]R_C$. ET rates should drop exponentially with the tunneling pathway length rather than the physical distance between donor and acceptor. The value of ϵ_C (0.6) implies an exponential decay constant of 0.73 $Å^{-1}$.

A hybrid approach to the electronic coupling problem in proteins has been introduced by Siddarth and Marcus (*10–12*). In this method, a pathway-searching algorithm is combined with extended Hückel (EH) calculations on a subset of amino acids in the protein. As with the TP model, a search of the protein structure for important residues reduces the complexity of the electronic structure problem. These residues are then used as a basis in an EH calculation of the electronic coupling matrix element T_{DA}. The method is not expected to produce highly accurate absolute values of T_{DA}, but it is expected to describe relative values of this matrix element.

Cytochrome c

Our approach to studying the effects of the inhomogeneous protein medium on ET rates is to probe different sections of a single protein at many predefined distances and intervening structures by modifying surface sites with redox-active probes (*5*). We have completed measurements of intramolecular electron transfer on six different ruthenium derivatives of wild-type and mutant cytochromes (*33–36*). Each derivative contains a histidine residue that is covalently modified with a bis(2,2′-bipyridine)imidazoleruthenium complex. The histidine residues are either naturally occurring (His-33 of horse cytochrome *c* and His-39 of *Candida krusei* cytochrome *c*), introduced semisynthetically (His-72 of horse cytochrome *c*), or genetically engineered (His-58, His-62, and His-66 of *Saccharomyces cerevisiae* iso-1-cytochrome *c*).

The rates of intramolecular $Fe^{2+} \rightarrow Ru^{3+}$ electron transfer were measured by a flash-quench technique (*33*); the results are listed in Table I along with other pertinent ET parameters. Measured ET rates span 2

Table I. ET Parameters for Ru(His-*X*)-Modified Cytochromes *c* and Human Myoglobin

Type of Protein	X^a	k_{ET} (s^{-1})	$-\Delta G°$ (eV)	k_{max} (s^{-1})	T_{DA} (cm^{-1})	d (Å)	σ_l (Å)
Cytochrome c^b	39	3.2×10^6	0.74	3.3×10^6	1.1×10^{-1}	12.3	19.6
	33	2.6×10^6	0.74	2.7×10^6	9.7×10^{-2}	11.1	19.5
	66	1.0×10^6	0.77	1.1×10^6	6.0×10^{-2}	13.3	19.6
	72	9.0×10^5	0.74	9.4×10^5	5.7×10^{-2}	8.4	24.6
	58	5.2×10^4	0.69	6.0×10^4	1.4×10^{-2}	13.2	21.4
	62	1.0×10^4	0.74	1.0×10^4	5.9×10^{-3}	14.5	28.8
Human myoglobin^c	70	1.6×10^7	0.82	7.2×10^7	5.6×10^{-1}	9.5	24.3
	48	7.2×10^4	0.82	3.3×10^5	3.8×10^{-2}	12.7	37.5
		1.1×10^5	0.96				
	83	4.0×10^2	0.82	1.8×10^3	2.8×10^{-3}	15.5	43.5
		7.3×10^2	0.96				

[a] Cytochrome *c* was modified with $Ru(bpy)_2(im)(His\text{-}X)$ (*33–36*). Human myoglobin was modified with $Ru(NH_3)_5(His\text{-}X)$ (*42*). *X* is the position of the histidine in the amino acid sequence of the protein.

[b] Values of k_{ET} correspond to $Fe^{2+} \rightarrow Ru(bpy)_2(im)(His\text{-}X)^{3+}$ ET; k_{max} and T_{DA} were estimated by using $\lambda = 0.8$ eV (λ is the reorganization energy for the ET reaction).

[c] Values of k_{ET} correspond to $*ZnP \rightarrow Ru(NH_3)_5(His\text{-}X)^{3+}$ (upper) and $Ru(NH_3)_5(His\text{-}X)^{2+} \rightarrow ZnP^+$ (lower) ET; k_{max} and T_{DA} were estimated by using $\lambda = 1.3$ eV.

orders of magnitude and a 6-Å range of edge–edge donor–acceptor separations. Maximum ET rates, k_{max}, were estimated by a classical expression for FC ($\lambda = 0.8$ eV) (*23*).

The 1DSB model [using Dutton's parameters, $\beta = 1.4$ Å^{-1} and $k_{max}(R = 3\ \text{Å} = R_0) = 10^{13}\ \text{s}^{-1}$] consistently predicts ET rates faster than are observed, and k_{max} is not strongly correlated with R (Figure 1) (*34–36*). The systematic deviations could be corrected by choosing a

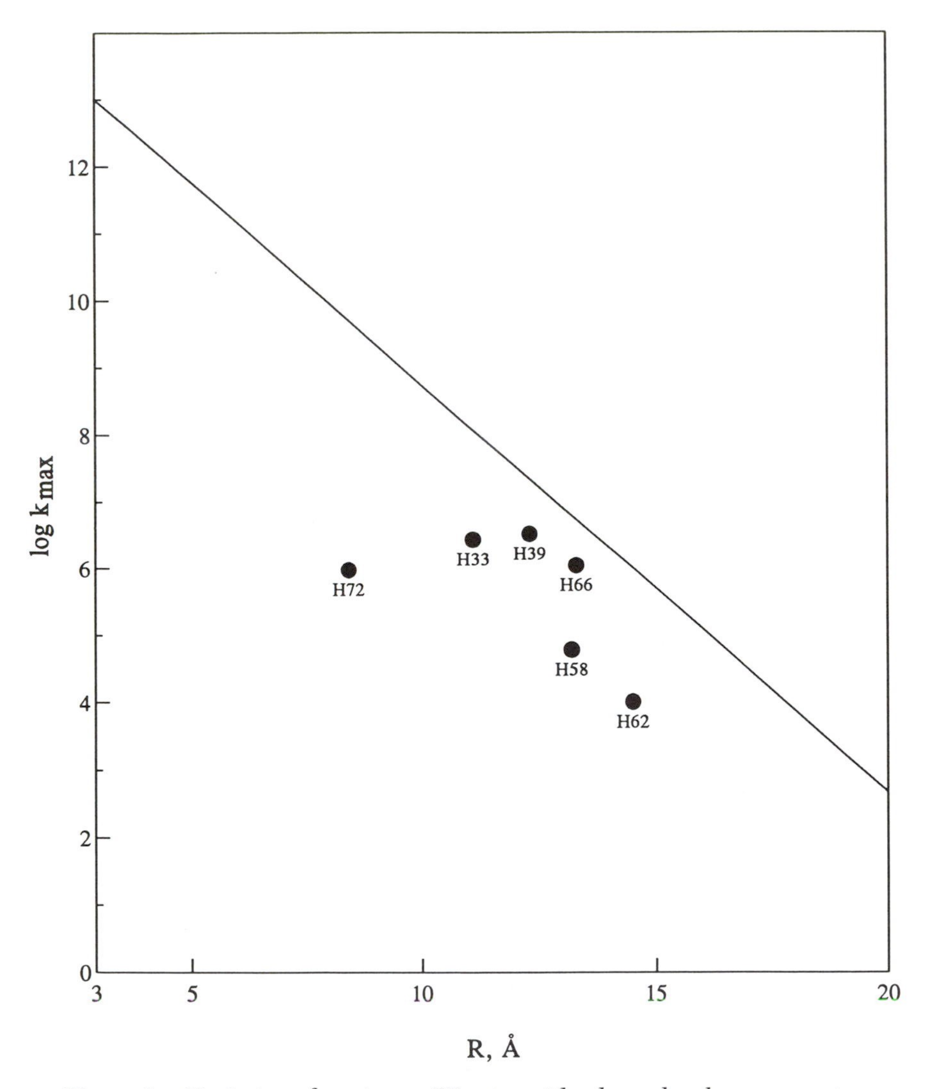

Figure 1. Variation of maximum ET rates with edge–edge donor–acceptor distance in Ru(bpy)$_2$(im)(HisX)-modified cytochrome c *(*X *= 33, 39, 58, 62, 66, and 72).*

more appropriate distance measure (e.g., metal-to-metal rather than edge-to-edge) or, equivalently, a smaller value of k_{max} at close contact. The scatter, however, reflects a weakness in the 1DSB model, most likely its failure to account for the inhomogeneous nature of the intervening medium.

Explicit inclusion of the protein structural details, even at the relatively basic level of the TP model, leads to a noticeable improvement over the 1DSB model. A plot of log k_{max} versus σ_1 appears in Figure 2. The solid line is the best fit to the data with β' constrained to a value of 0.73 $Å^{-1}$. Clearly, the ET rates correlate much better with σ_1 than with R, a result indicating that the structural details of the intervening protein medium are important in determining donor–acceptor coupling strengths.

The results of the TP calculation also can be represented as a plot of log k_{max}(experimental) versus log k_{max}(calculated) (Figure 3). The TP model does not calculate absolute coupling strengths; values of k_{max}(calculated) were obtained from the intercept of the line in Figure 2. The TP model does a very good job of describing k_{max} in four of the six ET rates for the Ru-modified cytochromes and is within about 1 order of magnitude for the other two systems. Given that the TP model assumes only three types of interactions, the agreement is remarkable. The simplicity of the model may be responsible, in part, for some of the deviations apparent in Figures 2 and 3. The assumption that a single pathway dominates the coupling between redox sites may also contribute to the discrepancies.

Direct calculation of T_{DA} at the extended Hückel level provides a good correlation of k_{max}(experimental) with k_{max}(calculated), although only four of the six modified cytochromes have been analyzed (*11*). Again, scaling of the calculated values is required to eliminate a systematic deviation from experimental values. Figure 4 compares the dominant pathways identified in the EH method to those found using the TP algorithm. Both models find the same link between His-33 and the heme: direct coupling along the peptide from His-33 to Pro-30, then a hydrogen-bonded contact between the carboxyl oxygen of Pro-30 and the delta nitrogen of His-18. Substantially different pathways are found, however, for the His-39, His-62, and His-72 derivatives. The discrepancies could be the result of the different criteria used in determining couplings between groups. The large differences in the His-72 pathways are, nevertheless, surprising.

The homogeneous barrier model for electronic couplings clearly is not adequate for cytochrome *c*, and a model that includes the explicit structure of the intervening medium is needed to provide a semiquantitative description of the tunneling process. Simple 1DSB models capture the overall decay of coupling for a very large range of distances

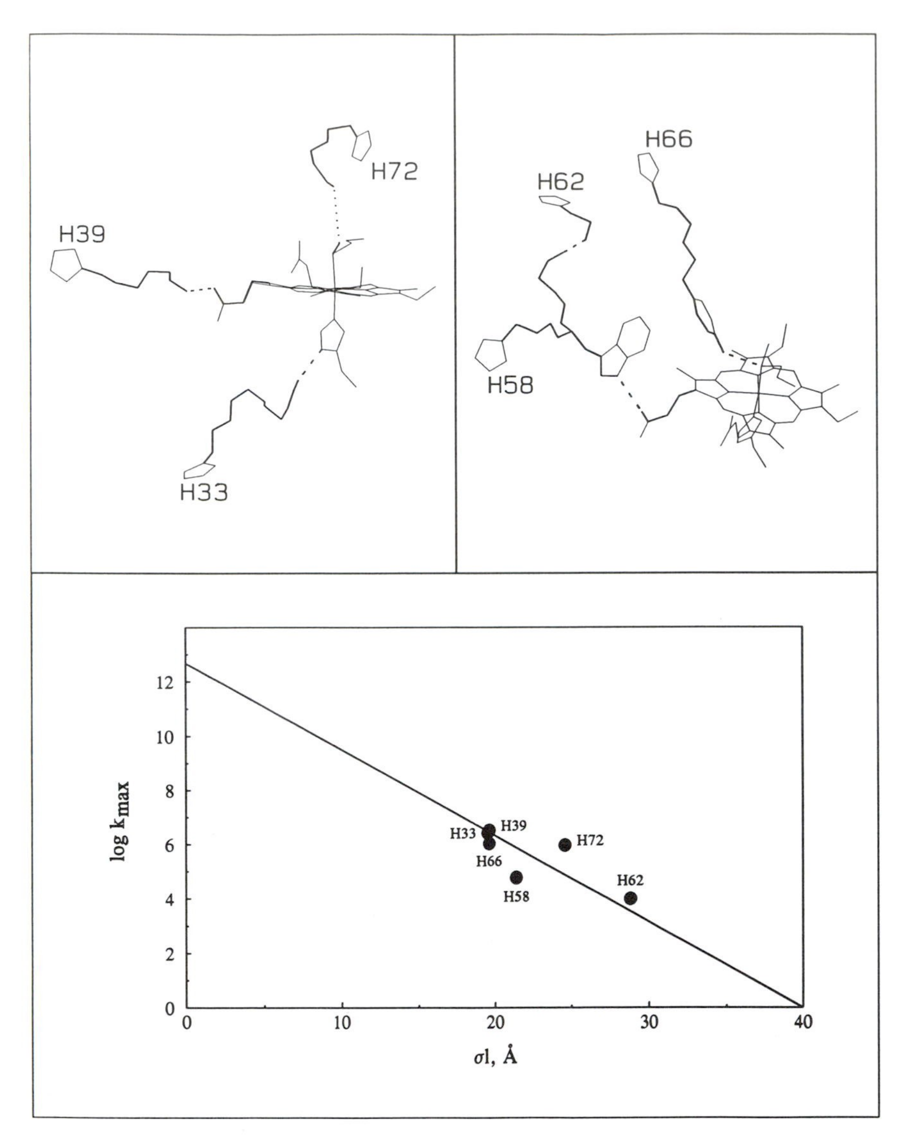

Figure 2. Top: Dominant tunneling pathways in $Ru(bpy)_2(im)(HisX)$-modified cytochrome c *(X = 33, 39, 58, 62, 66, and 72). Bottom: Variation of maximum ET rates with tunneling length (σ_l).*

and rates; they do not, however, account for the observed data scatter arising from the three-dimensional structures of the cytochromes that have been examined (*28*). The TP model contains a sufficiently detailed description of the medium to account for the considerable rate differences in the Ru-labeled cytochromes that are inconsistent with 1DSB models. The EH method also provides good agreement with experiment.

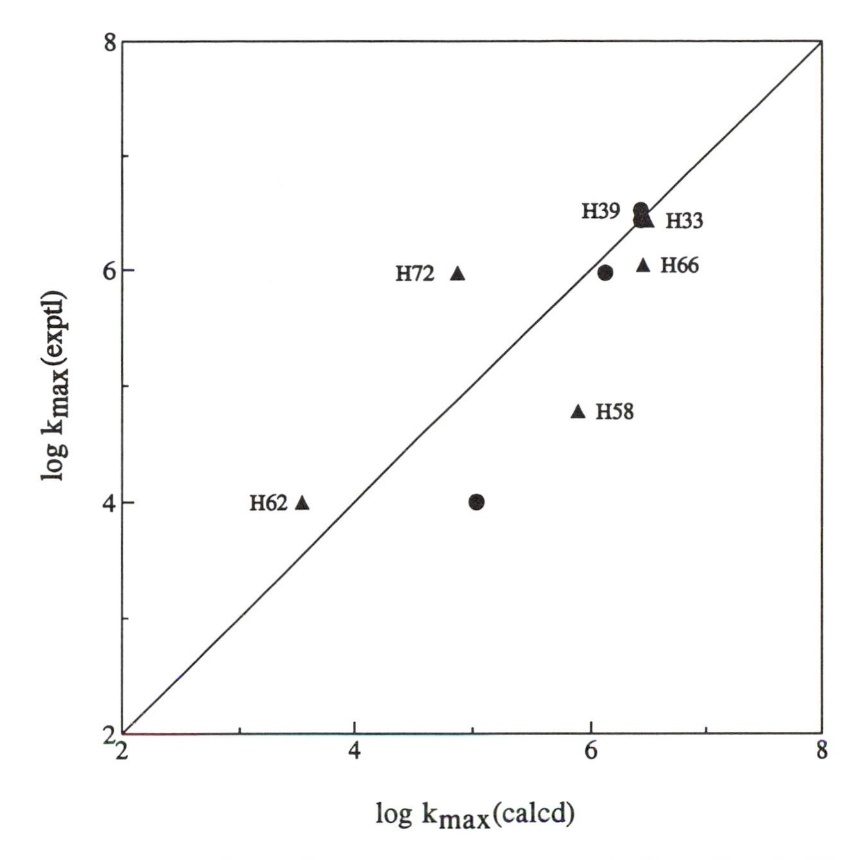

Figure 3. Correlation of calculated and experimental values of k_{max} *for ET in Ru(bpy)$_2$(im)(HisX)-modified cytochrome* c. *Key: triangles, TP calculation; and circles, EH calculation.*

In both cases, the data indicate that the structure of the intervening protein is absolutely critical in determining the rates of biological ET reactions.

A point of special interest is that the electronic couplings for intramolecular ET reactions in His-58 and His-66 cytochromes are not enhanced by aromatic residues (Trp-59 and Tyr-67) in the intervening media (*36*). The correlation of ET rates with σ_1 does not preclude a coupling role for the π orbitals of the aromatic groups in the pathway, but it does indicate that, in the Ru-modified cytochromes that we examined, they are no more efficient in mediating the coupling than is the σ-bonded framework. Hence, the presence of aromatic groups in the medium between redox sites does not necessarily result in faster ET than in a purely aliphatic medium (*15–20, 37–41*).

Gruschus and Kuki (*13, 14*) developed an inhomogeneous aperiodic lattice (IAL) model to calculate the charge resonance energy (T_{DA}) for

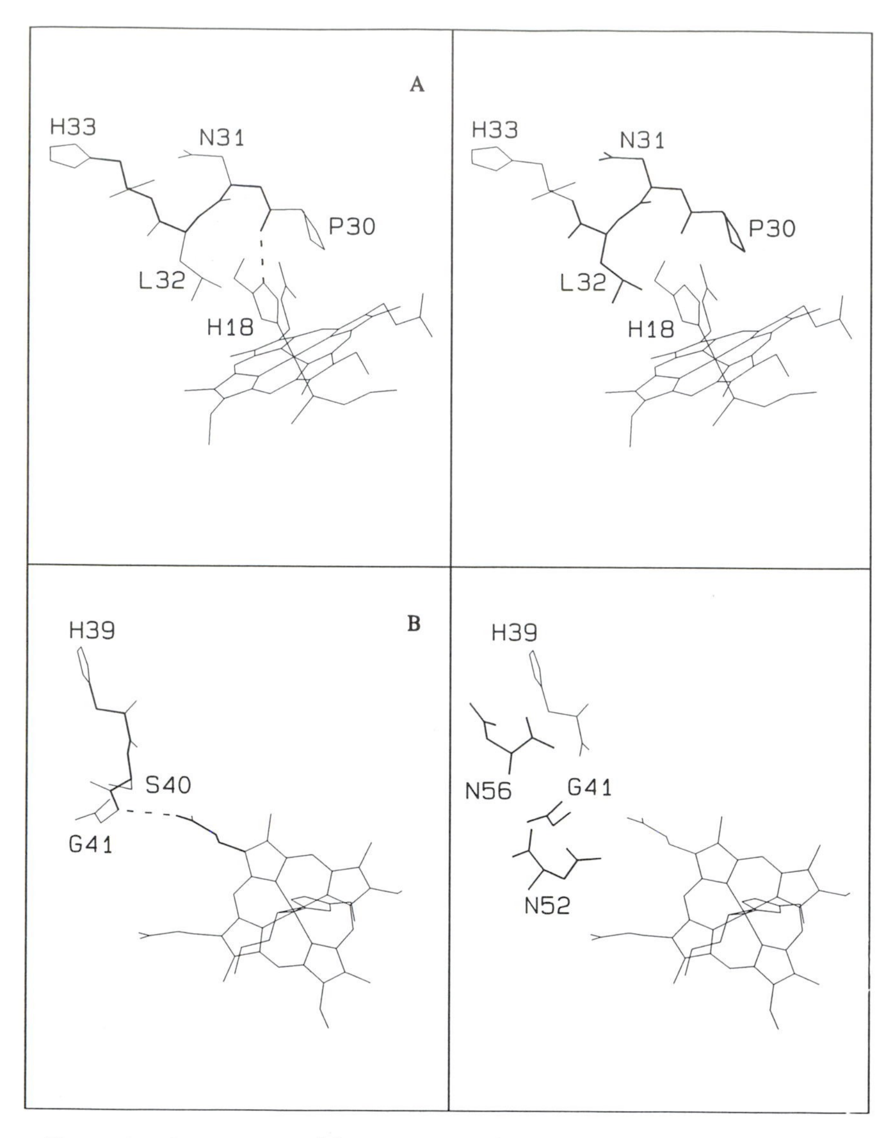

Figure 4. Comparison of dominant tunneling pathways (left) and residues identified in EH searches (right) of Ru(bpy)$_2$(im)(HisX)-modified cytochrome c [X = 33 *(A), 39 (B), 62 (C), and 72 (D)*].

long-range electron transfer in proteins. The IAL treatment depends upon very careful calibration of diagonal and off-diagonal energies for specific protein subunits. Armed with these subunit matrix elements, the donor–acceptor charge resonance energy is calculated by using the entire protein structure. The objective of the IAL approach is to produce

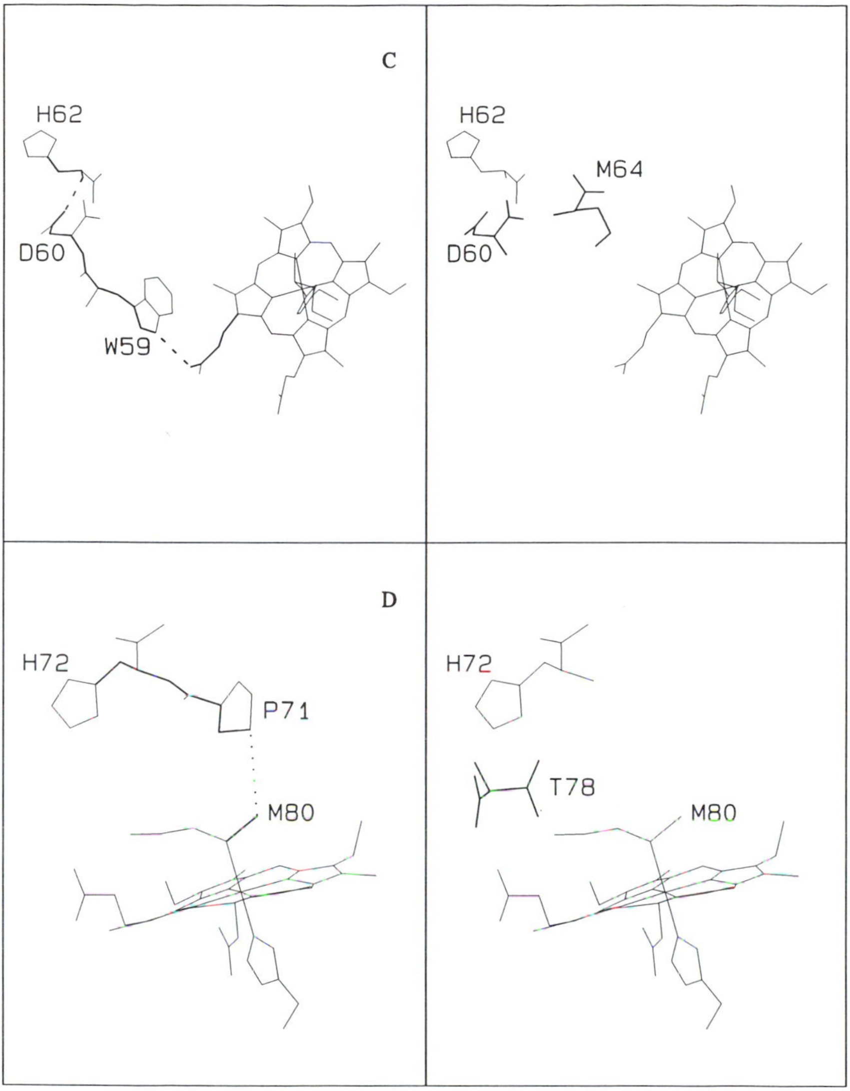

Figure 4. Continued.

accurate absolute values of T_{DA}. IAL calculations of charge resonance energies have been performed for three Ru-amine derivatives (His-33, His-39, and His-62) of Zn-substituted cytochrome *c* (*13*). The calculated values of $T_{DA}{}^2$ are all within an order of magnitude of the experimentally derived quantities. Quantum interference effects were particularly important in determining the overall coupling strength. In addition, all atoms within an ~10-Å-wide cross section between the redox sites made significant contributions to the superexchange matrix element.

Myoglobin

ET data obtained for Ru-modified zinc-substituted human myoglobins (Mb) are set out in Table I (*42*). The simple pathway-search algorithm finds many paths that contribute significantly to the donor–acceptor electronic coupling in each of the modified myoglobins. As expected for a protein system in which there is a high density of paths, the rates correlate with direct donor–acceptor separation. Although the dominant tunneling pathway significantly underestimates the coupling strength in each Mb derivative, a simple sum of best pathways (neglecting interferences) gives reasonably good agreement (Figure 5) (*42*). Good agreement with experiment is also obtained by employing the EH method (*43*); in general, more amino acids are used in the calculations than compose the dominant TP route (Figure 6). It is encouraging, however, that both the TP and EH analyses suggest that multiple pathways are important in mediating the electronic coupling in myoglobin.

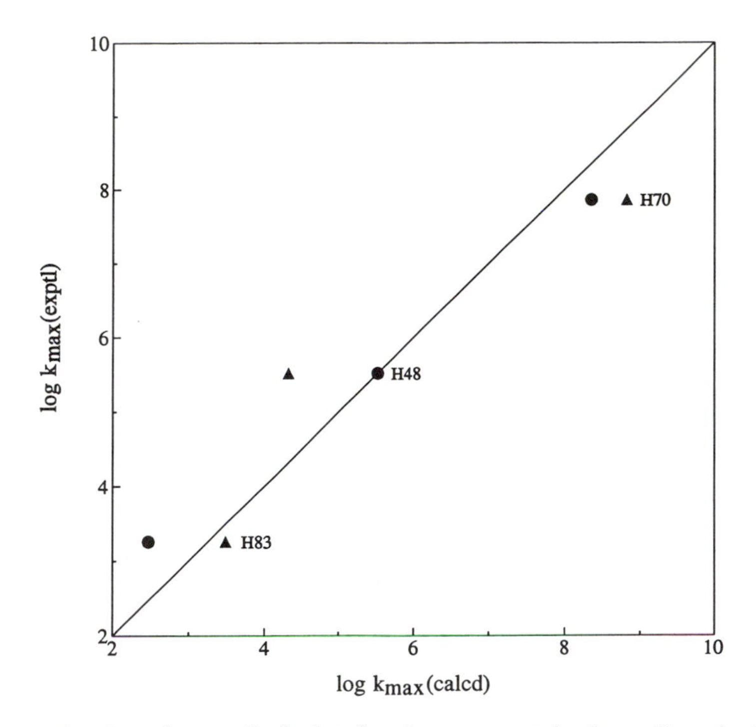

Figure 5. Correlation of calculated and experimental values of k_{max} *for ET in* $Ru(NH_3)_5(HisX)$*-modified human myoglobin. Key: triangles, multiple-path TP calculation; and circles, EH calculation.*

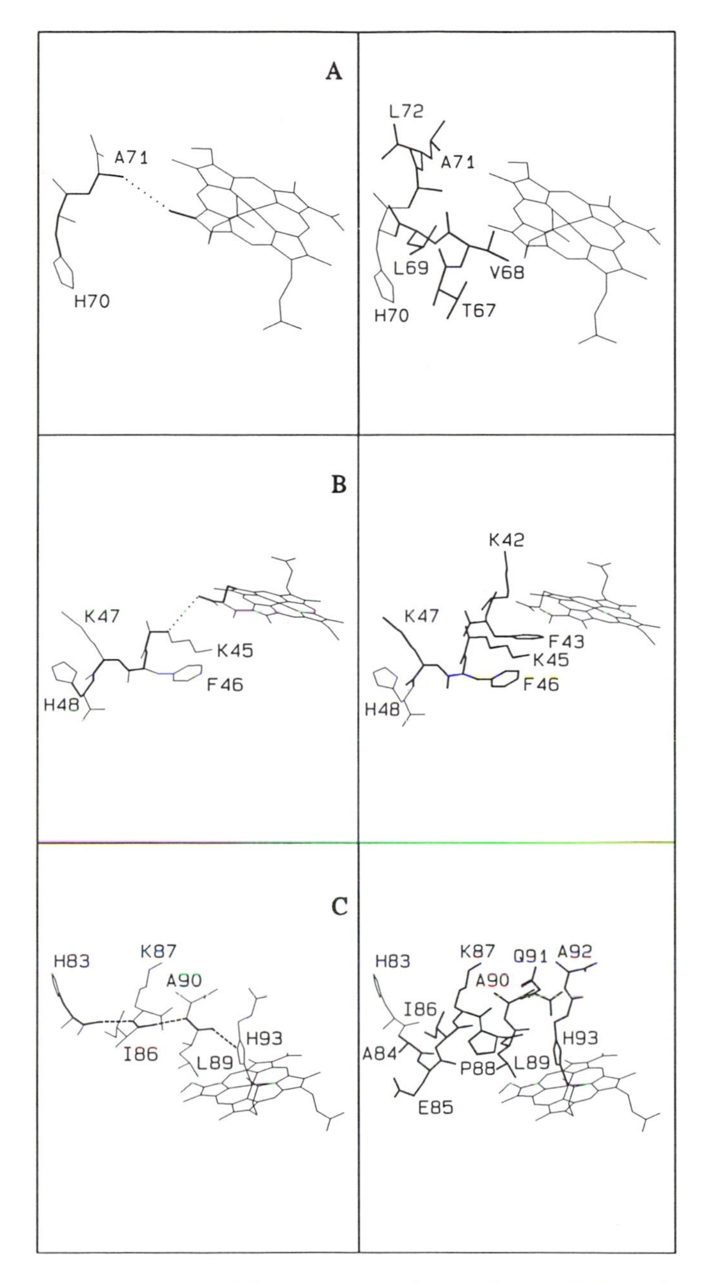

Figure 6. Comparison of dominant tunneling pathways (left) and residues identified in EH searches (right) of $Ru(NH_3)_5$(HisX)-modified Zn-substituted Mb [X = 70 (A), 48 (B), and 83 (C)].

Acknowledgments

We thank A. Kuki and R. A. Marcus for preprints of references 13 and 43, respectively. This research was supported by the National Science Foundation, the National Institutes of Health, the Department of Energy, and the Arnold and Mabel Beckman Foundation.

References

1. *Metal Ions in Biological Systems;* Sigel, H.; Sigel, A., Eds.; Dekker: New York, 1991; Vol. 27.
2. Closs, G. L.; Miller, J. R. *Science (Washington, D.C.)* **1988,** *240,* 440–447.
3. Jordan, K. D.; Paddon-Row, M. N. *Chem. Rev.* **1992,** *92,* 395–410.
4. Wasielewski, M. R. *Chem. Rev.* **1992,** *92,* 435–461.
5. Winkler, J. R.; Gray, H. B. *Chem. Rev.* **1992,** *92,* 369–379.
6. McLendon, G.; Hake, R. *Chem. Rev.* **1992,** *92,* 481–490.
7. McLendon, G.; Zhang, Q.; Wallin, S. A.; Miller, R. M.; Billstone, V.; Spears, K. G.; Hoffman, B. M. *J. Am. Chem. Soc.* **1993,** *115,* 3665–3669.
8. Wallin, S. A.; Stemp, E. D. A.; Everest, A. M.; Nocek, J. M.; Netzel, T. L.; Hoffman, B. M. *J. Am. Chem. Soc.* **1991,** *113,* 1842–1844.
9. Onuchic, J. N.; Beratan, D. N.; Winkler, J. R.; Gray, H. B. *Annu. Rev. Biophys. Biomol. Struct.* **1992,** *21,* 349–377.
10. Siddarth, P.; Marcus, R. A. *J. Phys. Chem.* **1990,** *94,* 2985–2989.
11. Siddarth, P.; Marcus, R. A. *J. Phys. Chem.* **1990,** *94,* 8430–8434.
12. Siddarth, P.; Marcus, R. A. *J. Phys. Chem.* **1992,** *96,* 3213–3217.
13. Gruschus, J. M.; Kuki, A. *J. Phys. Chem.* **1993,** *97,* 5581–5593.
14. Gruschus, J. M.; Kuki, A. *Chem. Phys. Lett.* **1992,** *192,* 205–212.
15. Christensen, H. E. M.; Conrad, L. S.; Mikkelsen, K. V.; Nielsen, M. K.; Ulstrup, J. *Inorg. Chem.* **1990,** *29,* 2808–2816.
16. Christensen, H. E. M.; Conrad, L. S.; Hammerstad-Pedersen, J. M.; Ulstrup, J. *FEBS Lett.* **1992,** *296,* 141–144.
17. Christensen, H. E. M.; Conrad, L. S.; Mikkelsen, K. V.; Ulstrup, J. *J. Phys. Chem.* **1992,** *96,* 4451–4454.
18. Broo, A. *Chem. Phys.* **1993,** *169,* 135–150.
19. Broo, A. *Chem. Phys.* **1993,** *169,* 152–163.
20. Broo, A.; Larsson, S. *J. Phys. Chem.* **1991,** *95,* 4925–4928.
21. Larsson, S.; Broo, A.; Kallebring, B.; Volosov, A. *Int. J. Quant. Chem.* **1988,** *S15,* 1–22.
22. Northrup, S. H.; Boles, J. O.; Reynolds, J. C. L. *Science (Washington, D.C.)* **1988,** *241,* 67–70.
23. Marcus, R. A.; Sutin, N. *Biochim. Biophys. Acta* **1985,** *811,* 265–322.
24. Hopfield, J. J. *Proc. Natl. Acad. Sci. U.S.A.* **1974,** *71,* 3640–3644.
25. Jortner, J. *J. Chem. Phys.* **1976,** *64,* 4860–4867.
26. Moser, C. C.; Keske, J. M.; Warncke, K.; Farid, R. S.; Dutton, P. L. *Nature (London)* **1992,** *355,* 796–802.
27. Farid, R. S.; Moser, C. C.; Dutton, P. L. *Curr. Opin. Struct. Biol.* **1993,** *3,* 225–233.
28. Beratan, D. N.; Betts, J. N.; Onuchic, J. N. *Science (Washington, D.C.)* **1991,** *252,* 1285–1288.
29. Onuchic, J. N.; Andrade, P. C. P.; Beratan, D. N. *J. Chem. Phys.* **1991,** *95,* 1131–1138.
30. Onuchic, J. N.; Beratan, D. N. *J. Chem. Phys.* **1990,** *92,* 722–733.

31. Beratan, D. N.; Onuchic, J. N. *Photosynth. Res.* **1989,** *22,* 173–186.
32. Beratan, D. N.; Onuchic, J. N.; Hopfield, J. J. *J. Chem. Phys.* **1987,** *86,* 4488–4498.
33. Chang, I.-J.; Gray, H. B.; Winkler, J. R. *J. Am. Chem. Soc.* **1991,** *113,* 7056–7057.
34. Wuttke, D. S.; Bjerrum, M. J.; Winkler, J. R.; Gray, H. B. *Science (Washington, D.C.)* **1992,** *256,* 1007–1009.
35. Wuttke, D. S.; Bjerrum, M. J.; Chang, I.-J.; Winkler, J. R.; Gray, H. B. *Biochim. Biophys. Acta* **1992,** *1101,* 168–170.
36. Casimiro, D. R.; Richards, J. H.; Winkler, J. R.; Gray, H. B. *J. Phys. Chem.* **1993,** *97,* 13073–13077.
37. Farver, O.; Pecht, I. *J. Am. Chem. Soc.* **1992,** *114,* 5764–5767.
38. Farver, O.; Skov, L. K.; Pascher, T.; Karlsson, B. G.; Nordling, M.; Lundberg, L. G.; Vänngard, T.; Pecht, I. *Biochemistry* **1993,** *32,* 7317–7322.
39. Govindaraju, K.; Christensen, H. E. M.; Lloyd, E.; Olsen, M.; Salmon, G. A.; Tomkinson, N. P.; Sykes, A. G. *Inorg. Chem.* **1993,** *32,* 40–46.
40. Everest, A. M.; Wallin, S. A.; Stemp, E. D. A.; Nocek, J. M.; Mauk, A. G.; Hoffman, B. M. *J. Am. Chem. Soc.* **1991,** *113,* 4337–4338.
41. Bowler, B. E.; Meade, T. J.; Mayo, S. L.; Richards, J. H.; Gray, H. B. *J. Am. Chem. Soc.* **1989,** *111,* 8757–8759.
42. Casimiro, D. R.; Wong, L.-L.; Coln, J. L.; Zewert, T. E.; Richards, J. H.; Chang, I.-J.; Winkler, J. R.; Gray, H. B. *J. Am. Chem. Soc.* **1993,** *115,* 1485–1489.
43. Siddarth, P.; Marcus, R. A. *J. Phys. Chem.* **1993,** *97,* 13078–13082.

RECEIVED for review July 19, 1993. ACCEPTED revised manuscript December 10, 1993.

INDEXES

Author Index

Affiliation Index

Subject Index

D

E

Q

R

S

U

V

X

Z

Production: Paula M. Bérard
Acquisition: Rhonda Bitterli
Cover design: Neal Clodfelter and Cornithia Harris

Copy edited, typeset, and indexed by Tapsco, Akron, PA
Printed and bound by Maple Press, York, PA